AF790614

Emergence, Complexity and Computation

Volume 9

Series editors

Ivan Zelinka, Technical University of Ostrava, Ostrava, Czech Republic
e-mail: ivan.zelinka@vsb.cz

Andrew Adamatzky, University of the West of England, Bristol, United Kingdom
e-mail: adamatzky@gmail.com

Guanrong Chen, City University of Hong Kong, Hong Kong
e-mail: eegchen@cityu.edu.hk

Editorial Board

Ajith Abraham, MirLabs, USA
Ana Lucia C. Bazzan, Universidade Federal do Rio Grande do Sul, Porto Alegre RS Brasil
Juan C. Burguillo, University of Vigo, Spain
Sergej Čelikovský, Academy of Sciences of the Czech Republic, Czech Republic
Mohammed Chadli, University of Jules Verne, France
Emilio Corchado, University of Salamanca, Spain
Donald Davendra, Technical University of Ostrava, Czech Republic
Andrew Ilachinski, Center for Naval Analyses, USA
Jouni Lampinen, University of Vaasa, Finland
Martin Middendorf, University of Leipzig, Germany
Edward Ott, University of Maryland, USA
Linqiang Pan, Huazhong University of Science and Technology, Wuhan, China
Gheorghe Păun, Romanian Academy, Bucharest, Romania
Hendrik Richter, HTWK Leipzig University of Applied Sciences, Germany
Juan A. Rodriguez-Aguilar, IIIA-CSIC, Spain
Otto Rössler, Institute of Physical and Theoretical Chemistry, Tübingen, Germany
Vaclav Snasel, Technical University of Ostrava, Czech Republic
Ivo Vondrák, Technical University of Ostrava, Czech Republic
Hector Zenil, Karolisnka Institute, Sweden

For further volumes:
http://www.springer.com/series/10624

About this Series

The Emergence, Complexity and Computation (ECC) series publishes new developments, advancements and selected topics in the fields of complexity, computation and emergence. The series focuses on all aspects of reality-based computation approaches from an interdisciplinary point of view especially from applied sciences, biology, physics, or Chemistry. It presents new ideas and interdisciplinary insight on the mutual intersection of subareas of computation, complexity and emergence and its impact and limits to any computing based on physical limits (thermodynamic and quantum limits, Bremermann's limit, Seth Lloyd limits...) as well as algorithmic limits (Gödel's proof and its impact on calculation, algorithmic complexity, the Chaitin's Omega number and Kolmogorov complexity, non-traditional calculations like Turing machine process and its consequences,...) and limitations arising in artificial intelligence field. The topics are (but not limited to) membrane computing, DNA computing, immune computing, quantum computing, swarm computing, analogic computing, chaos computing and computing on the edge of chaos, computational aspects of dynamics of complex systems (systems with self-organization, multiagent systems, cellular automata, artificial life,...), emergence of complex systems and its computational aspects, and agent based computation. The main aim of this series it to discuss the above mentioned topics from an interdisciplinary point of view and present new ideas coming from mutual intersection of classical as well as modern methods of computation. Within the scope of the series are monographs, lecture notes, selected contributions from specialized conferences and workshops, special contribution from international experts.

Mikhail Prokopenko
Editor

Guided Self-Organization: Inception

Springer

Mikhail Prokopenko
CSIRO Computational Informatics
Epping
Australia

and

School of Physics
The University of Sydney
Australia

and

Department of Computing
Macquarie University
Australia

ISSN 2194-7287 ISSN 2194-7295 (electronic)
ISBN 978-3-642-53733-2 ISBN 978-3-642-53734-9 (eBook)
DOI 10.1007/978-3-642-53734-9
Springer Heidelberg New York Dordrecht London

Library of Congress Control Number: 2013956507

© Springer-Verlag Berlin Heidelberg 2014
This work is subject to copyright. All rights are reserved by the Publisher, whether the whole or part of the material is concerned, specifically the rights of translation, reprinting, reuse of illustrations, recitation, broadcasting, reproduction on microfilms or in any other physical way, and transmission or information storage and retrieval, electronic adaptation, computer software, or by similar or dissimilar methodology now known or hereafter developed. Exempted from this legal reservation are brief excerpts in connection with reviews or scholarly analysis or material supplied specifically for the purpose of being entered and executed on a computer system, for exclusive use by the purchaser of the work. Duplication of this publication or parts thereof is permitted only under the provisions of the Copyright Law of the Publisher's location, in its current version, and permission for use must always be obtained from Springer. Permissions for use may be obtained through RightsLink at the Copyright Clearance Center. Violations are liable to prosecution under the respective Copyright Law.
The use of general descriptive names, registered names, trademarks, service marks, etc. in this publication does not imply, even in the absence of a specific statement, that such names are exempt from the relevant protective laws and regulations and therefore free for general use.
While the advice and information in this book are believed to be true and accurate at the date of publication, neither the authors nor the editors nor the publisher can accept any legal responsibility for any errors or omissions that may be made. The publisher makes no warranty, express or implied, with respect to the material contained herein.

Printed on acid-free paper

Springer is part of Springer Science+Business Media (www.springer.com)

To Don Price and Tony Farmer

Preface

Studying how order is created out of interactions, despite a relentless flow of increasing entropy, is one of the most rewarding scientific experiences, whether one studies the convection pattern formation of Bénard cells, collective behaviour of a school of fish reacting to a predator, or a modular robot adapting its locomotion to a new terrain. Furthermore, finding ways to guide the processes that seemingly spontaneously self-organise, towards desirable outcomes is among the most complex engineering tasks. And identifying fundamental principles for *guided* self-organisation would probably make a profound theoretical and practical contribution with far-reaching implications for both science and engineering.

Guided Self-Organisation (GSO) is a fundamentally multi-disciplinary area, drawing on methods from computational, physical and biological domains, most notably from information theory, theory of computation, dynamical systems, machine learning, evolutionary biology, artificial life, statistical mechanics and thermodynamics, and graph theory. The chapters presented in this book are produced and reviewed by experts in these fields who have been contributing to GSO over the last few years, shaping its rich research agenda. The cross-disciplinary nature of GSO is described in the introductory chapter written by several pioneers of this field.

The attempts to formalise foundations of GSO highlighted essential dependencies between self-organising behaviour and system-environment interactions, such as sensorimotor loops, brain-body-environment dynamics, information cascades, etc. These interactions are likely to indicate some deep links between distributed computation and thermodynamics: e.g., external entropy production (exported to the exterior of an open system) is known to match an increase of predictability about the system, captured by a directional information transfer and contributing to local computation. This relationship points out that an increase in (self-)organisation or emergence of a new functionality within the system cannot occur without some entropy dissipated away, concurring with Schrödinger concept of negentropy. Crucially, these complementary dynamics are not only tightly coupled, but also provide a thermodynamic insight about how to guide a self-organising process. In other words, providing and controlling channels for the system's entropy flux with its surroundings is perhaps the most generic method for guiding an increase of organisation within the system. This, of course, may be realised in a multiplicity of forms, by specifying universal utility functions that shape a diffusive control strategy or an entropic search in an evolutionary landscape, setting task-dependent constraints that bind a "dissipative" exploration during adaptive learning, or limiting computation to suitable pathways toward specifically balanced regimes.

Sharing important perspectives with other vigorously developed areas presented some challenges for creating and maintaining a unique GSO image. An intrigue created out of this tension continues to strongly drive the research. The number of compelling questions that are only beginning to find their resolutions is still vast, and their diversity, to some extent reflected in this book, is staggering. Is a bacterium more autopoietic than a virus? Does (Shannon) information operate as a "currency of life"? Is there anything remaining "in between" the system and the union of its parts? Can spontaneous symmetry breaking lead to an adaptive behaviour? Is life more stable in two dimensions than in one? How to quantify morphological computation? How does a swarm use its collective memory? Are there evolutionary pressures towards neural network efficiency and small-world structures? Can we still create what we do not understand? ...

Let us conclude with several historical remarks. As a field of study, GSO emerged in a somewhat spontaneous fashion, following up on some earlier ideas such as "design for emergence" and "emergent functionality", advocated by Luc Steels in late 1980s. Nevertheless, in a truly "self-referential" way, this development was guided over the last decade by a number of pioneers who recognised importance of this research direction and started to coordinate their activities from within several research labs worldwide: Ralf Der and Nihat Ay from Max Planck Institute for Mathematics in the Sciences (Leipzig, Germany), Daniel Polani from University of Hertfordshire (UK), Larry Yaeger from Indiana University (USA) and Mikhail Prokopenko from CSIRO (Sydney, Australia). This coordination resulted in a series of International Workshops on GSO hosted by Sydney (2008 and 2012), Leipzig (2009), Bloomington (2010), and Hatfield (2011).

While a clearer picture of GSO was emerging, the effort was joined by a younger generation of scientists who have not only contributed to this pursuit but shaped its directions to a large extent — Joseph T. Lizier, Georg Martius, Christoph Salge, Virgil Griffith, among many others. Without a doubt, their work provides the strongest hope that GSO will develop into a well-recognised scientific field retaining both the vigour and inspiration of its inception years.

September, 2013, on the sidelines of GSO-2013 (Barcelona) Mikhail Prokopenko

Contents

Part I: Introduction

1 On the Cross-Disciplinary Nature of Guided Self-Organisation 3
Mikhail Prokopenko, Daniel Polani, Nihat Ay
1.1 Introduction 3
1.2 Background 4
1.3 Structure 6
1.4 Foundational Frameworks of GSO 7
1.5 Coordinated Behaviour and Learning within an Embodied Agent 9
1.6 Swarms and Networks of Agents 10
1.7 Conclusion 12
References 13

Part II: Foundational Frameworks

2 Information Measures of Complexity, Emergence, Self-organization, Homeostasis, and Autopoiesis 19
Nelson Fernández, Carlos Maldonado, Carlos Gershenson
2.1 Introduction 19
2.2 Background 20
2.2.1 Complexity 20
2.2.2 Emergence 21
2.2.3 Self-organization 21
2.2.4 Homeostasis 22
2.2.5 Autopoiesis 23
2.2.6 Information Theory 23
2.2.7 Random Boolean Networks 25
2.2.8 Limnology 25
2.3 Measures 26
2.3.1 Emergence 27

2.3.2 Self-organization ... 28
2.3.3 Complexity ... 29
2.3.4 Homeostasis ... 30
2.3.5 Autopoiesis ... 32
2.3.6 Multi-scale Profiles ... 33
2.4 Results ... 33
2.4.1 Random Boolean Networks ... 33
2.4.2 An Ecological System: An Arctic Lake ... 35
2.5 Discussion ... 43
2.5.1 Measures ... 43
2.5.2 Complexity As Balance or Entropy? ... 43
2.5.3 Fisher Information ... 46
2.5.4 Tsallis Entropy ... 46
2.5.5 Guided Self-Organization ... 46
2.5.6 Scales ... 47
2.5.7 Normalization ... 47
2.5.8 Autopoiesis and Requisite Variety ... 47
2.6 Conclusions ... 47
References ... 48

3 Generating Functionals for Guided Self-Organization ... 53
Claudius Gros
3.1 Controlling Complex Systems ... 53
3.2 Guiding Self-Organization ... 55
3.3 Generating Functionals ... 56
3.4 Equations of Motion ... 58
3.5 Adaptive Phase Space ... 60
3.6 Self-Organized Dynamical States ... 62
3.7 Discussion ... 64
References ... 65

4 Empowerment — An Introduction ... 67
Christoph Salge, Cornelius Glackin, Daniel Polani
4.1 Introduction ... 67
4.1.1 Overview ... 69
4.2 Related Work ... 69
4.2.1 Intrinsic Motivation ... 70
4.3 Empowerment Hypotheses ... 72
4.3.1 Behavioural Empowerment Hypothesis ... 72
4.3.2 Evolutionary Empowerment Hypothesis ... 74
4.3.3 AI Empowerment Hypothesis ... 74
4.4 Formalism ... 75
4.4.1 The Causal Interpretation of Empowerment ... 76
4.4.2 Empowerment in the Perception Action Loop ... 77
4.4.3 *n*-Step Empowerment ... 79
4.4.4 Context-Dependent Empowerment ... 79

4.4.5 Open vs. Closed-Loop Empowerment ... 81
4.4.6 Discrete Deterministic Empowerment ... 81
4.4.7 Non-deterministic Empowerment Calculation ... 82
4.5 Discrete Examples ... 83
4.5.1 Maze ... 83
4.5.2 Average Distance vs. Empowerment ... 83
4.5.3 Sensor and Actuator Selection ... 85
4.5.4 Horizon Extension ... 87
4.5.5 Impoverished Empowerment ... 88
4.5.6 Sensor and Actuator Evolution ... 89
4.5.7 Multi-agent Empowerment ... 91
4.6 Continuous Empowerment ... 92
4.6.1 Continuous Information Theory ... 93
4.6.2 Infinite Channel Capacity ... 94
4.6.3 Continuous Empowerment Approximation ... 95
4.6.4 Binning ... 95
4.6.5 Evaluation of Binning ... 96
4.6.6 Jung's Monte Carlo Integration ... 96
4.6.7 Evaluation of Monte Carlo Integration ... 98
4.6.8 Quasi-Linear Gaussian Approximation ... 98
4.6.9 MIMO Channel Capacity ... 100
4.6.10 Coloured Noise ... 101
4.6.11 Evaluation of QLG Empowerment ... 102
4.7 Continuous Examples ... 102
4.7.1 Pendulum ... 103
4.7.2 Action Selection ... 103
4.7.3 Resulting Control ... 104
4.7.4 Power Constraint ... 106
4.7.5 Model Acquisition ... 108
4.8 Conclusion ... 110
References ... 111

5 A Framework for the Local Information Dynamics of Distributed Computation in Complex Systems ... 115
Joseph T. Lizier, Mikhail Prokopenko, Albert Y. Zomaya
5.1 Introduction ... 115
5.2 Information-Theoretic Preliminaries ... 117
5.2.1 Fundamental Quantities ... 117
5.2.2 Measures for Time-Series Processes ... 118
5.2.3 Local Information-Theoretic Measures ... 120
5.3 Cellular Automata ... 121
5.3.1 Introduction to Cellular Automata ... 121
5.3.2 Computation in Cellular Automata ... 122
5.3.3 Examples of Distributed Computation in CAs ... 124

5.4 Information Storage ... 126
5.4.1 Excess Entropy As Total Information Storage ... 126
5.4.2 Local Excess Entropy ... 129
5.4.3 Active Information Storage ... 129
5.4.4 Local Information Storage Results ... 130
5.5 Information Transfer ... 138
5.5.1 Local Transfer Entropy ... 138
5.5.2 Total Information, Entropy Rate and Collective Information Transfer ... 142
5.5.3 Local Information Transfer Results ... 142
5.6 Information Modification ... 145
5.6.1 Local Separable Information ... 145
5.6.2 Local Separable Information Results ... 147
5.6.3 Outlook for Information Modification ... 148
5.7 Importance of Coherent Computation ... 149
5.8 Conclusion ... 152
References ... 153

6 Quantifying Synergistic Mutual Information ... 159
Virgil Griffith, Christof Koch
6.1 Introduction ... 159
6.1.1 Notation ... 160
6.1.2 Understanding PI-Diagrams ... 160
6.2 Information Can Be Redundant, Unique, or Synergistic ... 162
6.2.1 Example Rdn: Redundant Information ... 162
6.2.2 Example Unq: Unique Information ... 163
6.2.3 Example Xor: Synergistic Information ... 163
6.3 Two Examples Elucidating Properties of Synergy ... 164
6.3.1 Duplicating a Predictor Does Not Change Synergistic Information ... 164
6.3.2 Adding a New Predictor Can Decrease Synergy ... 166
6.4 Prior Measures of Synergy ... 167
6.4.1 $\mathrm{I}_{\max}$ synergy: $\mathcal{S}_{\max}(\mathbf{X} : Y)$ ... 167
6.4.2 WholeMinusSum Synergy: $\mathrm{WMS}(\mathbf{X} : Y)$ ... 167
6.4.3 Correlational Importance: $\Delta\mathrm{I}(\mathbf{X}; Y)$ ... 168
6.5 Synergistic Mutual Information ... 170
6.6 Properties of I_{VK} ... 172
6.7 Applying the Measures to Our Examples ... 173
6.8 Conclusion ... 174
Appendix ... 175
A Three Extra Examples ... 175
B Connecting Back to $\mathrm{I}_{\cap}$ ... 179
C Desired Properties of $\mathrm{I}_{\cup}$ ... 179
D Analytic Upperbound on $\mathrm{I}_{\mathrm{VK}}(\mathbf{X} : Y)$ ... 182
E Essential Proofs ... 184

E.1 State-Dependent I_{VK} and $\mathcal{S}_{VK}$ ... 184
E.2 Proof Duplicate Predictors Don't Increase Synergy ... 185
E.3 Proof of Bounds of $\mathcal{S}_{VK}(\mathbf{X}:Y)$ ... 185
F Algebraic Simplification of ΔI ... 188
References ... 189

Part III: Coordinated Behaviour and Learning within an Embodied Agent

7 On the Role of Embodiment for Self-Organizing Robots: Behavior As Broken Symmetry ... 193
Ralf Der
7.1 Introduction ... 193
7.2 Vehicles ... 195
7.2.1 Braitenbergs Idea ... 195
7.2.2 Autistic Vehicles ... 195
7.2.3 Symmetries ... 197
7.3 The Braitenberg Man—Fundamental Modes ... 197
7.3.1 The HUMANOID ... 197
7.3.2 A Fundamental Mode of the HUMANOID ... 198
7.4 Unsupervised Learning for Self-Organization ... 200
7.4.1 Learning Rules for Self-model and Control ... 200
7.4.2 Anti-Hebbian and Differential Hebbian Learning: A Productive Competition ... 201
7.4.3 Relation to Infomax Principles ... 202
7.5 Homeokinesis: Body Inspired Behavior ... 203
7.5.1 Principles of Action ... 204
7.6 Vehicles: Behavior As Broken Symmetry ... 205
7.6.1 Least Biased Initialization ... 205
7.6.2 Symmetry Breaking—A Rule of Thumb ... 206
7.6.3 The Autistic Vehicle: Fundamental Modes ... 206
7.6.4 Synergy of Learning and Physical State Dynamics ... 208
7.6.5 The Pattern Factory ... 210
7.6.6 Patterns As Expressions of Embodiment ... 210
7.6.7 Modes ... 211
7.7 The Looping HUMANOID ... 212
7.7.1 High Symmetry Motion Patterns ... 212
7.7.2 Exterioception May Guide Self-Organization ... 213
7.7.3 Starting in a Mode ... 215
7.8 The HEXAPOD ... 215
7.8.1 Modes ... 215
7.8.2 Perspectives for Guidance and Reinforcement Learning ... 217
7.9 Concluding Remarks ... 217
References ... 219

8 Robot Learning by Guided Self-Organization 223
Georg Martius, Ralf Der, J. Michael Herrmann
8.1 Introduction 223
8.2 Homeokinesis 224
8.2.1 Example of Emergent Behavior 227
8.2.2 Behavior and Critical Dynamics in High-Dimensional Cricket Robot 228
8.3 Guided Self-Organization 229
8.4 Guidance by Mild Supervision 232
8.4.1 Integration of Problem-Specific Error Functions 232
8.4.2 Direct Motor Teaching 233
8.4.3 Direct Sensor Teaching and Distal Learning 235
8.5 Self-Organized Interaction with the Environment 237
8.5.1 Integration of Vision into the Sensorimotor Loop 237
8.5.2 Guiding towards an Object 238
8.5.3 Emergent Behaviors 239
8.5.4 Robustness against Structural Changes 241
8.6 Reward-Driven Self-Organization 244
8.6.1 Reinforcement Learning and Guided Self-Organization 244
8.6.2 Modulation of Behavior in a Spherical Robot 245
8.7 Channeling Self-Organization 248
8.7.1 From Spontaneous to Guided Symmetry Breaking 249
8.7.2 Multiple Motor Relations 250
8.7.3 Guiding to Directed Locomotion 251
8.7.4 Scaling Properties 254
8.7.5 Coordination of Finger Movements for Grasping 255
8.8 Discussion 257
References 259

9 On the Causal Structure of the Sensorimotor Loop 261
Nihat Ay, Keyan Zahedi
9.1 Introduction 261
9.2 Causal Networks 262
9.2.1 The Definition of Causal Networks 262
9.2.2 The Causal Structure of the Sensorimotor Loop 266
9.3 Causal Effects 268
9.3.1 The Definition of Causal Effects 268
9.3.2 Identification of Causal Effects 271
9.3.3 Causal Effects in the Sensorimotor Loop 272
9.4 Information Flows 274
9.4.1 Information-Theoretic Preliminaries 274

9.4.2 Transfer Entropy and Causality 275
9.4.3 Information Flows in the Sensorimotor Loop 281
9.5 Predictive Information and Its Maximization – An Experimental Case Study 284
Appendix 291
References 293

10 Action Switching in Brain-Body-Environment Systems 295
Eran Agmon
10.1 Introduction 295
10.2 Ashby's Self-Organization in Brain-Body-Environment Systems 297
10.3 Beer's Adaptive Behavior Program 301
10.3.1 CTRNNs and Genetic Algorithms 302
10.3.2 Dynamical Systems Theory 304
10.3.3 A Simple Chemotaxis Agent 305
10.4 Action Switching 306
10.4.1 Evolving an Action Switcher 308
10.4.2 The Agent's Behavior and Dynamics 309
10.4.3 Discussion of Action Switching 314
10.5 The Prospect of Brain-Body-Environment Systems 315
References 316

11 Guided Self-Organization of Input-Driven Recurrent Neural Networks 319
Oliver Obst, Joschka Boedecker
11.1 Introduction 319
11.2 Assessing the Computational Power and Mechanisms of Information Processing of Reservoirs 322
11.2.1 Information-Theory Related Measures 323
11.2.2 Measures Related to Learning Theory 325
11.2.3 Measures Related to Dynamical Systems Theory 327
11.3 Improving Reservoir Information Processing Capabilities through Self-Organized Adaptation 328
11.3.1 SORN: Self-Organized Optimization Based on 3 Local Plasticity Mechanisms 329
11.3.2 Hierarchical Self-Organizing Reservoirs 331
11.3.3 Guided Self-Organization of Reservoir Information Transfer 332
11.4 Quantifying Task Complexity 334
11.5 Conclusion 336
References 337

Part IV: Swarms and Networks of Agents

12 Measuring Information Dynamics in Swarms 343
Jennifer M. Miller, X. Rosalind Wang, Joseph T. Lizier, Mikhail Prokopenko, Louis F. Rossi
12.1 Introduction 343
12.1.1 Background on Information Cascades 344
12.1.2 Motivation and Objectives 345
12.2 Three Zones Model for Swarms 346
12.3 Information Dynamics in Swarms 347
12.4 Results and Discussion 349
12.4.1 Variable-Speed Swarm Model 349
12.4.2 Constant-Speed Swarm Model 355
12.5 Conclusions 361
References 362

13 Guiding Designs of Self-Organizing Swarms: Interactive and Automated Approaches 365
Hiroki Sayama
13.1 Introduction 365
13.2 Model: Swarm Chemistry 367
13.3 Interactive Approach 370
13.3.1 User Experience 372
13.3.2 Design Quality 373
13.4 Automated Approach 375
13.4.1 Exploring Experimental Conditions 377
13.4.2 Quantifying Observed Evolutionary Dynamics 380
13.5 Conclusions 385
References 385

14 Mutual Information As a Task-Independent Utility Function for Evolutionary Robotics 389
Valerio Sperati, Vito Trianni, Stefano Nolfi
14.1 Introduction 389
14.2 Short Introduction to Information Theory 390
14.3 Related Work 392
14.4 Experimental Setup 394
14.4.1 The Robot and the Neural Controller 395
14.4.2 The Evolutionary Process 398
14.4.3 The Fitness Function 398
14.5 Results 400
14.5.1 Experiment E_l 400
14.5.2 Experiment E_d 405
14.6 Conclusion 410
References 411

15 Evolution of Complexity and Neural Topologies 415
Larry S. Yaeger
15.1 Introduction 415
15.2 Complexity 416
15.3 Simulation Software 419
15.4 Natural Selection vs. Random Drift 420
15.5 Data Generation and Acquisition 426
15.6 Complexity As a Fitness Function 427
15.7 Evolutionary Trends of Complexity 428
15.8 Evolutionary Trends of Network Topology 433
15.8.1 Clustering Coefficient 434
15.8.2 Characteristic Path Length 435
15.8.3 Global Efficiency 436
15.8.4 Small-World Index 438
15.9 Relating Neural Complexity to Network Topology 439
15.9.1 Clustering Coefficient 440
15.9.2 Characteristic Path Length 440
15.9.3 Global Efficiency 441
15.9.4 Small-World Index 442
15.10 Broader Applicability 444
15.11 Discussion and Conclusions 446
References 450

16 Clustering and Modularity in Self-Organized Networks 455
Somwrita Sarkar, Peter A. Robinson
16.1 Introduction 455
16.2 Modularity of Self-Organized Systems 456
16.3 Chapter Summary 457
16.4 Background 458
16.4.1 Spectra and Graph Structure 458
16.4.2 Spectral Clustering and Partitioning Approaches 459
16.4.3 Spectral Fingerprints of Modularity and Hierarchical Modularity: Adjacency Matrix 462
16.5 Detecting the Modular Structure 464
16.6 Discussion and Conclusion 466
References 467

Index 469

List of Contributors

Eran Agmon
Indiana University, 107 S. Indiana Avenue
Bloomington, Indiana 47405-7000, United States
e-mail: agmon.eran@gmail.com

Nihat Ay
Max Planck Institute for Mathematics in the Sciences, Inselstraße 22,
D-04103, Leipzig, Germany
e-mail: nay@mis.mpg.de
& Santa Fe Institute, 1399 Hyde Park Road, Santa Fe,
New Mexico 87501, United States

Joschka Boedecker
Machine Learning Lab, University of Freiburg, Freiburg, Germany
e-mail: jboedeck@informatik.uni-freiburg.de

Ralf Der
Max Planck Institute for Mathematics in the Sciences, Inselstraße 22,
D-04103, Leipzig, Germany
e-mail: ralfder@mis.mpg.de

Nelson Fernández
Laboratorio de Hidroinformática, Facultad de Ciencias Básicas
Univesidad de Pamplona, Colombia
e-mail: nelferpa@gmail.com
& Centro de Micro-electrónica y Sistemas Distribuidos
Universidad de los Andes, Mérida, Venezuela

Carlos Gershenson
Departamento de Ciencias de la Computación, Instituto de Investigaciones en Matemáticas Aplicadas y en Sistemas & Centro de Ciencias de la Complejidad Universidad Nacional Autónoma de México
e-mail: cgg@unam.mx

Cornelius Glackin
Adaptive Systems Research Group, School of Computer Science
University of Hertfordshire Hatfield, United Kingdom
e-mail: c.glackin2@herts.ac.uk

Virgil Griffith
Computation and Neural Systems, California Institute of Technology, Pasadena, California 91125, United States
e-mail: virgil@caltech.edu

Claudius Gros
Institute for Theoretical Physics, Goethe University Frankfurt, Germany
e-mail: gros@itp.uni-frankfurt.de

J. Michael Herrmann
Bernstein Center for Computational Neuroscience, Am Faßberg 17, 37077, Göttingen, Germany & Institute for Perception, Action and Behaviour, School of Informatics, University of Edinburgh 10 Crichton St, Edinburgh, EH8 9AB Scotland, United Kingdom
e-mail: michael.herrmann@ed.ac.uk

Christof Koch
Computation and Neural Systems, California Institute of Technology, Pasadena, California 91125, United States
e-mail: koch@klab.caltech.edu
& Allen Institute for Brain Science Seattle, WA 98103 United States

Joseph T. Lizier
CSIRO Computational Informatics, PO Box 76, Epping, NSW 1710, Australia
& School of Information Technologies, The University of Sydney
NSW 2006, Australia
& Max Planck Institute for Mathematics in the Sciences, Inselstraße 22, D-04103, Leipzig, Germany

Carlos Maldonado
Facultad de Ciencias & Departamento de Ciencias de la Computación, Instituto de Investigaciones en Matemáticas Aplicadas y en Sistemas, Universidad Nacional Autónoma de México
e-mail: cdmc89@gmail.com

Georg Martius
Max Planck Institute for Mathematics in the Sciences, Inselstraße 22,
D-04103, Leipzig, Germany
e-mail: martius@mis.mpg.de

Jennifer M. Miller
Department of Mathematics, Trinity College, Hartford,
Connecticut 06106, United States
e-mail: jennifer.miller@trincoll.edu

Stefano Nolfi
Laboratory of Autonomous Robotics and Artificial Life, Institute of Cognitive Sciences and Technologies, CNR via S. Martino della Battaglia,
44 - 00185 Rome, Italy
e-mail: stefano.nolfi@istc.cnr.it

Oliver Obst
CSIRO Computational Informatics
PO Box 76, Epping, NSW 1710, Australia
e-mail: oliver.obst@csiro.au

Daniel Polani
Adaptive Systems Research Group, School of Computer Science
University of Hertfordshire, Hatfield, United Kingdom
e-mail: daniel.polani@gmail.com

Mikhail Prokopenko
CSIRO Computational Informatics
PO Box 76, Epping, NSW 1710, Australia
e-mail: mikhail.prokopenko@csiro.au
& School of Physics The University of Sydney NSW 2006, Australia
& Department of Computing Macquarie University
NSW 2109, Australia

Peter A. Robinson
Complex Systems Group, School of Physics
University of Sydney NSW 2006, Australia
e-mail: robinson@physics.usyd.edu.au
& Brain Dynamics Center, Sydney Medical School
University of Sydney, Westmead NSW 2145, Australia

Louis F. Rossi
Department of Mathematical Sciences, University of Delaware, 501 Ewing Hall,
Newark, Delaware 19716, United States
e-mail: rossi@math.udel.edu

Christoph Salge
Adaptive Systems Research Group, School of Computer Science, University of Hertfordshire, Hatfield, United Kingdom
e-mail: `christophsalge@gmail.com`

Somwrita Sarkar
Design Lab, Faculty of Architecture, Design, and Planning & Complex Systems Group, School of Physics, University of Sydney NSW 2006, Australia
e-mail: `sarkar@physics.usyd.edu.au`

Hiroki Sayama
Collective Dynamics of Complex Systems Research Group, Binghamton University, State University of New York Binghamton, New York 13902-6000, United States
e-mail: `sayama@binghamton.edu`

Valerio Sperati
Laboratory of Autonomous Robotics and Artificial Life, Institute of Cognitive Sciences and Technologies, CNR, via S. Martino della Battaglia, 44 - 00185 Rome, Italy
e-mail: `valerio.sperati@istc.cnr.it`

Vito Trianni
Laboratory of Autonomous Robotics and Artificial Life, Institute of Cognitive Sciences and Technologies, CNR, via S. Martino della Battaglia, 44 - 00185 Rome, Italy
e-mail: `vito.trianni@istc.cnr.it`

X. Rosalind Wang
CSIRO Computational Informatics
PO Box 76, Epping, NSW 1710, Australia
e-mail: `rosalind.wang@csiro.au`

Larry S. Yaeger
School of Informatics & Computing, Indiana University,
919 E. 10th St., Bloomington, Indiana 47408, United States
e-mail: `larry.yaeger@gmail.com`
& Google Inc. 1600 Amphitheatre Parkway,
Mountain View, CA 94043, United States

Keyan Zahedi
Max Planck Institute for Mathematics in the Sciences, Inselstraße 22, D-04103, Leipzig, Germany
e-mail: `zahedi@mis.mpg.de`

Albert Y. Zomaya
School of Information Technologies, The University of Sydney, NSW 2006, Australia
e-mail: `albert.zomaya@sydney.edu.au`

Part I
Introduction

Chapter 1
On the Cross-Disciplinary Nature of Guided Self-Organisation

Mikhail Prokopenko, Daniel Polani, and Nihat Ay

1.1 Introduction

Self-organisation is pervasive: neuronal ensembles self-organise into complex spatio-temporal spike patterns which facilitate synaptic plasticity and long-term consolidation of information; large-scale natural or social systems, as diverse as forest fires, landslides, or epidemics, produce spontaneous scale-invariant behaviour; robotic modules self-organise into coordinated motion patterns; individuals within a swarm achieve collective coherence out of isolated actions; and so on. Self-organisation is also valuable: the resultant increase in an internal organisation brings benefits to the (collective) organism, be it a learning brain, a co-evolving ecosystem, an adapting modular robot, or a re-configuring swarm. These benefits are typically realised in increased resilience to external disturbances, adaptivity to novel tasks, and scalability with respect to new challenges. However, self-organisation is difficult to engineer on demand: the intricate fabric of interactions within a self-organising system cannot follow a simple-minded blueprint and resists crude interventions.

Mikhail Prokopenko
CSIRO Computational Informatics PO Box 76, Epping, NSW 1710, Australia
e-mail: mikhail.prokopenko@csiro.au
& School of Physics, The University of Sydney NSW 2006, Australia
& Department of Computing Macquarie University NSW 2109, Australia

Daniel Polani
Adaptive Systems Research Group, School of Computer Science
University of Hertfordshire, Hatfield, United Kingdom
e-mail: daniel.polani@gmail.com

Nihat Ay
Max Planck Institute for Mathematics in the Sciences, Inselstraße 22, D-04103, Leipzig, Germany
e-mail: nay@mis.mpg.de
Santa Fe Institute, 1399 Hyde Park Road, Santa Fe, New Mexico 87501, United States

M. Prokopenko (ed.), *Guided Self-Organization: Inception*,
Emergence, Complexity and Computation 9,
DOI: 10.1007/978-3-642-53734-9_1, © Springer-Verlag Berlin Heidelberg 2014

Thus, the goal of *Guided Self-Organisation* (GSO) is to leverage the strengths of self-organisation while still being able to indirectly affect the outcome of the self-organising process. GSO typically has the following features: (i) an increase in organisation (structure and/or functionality) over some time; (ii) the local interactions are not *explicitly* guided by any external agent; (iii) task-independent objectives, shaped up by universal or intrinsic selection pressures, are combined with task-dependent constraints and costs.

This set of features is sufficiently general: it allows to include multiple drivers to guide a self-organising system/process, by treating these drivers as additional objective functions and/or constraints imposed on the system under consideration (Ay et al. 2011). These features also do not presume any specific modelling methodology. Nevertheless, several frameworks and concepts developed in this field over the last decade have been particularly useful, employing methods from

- information theory, e.g. (Polani et al. 2007; Polani 2009; Prokopenko et al. 2009; Martius et al. 2013),
- theory of computation, e.g. (Lizier et al. 2008b; Crutchfield 2009; Egri-Nagy and Nehaniv 2011; Dini et al. 2013),
- dynamical systems, e.g. (Gros 2008; Williams and Beer 2010; Der and Martius 2012; Beer 2013, 2014),
- statistical mechanics and thermodynamics, e.g. (Friston 2009; Prokopenko et al. 2011, 2013; Still et al. 2012; Wissner-Gross and Freer 2013), and
- graph theory, e.g. (Rubinov and Sporns 2010; Gershenson 2012; Piraveenan et al. 2009, 2012; Yaeger 2013).

The successes in applying these theories to GSO can be explained by their generic models that are able to express the most essential properties of the system in point, at multiple scales (local and global), and varying in both space and time. Crucially, these methods enable comparative analyses across domains, often allowing to transfer successful techniques across disciplines, without a major conceptual redesign.

1.2 Background

Many GSO approaches use the characterisation of a system-environment loop (e.g., sensorimotor or perception-action loop) in information-theoretic terms. These foundations can be derived from fundamental limits on information processing involving a task (Touchette and Lloyd 2000). This can also be put into perspective of exploratory (Still 2009) and task-relevant information processing costs (Tishby and Polani 2011). Interestingly, the formal expression arising from this is virtually identical to the free energy used in regular thermodynamics. For example, a free energy principle has been used to produce intrinsically self-motivated agent behaviours, by combining a variational Bayes formalism together with the hypothesis that evolved cognitive entities carry out a near-optimal Bayes inference (Friston 2009). Thus, the connections between information-theoretic and thermodynamic (or statistical-mechanical) models go beyond a simple convenience, but rather reflect on rich common dynamics underlying guided self-organisation in open systems.

The idea that living beings are information-processing systems and that the optimisation of these processes might provide an evolutionary advantage was formalised via excess entropy or predictive information (Bialek et al. 2001), a measure for a system's complexity that quantifies the total information of past experience that can be used for predicting future events. This approach provides both principled criteria for optimal agent learning as well as drivers generating structured behaviours resembling those of volitional decisions of organisms, without externally given tasks (Prokopenko et al. 2006b,a; Ay et al. 2008; Zahedi et al. 2010; Ay et al. 2012). In particular, an intrinsic selection pressure, captured via the generalised excess entropy, produced coordinated locomotion of a modular robot (Prokopenko et al. 2006a), while several explicit learning rules were derived in a task-free way by a gradient ascent on the predictive information (Zahedi et al. 2010; Ay et al. 2012). These learning rules were shown to have a Hebbian-like structure drawing additional parallels to biological systems.

Some of these results are related to Infomax principle: an optimisation principle for information processing systems, such as neural networks, which suggests to maximise (by choice or by learning) the average Shannon mutual information between inputs and outputs of the system (Linsker 1988). The optimisation is typically carried out subject to constraints and/or noise processes. For example, maximisation of the entropy of a neuron's output while the average output firing-rate is fixed results in intrinsic plasticity (Triesch 2005), producing complex dynamical phenomena such as neuronal self-regulation, etc. (Bell and Sejnowski 1995; Butko and Triesch 2005; Lazar et al. 2006; Lungarella and Sporns 2006).

It is well-recognised now that sensorimotor interaction and body morphology can induce statistical regularities and information structure both in sensory inputs and within the neural control architecture (Lungarella and Sporns 2006). The hypothesis that such statistical regularities result from the combined action of sensory and motor systems, given some body morphology, has been proposed and experimentally verified by Lungarella and Sporns, who critically pointed out that embodied systems actively seek information (stimuli) while engaging in behavior (Lungarella and Sporns 2006). In addition, this study has identified important patterns of information flow between sensors, neural units, and effectors, quantifying these with mutual information and transfer entropy where the latter measure quantifies the statistical coherence between systems evolving in time (Schreiber 2000).

One of the most representative examples motivated by InfoMax principle is the concept of *empowerment*. Empowerment was first introduced in (Klyubin et al. 2005a) and studied over the subsequent years: it measures the maximum quantity of Shannon information that an agent could *potentially* inject into the environment and recover via its sensors: it is the channel capacity of the exterior part of an agent's action-perception loop. In its simplest form, this turns out to be the maximum entropy that an agent has the potential to *controllably* produce in its environment. It also allows more sophisticated considerations which separate agent- and environment-generated noise, as well as account for the ability of the agent itself to profit from the entropy it had generated itself.

Entropy maximization-like approaches using empowerment were applied in a variety of scenarios, e.g., discovery of desirable "degrees of object manipulation" (Klyubin et al. 2005a); sensor evolution (Klyubin et al. 2005b); coordination of collective systems (Capdepuy et al. 2007); non-trivial "survival"-type control tasks, such as pole-balancing, acrobot and others, carried out without scenario-specific goals (Jung et al. 2011; Salge et al. 2012).

Recently, Wissner-Gross and Freer (2013) proposed a similar method to derive an entropic forcing mechanism as a precursor to a potential cognitive driver, aiming to demonstrate interesting and potentially biologically relevant behaviours. Different from the models discussed earlier, the entities governed by Wissner-Gross and Freer's dynamics are also essentially driven by direct external statistical physical principles and do not consider an explicit perception-action loop with dedicated sensors and actuators. Furthermore, the authors' arguments in favour of the mechanism point to generic cosmological speculations and, more importantly, to a fragile link via the Maximum Entropy Production Principle (MEPP). In the physical motivation, one is, however, currently still confronted with a conceptual gap: essential arguments in favour of the MEPP (Dewar 2003, 2005) have been found faulty (Grinstein and Linsker 2007). Thus, despite its attractiveness and repeated attempts in using MEPP to motivate the emergence of universal mechanisms generating self-organising processes, such as Wissner-Gross and Freer's entropic forcing, it must be considered an unresolved question whether such a derivation can be achieved without further assumptions.

In general, however, the rigorous links between adaptive behaviour, critical dynamics, and statistical patterns of information processing can and should be interpreted thermodynamically (Crooks 2007; Prokopenko et al. 2011). For example, a thermodynamic interpretation of transfer entropy shows that this quantity is proportional to the external entropy production by the system, attributed to a source of irreversibility (Prokopenko et al. 2013). Interestingly, transfer entropy was also shown to capture one of the three elements of distributed computation: communication (Lizier et al. 2008b), connecting GSO to another generic theory — the theory of computation (Langton 1990; Lizier et al. 2010, 2012c, 2013).

1.3 Structure

This book aims to present the state-of-the-art in the GSO field, by describing most of its subjects in sufficient detail and highlighting several fundamental interconnections.

Some of the chapters in this book follow presentations at The GSO-2012 Workshop, the fifth in the series, held in Sydney, 26-28 September 2012, which brought together invited experts and researchers in self-organising systems, with particular emphasis on foundations of GSO and the information dynamics of adaptive systems. Selected papers of previous Workshops were published in the special or topical issues on Guided Self-Organisation by the *Human Frontier Science Program Journal* (GSO-2008) (Prokopenko 2009), *Theory in Biosciences* (GSO-2009)

(Ay et al. 2011), *Advances in Complex Systems* (GSO-2010 and GSO-2011) (Polani et al. 2013). Following these workshops and the series of journal issues, it has become apparent that this is the right time to have a book that includes chapters not only focussing on particular topics presented at the most recent workshop, but also covering development of the topic over the last few years since the beginning of the workshop series.

The book captures the most representative approaches, includes in-depth reviews and descriptions of future research in the most important GSO areas. In short, it offers a comprehensive contemporary GSO perspective, aiming to become a mini-encyclopaedia for the early years of this emerging field. Hence, the title's emphasis: "Guided Self-Organization: Inception".

The book is structured around three main sections:

- foundational frameworks (including abstract models and theory of GSO);
- coordinated behaviour and learning within an embodied agent;
- swarms and networks of agents.

1.4 Foundational Frameworks of GSO

As mentioned in Section 1.1, several specific frameworks have been influential in this emerging field, using developing information-theoretic, computation-theoretic and dynamical systems methods and models. The chapters combined in this section describe some of these foundational methodologies. It is worth to point out that the deep theoretical links uncovered in these studies show a promise for the entire field, indicating a tangible possibility for a unifying basis: a theory of GSO.

The Chapter of Fernández et al. studies measures of emergence, self-organisation, complexity, homeostasis, and autopoiesis, using information theory. Many concepts used in the scientific study of complex systems have become so widespread that their misuse and misinterpretation have led to ambiguity and confusion. And so the purpose of this chapter is to clarify the meaning of these concepts with the aid of the proposed formal measures, derived from several proposed axioms. In a simplified version, (information) emergence is defined as the information a system or process produces. Defined in such a way, emergence becomes the opposite of self-organisation, while complexity represents their balance. Homeostasis can be seen as a measure of the stability of the system. Autopoiesis can be measured as the ratio between the complexity of a system and the complexity of its environment. The proposed measures can be applied at different scales, with multi-scale profiles, and the chapter illustrates the measures with simulations of random Boolean networks and an Arctic lake ecosystem.

The Chapter of Gros considers a generic question of how to control a complex and self-organising system, achieving stability of a default working regime with respect to possible external influences and statistical fluctuations. The author points out that one way to guide a dynamical system is to restrict its flow to a certain region in phase space, allowing for an otherwise unrestricted development within this bounded area of phase space. Thus, one possibility to control a GSO process is to formulate the targets in terms of statistical properties of the desired dynamical state. The chapter also offers some specific examples, by explicitly deriving equations of motions from generating functionals that incorporate polyhomeostatic and other targets. The approach of generating functionals is proposed as a general method for building increasingly complex dynamical systems and cognitive architectures.

The Chapter of Salge et al. offers a detailed review and discussion of the now increasingly maturing empowerment formalism. As mentioned in Section 1.2, empowerment is aimed at capturing intrinsic motivations which can be used to generate self-organizing behaviours in agents. The chapter presents different scenarios that highlight the universality of empowerment-driven algorithms, as well as posing several open questions: how to integrate explicit non-default goals into empowerment, how to manage its computational feasibility, and how to search for optimal solutions that lie behind the (local) empowerment's horizon.

Lizier et al. point out that the nature of distributed computation has often been described in terms of the component operations of universal computation: information storage, transfer and modification. They review the first complete framework that quantifies each of these individual information dynamics on a local scale within a system, and describe the manner in which they interact to create non-trivial computation where "the whole is greater than the sum of the parts". They further apply the framework to cellular automata (CA), a simple yet powerful model of distributed computation. In this application, the framework is demonstrated to be the first to provide quantitative evidence for several important conjectures about distributed computation in CA: that blinkers embody information storage, particles are information transfer agents, and particle collisions are information modification events. The framework is also used to investigate and contrast the computations conducted by several well-known CA, highlighting the importance of information coherence in complex computation. Their results provide important quantitative insights into the fundamental nature of distributed computation and the dynamics of complex systems, as well as impetus for the framework to be applied to the analysis and design of other systems.

Griffith and Koch further develop the notion of information modification — a crucial element of non-trivial distributed computation — by quantifying synergistic mutual information. Their chapter reviews several existing information-theoretic measures of synergy developed over the last two decades, and introduces a novel synergy measure. The proposed measure is defined as the difference between the whole and the *union* of its parts. The new and existing measures are compared with respect to a set of representative examples (a suite of binary circuits), demonstrating a superior performance of the new measure. The chapter also elucidates the reasons behind the success, pointing out that an existing alternative underestimates synergy by double-subtracting redundancies. In addition, an upper bound for the introduced synergy measure is analytically derived, addressing questions on the measure's computability. Similarly to Chapter by Fernández et al., this chapter presents the new synergy measure within an *axiomatic* framework.

The importance of an axiomatic argument is beginning to be recognised in this field, and we believe that novel measures and approaches are much more convincing when they are shown to satisfy a set of axioms, rather than simply capture some intuition. Axiomatic approaches have demonstrated enormous power in other branches of science by rendering intuitions precise. It therefore would be desirable to obtain a more robust understanding of the underlying concepts for the quickly growing field of GSO by identifying axiomatic principles that they should respect, and comparing the axioms rather than individual examples.

1.5 Coordinated Behaviour and Learning within an Embodied Agent

The investigations that take the agent-centric perspective on GSO have almost always considered various perception-action or sensorimotor loops, coupling the agent with its environment. While most of the methods presented in this section can potentially be transferable to other contexts, the focus on embodiment and cognitive agent's behaviour is clear. Moreover, the depth and breadth of the results presented in this section point to maturity of GSO research in this particular area.

The Chapter by Der is centred on two cornerstones of modern robotics and models of human and animal intelligence: embodiment and self-organisation. It introduces a new, strictly local, unsupervised learning rule for a neurally controlled robot that drives the system into self-organization, achieving global, whole-body motion patterns. Interestingly, the proposed rule balances a differential Hebbian and an anti-Hebbian learning mechanism: while the former drives the system to activity, the latter acts as a confinement, keeping the system under control. In addition, the chapter introduces the concept of fundamental modes of a closely coupled system under parsimonious control, highlighting the crucial role of spontaneous symmetry breaking. Specifically, it demonstrates that the unsupervised learning rule induces a variety of behaviors, interpreting the resultant spatio-temporal patterns as broken symmetries.

The following chapter by Martius et al. continues the theme of this section, focussing on the question of how a robot can be equipped with an internal drive for innovation that may serve as a path for an open ended, self-determined development. Building on earlier work in homeokinetic learning, it presents several mechanisms for guided self-organisation of robot behavior. While homeokinesis bootstraps the exploration process of embodied systems, leading to self-organization of various behavioural patterns, the proposed guiding mechanism by cross-motor "teaching" signals produces a goal-oriented behavior. Importantly, the balance between self-organisation and guidance forces can be adjusted with a single parameter. The findings are exemplified by a number of case studies, and complemented by a comparative analysis of contemporary approaches to learning of autonomous robot behavior.

Ay and Zahedi review basic concepts of Pearl's causality theory. Within this theory, the notion of experimental intervention is essential and provides a formal basis for the definition and the study of causal effects that can be identified from an agent-centric perspective. The identification of casual effects based on observational data represents a particularly important problem within this study. While it is well-known that, given appropriate structural information, a causal effect can be identified based on solely observational data, the study demonstrates this in the context of the agent's sensorimotor loop, considering various kinds of information flows. Crucially, it points out that if structural information is not available, it is still possible to infer important properties of the underlying causal structure. The chapter also discusses transfer entropy and predictive information relating these measures to coordinated behaviour and morphological computation. The framework developed in this chapter may also provide a perspective on results presented in other chapters. This perspective is centered on the notion of world model which plays an important role in evaluating causal effects in the sensorimotor loop, being also an essential object within both the empowerment approach to behavior, as argued in Chapter by Salge et al., and the learning processes presented in Chapters by Der and Martius et al.

The Chapter by Agmon argues that living systems need to be considered as organisms embedded within a complex environment presenting numerous diverse obstacles. In order to navigate such an environment, organisms must exhibit diverse, complex and adaptive behaviours. Specifically, they must use flexible behavioral strategies that rapidly integrate and coordinate multiple possible actions. The proposed approach models behaviour from a dynamical systems perspective, enabling a study of dynamical strategies required for efficient transitions between such coordinated actions. Having developed an understanding of this dynamical coordination via appropriate simulations, the chapter proceeds to identifications of some key action variables and general patterns, proposing a framework for modelling action switching in terms of dynamical systems.

Obst and Boedecker utilise the framework of information dynamics, that was presented in detail in Chapter by Lizier et al., in evaluating different computational properties of input-driven recurrent neural networks (RNNs), and reservoir computing networks in particular. In addition, some of the methods are drawn from statistical learning theory and dynamical systems theory, sharing the motivation expressed in the previous chapter (by Agmon), and highlighting once more that dynamic aspects of information processing in self-organising systems can be studied from different complementary perspectives. Interestingly, the identified information-processing elements are argued to be potentially useful as target signals to guide self-organised optimisation procedures — and this aspect resonates strongly with the motivation outlined in chapters by Martius et al., as well as the Chapter by Der. For example, a learning goal can be specified in terms of information transfer between input and desired output, guiding local adjustments to self-recurrence of each reservoir unit.

1.6 Swarms and Networks of Agents

Swarm and network dynamics provide one of the most striking visual examples of self-organisation. Importantly, these self-organising dynamics bring about evident collective benefits. And so our story would be incomplete without the cases centred on collective dynamics. It is also no longer surprising that these studies are carried out within well-motivated frameworks built around theories that we have encountered already.

Miller et al. utilise a novel information-theoretic characterisation of dynamics within swarms, in the context of GSO, through explicitly measuring the extent of collective communications and tracing collective memory. The details of the required information-theoretic treatment were presented in The Chapter by Lizier et al. This study follows the argument that collective communication and memory are two necessary elements of distributed computation, and study these elements in two different models (one being more constrained/guided than the other). The approach deals with both global and local information dynamics ultimately discovering diverse ways in which an individual's location within the group is related to its information processing role. The comparative analysis shows that there are significant differences between the information dynamics that emerge from the two models: the constrained system is observed to have lower peaks of collective memory and communications, resulting in less-formed and delayed information cascades. In other words, it is harder for a constrained system to self-organise into a coherent, swarming state.

The next chapter by Sayama continues the topic of particle swarm self-organisation, extending the scope to heterogeneous swarms which are particularly rich in their dynamics rendering their design difficult in a traditional top-down manner. This problem is addressed by two complementary approaches: employing and enhancing the methods of interactive evolutionary design, and spontaneously evolving self-organising swarms within an artificial ecosystem. One of the key messages of this thought-provoking chapter is that designs obtained through open-ended evolutionary processes "may have a potential to be more creative and innovative than those produced through optimization for explicit selection criteria".

Sperati et al. investigate the use of information-theoretic concepts as task-independent utility functions (e.g., fitness functions or reward/error measures) in evolving and/or adapting behaviour of mobile robots. This chapter utilises the information-theoretic utility functions to drive the evolution of coordinated behaviours in groups of homogeneous robots. The argument is that maximising mutual information results in evolved controllers that produce structured coordinated behaviours. Importantly, the proposed methodology is verified in the swarm robotics context, which is particularly challenging because of the multi-scale relationships within the self-organising system: (a) between control rules of a single robot and its individual behaviour, and (b) between interacting individuals and the global pattern self-organising at the swarm level.

The Chapter by Yaeger studies and experiments with mechanisms that give rise to complexity, using the Polyworld simulation environment, an evolutionary model of a computational ecosystem, populated by haploid agents with a suite of primitive, neurally controlled behaviors. The experiments incorporate an information-theoretic measure of neural complexity to guide agent evolution and behavior, by using complexity as a fitness function. The results demonstrate a correlation between behavioral adaptation and the employed measures of neural complexity and graph topology. The study suggests that it is reasonable to expect that the presented simulation results on open-ended evolution and dynamical complexity may apply to biological systems.

As an example of a scenario which is relevant to real-world decentralised applications, Sarkar and Robinson study large scale evolving peer-to-peer networks, tracing their modularity properties over time by spectral methods. Once some optimisation or evolutionary rules are specified at the design stage, the resultant evolution of the system can be charted using the spectra. Importantly, if modularity or hierarchical organisation appears, the spectra can classify the rules that lead to the self-organisation of modularity.

1.7 Conclusion

The last decade witnessed the emergence of several generic frameworks and principles unifying the research in Guided Self-Organisation: (a) Various information-processing limits and costs need to be accounted by both universal objectives and task-relevant constraints. (b) Different elements of distributed computation, captured information-theoretically as distinct axes of complexity, may be used to filter (and evolve for) coherent spatio-temporal computing structures. (c) Restricting the system's dynamics to a certain region in phase space may guide the self-organisation by specifying the targets via statistical properties of the desired dynamical state or critical regime. (d) Identifying and tracing graph-theoretical properties of networks under consideration may classify and predict their evolution and growth.

Equipped with these guidelines, designers may consider diverse representative application scenarios. For instance, taking the agent-centric perspective on GSO allows one to relate embodiment and self-organisation, deriving learning rules and guiding signals that produce coordinated behaviour. This behaviour can often be related to fundamental concepts developed across several disciplines, e.g., morphological computation, symmetry breaking, reservoir computing, information cascades, interactive evolution, etc. It is evident that most of the studies benefit from being formulated in terms of universal principles, such as the maximisation of predictive information, transfer entropy, neural complexity, empowerment, open-ended evolution, or driving towards some spatio-temporal structure within the system's dynamical state-space. This is not surprising, as these connections are based on strong theoretical foundations. For example, symmetry breaking is a known phenomenon occurring at phase transitions and thus, guiding the self-organising system towards a critical regime is likely produce rich behaviours. Similarly, a self-organising swarm is expected to pass through phases of high collective coherence, and so considering it as a system collectively "computing" its next states allows to explicitly trace information cascades and collective memory.

While most of the highlighted approaches and unifying principles are being refined, their contributions to numerous applications and case studies indicate some maturity of the field. And so we believe that the field of GSO that is going to follow these inception years will grow on strong foundations, continuing its cross-disciplinary development and making significant impact on both the theory and practice of self-organising systems.

Acknowledgements. Mikhail Prokopenko is grateful for travel support from The Max Planck Institute for Mathematics in the Sciences (Leipzig, Germany) in 2013. Daniel Polani was partly supported by the European Commission via the CORBYS (Cognitive Control Framework for Robotic Systems) project under contract FP7 ICT-270219. The views expressed in this book are those of its editors and authors, and not necessarily those of the consortium.

References

Ay, N., Bernigau, H., Der, R., Prokopenko, M.: Information-driven self-organization: the dynamical system approach to autonomous robot behavior. Theory in Biosciences 131, 161–179 (2012)

Ay, N., Bertschinger, N., Der, R., Güttler, F., Olbrich, E.: Predictive information and explorative behavior of autonomous robots. European Journal of Physics B 63, 329–339 (2008)

Ay, N., Der, R., Prokopenko, M.: Guided self-organization: perception-action loops of embodied systems. Theory in Biosciences, 1–3 (2011)

Beer, R.: Dynamical systems and embedded cognition, ch. 12. Cambridge University Press (2013)

Beer, R.: Dynamical analysis of evolved agents: A primer. MIT Press (2014)

Bell, A.J., Sejnowski, T.J.: An information-maximisation approach to blind separation and blind deconvolution. Neural Computation 7, 1129–1159 (1995)

Bialek, W., Nemenman, I., Tishby, N.: Predictability, complexity and learning. Neural Computation 13, 2409–2463 (2001)

Butko, N.J., Triesch, J.: Exploring the role of intrinsic plasticity for the learning of sensory representations. In: ESANN 2006 Proceedings - 14th European Symposium on Artificial Neural Networks Bruges. Neurocomputing., pp. 467–472 (2005)

Capdepuy, P., Polani, D., Nehaniv, C.: Maximization of potential information flow as a universal utility for collective behaviour. In: 2007 IEEE Symposium on Artificial Life, pp. 207–213. IEEE (2007)

Crooks, G.: Measuring thermodynamic length. Physical Review Letters, 99(10), 100602+ (2007)

Crutchfield, J.P.: Computational mechanics: Natural computation and self-organization. In: Calude, C.S., Costa, J.F., Dershowitz, N., Freire, E., Rozenberg, G. (eds.) UC 2009. LNCS, vol. 5715, pp. 3–3. Springer, Heidelberg (2009)

Der, R., Martius, G.: The Playful Machine - Theoretical Foundation and Practical Realization of Self-Organizing Robots. Springer (2012)

Dewar, R.: Information theory explanation of the fluctuation theorem, maximum entropy production and self-organized criticality in non-equilibrium stationary states. J. Phys. A: Math. Gen. 36(3), 631–641 (2003)

Dewar, R.: Maximum entropy production and the fluctuation theorem. J. Phys. A: Math. Gen. 38, 371–381 (2005)

Dini, P., Nehaniv, C.L., Egri-Nagy, A., Schilstra, M.J.: Exploring the concept of interaction computing through the discrete algebraic analysis of the Belousov-Zhabotinsky reaction. Biosystems 112(2), 145–162 (2013)

Egri-Nagy, A., Nehaniv, C.L.: Symmetries of automata. In: Dömösi, P., Iván, S. (eds.) AFL, p. 391 (2011)

Friston, K.: The free-energy principle: a rough guide to the brain? Trends Cogn. Sci. 13(7), 293–301 (2009)

Gershenson, C.: Guiding the self-organization of random boolean networks. Theory in Biosciences 131(3), 181–191 (2012)
Grinstein, G., Linsker, R.: Comments on a derivation and application of the 'maximum entropy production' principle. J. Phys. A: Math. Theor. 40, 9717–9720 (2007)
Gros, C.: Complex and adaptive dynamical systems: a primer. Springer, Heidelberg (2008)
Jung, T., Polani, D., Stone, P.: Empowerment for continuous agent-environment systems. Adaptive Behaviour 19(1), 16–39 (2011)
Klyubin, A.S., Polani, D., Nehaniv, C.L.: All else being equal be empowered. In: Capcarrère, M.S., Freitas, A.A., Bentley, P.J., Johnson, C.G., Timmis, J. (eds.) ECAL 2005. LNCS (LNAI), vol. 3630, pp. 744–753. Springer, Heidelberg (2005a)
Klyubin, A.S., Polani, D., Nehaniv, C.L.: Empowerment: a universal agent-centric measure of control. In: The 2005 IEEE Congress on Evolutionary Computation, vol. 1, pp. 128–135. IEEE (2005b)
Langton, C.G.: Computation at the edge of chaos: phase transitions and emergent computation. Physica D 42(1-3), 12–37 (1990)
Lazar, A., Pipa, G., Triesch, J.: The combination of STDP and intrinsic plasticity yields complex dynamics in recurrent spiking networks. In: ESANN 2006 Proceedings - 14th European Symposium on Artificial Neural Networks Bruges, pp. 647–652 (2006)
Linsker, R.: Self-organization in a perceptual network. Computer 21(3), 105–117 (1988)
Lizier, J.T., Flecker, B., Williams, P.L.: Towards a synergy-based approach to measuring information modification. In: IEEE Symposium Series on Computational Intelligence (SSCI 2013) — IEEE Symposium on Artificial Life, Singapore. IEEE Press (April 2013)
Lizier, J.T., Prokopenko, M., Zomaya, A.Y.: Local information transfer as a spatiotemporal filter for complex systems. Physical Review E 77(2), 026110 (2008)
Lizier, J.T., Prokopenko, M., Zomaya, A.Y.: Information modification and particle collisions in distributed computation. Chaos 20(3), 037109 (2010)
Lizier, J.T., Prokopenko, M., Zomaya, A.Y.: Local measures of information storage in complex distributed computation. Information Sciences 208, 39–54 (2012)
Lungarella, M., Sporns, O.: Mapping information flow in sensorimotor networks. PLoS Comput. Biol. 2(10), e144 (2006)
Martius, G., Der, R., Ay, N.: Information driven self-organization of complex robotic behaviors. PLoS ONE 8(5), e63400 (2013)
Piraveenan, M., Prokopenko, M., Zomaya, A.Y.: Local assortativity and growth of Internet. European Physical Journal B 70(2), 275–285 (2009)
Piraveenan, M., Prokopenko, M., Zomaya, A.Y.: Assortative mixing in directed biological networks. IEEE/ACM Trans. Comput. Biology Bioinform. 9(1), 66–78 (2012)
Polani, D.: Information: currency of life? HFSP Journal 3(5), 307–316 (2009)
Polani, D., Prokopenko, M., Yaeger, L.S.: Information and self-organization of behavior. Advances in Complex Systems (ACS) 16(02) (2013)
Polani, D., Sporns, O., Lungarella, M.: How information and embodiment shape intelligent information processing. In: Lungarella, M., Iida, F., Bongard, J.C., Pfeifer, R. (eds.) 50 Years of Aritficial Intelligence. LNCS (LNAI), vol. 4850, pp. 99–111. Springer, Heidelberg (2007)
Prokopenko, M.: Guided self-organization. HFSP Journal 3(5), 287–289 (2009)
Prokopenko, M., Boschetti, F., Ryan, A.J.: An information-theoretic primer on complexity, self-organization, and emergence. Complexity 15(1), 11–28 (2009)

Prokopenko, M., Gerasimov, V., Tanev, I.: Evolving spatiotemporal coordination in a modular robotic system. In: Nolfi, S., Baldassarre, G., Calabretta, R., Hallam, J.C.T., Marocco, D., Meyer, J.-A., Miglino, O., Parisi, D. (eds.) SAB 2006. LNCS (LNAI), vol. 4095, pp. 558–569. Springer, Heidelberg (2006a)

Prokopenko, M., Gerasimov, V., Tanev, I.: Measuring spatiotemporal coordination in a modular robotic system. In: Rocha, L.M., Yaeger, L.S., Bedau, M.A., Floreano, D., Goldstone, R.L., Vespignani, A. (eds.) Proceedings of the 10th International Conference on the Simulation and Synthesis of Living Systems (ALifeX), Bloomington, Indiana, USA, pp. 185–191. MIT Press (2006b)

Prokopenko, M., Lizier, J.T., Obst, O., Wang, X.R.: Relating Fisher information to order parameters. Physical Review E 84(4), 041116 (2011)

Prokopenko, M., Lizier, J.T., Price, D.C.: On thermodynamic interpretation of transfer entropy. Entropy 15(2), 524–543 (2013)

Rubinov, M., Sporns, O.: Complex network measures of brain connectivity: Uses and interpretations. NeuroImage 52(3), 1059–1069 (2010)

Salge, C., Glackin, C., Polani, D.: Approximation of empowerment in the continuous domain. Advances in Complex Systems 16(1/2), 1250079 (2012)

Schreiber, T.: Measuring information transfer. Physical Review Letters 85(2), 461–464 (2000)

Still, S.: Information-theoretic approach to interactive learning. EPL (Europhysics Letters) 85(2), 28005–28010 (2009)

Still, S., Sivak, D.A., Bell, A.J., Crooks, G.E.: Thermodynamics of prediction. Phys. Rev. Lett. 109, 120604 (2012)

Tishby, N., Polani, D.: Information theory of decisions and actions. In: Cutsuridis, V., Hussain, A., Taylor, J. (eds.) Perception-Action Cycle: Models, Architecture and Hardware, pp. 601–636. Springer (2011)

Touchette, H., Lloyd, S.: Information-theoretic limits of control. Phys. Rev. Lett. 84, 1156 (2000)

Triesch, J.: A gradient rule for the plasticity of a neuron's intrinsic excitability. In: Duch, W., Kacprzyk, J., Oja, E., Zadrożny, S. (eds.) ICANN 2005. LNCS, vol. 3696, pp. 65–70. Springer, Heidelberg (2005)

Williams, P.L., Beer, R.D.: Information dynamics of evolved agents. In: Doncieux, S., Girard, B., Guillot, A., Hallam, J., Meyer, J.-A., Mouret, J.-B. (eds.) SAB 2010. LNCS, vol. 6226, pp. 38–49. Springer, Heidelberg (2010)

Wissner-Gross, A.D., Freer, C.E.: Causal entropic forces. Phys. Rev. Lett. 110, 168702 (2013)

Yaeger, L.S.: Identifying neural network topologies that foster dynamical complexity. Advances in Complex Systems (ACS) 16(02) (2013)

Zahedi, K., Ay, N., Der, R.: Higher coordination with less control – A result of information maximization in the sensorimotor loop. Adaptive Behavior 18(3-4), 338–355 (2010)

[illegible]

Part II
Foundational Frameworks

Chapter 2
Information Measures of Complexity, Emergence, Self-organization, Homeostasis, and Autopoiesis

Nelson Fernández, Carlos Maldonado, and Carlos Gershenson

2.1 Introduction

In recent decades, the scientific study of complex systems (Bar-Yam 1997; Mitchell 2009) has demanded a paradigm shift in our worldviews (Gershenson et al. 2007; Heylighen et al. 2007). Traditionally, science has been reductionistic. Still, complexity occurs when components are difficult to separate, due to relevant *interactions*. These interactions are relevant because they generate novel information which determines the future of systems. This fact has several implications (Gershenson 2013).

Nelson Fernández
Laboratorio de Hidroinformática, Facultad de Ciencias Básicas
Univesidad de Pamplona, Colombia
e-mail: nelferpa@gmail.com
http://unipamplona.academia.edu/NelsonFernandez

Nelson Fernández
Centro de Micro-electrónica y Sistemas Distribuidos,
Universidad de los Andes, Mérida, Venezuela

Carlos Maldonado
Facultad de Ciencias
Universidad Nacional Autónoma de México
e-mail: cdmc89@gmail.com

Carlos Maldonado · Carlos Gershenson
Departamento de Ciencias de la Computación,
Instituto de Investigaciones en Matemáticas Aplicadas y en Sistemas,
Universidad Nacional Autónoma de México

Carlos Gershenson
Centro de Ciencias de la Complejidad
Universidad Nacional Autónoma de México
e-mail: cgg@unam.mx
http://turing.iimas.unam.mx/~cgg

M. Prokopenko (ed.), *Guided Self-Organization: Inception*,
Emergence, Complexity and Computation 9,
DOI: 10.1007/978-3-642-53734-9_2, © Springer-Verlag Berlin Heidelberg 2014

A key implication: reductionism—the most popular approach in science—is not appropriate for studying complex systems, as it attempts to simplify and separate in order to predict. Novel information generated by interactions limits prediction, as it is not included in initial or boundary conditions. It implies computational irreducibility (Wolfram 2002), i.e. one has to reach a certain state before knowing it will be reached. In other words, *a priori* assumptions are of limited use, since the precise future of complex systems is known only *a posteriori*. This does not imply that the future is random, it just implies that the degree to which the future can be predicted is inherently limited.

It can be said that this novel information is *emergent*, since it is not in the components, but produced by their interactions. Interactions can also be used by components to *self-organize*, i.e. produce a global pattern from local dynamics. Interactions are also key for feedback control loops, which help systems regulate their internal states, an essential aspect of living systems.

We can see that reductionism is limited for describing such concepts as complexity, emergence, self-organization, and life. In the wake of the fall of reductionism as a dominant worldview (Morin 2007), a plethora of definitions, notions, and measures of these concepts has been proposed. Still, their diversity seems to have created more confusion than knowledge. In this chapter, we revise a proposal to ground measures of these concepts in information theory. This approach has several advantages:

- Measures are precise and formal.
- Measures are simple enough to be used and understood by people without a strong mathematical background.
- Measures can help clarify the meaning of the concepts they describe.
- Measures can be applied to any phenomenon, as anything can be described in terms of information (Gershenson 2012b).

This chapter is organized as follows: In the next section, background concepts are presented, covering briefly complexity, emergence, self-organization, homeostasis, autopoiesis, information theory, random Boolean networks, and limnology. Section 2.3 presents axioms and derives measures for emergence, self-organization, complexity, homeostasis and autopoiesis. To illustrate the measures, these are applied to two case studies in Section 2.4: random Boolean networks and an Arctic lake ecosystem. Discussion and conclusions close the chapter.

2.2 Background

2.2.1 Complexity

There are dozens of notions and measures of complexity, proposed in different areas with different purposes (Edmonds 1999; Lloyd 2001). Etymologically, complexity comes from the Latin *plexus*, which means interwoven. Thus, something complex is difficult to separate. This means that its components are interdependent, i.e. their future is partly determined by their *interactions* (Gershenson 2013). Thus, studying the components in isolation—as reductionistic approaches attempt—is not sufficient to describe the dynamics of complex systems.

Nevertheless, it would be useful to have global measures of complexity, just as temperature characterizes the properties of kinetic energy of molecules or photons. Each component can have a different kinetic energy, but the statistical average is represented in the temperature. For complex systems, particular interactions between components can be different, but we can say that complexity measures should represent the type of interactions between components, just as Lyapunov exponents characterize different dynamical regimes.

A useful measure of complexity should enable us to answer questions such as: Is a desert more or less complex than a tundra? What is the complexity of different influenza outbreaks? Which organisms are more complex: predators or preys; parasites or hosts; individual or social? What is the required complexity of a company to face the complexity of a market[1]?

Moreover, with the recent scandalous increase of data availability in most domains, we urgently need measures to make sense of it.

2.2.2 *Emergence*

Emergence has probably been one of the most misused concepts in recent decades. The reasons for this misuse are varied and include: polysemy (multiple meanings), buzzwording, confusion, hand waving, Platonism, and even mysticism. Still, the concept of emergence can be clearly defined and understood (Anderson 1972). The properties of a system are emergent if they are not present in their components. In other words, global properties which are produced by local interactions are emergent. For example, the temperature of a gas can be said to be emergent (Shalizi 2001), since the molecules do not possess such a property: it is a property of the collective. In a broad an informal way, emergence can be seen as differences in phenomena as they are observed at different scales (Prokopenko et al. 2009).

Some might perceive difficulties in describing phenomena at different scales (Gershenson 2013), but this is a consequence of attempting to find a single "true" description of phenomena. Phenomena do not depend on the descriptions we have of them, and we can have several different descriptions of the same phenomenon. It is more informative to handle several descriptions at once, and actually it is necessary when studying emergence and complex systems.

2.2.3 *Self-organization*

Self-organization has been used to describe swarms, flocks, traffic, and many other systems where the local interactions lead to a global pattern or behavior (Camazine et al. 2003; Gershenson 2007). Intuitively, self-organization implies that a system

[1] This question is related to the law of requisite variety (Ashby 1956).

increases its own organization. This leads to the problems of defining organization, system, and self. Moreover, as Ashby showed (1947b), almost any dynamical system can be seen as self-organizing: if it has an attractor, and we decide to call that attractor "organized", then the system dynamics will tend to it, thus increasing by itself its own organization. If we can describe almost any system as self-organizing, the question is not whether a system *is* self-organizing or not, but rather, when is it useful to describe a system as self-organizing (Gershenson and Heylighen 2003)?

In any case, it is convenient to have a measure of self-organization which can capture the nature of local dynamics at a global scale. This is especially relevant for the nascent field of guided self-organization (GSO) (Prokopenko 2009; Ay et al. 2012; Polani et al. 2013). GSO can be described as *the steering of the self-organizing dynamics of a system towards a desired configuration* (Gershenson 2012a). This desired configuration will not always be the natural attractor of a controlled system. The mechanisms for guiding the dynamics and the design of such mechanisms will benefit from measures characterizing the dynamics of systems in a precise and concise way.

2.2.4 Homeostasis

Originally, the concept of homeostasis was developed to describe internal and physiological regulation of bodily functions, such as temperature or glucose levels. Probably the first person to recognize the internal maintenance of a near-constant environment as a condition for life was Bernard (1859). Subsequently, Canon (1932) coined the term homeostasis from the Greek hómoios (similar) and stasis (standing still). Cannon defined homeostasis as the ability of an organism to maintain steady states of operation during internal and external changes. Homeostasis does not imply an immobile or a stagnant state. Although some conditions may vary, the main properties of an organism are maintained.

Later, the British cybernetician William R. Ashby proposed, in an alternative form, that homeostasis implicates an adaptive reaction to maintain "essential variables" within a range (Ashby 1947a, 1960). In order to explain the generation of behavior and learning in machines and living systems, Ashby also contributed by linking the concepts of ultrastability and homeostatic adaptation (Di Paolo 2000). Ultrastability refers to the normal operation of the system within a "viability zone" to deal with environmental changes. This viability zone is defined by the lower and upper bounds of the essential variables. If the value of variables crosses the limits of its viability zone, the system has a chance of finding new parameters that make the challenged variables return to their viability zone.

A dynamical system has a high homeostatic capacity if it is able to maintain its dynamics close to a certain state or states (attractors). As explained above, when perturbations or environmental changes occur, the system adapts to face the changes within the viability zone, that is, without the system "breaking" (Ashby 1947a). Homeostasis can be seen as a dynamic process of self-regulation and adaptation by which systems adapt their behavior over time (Williams 2006). The homeostasis concept can be applied to different fields beyond life sciences and is also closely related to self-organization and to robustness (Wagner 2005; Jen 2005).

2.2.5 Autopoiesis

Autopoiesis comes from the Greek *auto* (self) and *poiesis* (creation, production) and was proposed as a concept to define the living. According to Maturana (2011), the notion of autopoiesis was created to connote and describe the molecular processes taking place in the realization of living beings as autonomous entities. However, this meaning of the word autopoiesis, which was used to describe closed networks of molecular production, was chosen only until 1970 (Maturana and Varela 1980). This notion arises from a series of questions, related to the internal dynamics of living systems, which Maturana began considering in the 1960s, such as: "What should be the constitution of a system so that I see a living system as a result of its operation?", "What kind of systems or entities are living systems?", and another question that a student asked Maturana: "What happened three billion eight hundred million years ago so that you can now say that living systems began then?"

In the context of autopoiesis, living beings occur as discrete autonomous dynamic molecular autopoietic entities. These entities are in a continuous realization of their self-production. Thus, autopoiesis describes the internal dynamics of a living system in the molecular domain. Maturana notices that living beings are dynamical systems in continuous change. Interactions between elements of an autopoietic system regulate the production and regeneration of the system's components, having the potential to develop, preserve, and produce their own organization (Varela et al. 1974).

For example, a bacterium may produce another bacterium by cellular division, while a virus requires a host cell to produce another virus. The production of the new bacterium is made by the interactions between the elements of another bacterium. The production of a new virus depends on interactions between elements of an external system. Thus, it can be said that a bacterium is more autopoietic than a virus. In this sense, autopoiesis is much related to autonomy (Ruiz-Mirazo and Moreno 2004). Autonomy is always limited in open systems, as their states depend on environmental interactions. However, differences in autonomy can be clearly identified, just like in the previous example.

The concept of autopoiesis has been extended to other areas beyond biology (Luisi 2003; Seidl 2004; Froese and Stewart 2010), although no formal measure had been proposed so far.

2.2.6 Information Theory

Information has had a most interesting history (Gleick 2011). Information theory was created by Claude Shannon in 1948 in the context of telecommunications.

He analyzed whether it was possible to reconstruct data transmitted across a noisy channel. In his model, **information** is represented as a string $X = x_0x_1...$ where each x_i is a symbol from a finite set of symbols $\mathcal{A}$ called the **alphabet**. Moreover, each symbol in the alphabet has a given probability $P(x)$ of occurring in the string. Common symbols will have a high $P(x)$ while infrequent symbols will have a low $P(x)$.

Shannon was interested in a function to measure how much information a process "produces". Quoting Shannon (1948)[2]:

> Suppose we have a set of possible events whose probabilities of occurrence are $p_1, p_2, ..., p_n$. These probabilities are known but that is all we know about the event that might occur. Can we find a measure of how much "choice" is involved in the selection of the event or how uncertain we are of the outcome? If there is such a measure, say $(p_1, p_2, ..., p_n)$ it is reasonable to require of it the following properties:
>
> 1. I should be continuous in each p_i.
> 2. If all the p_i are equal, $p_i = 1/n$, then I should be a monotonic increasing function of n. With equally n likely events there is more choice, or uncertainty, when there are more possible events.
> 3. If a choice be broken down into two successive choices, the original I should be the weighted sum of the individual values of I.

With these few *axioms*, Shannon demonstrates that the only function I satisfying the three above is of the form:

$$I = -K \sum_{i=i}^{n} p_i \log p_i, \tag{2.1}$$

where K is a positive constant.

For example, if we have a string '0001000100010001...', we can estimate $P(0) = 0.75$ and $P(1) = 0.25$, then $I = -(0.75 \cdot \log 0.75 + 0.25 \cdot \log 0.25)$. If we use $K = 1$ and a base 2 logarithm, then $I \approx 0.811$.

Shannon used H to describe information (we are using I) because he was thinking in the Boltzmann's H theorem[3] when he developed the theory. Therefore, he called equation 2.1 the entropy of the set of probabilities $p_1, p_2, ..., p_n$. In modern words, I is a function of a random variable X.

The unit of information is the bit (*bi*nary digi*t*). One bit represents the information gained when a binary random variable becomes known. However, since equation 2.1 is a sum of probabilities, Shannon's information is a unitless measure.

More details about information theory in general can be found in Ash (1990), while a primer on information theory related to complexity, self-organization, and emergence is found in Prokopenko et al. (2009).

[2] We replaced Shannon's H for I.

[3] The Boltzmann H theorem is given in the thermodynamic context. It states that the entropy of an ideal gas increases in an irreversible process. This might be also the reason why he required the second property.

2.2.7 Random Boolean Networks

Random Boolean networks (RBNs) are abstract computational models, originally proposed to study genetic regulatory networks (Kauffman 1969, 1993). However, being general models, their study and use has expanded beyond biology (Aldana-González et al. 2003; Gershenson 2004, 2012a).

A RBN is formed by N nodes linked by K connections[4]. Each node has a Boolean state, i.e. zero or one. The future state of each node is determined by the current states of the nodes that link to it and a lookup table which specifies how the update will take place. The connectivity (which nodes affect which) and the lookup tables (how nodes affect their states) are usually generated randomly for a network, but remain fixed during its dynamics.

RBNs have been found to have three different dynamical regimes, which have been studied extensively (Gershenson 2004):

Ordered. Most nodes are static, RBNs are robust to perturbations.
Chaotic. Most nodes are changing, RBNs are fragile to perturbations.
Critical. Some nodes are changing, RBNs have adaptive potential.

Different parameters and properties determine the regime, which can be used to guide a particular RBN towards a desired regime (Gershenson 2012a).

It can be said that the critical regime balances the robustness of the ordered phase and the changeability of the chaotic phase. It has been argued that computation and life require this balance to be able to compute and adapt (Langton 1990; Kauffman 1993).

RBNs will be used in Section 2.4.1 to illustrate the measures proposed in the next section.

2.2.8 Limnology

Lakes are studied by limnology. Lakes can be divided in different zones, as shown in Figure 2.1: (i) The macrophyte zone, composed mainly of aquatic plants, which are rooted, floating or submerged. (ii) The planktonic zone corresponds to the open surface waters; away from the shore in which organisms passively float and drift (phyto and zooplankton). Planktonic organisms are incapable of swimming against a current. However, some of them are motile. (iii) The benthic zone is the lowest level of a body of water related with the substratum, including the sediment surface and subsurface layers. (iv) The mixing zone is where the exchange of water from planktonic and benthic zones occurs.

At different zones, one or more components or subsystems can be an assessment for the ecosystem dynamics. For our case study to be presented in Section 2.4.2, we considered three components: physiochemical, limiting nutrients and photosynthetic biomass for the planktonic and benthic zones.

[4] This K is different from the constant used in equation 2.1.

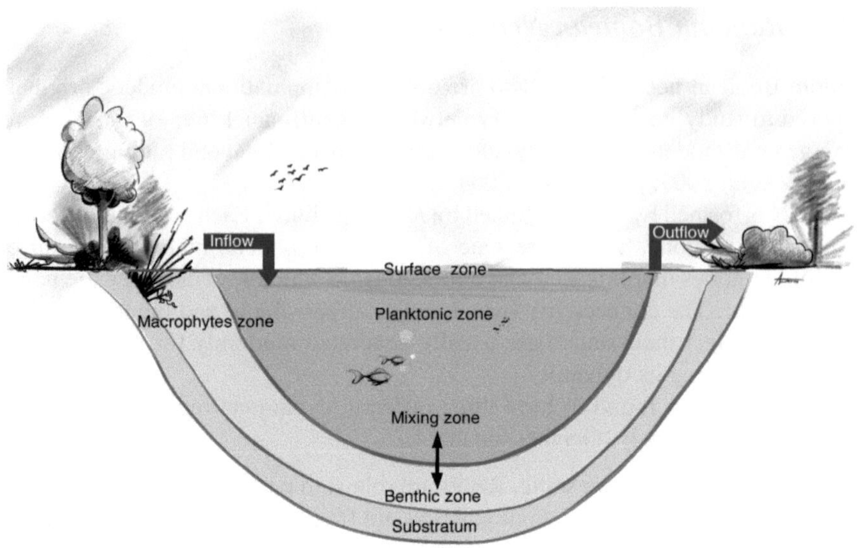

Fig. 2.1 Zones of lakes studied in limnology

The physiochemical component refers to the chemical composition of water. It is affected by various conditions and processes such as geological nature, the water cycle, dispersion, dilution, solute and solids generation (e.g. photosynthesis), and sedimentation. In this component, we highlight two water variables that are important for the aquatic life: (i) the pH equilibrium that affects, among others, the interchange of elements between the organism and its environment and (ii) the temperature regulation that is supported in the specific heat of the water.

Related to the physiochemical component, limiting nutrients which are basic for photosynthesis are associated with the biogeochemical cycles of nitrogen, carbon, and phosphorous. These cycles permit the adsorption of gases into the water or the dilution of some limiting nutrients.

In addition, among limnetic biota, photoautotrophic biomass is the basis for the trophic web establishment. The term autotrophs is used for organisms that increase their mass through the accumulation of proteins which they manufacture, mainly from inorganic radicals (Stumm 2004). This type of organisms can be found at the planktonic and benthic zones.

The previous basic limnology concepts will be useful to follow the case study of an Arctic lake, presented in Section 2.4.2.

2.3 Measures

We have recently proposed earlier versions of the measures presented in this chapter (Fernández et al. 2012; Gershenson and Fernández 2012). The ones presented here are more refined and are based on axioms. The benefit of using axioms is that the discussion is not taken so much at the level of the measures, but at the level of the presuppositions or the properties we want measures to have.

A comparison of the proposed measures with others can be found in Gershenson and Fernández (2012). It is worth noting that all of the proposed measures are unitless.

2.3.1 Emergence

We mentioned that emergence refers to properties of a phenomenon which are present at one scale and are not at another scale. Scales can be temporal or spatial. If we describe phenomena in terms of information, in order to have "new" information, "old" information has to be *transformed*. This transformation can be dynamic, static, active, or stigmergic (Gershenson 2012b). For example, new information is produced when a dynamical system changes its behavior, but also when a description of a system changes. Concerning the first case, approaches measuring the difference between past and future states have been proposed, e.g. (Shalizi and Crutchfield 2001). We can call this *dynamic emergence*. Concerning the second case, approaches measuring differences between scales have been used, e.g. (Shalizi 2001; Holzer and De Meer 2011). We can call this *scale emergence*. Even when there are differences between dynamic and scale emergencies, both can be seen as new information being produced. In the first case, dynamics produce new information. In the second case, the change of description produces new information. Thus, *information emergence E* includes both dynamic emergence and scale emergence. If we recall, Shannon proposed a quantity which measures how much information a process "produces". Therefore, we can say that emergence is the same as Shannon's information I. From now on, we will consider the emergence of a process E as the information I and we will use the base two logarithm.

$$E = I. \tag{2.2}$$

We now revise that the intuitive idea of emergence fulfills the three basic notions (axioms) that Shannon used to derive I (Shannon's H). For the continuity axiom, it is expected of a measure not to give big jumps when small changes are made. The second axiom will be harder to show. It states that if we consider an auxiliary function i which is the I function when there are n events with the same probability $1/n$ then the function i is monotonic increasing. If we have the same configuration for emergence, then we could think the process to be with equally likelihood in any of n available states. If something happens and now the process can be in $n+k$ equally likely states we can say that the process has had emergence, since now we need more information to know in which state the process is. For the third axiom, we need to find a way to figure out how is that we can 'split' the process. Lets recall that the third property required by Shannon is that if a choice can be broken into two different choices, the original I should be the average of the other two I. In a process, we can think the choices as a fraction of the process that we are currently observing. For this purpose, we can make a partition[5] of the domain, in our case, we get two subsets whose intersection is the null set and whose union is the full original set. After this, we compute the I function for each. Since we observe two different parts of a process and in each observation we get the average[6] new information required to describe the (partial) process, then it makes sense to take the average of both when observing the full process.

E, as well as I, is a probabilistic measure. $E = 1$ means that when any random binary variable becomes known, one bit of information emerges. If $E = 0$, then no new information will emerge, even as random binary variables become "known" (they are known beforehand). Again, we emphasize that emergence can take place at the level of a phenomenon observed or at the level of the description of the phenomenon observed. Either can produce novel information.

2.3.1.1 Multiple Scales

When Shannon defined equation 2.1, he included K which is a positive constant. This is important because we will change the value of K to normalize a measure onto the $[0,1]$ interval. The value of K will depend on the length of the finite alphabet $\mathcal{A}$ we use. In the particular Boolean case when we have the alphabet $\mathcal{A} = \{0,1\}$ with length $|\mathcal{A}| = 2$. Then the value $K = 1$ will normalize the measure to the interval $[0,1]$. Because of the relevance of the binary notation in computer science and other applications, we will often use the Boolean alphabet. Nevertheless, we can compute the entropy for alphabets with different lengths. We only have to consider the equation

$$K = \frac{1}{\log_2 b}, \tag{2.3}$$

where b is the length of the alphabet we use. In this way we will normalize E and measures derived from it, having a maximum of 1 and a minimum of 0.

For example, consider the string in base 4 '0133013301330133...'. We can estimate $P(0) = P(1) = 0.25$, $P(2) = 0$, and $P(3) = 0.5$. Following equation 2.1, we have $I = -K(0.25 \cdot \log 0.25 + 0.25 \cdot \log 0.25 + 0 + 0.5 \cdot \log 0.5)$. Since $b = 4$, $K = \frac{1}{\log_2 4} = 0.5$. Thus, we obtain a normalized $I = 0.75$.

2.3.2 *Self-organization*

Self-organization has been correlated with an increase in order, i.e. a reduction of entropy (Gershenson and Heylighen 2003). If emergence implies an increase of information, which is analogous to entropy and disorder, self-organization should be anti-correlated with emergence.

---

[5] We are using the set theory partition, we could have any finite number of partitions where the intersection of all of them is the null set and whose union is the original set.

[6] When there are more than two subsets in the partition, we can make a weighted average. A sort of expectation where the distribution probability is given by the nature of the process.

A measure of self-organization S should be a function $S : \Sigma \to \mathbb{R}$ (where $\Sigma = \mathcal{A}^{\mathbb{N}}$) with the following properties:

1. The range of S is the real interval $[0,1]$
2. $S(X) = 1$ if and only if X is deterministic, i.e. we know beforehand the value of the process.
3. $S(X) = 0$ if and only if X has a uniform distribution, i.e. any state of the process is equally likely.
4. $S(X)$ has a negative correlation with emergence E.

We propose as the measure

$$S = 1 - I = 1 - E \tag{2.4}$$

It is straightforward to check that this function fulfills the axioms stated. Nevertheless it is not unique. However, it is the only affine (linear) function which fulfills the axioms. For simplicity, we propose the use of 2.4 as a measure of self-organization.

$S = 1$ means that there is maximum order, i.e. no new information is produced ($I = E = 0$). On the other extreme, $S = 0$ when there is no order at all, i.e. when any random variable becomes known, information is produced/emerges ($I = E = 1$). When $S = 1$, maximum order, dynamics do not produce novel information, so the future is completely known from the past. On the other hand, when $S = 0$, minimum order, no past information tells us anything about future information.

Note that equation 2.4 makes no distinction on whether the order is produced by the system (self) or by its environment. Thus, S would have a high value in systems with a high organization, independently on whether this is a product of local interactions or imposed externally. This distinction can be easily made describing the detailed behavior of systems, but is difficult and unnecessary to capture with a general probabilistic measure such as S. As an analogy, one can measure the temperature of a substance, but temperature does not differentiate (and does not need to differentiate) between substances which are heated or cooled from the outside and substances whose temperature is dependent mainly on internal chemical reactions.

2.3.3 Complexity

Following Lopez-Ruiz et al. (1995), we can define complexity C as the balance between change (chaos) and stability (order). We have just defined such measures: emergence and self-organization. The complexity function $C : \Sigma \to \mathbb{R}$ should have the following properties:

1. The range is the real interval $[0,1]$.
2. $C = 1$ if and only if $S = E$.
3. $C = 0$ if and only if $S = 0$ or $E = 0$.

It is natural to consider the product of S and I to satisfy the last two requirements. We propose:

$$C = 4 \cdot E \cdot S. \tag{2.5}$$

Where the constant 4 is added to normalize the measure to $[0,1]$, fulfilling the first axiom. C can also be represented in terms of I as:

$$C = 4 \cdot I \cdot (1 - I). \tag{2.6}$$

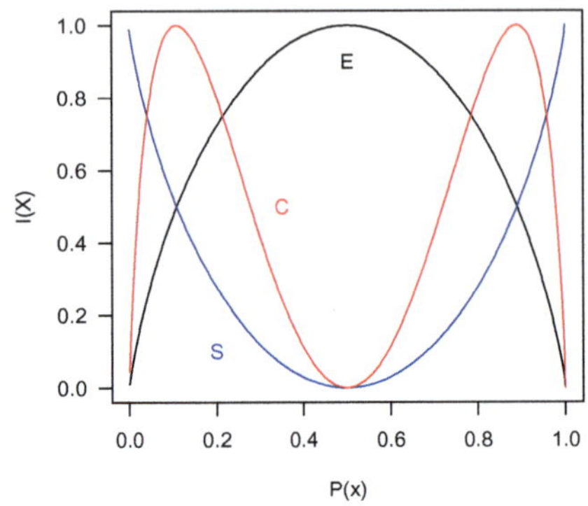

Fig. 2.2 Emergence E, self-organization S, and complexity C

Figure 2.2 plots the measures proposed so far for different values of $P(x)$. It can be seen that E is maximal when $P(x) = 0.5$ and minimal when $P(x) = 0$ or $P(x) = 1$. The opposite holds for S: it is minimal when $P(x) = 0.5$ and maximal when $P(x) = 0$ or $P(x) = 1$. C is minimal when S or E are minimal, i.e. $P(x) = 0$, $P(x) = 0.5$, or $P(x) = 1$. C is maximal when $E = S = 0.5$, which occurs when $P(x) \approx 0.11$ or $P(x) \approx 0.89$.

Shannon information can be seen as a balance of zeros and ones (maximal when $P(0) = P(1) = 0.5$), while C can be seen as a balance of E and S (maximal when $E = S = 0.5$).

2.3.4 Homeostasis

The previous three measures (E, S, and C) study how single variables change in time. To calculate the measures for a system, one can plot the histogram or simply average the measures for all variables in a system. For homeostasis H, we are interested on how all variables of a system change or not in time. Table 2.1 shows this difference: E, S, and C focusses on time series of *variables* (columns), while H focusses on *states* (rows).

Let $X = x_1x_2x_3...x_n$ represent the *state* of a system of n variables (i.e. a row in Table 2.1). If the system has a high homeostasis, we would expect that its states do

Table 2.1 Difference between observing single variables in time (columns) and several variables at one time (rows)

	X	Y	Z
$t=m-2$	x_{m-2}	y_{m-2}	z_{m-2}
$t=m-1$	x_{m-1}	y_{m-1}	z_{m-1}
$t=m$	x_m	y_m	z_m

not change too much in time. The homeostasis function $H:\Sigma\times\Sigma\rightarrow\mathbb{R}$ should have the following properties:

1. The range is the real interval $[0,1]$.
2. $H=1$ if and only if for states X and X', $X=X'$, i.e. there is no change in time.
3. $H=0$ if and only if $\forall i, x_i \neq x_i'$, i.e. all variables in the system changed.

A useful function for comparing strings of equal length is the Hamming distance. The Hamming distance d measures the percentage of different symbols in two strings X and X'. For binary strings, it can be calculated with the XOR function ($\oplus$). Its normalization bounds the Hamming distance to the interval $[0,1]$:

$$d(X,X') = \frac{\sum\limits_{i\in\{0,\ldots,|X|\}} x_i \oplus x_i'}{|X|}. \tag{2.7}$$

d measures the fraction of different symbols between X and X'. For the Boolean case, $d=0 \iff X=X'$ and $d=1 \iff X=\neg X'$, while X and X' are uncorrelated $\iff d\approx 0.5$.

We can use the inverse of d to define h:

$$h(X^t,X^{t+1}) = 1-d(X^t,X^{t+1}), \tag{2.8}$$

which clearly fulfills the desired properties of homeostasis between two states.

To measure the homeostasis of a system in time, we can generalize:

$$H = \frac{1}{m-1}\sum_{t=0}^{m-1} h(X^t,X^{t+1}), \tag{2.9}$$

where m is the total number of time steps being evaluated. H will be simply the average of different h from $t=0$ to $t=m-1$. As well as the previous measures based on I, H is a unitless measure.

When H is measured at higher scales, it can capture periodic dynamics. For example, let us have a system with $n=2$ variables and a cycle of period 2: $11\rightarrow 00\rightarrow 11$. H for base 2 will be minimal, since every time step all variables change, i.e. ones turn into zeros or zeros turn into ones. However, if we measure H in base 4, then we will be actually comparing pairs of states, since to make one time step in base 4 we take two binary time steps. Thus, in base 4 the attractor becomes

$22 \to 22$, and $H = 1$. The same applies for higher bases. An example of the usefulness of measuring H at multiple scales in elementary cellular automata is explained in Gershenson and Fernández (2012).

2.3.5 Autopoiesis

Let $\bar{X}$ represent the trajectories of the variables of a system and $\bar{Y}$ represent the trajectories of the variables of the environment of the system. A measure of autopoiesis $A : \Sigma \times \Sigma \to \mathbb{R}$ should have the following properties:

1. $A \geq 0$.
2. A should reflect the independence of $\bar{X}$ over $\bar{Y}$. This implies:
   a. $A > A' \iff \bar{X}$ produces more of its own information than $\bar{X}'$ for a given $\bar{Y}$.
   b. $A > A' \iff \bar{X}$ produces more of its own information in $\bar{Y}$ than in $\bar{Y}'$.
   c. $A = A' \iff \bar{X}$ produces as much of its own information than $\bar{X}'$ for a given $\bar{Y}$.
   d. $A = A' \iff \bar{X}$ produces as much of its own information in $\bar{Y}$ than in $\bar{Y}'$.
   e. $A = 0$ if all of the information in $\bar{X}$ is produced by $\bar{Y}$.

Following the classification of types of information transformation proposed in Gershenson (2012b), dynamic and static transformations are internal (a system producing its own information), while active and stigmergic transformations are external (information produced by another system).

It is problematic to define in a general and direct way how some information depends on other information, as causality can be confounded with co-occurrence. For this reason, measures such as mutual information are not suitable for measuring A.

As it has been proposed, adaptive systems require a high C in order to be able to cope with changes of its environment while at the same time maintaining their integrity (Langton 1990; Kauffman 1993). If $\bar{X}$ had a high E, then it would not be able to produce the same patterns for different $\bar{Y}$. With a high S, $\bar{X}$ would not be able to adapt to changes in $\bar{Y}$. Therefore, we propose:

$$A = \frac{C(\bar{X})}{C(\bar{Y})}. \tag{2.10}$$

If $C(\bar{X}) = 0$, then either $\bar{X}$ is static ($E(\bar{X}) = 0$) or pseudorandom ($S(\bar{X}) = 0$). This implies that any pattern (complexity) which could be observed in $\bar{X}$ (if any) should come from $\bar{Y}$. This case gives a minimal A. On the other hand, if $C(\bar{Y}) = 0$, it implies that any pattern (if any) in $\bar{X}$ should come from itself. This case gives a maximal $A = \infty$. A particular case occurs if $C(\bar{X}) = 0$ and $C(\bar{Y}) = 0$. A becomes undefined. But how can we say something about autopoiesis if we are comparing two systems which are either without variations ($S = 1$) or pseudorandom ($E = 1$)? This case *should* be undefined. The rest of the properties are evidently fulfilled by equation 2.10. This is certainly not the unique function to fulfill the desired axioms. The exploration of alternatives requires further study.

Since A represents a ratio of probabilities, it is a unitless measure. $A \in [0, \infty)$, although it could be mapped to $[0, 1)$ using a function such as $f(A) = \frac{A}{1+A}$. We do not normalize A because it is useful to distinguish $A > 1$ and $A < 1$ (see Section 2.5.8).

2.3.6 *Multi-scale Profiles*

Bar-Yam (2004) proposed the "complexity profile", which plots the complexity of systems depending on the scale at which they are observed. This allows to compare how a measure changes with scale. For example, the σ profile compares the "satisfaction" of systems at different scales to study organization, evolution and cooperation (Gershenson 2011).

In a similar way, multi-scale profiles can be used for each of the measures proposed, giving further insights about the dynamics of a system than measuring them at a single scale. This is clearly seen, for example, with different types of elementary cellular automata (Gershenson and Fernández 2012).

2.4 Results

In this section we apply the measures proposed in the previous section to two case studies: random Boolean networks and an aquatic ecosystem. A further case, elementary cellular automata, can be found in Gershenson and Fernández (2012).

2.4.1 *Random Boolean Networks*

Results show averages of 1000 RBNs, where 1000 steps were run from a random initial state and E, S, C and H were calculated from data generated in 1000 additional steps.

R (R Project Contributors 2012) was used with packages *BoolNet* (Müssel et al. 2010) and *entropy* (Hausser and Strimmer 2012).

Figure 2.3 shows results for RBNs with 100 nodes, as the connectivity K varies. For low K, there is high S and H, and a low E and C. This reflects the ordered regime of RBNs, where there is high robustness and few changes. Thus, it can be said that there is few or no information emerging and there is a high degree of self-organization and homeostasis. For high K, there is high E, low S and C, and uncorrelated $H \approx 0.5$. This reflects the chaotic regime of RBNs, where there is high fragility and many changes. Almost every bit (a new state for most nodes) carries novel emergent information, and this constant change implies low organization and complexity. For medium connectivities ($2 \leq K \leq 3$), there is a balance between E and S, leading to a high C. This corresponds to the critical regime of RBNs, which has been associated with complexity and the possibility of life (Kauffman 2000).

As for autopoiesis, to model a system and its environment, we coupled two RBNs: One "internal" RBN with N_i nodes and K_i average connections and one "external" with N_e nodes and K_e average connections. A "coupled" RBN is considered

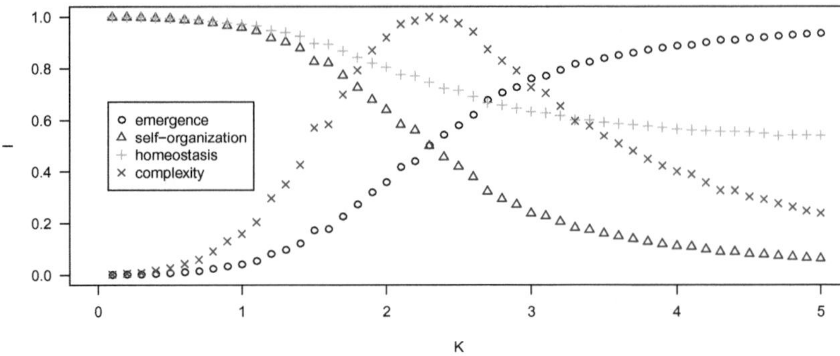

Fig. 2.3 Averages for 1000 RBNs, $N = 100$ nodes and varying average connectivity K (Gershenson and Fernández 2012)

with $N_c = N_i + N_e$ nodes and K_i connections. At every time step, the external RBN evolves independently. However, its state is copied to the N_e nodes representing it in the coupled RBN, which now evolves depending partly on the external RBN. Thus, the N_i nodes in the coupled RBN representing the internal RBN may be affected by the dynamics of the external RBN, but not vice versa. The C of each node is calculated and averaged separately for each network, obtaining an internal complexity C_i and an external complexity C_e.

Figure 2.4 and Table 2.2 show results for $N_e = 96$ and $N_i = 32$ for different combinations of K_e and K_i.

Table 2.2 A averages for 50 sets $N_e = 96, N_i = 32$. Same results as those shown in Figure 2.4.

$K_i \backslash K_e$	1	2	3	4	5
1	0.4464025	0.5151070	0.7526248	1.6460345	3.4081967
2	1.6043330	0.9586809	1.1379227	2.0669794	3.2473729
3	2.4965328	0.9999926	0.9355231	1.3604272	2.6283798
4	2.1476247	0.7249803	0.6151742	0.8055051	1.3890630
5	1.8969094	0.4760027	0.3871875	0.4755580	0.8648389

As it was shown in Figure 2.3, C changes with K, so it is expected to have $A \approx 1$ when $K_i \approx K_e$. When C_e is high ($K_e = 2$ or $K_e = 3$), then the environment dominates the patterns of the system, yielding $A < 1$. When C_e is low ($K_e < 2$ or $K_e > 3$), the patterns produced by the system are not affected that much by its environment, thus $A > 1$, as long as $K_i < K_e$ (otherwise the system is more chaotic that its environment, and so complex patterns have to come from outside).

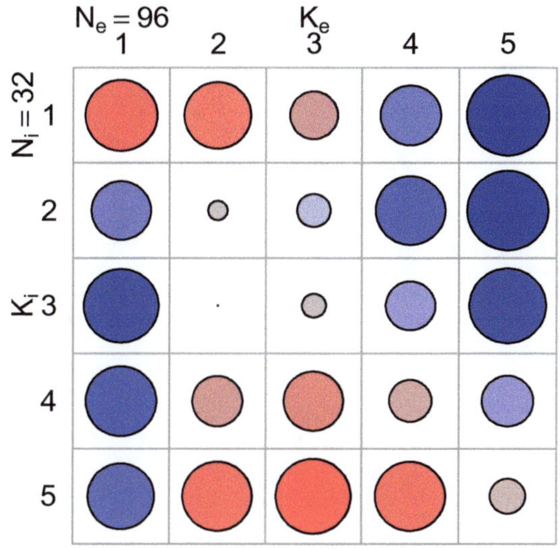

Fig. 2.4 A averages for 50 sets $N_e = 96, N_i = 32$. Values $A < 1$ are red while $A > 1$ are blue. Size of circles indicate how far A is from $A = 1$. Numerical values shown in Table 2.2.

A does not try to measure how much information emerges internally or externally, but how much the patterns are internally or externally produced. A high E means that there is no pattern, as there is constant change. A high S implies a static pattern. A high C reflects complex patterns. We are interested in A measuring the ratio of the complexity of patterns being produced by a system *compared* to the complexity of patterns produced by its environment.

2.4.2 An Ecological System: An Arctic Lake

The data from an Artic lake model used in this section was obtained using The Aquatic Ecosystem Simulator (Randerson and Bowker 2008).

In general, Arctic lake systems are classified as oligotrophic due to their low primary production, represented in chlorophyll values of 0.8-2.1 mg/m3. The lakeÕs water column, or limnetic zone, is well-mixed; this means that there are no stratifications (layers with different temperatures). During winter (October to March), the surface of the lake is ice covered. During summer (April to September), ice melts and the water flow and evaporation increase, as shown in Figure 2.5a. Consequently, the two climatic periods (winter and summer) in the Arctic region cause a typical hydrologic behavior in lakes as the one shown in Figure 2.5b. This hydrologic behavior influences the physiochemical subsystem of the lake.

Table 2.3 and Figure 2.6 show the variables and daily data we obtained from the Arctic lake simulation. The model used is deterministic, so there is no variation in

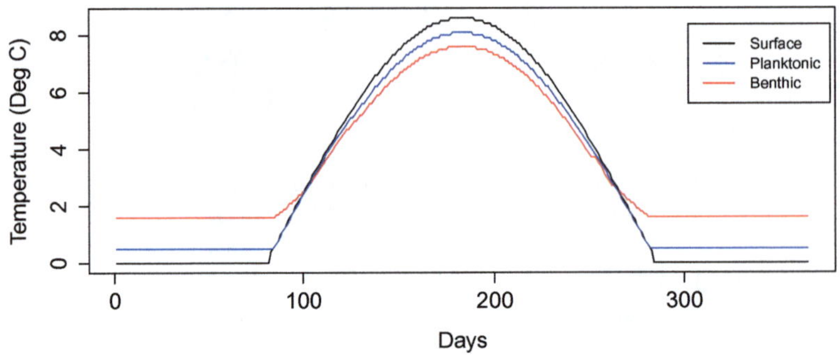

(a)

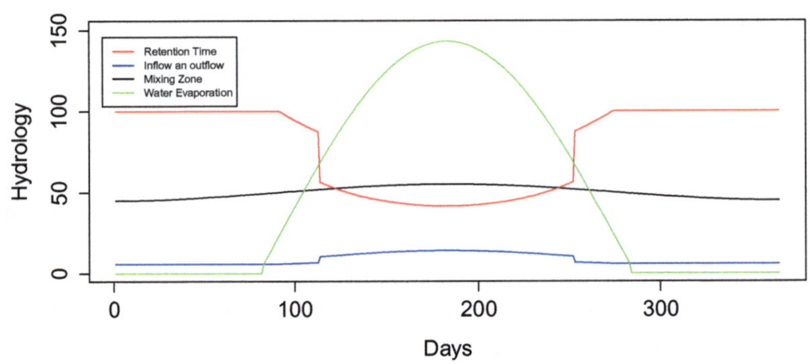

(b)

Fig. 2.5 (A) Climatic and (B) hydraulic regimes of Arctic lakes

different simulation runs. Figure 2.6 depicts a high dispersion for the following variables: temperature (T) and light (L) at the three zones of the Arctic lake (surface=S, planktonic=P and benthic=B), inflow and outflow (IO), retention time (RT) and evaporation (Ev). Ev is the variable with the highest dispersion.

Observing RT and IO in logarithmic scale, we can see that their values are located at the extremes, but their range is not long. Consequently, these variables have considerable variability in a short range. However, the ranges of the other variables do not reflect large changes. This situation complicates the interpretation and comparison of the physiochemical dynamics. To attend this situation, we normalize the data to base b of all points x of all variables X with the following equation:

Table 2.3 Physiochemical variables considered in the Arctic lake model

Variable	Units	Acronym	Max	Min	Median	Mean	std. dev.
Surface Light	MJ/m2/day	*SL*	30	1	5.1	11.06	11.27
Planktonic Ligth	MJ/m2/day	*PL*	28.2	1	4.9	10.46	10.57
Benthic Light	MJ/m2/day	*BL*	24.9	0.9	4.7	9.34	9.33
Surface Temperature	Deg C	*ST*	8.6	0	1.5	3.04	3.34
Planktonic Temperature	Deg C	*PT*	8.1	0.5	1.4	3.1	2.94
Benthic Temperature	Deg C	*BT*	7.6	1.6	2	3.5	2.29
Inflow and Outflow	m3/sec	*IO*	13.9	5.8	5.8	8.44	3.34
Retention Time	days	*RT*	100	41.7	99.8	78.75	25.7
Evaporation	m3/day	*Ev*	14325	0	2436.4	5065.94	5573.99
Zone Mixing	%/day	*ZM*	55	45	50	50	3.54
Inflow Conductivity	uS/cm	*ICd*	427	370.8	391.4	396.96	17.29
Planktonic Conductivity	uS/cm	*PCd*	650.1	547.6	567.1	585.25	38.55
Benthic Conductivity	uS/cm	*BCd*	668.4	560.7	580.4	600.32	40.84
Surface Oxygen	mg/litre	*SO2*	14.5	11.7	13.9	13.46	1.12
Planktonic Oxygen	mg/litre	*PO2*	13.1	10.5	12.6	12.15	1.02
Benthic Oxygen	mg/litre	*BO2*	13	9.4	12.5	11.62	1.51
Sediment Oxygen	mg/litre	*SdO2*	12.9	8.3	12.4	11.1	2.02
Inflow pH	ph Units	*IpH*	6.4	6	6.2	6.2	0.15
Planktonic pH	ph Units	*PpH*	6.7	6..40	6.6	6.57	0.09
Benthic pH	ph Units	*BpH*	6.6	6.4	6.5	6.52	0.07

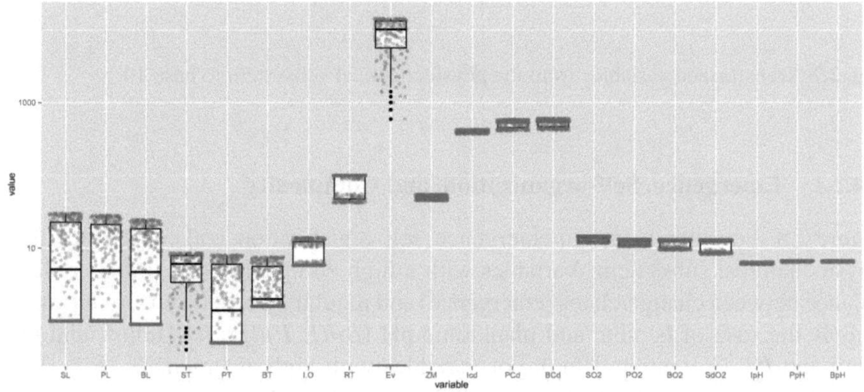

Fig. 2.6 Boxplots of variables from the physiochemical subsystem. Abbreviations expanded in Table 2.3.

$$f(x) = \left\lfloor b \cdot \frac{x - \min X}{\max X - \min X} \right\rfloor, \tag{2.11}$$

where $\lfloor x \rfloor$ is the floor function of x.

Once all variables are in transformed into a finite alphabet, in this case, base 10 ($b = 10$), we can calculate emergence, self-organization, complexity, homeostasis and autopoiesis. Figure 2.7 depicts the number of points in each of the ten classes and shows the distribution of the values for each variable. Based on this distribution, the behavior for variables can be easily described and compared. Variables with a more homogeneous distribution will produce more information, yielding higher values of emergence. Variables with a more heterogeneous distribution will produce higher self-organization values. The complexity of variables is not easy to deduce from Figure 2.7.

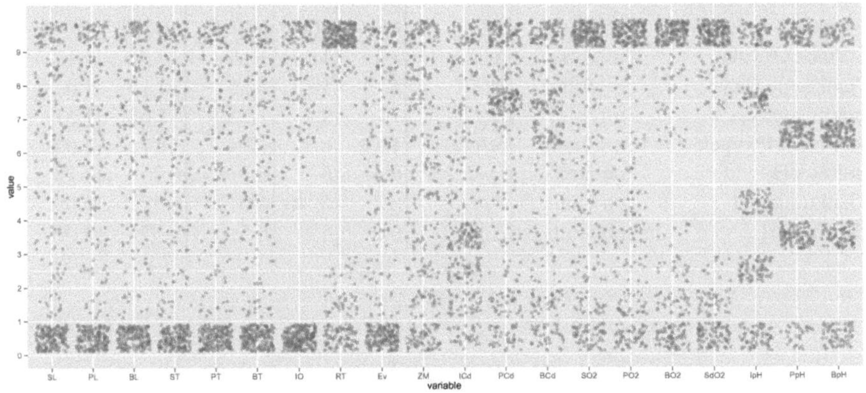

Fig. 2.7 Transformed variables from the physiochemical subsystem to base 10

2.4.2.1 Emergence, Self-organization, and Complexity

Figure 2.8 shows the values of emergence, self-organization, and complexity of the physiochemical subsystem. Variables with a high complexity $C \in [0.8, 1]$ reflect a balance between change/chaos (emergence) and regularity/order (self-organization). This is the case of benthic and planktonic pH (BpH; PpH), IO (Inflow and Outflow) and RT (Retention Time). For variables with high emergencies ($E > 0.92$), like Inflow Conductivity (ICd) and Zone Mixing (ZM), their change in time is constant; a necessary condition for exhibiting chaos. For the rest of the variables, self-organization values are low ($S < 0.32$), reflecting low regularity. It is interesting to notice that in this system there are no variables with a high self-organization nor low emergence.

Since $E, S, C \in [0, 1]$, these measures can be categorized into five categories as shown in Table 2.4. These categories are described on the basis of the range value, the color and the adjective in a scale from very high to very low. This categorization

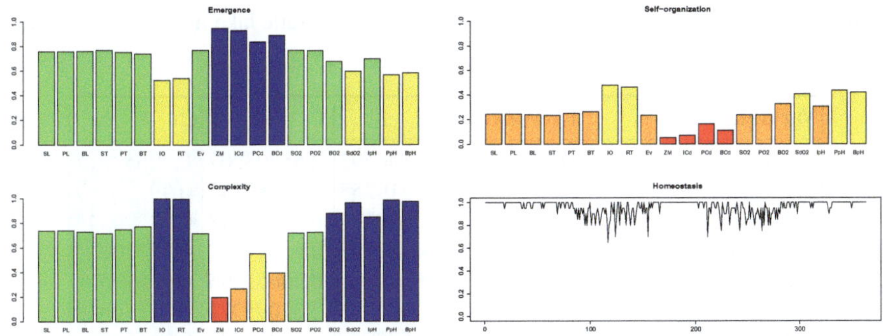

Fig. 2.8 E, S, and C of physiochemical variables of the Arctic lake model (also shown in Table 2.5) and daily variations of homeostasis H during a simulated year

Table 2.4 Categories for classifying E, S, and C

Category	Very High	High	Fair	Low	Very Low
Range	$[0.8,1]$	$[0.6,0.8)$	$[0.4,0.6)$	$[0.2,0.4)$	$[0,0.2)$
Color	Blue	Green	Yellow	Orange	Red

is inspired on the categories for Colombian water pollution indices. These indices were proposed by Ramírez et al. (2003) and evaluated in Fernández et al. (2005).

Table 2.5 shows results of E, S, and C using the categories just mentioned.

From Table 2.5 and a principal component analysis (not shown), we can divide the values obtained in complexity categories as follows:

Very High Complexity: $C \in [0.8,1]$. The following variables balance self-organization and emergence: benthic and planktonic pH (BpH, PpH), inflow and outflow (IO), and retention time (RT). It is remarkable that the increasing of the hydrological regime during summer is related in an inverse way with the dissolved oxygen ($SO2$; $BO2$). It means that an increased flow causes oxygen depletion. Benthic Oxygen ($BO2$) and Inflow Ph (IpH) show the lowest levels of the category. Between both, there is a negative correlation: a doubling of IpH is associated with a decline of $BO2$ in 40%.

High Complexity: $C \in [0.6,0.8)$. This group includes 11 of the 21 variables and involves a high E and a low S. These 11 variables that showed more chaotic than ordered states are highly influenced by the solar radiation that defines the winter and summer seasons, as well as the hydrological cycle. These variables were: Oxygen ($PO2$, $SO2$); surface, planktonic and benthic temperature (ST, PT, BT); conductivity (ICd, PCd, BCd); planktonic and benthic light (PL,BL); and evaporation (Ev).

Very Low Complexity: $C \in [0,0.2)$. In this group, E is very high, and S is very low. This category includes the inflow conductivity (ICd) and water mixing

Table 2.5 *E*, *S*, and *C* of physiochemical variables of the Arctic lake model. Also shown in Figure 2.2

Variable	**Acronym**	*E*	*S*	*C*
Benthic pH	*BpH*	0.44196793	0.55803207	0.98652912
In and Outflow	*IO*	0.52310253	0.47689747	0.99786509
Retention Time	*RT*	0.53890552	0.46109448	0.99394544
Planktonic pH	*PpH*	0.54122993	0.45877007	0.99320037
Sediment Oxygen	*SdO2*	0.59328705	0.40671295	0.96519011
Benthic Oxygen	*BO2*	0.67904928	0.32095072	0.87176542
Inflow pH	*IpH*	0.69570975	0.30429025	0.84679077
Benthic Temperature	*BT*	0.72661539	0.27338461	0.79458186
Planktonic Temperature	*PT*	0.75293885	0.24706115	0.74408774
Planktonic Light	*PL*	0.75582978	0.24417022	0.7382045
Surface Light	*SL*	0.75591484	0.24408516	0.73803038
Benthic Light	*BL*	0.76306133	0.23693867	0.72319494
Surface Oxygen	*SO2*	0.76509182	0.23490818	0.71890531
Surface Temperature	*ST*	0.76642121	0.23357879	0.71607895
Evaporation	*Ev*	0.76676234	0.23323766	0.71535142
Planktonic Oxygen	*PO2*	0.76887287	0.23112713	0.71082953
Benthic Conductivity	*BCd*	0.77974428	0.22025572	0.68697255
Planktonic Conductivity	*PCd*	0.78604873	0.21395127	0.6727045
Inflow Conductivity	*ICd*	0.92845597	0.07154403	0.26570192
Zone Mixing	*ZM*	0.94809050	0.0519095	0.1968596

variance (*ZM*). Both are high and directly correlated; it means that an increase of the mixing percentage between planktonic and benthic zones is associated with an increase of inflow conductivity.

2.4.2.2 Homeostasis

The homeostasis was calculated by comparing the daily values of all variables, representing the state of the Arctic subsystem. The temporal timescale is very important, because H can vary considerably if we compare states every minute or every month.

The h values have a mean (H) of 0.95739726 and a standard deviation of 0.064850247. The minimum h is 0.60 and the maximum h is 1.0. In an annual cycle, homeostasis shows four different patterns, as shown in Figure 2.8, which correspond with the seasonal variations between winter and summer. These four periods show scattered values of homeostasis as the result of transitions between winter and summer and winter back again. The winter period (first and last days of the year) has very high h levels (1 or close to 1) and starts from day 212 and goes to day 87. In this period, the winter conditions such as low light level, temperature, maximum time retention due to ice covering, low inflow and outflow, water mixing interchange between planktonic and benthic zones, low conductivities and pHs and high oxygen are present. The second, third and fourth periods correspond to summer. The second period starts with an increase of benthic pH, zone mixing, and inflow-outflow variables. Between days 83 and 154, this period is characterized for extreme fluctuations as a result of an increase in temperature and light. Homeostasis fluctuates and reaches a minimum of 0.6 in day 116. At the end of this period, the evaporation and zone mixing increase, while oxygen decreases in the benthic zone and sediment. The third period (days 155 to 162) reflects the stabilization of the summer conditions; It means maximum evaporation, temperature, light, mixing zone, conductivity and pH and the lowest oxygen level. Homeostasis is maximal again for this period. The fourth period (days 163-211), which has h fluctuations near 0.9, corresponds to the transition of summer to winter conditions.

As it can be seen, using h, periodic or seasonal dynamics can be followed and studied.

2.4.2.3 Autopoiesis

Autopoiesis was measured for three components (subsystems) at the planktonic and benthic zones of the Arctic lake. These were physiochemical, limiting nutrients and biomass. They include the variables and organisms related in Table 2.6.

Table 2.6 Variables and organisms used for calculating autopoiesis

Component	Planktonic zone	Benthic zone
Physiochemical	Light, Temperature, Conductivity, Oxygen, pH.	Light, Temperature, Conductivity, Oxygen, Sediment Oxygen, pH.
Limiting Nutrients	Silicates, Nitrates, Phosphates, Carbon Dioxide.	Silicates, Nitrates, Phosphates, Carbon Dioxide.
Biomass	Diatoms, Cyanobacteria, Green Algae, Chlorophyta.	Diatoms, Cyanobacteria, Green Algae.

According to the complexity categories established in Table 2.4, the planktonic and benthic components have been classified in the following categories: limiting nutrient variables in the low complexity category ($C \in [0.2, 0.4)$; orange color), planktonic physiochemical variables in the high complexity category ($C \in [0.6, 0.8)$; green color) and biomass and benthic physiochemical variables in the very high complexity category ($C \in [0.8, 1]$; blue color). A comparison of the complexity level for each subsystem of each zone (averaging their respective variables) is depicted in Figure 2.9.

In order to compare the autonomy of each group of variables, equation 2.10 was applied to the complexity data, as shown in Figure 2.10. For the planktonic and benthic zones, we calculated the autopoiesis of the biomass elements in relation to limiting nutrient and physiochemical variables. All A values are greater than 1. This means that the variables related to living systems have a greater complexity than

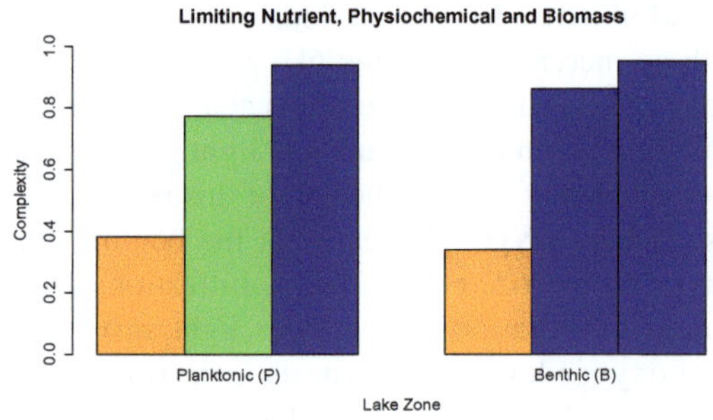

Fig. 2.9 *C* of planktonic and benthic components

the variables related to their environment, represented by the limiting nutrient and physiochemical variables. While we can say that some physiochemical variables, including limiting nutrients have more or fewer effects on the planktonic and benthic biomass, we can also estimate that planktonic and benthic biomass are more autonomous compared to their physiochemical and nutrient environments. The very high values of complexity of biomass imply that these living systems can adapt to the changes of their environments because of the balance between emergence and self-organization that they have.

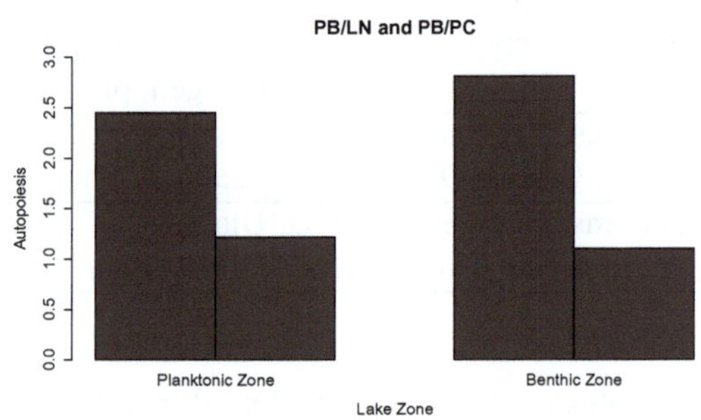

Fig. 2.10 *A* of biomass depending on limiting nutrients and physiochemical components

2.4.2.4 Multiple Scales

The previous analysis of the Arctic lake was performed using base ten. We obtained the measures for the same data using bases $2^i, \forall i \in [1..6]$, as shown in Figures 2.11 and 2.12.

For base 2 (Figure 2.11a), there is a very high *E* for all variables, as the richness of the dynamics cannot be captured by only two values. Thus, *S* and *C* are low. Base 8 (Figure 2.11c) gives results very similar to those of base 10 (Figure 2.8), indicating that the measures are not sensitive to slight changes of base. Base 4 (Figure 2.11b) gives intermediate values between base 2 and base 8. Results for bases 16, 32, and 64 (Figure 2.12) are very similar to those of base 10 and 8, showing that the choice of base is relevant but not a sensitive parameter.

As more diversity is possible with higher bases, homeostasis values decrease with base. Still, the different periods of the year can be identified at all scales, with different levels of detail.

In the case of the Arctic lake model, studying the dynamics with a single base, i.e. at a single scale, can be very informative. However, studying the same phenomena at multiple scales can give further insights, independently on whether the measures change or not with scale.

2.5 Discussion

2.5.1 Measures

The proposed measures characterize the different configurations and dynamics that elements of complex systems acquire through their interactions. Just like temperature averages the kinetic energy of molecules, much information is lost in the averaging, as the description of phenomena changes scale. The measures are probabilistic (except for H) and they all rely on statistical samples[7]. Thus, the caveats of statistics and probability should be taken into consideration when using the proposed measures. Still, these measures capture the properties and tendencies *of a system*, that is why the scale at which they are described is important. They will not indicate which element interacted with which element, how and when. If we are interested in the properties and tendencies of the elements, we can change scale and apply the measures there. Still, we have to be aware that the measures are averaging—and thus simplifying—the phenomena they describe. Whether *relevant* information is lost on the averaging depends not only on the phenomenon, but on what kind of information we are interested in, i.e. relevance is also partially dependent on the observer (Gershenson 2002).

2.5.2 Complexity As Balance or Entropy?

Some approaches relate complexity with a high entropy, i.e. information content (Bar-Yam 2004; Delahaye and Zenil 2007). Just as chaos should not be confused with complexity (Gershenson 2013), a very high entropy (high emergence E) implies too much change, where complex patterns are destroyed. On the other hand, very low entropy (high self-organization S), prevents complex patterns from emerging. As it has been proposed by several authors, complexity can be seen as balance between order and disorder (Langton 1990; Kauffman 1993; Lopez-Ruiz et al. 1995), and thus, it is logical to postulate C as a balance of E and S.

[7] This is also the reason for why all measures are unitless.

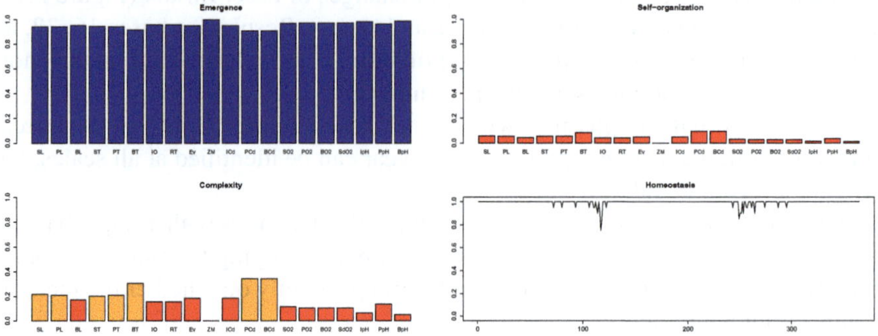

(a) . Base 2.

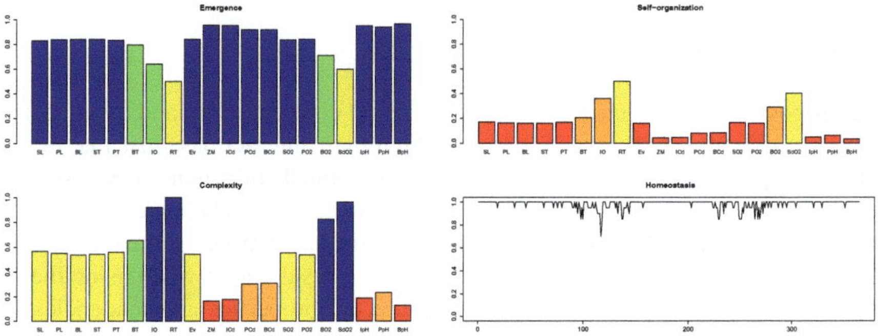

(b) . Base 4.

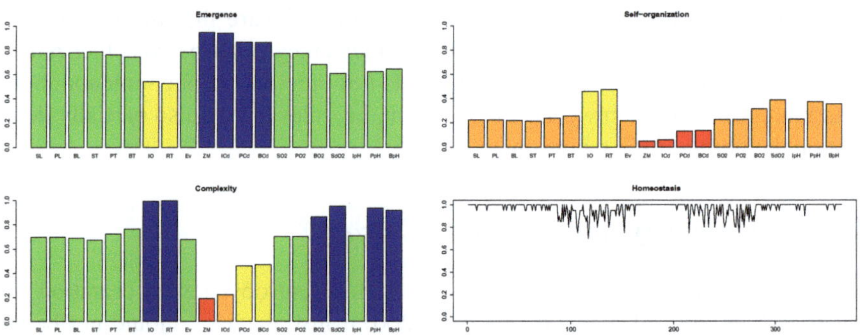

(c) . Base 8.

Fig. 2.11 Emergence, Self-organization, Complexity, and Homeostasis for Arctic lake model at multiple scales: 2, 4, and 8

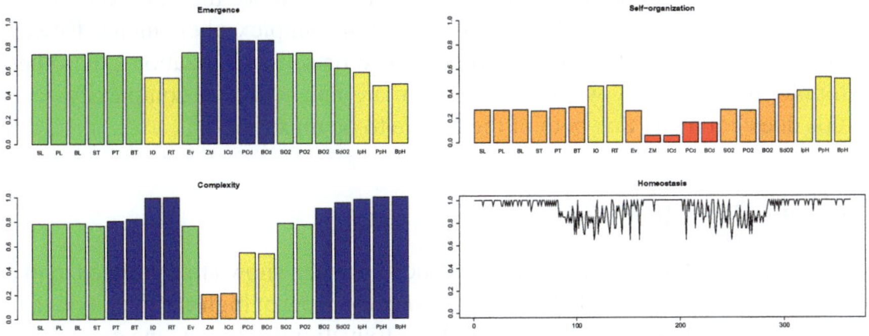

(a) . Base 16.

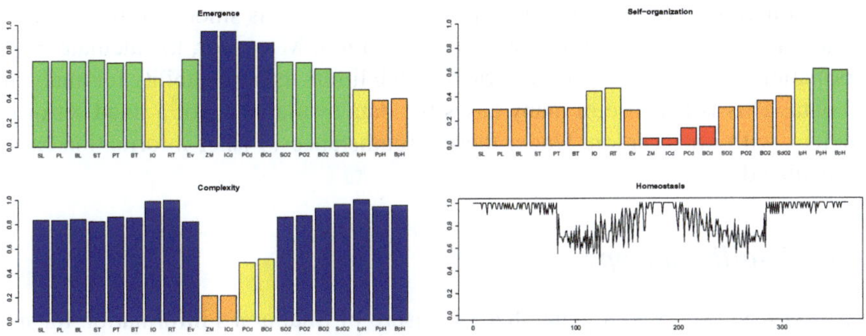

(b) . Base 32.

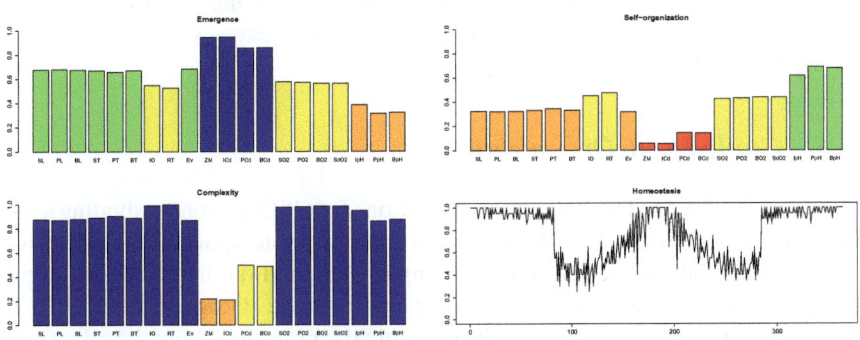

(c) . Base 64.

Fig. 2.12 Emergence, Self-organization, Complexity, and Homeostasis for Arctic lake model at multiple scales: 16, 32, and 64

It might seem contradictory to define emergence as the opposite of self-organization, as they are both present in several complex phenomena. However, when one takes one to the extreme (emergence or self-organization), the other is negligible. It is precisely when both of them are balanced that complexity occurs, but this does not mean that both of them have to be maximal.

2.5.3 Fisher Information

C is correlated with Fisher information, which has been shown to be related to phase transitions (Prokopenko et al. 2011). Following the view of high complexity as a balance, it is natural that C is maximal at phase transitions, which is the case for both C and Fisher information. However, the steepness of Fisher information is much higher than that of C. It is appropriate for determining phase transitions, but it makes little distinction of dynamics farther from transitions. C is smoother, so it can represent dynamical change in a more gradual fashion. Moreover, to calculate Fisher information, a parameter must be varied, which limits its applicability for analyzing real data. This is because in many cases the data available is for a fixed set of parameters, with no variation. Under these circumstances, Fisher information cannot be calculated.

2.5.4 Tsallis Entropy

Tsallis (1988) proposed a generalized measure of Shannon's information for non-ergodic systems. This measure has been correlated with complexity (Tsallis 2002; Gell-Mann and Tsallis 2004). On the one hand, it would be interesting to compare Tsallis entropy with C for different systems. On the other hand, it would be worth exploring what occurs when I is replaced with Tsallis entropy in E (eq. 2.2) and how this affects S, C, and A at multiple scales.

2.5.5 Guided Self-Organization

The measures proposed have several implications for GSO, beyond providing a measure of self-organization. In order to guide a complex system, one has to detect what kind of dynamical regime it has. Depending on this, and on the desired configuration for the system, different interventions can be made (Gershenson 2012a). The measures can inform directly about the dynamical regime and about the effect of the intervention.

For example, if we want to have a system with a high complexity, first we need to measure what is its actual complexity. If it is not the desired one, then the dynamics can be guided. But we also have to measure the complexity during the guiding process, to evaluate the effectiveness of the intervention.

2.5.6 *Scales*

The proposed measures can be applied at different scales, with drastic outcomes. For example, the string '1010101010' will have $E = 1$ in base 2, as $P(0) = P(1) = 0.5$. However, in base 4, each symbol pair is transformed into a single symbol, so the string is transformed to '22222', and thus $P(2) = 1$ and $P(0) = P(1) = P(3) = 0$, giving $E = 0$. Which scale(s) should be used is a question that has to be decided and justified. Multiscale profiles can be helpful in visualizing how the measures change with scale.

2.5.7 *Normalization*

For treating continuous data, we used equation 2.11 to normalize to a finite alphabet, which is equally distributed. Clustering methods could also be used to process data into finite categories. Still, an issue might arise for either case: if the available data does not represent the total range of possible values of a variable, e.g. data $\in [4.5, 5.5]$ but the variable $\in [0, 10]$. If we consider $b = 10$, then equation 2.11 would produce ten categories for the available data, which might be homogeneously distributed and this give a high E. However, if we considered the variable range for equation 2.11, it would categorize the available data in only two categories, leading to a low E. This problem is similar to the one of scales. We suggest to use the viability zone of a variable when known to normalize variables.

2.5.8 *Autopoiesis and Requisite Variety*

Ashby's Law of Requisite Variety (Ashby 1956) states that an active controller requires as much variety (number of states) as that of the controlled system to be stable. For example, if a system can be in four different states, its controller must be able to discriminate between those four states in order to regulate the dynamics of the system.

The proposed measure of autopoiesis is related to the law of requisite variety, as a system with a $A > 1$ must have a higher complexity (variety) than its environment, also reflecting its autonomy. Thus, a successful controller should have $A > 1$ (at multiple scales (Gershenson 2011)), although the controller will be more efficient if $A \to 1$, assuming that higher complexities have higher costs.

2.6 Conclusions

We reviewed measures of emergence, self-organization, complexity, homeostasis, and autopoiesis based on information theory. Axioms were postulated for each measure and equations were derived form them. Having in mind that there are several different measures already proposed (Prokopenko et al. 2009; Gershenson and Fernández 2012), this approach allows us to evaluate the axioms underlying the measures, as opposed to trying to compare different measures without a common ground.

The generality and usefulness of the proposed measures will be evaluated gradually, as these are applied to different systems. These can be abstract (e.g. Turing machines (Delahaye and Zenil 2007, 2012), ε-machines (Shalizi and Crutchfield 2001; Görnerup and Crutchfield 2008)), biological (ecosystems, organisms), economic, social or technological (Helbing 2011).

The potential benefits of general measures as the ones proposed here are manifold. Even if with time more appropriate measures are found, aiming at the goal of finding general measures which can characterize complexity, emergence, self-organization, homeostasis, autopoiesis, and related concepts for any observable system is a necessary step to take.

Acknowledgements. We should like to thank Nihat Ay, Ragnar Behncke, Andrés Burgos, Paul Bourgine, Niccolo Capanni, Wilfried Elmenreich, Tom Froese, Virgil Griffith, Joseph Lizier, Luis Miramontes-Hercog, Roberto Murcio, Oliver Obst, Daniel Polani, Mikhail Prokopenko, Robert Ulanowicz, and X. Rosalind Wang, for interesting critiques, discussions, and suggestions. Diego Lizcano and Cristian Villate helped with programming tasks. Figure 2.1 was made by Albán Blanco. C. G. was partially supported by SNI membership 47907 of CONACyT, Mexico.

References

Aldana-González, M., Coppersmith, S., Kadanoff, L.P.: Boolean dynamics with random couplings. In: Kaplan, E., Marsden, J.E., Sreenivasan, K.R. (eds.) Perspectives and Problems in Nonlinear Science. A Celebratory Volume in Honor of Lawrence Sirovich. Applied Mathematical Sciences Series, Springer, Berlin (2003)

Anderson, P.W.: More is different. Science 177, 393–396 (1972)

Ash, R.B.: Information Theory. Dover Publications, Inc. (1990)

Ashby, W.R.: The nervous system as physical machine: With special reference to the origin of adaptive behavior. Mind 56(221), 44–59 (1947a)

Ashby, W.R.: Principles of the self-organizing dynamic system. Journal of General Psychology 37, 125–128 (1947b)

Ashby, W.R.: An Introduction to Cybernetics. Chapman & Hall, London (1956)

Ashby, W.R.: Design for a brain: The origin of adaptive behaviour, 2nd edn. Chapman & Hall, London (1960)

Ay, N., Der, R., Prokopenko, M.: Guided self-organization: perception–action loops of embodied systems. Theory in Biosciences 131(3), 125–127 (2012)

Bar-Yam, Y.: Dynamics of Complex Systems. Studies in Nonlinearity. Westview Press, Boulder (1997)

Bar-Yam, Y.: Multiscale variety in complex systems. Complexity 9(4), 37–45 (2004)

Bernard, C.: Leçons sur les propriétés physiologiques et les alterations pathologiques des liquides de l'organisme, Paris (1859)

Camazine, S., Deneubourg, J.-L., Franks, N.R., Sneyd, J., Theraulaz, G., Bonabeau, E.: Self-Organization in Biological Systems. Princeton University Press, Princeton (2003)

Cannon, W.: The wisdom of the body. WW Norton & Co., New York (1932)

Delahaye, J.-P., Zenil, H.: On the Kolmogorov-Chaitin complexity for short sequences. In: Calude, C.S. (ed.) Randomness and Complexity: From Leibniz to Chaitin, p. 123. World Scientific, Singapore (2007)

Delahaye, J.-P., Zenil, H.: Numerical evaluation of algorithmic complexity for short strings: A glance into the innermost structure of randomness. Applied Mathematics and Computation 219(1), 63–77 (2012)

Di Paolo, E.A.: Homeostatic adaptation to inversion of the visual field and other sensorimotor disruptions. In: Meyer, J.-A., Berthoz, A., Floreano, D., Roitblat, H., Wilson, S.W. (eds.) From animals to animats 6: Proceedings of the 6th International Conference on the Simulation of Adaptive Behavior, pp. 440–449. MIT Press (2000)

Edmonds, B.: Syntactic Measures of Complexity. PhD thesis, University of Manchester, Manchester, UK (1999)

Fernández, N., Aguilar, J., Gershenson, C., Terán, O.: Sistemas dinámicos como redes computacionales de agentes para la evaluación de sus propiedades emergentes. In: II Simposio Científico y Tecnológico en Computación SCTC 2012, Universidad Central de Venezuela (2012)

Fernández, N., Ramírez, A., Solano, F.: Physico-chemical water quality indices. BISTUA 2, 19–30 (2005)

Froese, T., Stewart, J.: Life After Ashby: Ultrastability and the Autopoietic Foundations of Biological Autonomy. Cybernetics and Human Knowing 17(4), 7–50 (2010)

Gell-Mann, M., Tsallis, C. (eds.): Nonextensive Entropy - Interdisciplinary Applications. Oxford University Press (2004)

Gershenson, C.: Contextuality: A philosophical paradigm, with applications to philosophy of cognitive science. POCS Essay, COGS, University of Sussex (2002)

Gershenson, C.: Introduction to random Boolean networks. In: Bedau, M., Husbands, P., Hutton, T., Kumar, S., Suzuki, H. (eds.) Workshop and Tutorial Proceedings, Ninth International Conference on the Simulation and Synthesis of Living Systems (ALife IX), Boston, MA, pp. 160–173 (2004)

Gershenson, C.: Design and Control of Self-organizing Systems. CopIt Arxives, Mexico (2007), `http://tinyurl.com/DCSOS2007`

Gershenson, C.: The sigma profile: A formal tool to study organization and its evolution at multiple scales. Complexity 16(5), 37–44 (2011)

Gershenson, C.: Guiding the self-organization of random Boolean networks. Theory in Biosciences 131(3), 181–191 (2012a)

Gershenson, C.: The world as evolving information. In: Minai, A., Braha, D., Bar-Yam, Y. (eds.) Unifying Themes in Complex Systems, vol. VII, pp. 100–115. Springer, Heidelberg (2012b)

Gershenson, C.: The implications of interactions for science and philosophy. In: Foundations of Science, Early View (2013)

Gershenson, C., Aerts, D., Edmonds, B. (eds.): Philosophy and Complexity. Worldviews, Science and Us. World Scientific, Singapore (2007)

Gershenson, C., Fernández, N.: Complexity and information: Measuring emergence, self-organization, and homeostasis at multiple scales. Complexity 18(2), 29–44 (2012)

Gershenson, C., Heylighen, F.: When can we call a system self-organizing? In: Banzhaf, W., Ziegler, J., Christaller, T., Dittrich, P., Kim, J.T. (eds.) ECAL 2003. LNCS (LNAI), vol. 2801, pp. 606–614. Springer, Heidelberg (2003)

Gleick, J.: The information: A history, a theory, a flood. Pantheon, New York (2011)

Görnerup, O., Crutchfield, J.P.: Hierarchical self-organization in the finitary process soup. Artificial Life 14(3), 245–254 (2008)

Hausser, J., Strimmer, K.: R package 'entropy'. v. 1.1.7 (2012)
Helbing, D.: FuturICT - new science and technology to manage our complex, strongly connected world. arXiv:1108.6131 (2011)
Heylighen, F., Cilliers, P., Gershenson, C.: Complexity and philosophy. In: Bogg, J., Geyer, R. (eds.) Complexity, Science and Society, pp. 117–134. Radcliffe Publishing, Oxford (2007)
Holzer, R., De Meer, H.: Methods for approximations of quantitative measures in self-organizing systems. In: Bettstetter, C., Gershenson, C. (eds.) IWSOS 2011. LNCS, vol. 6557, pp. 1–15. Springer, Heidelberg (2011)
Jen, E. (ed.): Robust Design: A Repertoire of Biological, Ecological, and Engineering Case Studies. Santa Fe Institute Studies on the Sciences of Complexity. Oxford University Press, Oxford (2005)
Kauffman, S.A.: Metabolic stability and epigenesis in randomly constructed genetic nets. Journal of Theoretical Biology 22, 437–467 (1969)
Kauffman, S.A.: The Origins of Order. Oxford University Press, Oxford (1993)
Kauffman, S.A.: Investigations. Oxford University Press, Oxford (2000)
Langton, C.: Computation at the edge of chaos: Phase transitions and emergent computation. Physica D 42, 12–37 (1990)
Lloyd, S.: Measures of complexity: a non-exhaustive list. Department of Mechanical Engineering. Massachusetts Institute of Technology (2001)
Lopez-Ruiz, R., Mancini, H.L., Calbet, X.: A statistical measure of complexity. Physics Letters A 209(5-6), 321–326 (1995)
Luisi, P.L.: Autopoiesis: a review and a reappraisal. Naturwissenschaften 90(2), 49–59 (2003)
Maturana, H.: Ultrastability...Autopoiesis? Reflective Response to Tom Froese and John Stewart. Cybernetics Human Knowing 18(1-2), 143–152 (2011)
Maturana, H., Varela, F.: Autopoiesis and Cognition: The realization of living. Reidel Publishing Company, Dordrecht (1980)
Mitchell, M.: Complexity: A Guided Tour. Oxford University Press, Oxford (2009)
Morin, E.: Restricted complexity, general complexity. In: Gershenson, C., Aerts, D., Edmonds, B. (eds.) Philosophy and Complexity, Worldviews, Science and Us, pp. 5–29. World Scientific, Singapore (2007), Translated from French by Carlos Gershenson
Müssel, C., Hopfensitz, M., Kestler, H.A.: BoolNet – an R package for generation, reconstruction and analysis of Boolean networks. Bioinformatics 26(10), 1378–1380 (2010)
Polani, D., Prokopenko, M., Yaeger, L.S.: Information and self-organization of behavior. Advances in Complex Systems 16(2&3), 1303001 (2013)
Prokopenko, M.: Guided self-organization. HFSP Journal 3(5), 287–289 (2009)
Prokopenko, M., Boschetti, F., Ryan, A.J.: An information-theoretic primer on complexity, self-organisation and emergence. Complexity 15(1), 11–28 (2009)
Prokopenko, M., Lizier, J.T., Obst, O., Wang, X.R.: Relating Fisher information to order parameters. Phys. Rev. E 84, 041116 (2011)
Project Contributors, R.: The R project for statistical computing (2012)
Ramírez, A., Restrepo, R., Fernández, N.: Evaluación de impactos ambientales causados por vertimientos sobre aguas continentales. Ambiente y Desarrollo 2, 56–80 (2003)
Randerson, P., Bowker, D.: Aquatic Ecosystem Simulator (AES) — a learning resource for biological science students (2008)
Ruiz-Mirazo, K., Moreno, A.: Basic autonomy as a fundamnental step in the synthesis of life. Artificial Life 10(3), 235–259 (2004)
Seidl, D.: Luhmann's theory of autopoietic social systems. Technical Report 2004-2, Ludwig-Maximilians-Universität München. Munich Business Research paper (2004)

Shalizi, C., Crutchfield, J.: Computational mechanics: Pattern and prediction, structure and simplicity. Journal of Statistical Physics 104, 816–879 (2001)
Shalizi, C.R.: Causal Architecture, Complexity and Self-Organization in Time Series and Cellular Automata. PhD thesis, University of Wisconsin at Madison (2001)
Shannon, C.E.: A mathematical theory of communication. Bell System Technical Journal 27, 379–423, 623–656 (1948)
Stumm, W.: Chemical Processes Regulating the Composition of Lake Waters. In: O'Sullivan, P., Reynolds, C. (eds.) The Lakes Handbook Vol 1. Limonlogy and Limnetic Ecology, ch. 5, pp. 79–106. Blackwell Science Ltd., Malden (2004)
Tsallis, C.: Possible generalization of Boltzmann-Gibbs statistics. Journal of Statistical Physics 52(1-2), 479–487 (1988)
Tsallis, C.: Entropic nonextensivity: a possible measure of complexity. Chaos, Solitons & Fractals 13(3), 371–391 (2002)
Varela, F.G., Maturana, H.R., Uribe, R.: Autopoiesis: The organization of living systems, its characterization and a model. Biosystems 5(4), 187–196 (1974)
Wagner, A.: Robustness and Evolvability in Living Systems. Princeton University Press, Princeton (2005)
Williams, H.T.P.: Homeostatic adaptive networks. PhD thesis, University of Leeds (2006)
Wolfram, S.: A New Kind of Science. Wolfram Media, Champaign (2002)

Chapter 3
Generating Functionals for Guided Self-Organization

Claudius Gros

3.1 Controlling Complex Systems

One may take it as a running joke, that complex systems are complex since they are complex. It is however important to realize, this being said, that complex systems come in a large varieties, and in many complexity classes, ranging from relatively simple to extraordinary complex. One may distinguish in this context between *classical* and *modern* complex system theory. In the classical approach one would typically study a standardized model, like the Lorentz model or the logistic map, being described usually by maximally a handful of variables and parameters (Gros 2008). Many real-world systems are however characterized by a very large number of variables and control parameters, especially when it comes to biological and cognitive systems. It has been noted, in this context, that scientific progress may generically be dealing with complexity barriers of various severities, in far reaching areas like medicine and meteorology (Gros 2012b), when researching real-world natural or biological complex systems.

Generically, a complex system may be described by a set of first-order differential equations (or maps), like

$$\dot{x}_i = f_i(x_1, x_2, \ldots | \gamma_1, \gamma_2, \ldots) , \tag{3.1}$$

where the $\{x_i\}$ are the primary dynamical variables and the $\{\gamma_j\}$ the set of control parameters. Modern complex system theory has often to deal with the situation where the phase space of dynamical variables and parameters are both high dimensional. Everything in the macroscopic world, f.i. the brain, can be described by an appropriate set of equations of motion, like (3.1), and we are hence confronted with two types of control problems:

Claudius Gros
Institute for Theoretical Physics, Goethe University Frankfurt, Germany
e-mail: gros@itp.uni-frankfurt.de

M. Prokopenko (ed.), *Guided Self-Organization: Inception*,
Emergence, Complexity and Computation 9,
DOI: 10.1007/978-3-642-53734-9_3, © Springer-Verlag Berlin Heidelberg 2014

– How do we derive governing equations of type (3.1)?
– Given a set of equations of motion, like (3.1), how do we investigate its properties and understand the resulting behavior as a function of the control parameters?

At its core, we are interested here in how to generically control, in general terms, a complex and self-organizing system. A range of complementary approaches are commonly used in order to alleviate the control problem, we discuss here some of the most prominent (non mutually exclusive) approaches.

- *Delegation to Evolution*
  One is often interested, especially in biology and in the neurosciences, in biologically realistic models and simulations (Markram 2006). In this case both the functional form and the parametric dependences are taken from experiment. One may then expect, thanks to Darwinian selection, that the such constructed dynamical equation should exhibit meaningful behavior, replicating observations.
- *Exploring Phase Space*
  A complete understanding would correspond, within dynamical system theory, to a full control of both the qualitative behavior of the flow in phase space and of its dependency on the control parameters. Achieving this kind of complete control is clearly very desirable, but often extremely hard to achieve when dealing with large numbers of dynamical variables and control parameters, the typical situation in modern complex system theory. The exploration of phase space, typically through a combination of analytical and numerical investigations, is in any case an indispensable tool, even when only a small fraction of the overall phase space volume can be probed.
- *Classical Control Theory*
  Classical control theory deals with the objective to control a real-world system, like a rocket, such that a desired behavior is optimally achieved, in the wake of noise both in the sensor readings and in the action effectiveness (Leigh 2004). Classical control theory is of widespread use in engineering and for robot control (De Wit et al. 1996). Our present discussion deals however with the general control of working regimens of a self-organizing complex system; if we knew what the system is supposed to do, we would be done.
- *Diffusive Control*
  Neuromodulators (Marder 2012), like dopamine, serotonin, choline, norepinephrine, neuropeptides and neurohormons, act in the brain as messengers of a diffusive control system (Gros 2010, 2012a), controlling intrinsic and synaptic properties like neural gain and threshold, or synaptic plasticity. Diffusive control is needed to stabilize a desired working regime, a process also denoted as metalearning (Doya 2002), and to switch between different working regimes in order to achieve behavioral flexibility (Arnsten et al. 2012). Diffusive control is a very general strategy for controlling a complex system.
- *Generating Functionals*
  This is the subject we will develop here. One can achieve an improved understanding when considering classes of dynamical systems derived from superordinated functionals. In this case the equations of motion are not given a priori, but

derived from a generating principle. Here we will detail out how this approach leads to an alleviation of the control problem.

One needs to recall, coming back to the introductory statement, that there is no one-size-fits-all method for controlling complex systems (Frei and Serugendo 2011), as there are many kinds and varieties of complex systems. Here we will consider primarily systems made up of a potentially large number of similar functional units, as typical for neural networks. A related aspect of the generic control problem discussed above regards, in this context, the stability of a default working regime with regard to external influences and statistical fluctuations (Clarke 2007). This is particularly important in functional complex systems, such as an ecosystem (Holling 1973; May 2001), or cognitive architectures, the subject of our interest here.

3.2 Guiding Self-Organization

There is no strict scientific definition of what self-organization means or implies. It is however generally accepted to consider processes as self-organizing when a rich and structured dynamics results from a set of relatively simple evolution rules. The term self-organization is of widespread use (Haken 2006), ranging from classical non-equilibrium physical (Nicolis 1989) and biological (Camazine 2003) systems to the assembly of complex macromolecules (Lehn 2002); it is quite generally accepted that the foundations of life are based on self-organizing principles (Kauffman 1993). The brain in particular, possibly the most complex object presently known to humanity, is expected to result from a plethora of intertwined self-organizing processes (Kelso 1995), ranging from self-organized cognitive functions (Kohonen 1988) to self-organized critical states (Bak 1999; Chialvo 2010).

Self-organization is, per se, content free, having no semantic relevancy. The stars in a rotating galaxy, to give an example, may spontaneously organize into a set of distinct density waves, known as the arms of a spiral galaxy. Even though pretty to the eye, the spiral arms of the Milky Way do not serve any purpose; self-organization is in this case just a byproduct of Newton's law. The situation is however generically distinct for biological settings, or for man-made systems, where functionality is the key objective.

The design of functionality is of course a standard objective for the vast majority of man-made systems, and contrasts with the absence of functionality of natural phenomena. Here we are interested in self-organizing processes which are neither fully designed nor without any objective. There is a middleway, which has been denoted *"targeted self-organization"* (Gros 2008) or, alternatively, *"guided self-organization"* (Prokopenko 2009; Martius and Herrmann 2010).

designed	$\longrightarrow$	guided	$\longrightarrow$	natural

For a designed system the functionality is specified in detail in order to achieve optimal performance for a given task. The target for a self-organizing process is however presumed to be a generic principle, often based on information theoretical considerations, with the actual functionality arising indirectly through self-organizing processes. Targeted and guided self-organization are essentially identical terms, with guided self-organization having a somewhat broader breath. One could guide, for example, a dynamical system by restricting its flow to a certain region in phase space, allowing for an otherwise unrestricted development within this bounded area of phase space. Here we will neglect the differences in connotation between targeted and guided self-organization and use both terms interchangeably.

Let us come back at this point to the general formulation of a complex dynamical system through a set of parameterized first-order differential equations, as given by (3.1). The distinction between a parameter γ_j and a primary dynamical variable $x_k(t)$ is a question of time scales.

$$\left.\begin{array}{l}\dot{x}_k\ :\ \text{fast}\\ \dot{\gamma}_k\ :\ \text{slow}\end{array}\right\}\quad \text{time evolution}$$

The flow $(x_1(t), x_2(t), \ldots)$ of the primary dynamics is taking place in the slowly changing environment of parameter space, defining the adiabatic background. The slow adaption of parameters is what controls in the end the working regime of a dynamical system, and is also denoted sometimes as metalearning (Vilalta and Drissi 2002). Not all parameters can be involved in metalearning, a small but finite set of core parameters $\{\gamma'_j\} \in \{\gamma_k\}$ must be constant and immutable,

$$\dot{\gamma}'_j = 0\,.$$

This set of core parameters is what defines in the end the system. One has achieved a dimensional reduction of the control problem if the number $|\{\gamma'_j\}|$ of core parameters is small. This is the aim of guided self-organization, that a concise set of core parameters controls the development and the dynamical properties of a system, with quantitative tuning of the values of the control parameters inducing modifications of the system's characteristics, both on a quantitative and a qualitative level.

3.3 Generating Functionals

There are two principle venues on how to express guiding principles for dynamical systems, implicitly or explicitly. In analogy, one can implement conservation laws in physics by writing down directly appropriate equations of motion, demonstrating that, e.g., energy is conserved. In this case energy conservation is implicitly present in the formulation of the dynamical system. Alternatively one may consider directly a time independent Lagrange function, a condition which explicitly guarantees energy conservation for the respective Lagrange equations of motion. Here we will concentrate on the second approach, the explicit derivation of equations of motion for targeted self-organization through appropriate generating functionals.

The term generating functional has a wide range of connotations in the sciences. The action functional in classical mechanics and quantum field theory is a prominent example from physics, the generating functional $\sum_k p_k x^k$ for a distribution function p_k (with $p_k \geq 0$ and $\sum_k p_k = 1$) another from information theory. In the neurosciences it is custom to speak of objective functions (Intrator and Cooper 1992; Goodhill and Sejnowski 1997) instead of generating functionals.

As a first example we consider a simple energy functional

$$E(\{x_k\}) = \frac{\Gamma}{2}\sum_k x_k^2 - \frac{1}{2}\sum_{kl} y(x_k)w_{kl}y(x_l), \tag{3.2}$$

which is suitable for a network of neurons with membrane potential x_k and firing rate $y(x_k)$. Here $y(x)$ is the sigmoidal transfer function

$$y(x) = \frac{1}{1+\mathrm{e}^{a(b-x)}}, \tag{3.3}$$

parameterized by the gain a and the threshold b. The w_{ij} in (3.2) will turn into the synaptic weights, as we will show lateron, and Γ into a relaxation rate. Concerning the terminology, one could consider $E(\{x_k\})$ also to be an energy function (instead of a functional), being a function of the individual x_k. Here we use the term energy functional, for the functional dependence on the vector $\mathbf{x} = (x_1, x_2, \ldots)$ of membrane potentials.

For our second example we consider a general functional based on the principle of polyhomeostasis (Marković and Gros 2010). One speaks of a homeostatic feedback loop when a target value for a single scalar quantity is to be achieved. Life per se is based on homeostasis, the concentrations of a plethora of biological relevant substances, minerals and hormones need to be regulated, together with a vast number of physical properties, like the body temperature or the heart beating frequency. Polyhomeostasis is, in contrast, typically necessary for time allocation problems.

The problem of allocating time for various tasks constitutes the foundation of behavior. Every living being needs to decide how much time to spend, relatively, on vitally important behaviors, like foraging, resting, exploring or socializing. Maximizing only a single of the possible behavioral patterns would be counterproductive, only a suitable mix of behaviors, as an average over time, is optimal. Mathematically this goal is equivalent to optimizing a distribution function, hence the term polyhomeostasis, in contrast to the case of homeostasis, corresponding to the optimization of a single scalar quantity.

All a neuron can do, at any given moment, is to fire or not to fire, a typical time allocation problem. The generic functional

$$F[p] = \int p(y)f(p(y)))dy \tag{3.4}$$

of the firing rate distribution

$$p(y) = \frac{1}{T}\int_0^T \delta(y - y(t-\tau))\,d\tau \tag{3.5}$$

is an example of the polyhomeostatic principle. Minimizing $F[p]$ corresponds to optimizing a given function $f(p)$ of the neural activity distribution $p(y)$. The resulting adaption rates will then influence the timeline $y(t)$ of the neural activity. This is an example of guided self-organization, since the target functional is expressed in terms of general statistical properties of the dynamical flow, independently of an eventual semantic content. The explicit form and derivation of the adaption rates will be discussed further below, both for the polyhomeostatic functional (3.4) and for the energy functional (3.2).

3.4 Equations of Motion

There are several venues for deriving equations of motions from a given target functional. One uses variational calculus, within classical mechanics, when deriving the Lagrange equations of motion. In classical mechanics the target functional, the action, needs to be stationary with respect to an arbitrary variation of the trajectory. Here we will consider instead generic objective functions which are to be minimized.

Minimizing an objective function is a very generic task for which a wide range of methods and algorithms have been developed (Papadimitriou and Steiglitz 1998; Goldberg 1989; Kennedy and Eberhart 1995). Here we are however interested in a different aspect. Our aim is not to actually find the global minimum of a given objective function, or any stationary point, which is not of interest per se. Objective functions serve as a guiding principle and equations of motion induced by minimizing a given objective function will tend to minimize it. Other driving influence will however in general compete with this goal and it is this very competition which may result in complex and novel dynamical states.

For an objective function which is an explicit function of the dynamical variable, like the energy functional (3.2), the equations of motion just correspond to the downhill flow within the energy landscape,

$$\dot{x}_j = -\frac{1}{T_e}\frac{\partial}{\partial x_j}E(\{x_k\}) , \tag{3.6}$$

where the timescale T_e of the flow in normally set to unity, $T_e = 1$. In our case we obtain

$$\dot{x}_k = -\Gamma x_k + a_k y_k(1-y_k)\sum_j w_{kj} y_j , \tag{3.7}$$

where we have used (3.3) and

$$y'(x) = \frac{\partial y}{\partial x} = ay(1-y) . \tag{3.8}$$

The dynamical system (3.7) just corresponds to a network of leaky integrators (Hopfield 1982, 1984), with the x_k and y_k corresponding to the membrane potential and the mean neural firing rate respectively. The neurons are coupled through the weight matrix w_{kj}, the synaptic weights. The term $a_k y_k(1-y_k)$ in front of the inter-neural coupling is present only when deriving (3.7) from the energy functional (Linkerhand and Gros 2012a), and not when formulating equivalent neural updating rules directly from neurobiological considerations (Olshausen et al. 1993).

The polyhomeostatic functional (3.4) is used to derive adaption rules for the intrinsic parameters a_i and b_i of the transfer function (3.3). The lack of an explicit dependence on either a_i or b_i rules out adaption rules like $\dot{a}_i \propto -\partial F[p]/\partial a_i$, which would be the equivalent to (3.6). It is however possible to derive implicit adaption rules, for which the minimization of the objective functions $F[p]$ is performed stochastically in the sense that the time-averaged firing rate $p(y)$ is sampled along the flow during the time evolution. For this purpose we change variables and rewrite the generating functional

$$F[p] = \int p(x) f(p(y)/y')) dx, \qquad p(y)dy = p(x)dx \tag{3.9}$$

as an expectation value over the distribution $p(x)$ of the membrane potential x, the input. The transfer function (3.3) maps the input of a neuron to its output and adaption rules for the intrinsic parameters should hence not depend explicitly on the actual distribution $p(x)$ of the input, they should be universal in the sense that the intrinsic adaption rules should abstract from the actual semantic content of the information being processed. Noting that $p(x)$ does not depend explicitly on the gain a and the threshold b, we have

$$\frac{\partial}{\partial \theta} F[p] = \int dx\, p(x) \frac{\partial}{\partial \theta} f(p(y)/y')), \qquad \theta = a, b\,. \tag{3.10}$$

For the overall global minimum of $F[p]$ the weighting with respect to the input distribution $p(x)$ would be needed to be taken into account. As we are however interested only in adaption rules abstracting from the actual form of the input distribution, and noting that $p(x) \geq 0$ is positive definite, we demand that the adaption process should lead to a uniform minimization of the kernel of (3.10),

$$\dot{\theta} = -\varepsilon_\theta \frac{\partial}{\partial \theta} f(p(y)/y')), \qquad \theta = a, b\,, \tag{3.11}$$

where ε_θ are the respective adaption rates. The adaption process should generally be slow, as typical for metalearning, and the adaption rates ε_a and ε_b small. In this case the updating rule (3.11) will statistically sample the input distribution $p(x)$, as an average over time, and become equivalent with (3.10).

The adaption rates (3.11) are generic and need to be concretized for a specific polyhomeostatic function $f(p)$. A straightforward target functional for the problem of allocating time is to consider a target distribution function $q(y)$ for the neural firing rate. In this case the functional

$$F[p] = \int p(y) f(p(y)) dy, \qquad f(p) = \ln(p/q) \tag{3.12}$$

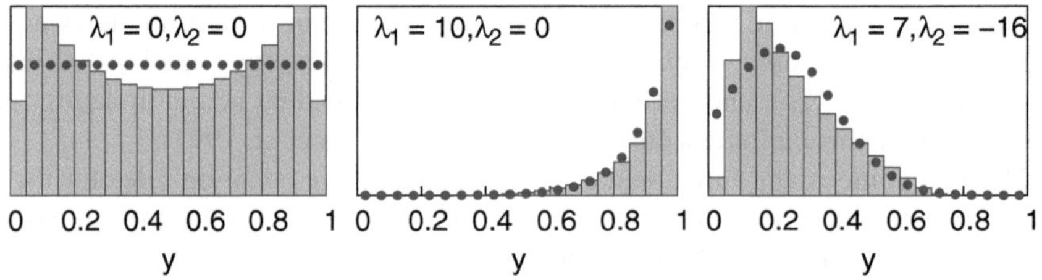

Fig. 3.1 The results of the intrinsic adaption rules (3.14) and (3.15) for the time averaged firing rate distribution (boxes, see Eq. (3.5)) of a single neuron driven by a white-noise input and for several information maximizing target distributions (points, see Eq. (3.13))

corresponds to the Kullback-Leibler divergence (Gros 2008), which is a positive definite measure for the similarity of two distribution functions p and q. The Kullback-Leibler divergence is minimal whenever $p(y)$ and $q(y)$ are as similar as possible, within the configuration of all dynamically realizable firing rate distributions $p(y)$.

The target firing rate distribution $q(y)$ could be any positive and normalized distribution function. Here we demand that $q(y)$ should maximize Shannon's information entropy $-q\ln(q)$, which can be achieved using variational calculus:

$$0 = -\delta \int q\left[\ln(q) - \lambda_1 y - \lambda_2 y^2\right] dy, \qquad q(y) \propto e^{\lambda_1 y + \lambda_2 y^2} . \tag{3.13}$$

Here λ_1/λ_2 are suitable Lagrange parameters enforcing a given mean/variance. The flat distribution $\lambda_1 = \lambda_2 = 0$ maximizes information entropy in the absence of any constraint. Using (3.11) and $y' = ay(1-y)$, see Eq. (3.8), we obtain then the adaption rules (Triesch 2005, 2007; Marković and Gros 2010; Linkerhand and Gros 2012b)

$$\dot{a} = \varepsilon_a \left(\frac{1}{a} + (x-b)\,\Delta\tilde{\theta} \right) \tag{3.14}$$

$$\dot{b} = \varepsilon_b\,(-a)\,\Delta\tilde{\theta}, \qquad \Delta\tilde{\theta} = (1-2y) + y(1-y)\left[\lambda_1 + 2\lambda_2 y\right] . \tag{3.15}$$

In Fig. 3.1 we present the results for a single polyhomeostatically adapting neuron, driven by white noise, for various target distributions $q(y)$. Note that there are only two intrinsic parameters, the threshold b and the gain a, to be optimized and that the transfer function (3.3) can hence not change, during the adaption process, its functional form arbitrarily. The firing rate distribution $p(y)$ approximates, considering this limitation, the target distribution $q(y)$ remarkably well, an exemplification of the principle of targeted self-organization.

3.5 Adaptive Phase Space

It is illuminating to investigate somewhat in detail the behavior of the adaption process in the phase space (a,b) of the intrinsic adaption parameters, and to study individual trajectories $(a(t),b(t))$. In Fig. 3.2 we present a selection of trajectories

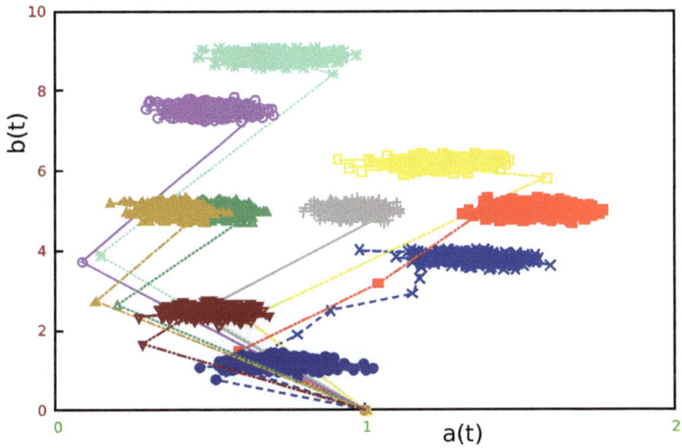

Fig. 3.2 Sample trajectories $(a(t), b(t))$ resulting from the intrinsic adaption rules (3.14) and (3.15), color coded for various parameters λ_1 and λ_2 of the target distributions $q(t)$, compare Eq. (3.13). All trajectories start at $(a(0), b(0)) = (1, 0)$ and then settle into distinct regions of phase space, where they perform a confined stochastic walk, due to the white-noise input.

for distinct realizations of the target distribution $q(y)$, as given by Eq. (3.13). The neuron is driven by a white noise input, the starting gain and threshold are $a = 1$ and $b = 0$, for all trajectories. After a relatively fast initial transient the intrinsic parameters settle to distinct respective regions in the phase space, where they perform a stochastic motion, reflecting the white-noise character of the driving input. Three of the resulting firing rate distributions $p(y)$ are shown in Fig. 3.1.

The target distribution $q(y)$, see Eq. (3.13), can be selected to be bimodal, which is generally the case for inverse Gaussians having $\lambda_1 < 0$ and $\lambda_2 > 0$. In Fig. 3.3 we present the adaptive walk through phase space $(a(t), b(t))$ for a bimodal target distribution having $\lambda_1 = -20$ and $\lambda_2 = 18.5$ and for various adaption rates $\varepsilon_a = \varepsilon_b$. When the adaption process is very slow, viz for small ε_a and ε_b the system average over extended periods of the stochastic input and the dynamics becomes smooth (Linkerhand and Gros 2012b), fluctuating with a reduced amplitude around a certain target region in phase space, just as illustrated in Fig. 3.2.

For a bimodal target distribution $q(y)$ there may however be two local minima in adaptive space, since the transfer function (3.3) is always monotonic. For any given pair of intrinsic parameters the system can hence approximate well only one of the two peaks of a bimodal transfer function. For small adaption it remains trapped in one of the local minima, but larger adaption rates ε_a and ε_b will lead to an enhanced sensibility with respect to the stochastic driving, inducing stochastic tipping transitions between the two local minima. This is a striking realization of the principle of guided self-organization.

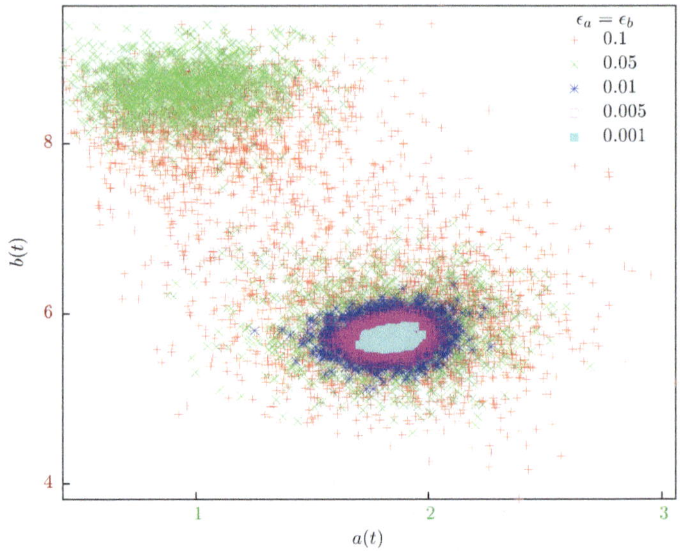

Fig. 3.3 Sample trajectories $(a(t), b(t))$ resulting from the intrinsic adaption rules (3.14) and (3.15), color coded for various adaption parameters $\varepsilon_a = \varepsilon_b$, as given in the legend. The single neuron is driven by white noise and the target distribution, see Eq. (3.13) is bimodal, parameterized by $\lambda_1 = -20$ and $\lambda_2 = 18.5$. For moderate large adaption rates the system is able to make stochastically driven tipping transitions between two local minima (Linkerhand and Gros 2012b).

3.6 Self-Organized Dynamical States

As a second example for the functioning of polyhomeostatic optimization we consider a network of N randomly interconnected neurons,

$$x_k = \sum_{j \neq k} w_{kj} y_j$$

which corresponds to (3.7) in the anti-adiabatic limit $\Gamma \to \infty$ (and without the factor $y(1-y)$). For the synaptic weights we select

$$w_{ij} = \begin{cases} +1/\sqrt{K} & \text{with probability} \quad \rho_{exc} \\ -1/\sqrt{K} & \text{with probability} \quad 1-\rho_{exc} \end{cases}, \tag{3.16}$$

where K is the in-degree. The system is balanced for $\rho_{exc} = 1/2$. As a second control parameter, besides the fraction ρ_{exc} of excitatory links, we consider the average target activity μ,

$$\mu = \int y q(y) dy, \qquad \int q(y) dy = 1, \tag{3.17}$$

which is taken to be uniform, viz identical for all sites.

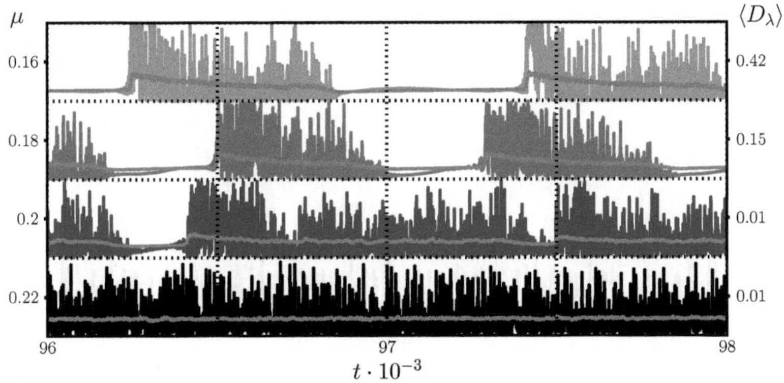

Fig. 3.4 For a network of $N = 1000$ adapting neurons, according to Eqs. (3.14) and (3.15), the activity of a randomly selected neuron and the average neural activity (green line). The network is balanced, with as many excitatory and inhibitory links, randomly selected according to Eq. (3.16). Shown are results for various target mean activities μ, see Eq. (3.17). The right-hand axis is not a scale, the numbers are the values of the network-averaged Kullback-Leibler divergence $\langle D_\lambda \rangle$, as defined by Eq. (3.12). One observes that the mean target activity μ entering the polyhomeostatic generating functional acts as a parameter controlling the resulting self-organized dynamical state (Marković and Gros 2012).

In Fig. 3.4 we present the results for a balanced network with $N = 1000$ adapting neurons, and an in-degree of $K = 100$. Shown are both the activity of a single, randomly selected site and the average activity, averaged over all sites. We notice that the network enters distinct dynamical states, as a function of the mean target activity μ (Marković and Gros 2010, 2012). For intermediate target activity levels the dynamics is chaotic, for smaller mean activities μ a regime with intermittent bursts is observed. One has hence the possibility to tune the self-organized dynamical state through the target set by the polyhomeostatic generating functional, an example of targeted self-organization. Interestingly the overall value of the network-averaged Kullback-Leibler divergence is minimal in the chaotic state.

In Fig. 3.5 we present the results for the same network of $N = 1000$ sites as in Fig. 3.4, but this time the network is not balanced, $\rho_{exc} > 1/2$. The mean target firing-rate activity is kept constant at $\mu = 0.3$. For larger values of ρ_{exc} the network synchronizes, not surprisingly, as a result of the predominance of positive feedback loops. For values of ρ_{exc} close to the balanced state, the system is chaotic, with a large variability around a partly synchronized state in between. One can regard ρ_{exc} as a controlling parameter of the energy functional (3.2), which hence allows to guide the self-organization of the resulting dynamical state. The value of the Kullback-Leibler divergence is, again, lowest in the chaotic state, which explores phase space best.

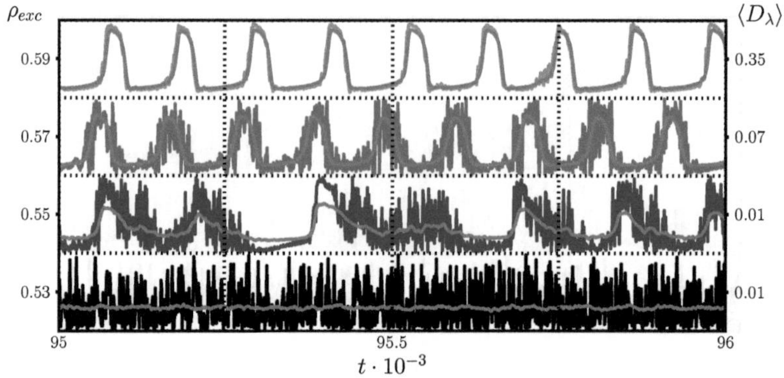

Fig. 3.5 For networks containing $N = 1000$ adapting neurons with an in-degree $K = 100$ and a target mean activity $\mu = 0.3$, see Eq. (3.17), the activity of a randomly selected neuron and the average neural activity (green line). The networks are not balanced, having a slight excess ρ_{exc} of randomly selected excitatory links see Eq. (3.16). Also given (on the right) are the respective values of the network-averaged Kullback-Leibler divergence $\langle D_{\lambda} \rangle$, as defined by Eq. (3.12). The network shows a transition between chaos and synchronization, as a function of ρ_{exc} (Marković and Gros 2012).

3.7 Discussion

A self-organizing process may be guided by presenting to the system one or more targets. If these targets are very concrete they may destroy the self-organizing character of the process, resulting in a driving force. One possibility to achieve a gentle way of controlling a self-organizing process is to formulate the targets in terms of statistical properties of the desired dynamical state, with a basic example being the time-average distribution function of activities. Optimizing the distribution of activities is an example of a time-allocation problem, which is intrinsically of polyhomeostatic nature.

A given set of goals may be achieved be a range of different tools, for example using evolutionary algorithms. In this treatise we have discussed the perspective, together with concrete examples, of explicitly deriving equations of motions from generating functionals incorporating polyhomeostatic and other targets. We believe that this approach offers several advantages. Having explicit time evolution equation at hand is, in our view, mandatory for time-efficient simulations and applications. Generating functionals can furthermore be seen as a route for solving the control problem, as they offer a substantial dimensional reduction in the number of free parameters. This is a particularily attractive feature, in view of the raising appreciation that the neuromodulator control system in the brain tunes the relative stability of a wide range of possible dynamical operative states of the affected downstream circuits.

From an alternative perspective one may view generating functionals also as a middleway between the study of simplified model systems and biological realistic simulations.

simple model systems $\longrightarrow$ generating functionals $\longrightarrow$ detailed / realistic simulations

Model systems may constitute important reference models, for understanding and developing key concepts and methods. Detailed simulations are, at the other extreme, often indispensable for obtaining a realistic comparison with experimental data, having however the drawback that an in-depth understanding is in general not achievable. We propose generating functionals as a venue for building increasingly complex dynamical systems and cognitive architectures, a venue which allows for the control of the operating modi of the system by tuning a limited number of high-level control parameters incorporating the targets of the respective generating functionals.

References

Arnsten, A., Wang, M., Paspalas, C.: Neuromodulation of thought: flexibilities and vulnerabilities in prefrontal cortical network synapses. Neuron 76, 223–239 (2012)

Bak, P.: How nature works: the science of self-organized criticality. Springer (1999)

Camazine, S., Deneubourg, J., Franks, N., Sneyd, J., Theraula, G., Bonabeau, E.: Self-organization in biological systems. Princeton University Press (2003)

Chialvo, D.: Emergent complex neural dynamics. Nature Physics 6(10), 744–750 (2010)

Clarke, B.: Stability of complex reaction networks. Wiley (2007)

De Wit, C., Bastin, G., Siciliano, B.: Theory of robot control. Springer, New York (1996)

Doya, K.: Metalearning and neuromodulation. Neural Networks 15, 495–506 (2002)

Frei, R., Serugendo, G.D.M.: Advances in complexity engineering. International Journal of Bio-Inspired Computation 3(4), 199–212 (2011)

Goldberg, D.: Genetic algorithms in search, optimization, and machine learning (1989)

Goodhill, G., Sejnowski, T.: A unifying objective function for topographic mappings. Neural Computation 9(6), 1291–1303 (1997)

Gros, C.: Complex and adaptive dynamical systems: A primer. Springer (2008)

Gros, C.: Cognition and emotion: perspectives of a closing gap. Cognitive Computation 2, 78–85 (2010)

Gros, C.: Emotional control–conditio sine qua non for advanced artificial intelligences? In: Müller, V. (ed.) Philosophy and Theory of Artificial Intelligence. Springer (2012a)

Gros, C.: Pushing the complexity barrier: diminishing returns in the sciences. Complex Systems 21, 183 (2012b)

Haken, H.: Information and self-organization: A macroscopic approach to complex systems, vol. 40. Springer (2006)

Holling, C.: Resilience and stability of ecological systems. Annual Review of Ecology and Systematics, 1–23 (1973)

Hopfield, J.: Neural networks and physical systems with emergent collective computational abilities. Proceedings of the National Academy of Sciences 79(8), 2554–2558 (1982)

Hopfield, J.: Neurons with graded response have collective computational properties like those of two-state neurons. Proceedings of the National Academy of Sciences 81(10), 3088–3092 (1984)

Intrator, N., Cooper, L.: Objective function formulation of the bcm theory of visual cortical plasticity: Statistical connections, stability conditions. Neural Networks 5(1), 3–17 (1992)

Kauffman, S.: The origins of order: Self-organization and selection in evolution. Oxford University Press (1993)
Kelso, J.: Dynamic patterns: The self-organization of brain and behavior. MIT press (1995)
Kennedy, J., Eberhart, R.: Particle swarm optimization. In: Proceedings of the IEEE International Conference on Neural Networks, vol. 4, pp. 1942–1948. IEEE (1995)
Kohonen, T.: Self-organization and associative memory. Springer (1988)
Lehn, J.: Toward self-organization and complex matter. Science 295(5564), 2400–2403 (2002)
Leigh, J.: Control theory. Peter Peregrinus Limited (2004)
Linkerhand, M., Gros, C.: Generating functionals for autonomous latching dynamics in attractor relict networks. arXiv preprint arXiv:1212.5054 (2012a)
Linkerhand, M., Gros, C.: Self-organized stochastic tipping in slow-fast dynamical systems. arXiv preprint arXiv:1207.2928 (2012b)
Marder, E.: Neuromodulation of neuronal circuits: Back to the future. Neuron 76, 1–11 (2012)
Marković, D., Gros, C.: Self-organized chaos through polyhomeostatic optimization. Physical Review Letters 105, 68702 (2010)
Marković, D., Gros, C.: Intrinsic adaptation in autonomous recurrent neural networks. Neural Computation 24(2), 523–540 (2012)
Markram, H.: The blue brain project. Nature Reviews Neuroscience 7, 153–160 (2006)
Martius, G., Herrmann, J.M.: Taming the beast: Guided self-organization of behavior in autonomous robots. In: Doncieux, S., Girard, B., Guillot, A., Hallam, J., Meyer, J.-A., Mouret, J.-B. (eds.) SAB 2010. LNCS, vol. 6226, pp. 50–61. Springer, Heidelberg (2010)
May, R.: Stability and complexity in model ecosystems. Princeton University Press (2001)
Nicolis, G.: Physics of far-from-equilibrium systems and self-organisation. The New Physics 11, 316–347 (1989)
Olshausen, B., Anderson, C., Van Essen, D.: A neurobiological model of visual attention and invariant pattern recognition based on dynamic routing of information. The Journal of Neuroscience 13(11), 4700–4719 (1993)
Papadimitriou, C., Steiglitz, K.: Combinatorial optimization: algorithms and complexity. Dover Publications (1998)
Prokopenko, M.: Guided self-organization. HFSP Journal 3, 287–289 (2009)
Triesch, J.: A gradient rule for the plasticity of a neuron's intrinsic excitability. In: Duch, W., Kacprzyk, J., Oja, E., Zadrożny, S. (eds.) ICANN 2005. LNCS, vol. 3696, pp. 65–70. Springer, Heidelberg (2005)
Triesch, J.: Synergies between intrinsic and synaptic plasticity mechanisms. Neural Computation 19(4), 885–909 (2007)
Vilalta, R., Drissi, Y.: A perspective view and survey of meta-learning. Artificial Intelligence Review 18(2), 77–95 (2002)

Chapter 4
Empowerment — An Introduction

Christoph Salge, Cornelius Glackin, and Daniel Polani

4.1 Introduction

Is it better for you to own a corkscrew or not? If asked, you as a human being would likely say "yes", but more importantly, you are somehow able to make this decision. You are able to decide this, even if your current acute problems or task do not include opening a wine bottle. Similarly, it is also unlikely that you evaluated several possible trajectories your life could take and looked at them with and without a corkscrew, and then measured your survival or reproductive fitness in each. When you, as a human cognitive agent, made this decision, you were likely relying on a behavioural "proxy", an internal motivation that abstracts the problem of evaluating a decision impact on your overall life, but evaluating it in regard to some simple fitness function. One example would be the idea of curiosity, urging you to act so that your experience new sensations and learn about the environment. On average, this should lead to better and richer models of the world, which give you a better chance of reaching your ultimate goals of survival and reproduction.

But how about questions such as, would you rather be rich than poor, sick or healthy, imprisoned or free? While each options offers some interesting new experience, there seems to be a consensus that rich, healthy and free is a preferable choice. We think that all these examples, in addition to the question of tool ownership above, share a common element of preparedness. Everything else being equal it is preferable to be prepared, to keep ones options open or to be in a state where ones actions have the greatest influence on ones direct environment.

The concept of *Empowerment*, in a nutshell, is an attempt at formalizing and quantifying these degrees of freedom (or options) that an organism or agent has as a proxy for "preparedness"; preparedness, in turn, is considered a proxy for prospective fitness via the hypothesis that preparedness would be a good indicator to

Christoph Salge · Cornelius Glackin · Daniel Polani
Adaptive Systems Research Group, School of Computer Science,
University of Hertfordshire, Hatfield, United Kingdom
e-mail: {christophsalge,daniel.polani}@gmail.com,
c.glackin2@herts.ac.uk

M. Prokopenko (ed.), *Guided Self-Organization: Inception*,
Emergence, Complexity and Computation 9,
DOI: 10.1007/978-3-642-53734-9_4, © Springer-Verlag Berlin Heidelberg 2014

distinguish promising from less promising regions in the prospective fitness landscape, without actually having to evaluate the full fitness landscape.

Empowerment aims to reformulate the options or degrees of freedom that an agent has as the agent's control over its environment; and not only of its control — to be reproducible, the agent needs to be aware of its control influence and sense it. Thus, empowerment is a measure of both the control an agent has over its environment, as well as its ability to sense this control. Note that this already hints at two different perspectives to evaluate the empowerment of an agent. From the agent perspective empowerment can be a tool for decision making, serving as a behavioural proxy for the agent. This empowerment value can be skewed by the quality of the agent world model, so it should be more accurately described as the agent's approximation of its own empowerment, based on its world model. The actual empowerment depends both on the agent's embodiment, and the world the agent is situated in. More precisely, there is a specific empowerment value for the current state of the world (the agent's current empowerment), and there is an averaged value over all possible states of the environment, weighted by their probability (the agent's average empowerment).

Empowerment, as introduced by Klyubin et al. (2005a,b), aims to formalize the combined notion of an agent controlling its environment and sensing this control in the language of information theory. The idea behind this is that this should provide us with a utility function that is inherently *local*, *universal* and *task-independent*.

1. *Local* means that the knowledge of the local dynamics of the agent is enough to compute it, and that it is not necessary to know the whole system to determine one's empowerment. Ideally, the information that the agent itself can acquire should be enough.
2. *Universal* means that it should be possible to apply empowerment "universally" to every possible agent-world interaction. This is achieved by expressing it in the language of information theory and thus making it applicable for any system that can be probabilistically expressed.
   For instance, even if an agent completely changes its morphology, it is still possible to compute a comparable empowerment value. Klyubin et al. (2005b) gave the examples of money in a bank account, of social status in a group of chimpanzees, and of sugar concentration around a bacterium as different scenarios, all as examples which would be treated uniformly by the empowerment formalism.
3. *Task-independent* means that empowerment is not evaluated in regard to a specific goal or external reward state. Instead, empowerment is determined by the agent's embodiment in the world. In particular, apart from minor niche-dependent parameters, the empowerment formalism should have the very same structure in most situations.

More concretely, the proposed formulation of empowerment defines it via the concept of potential information flow, or channel capacity, between an agent's actuator state at earlier times and their sensor state at a later time. The idea behind this is that empowerment would quantify how much an agent can reliably and perceptibly influence the world.

4.1.1 Overview

Since its original inception by Klyubin et al. (2005a,b) in 2005, several papers have been published about empowerment, both further developing the formalism, and demonstrating a variety of behaviours in different scenarios. Our aim here is to both present an overview of what has been done so far, and to provide readers new to empowerment with an easy entry point to the current state-of-the-art in the field. Due to the amount of content, some ideas and results are only reported in abstract form, and we would refer interested reader to the cited papers, where models and experiments are explained in greater detail.

Throughout the text we also tried to identify the open problems and questions that we currently see, and we put a certain emphasis on the parameters that affect empowerment. While empowerment is defined in a generic and general way, the review of the literature shows that there are still several choices one can take on how to exactly apply empowerment, and which can affect the outcome of the computation.

The remaining paper is structured as follows. First, after briefly outlining the related work previous to empowerment, we will spell out the different empowerment hypotheses motivating the research in empowerment. This will allow us to locate empowerment in relation to different fields, and also makes it easier to see how and where insights from the empowerment formalism apply to other areas.

The next section then focusses on discrete empowerment, first, in Sec. 4.4 introducing the formalism, and then, in Sec. 4.5, describing several different examples, showcasing the genericity of the approach.

Section 4.6 then deals with empowerment in continuous settings, which is currently not as far developed and sees vigorous activity. Here we will discuss the necessity for suitable approximations, and outline the current technical challenges to provide good but fast approximations for empowerment in the continuous domain.

4.2 Related Work

Empowerment is based on and connects to several fields of scientific inquiry. One foundational idea for empowerment is to apply information theory to living, biological systems. (Gibson James 1979) points out the importance of information in embodied cognition, and earlier work (Barlow 1959; Attneave 1954) investigates the informational redundancy in an agent's sensors. Later research (Atick 1992) based on this identifies the importance of *information bottlenecks* for the compression of redundancies, which are later formalized in information theoretic terms (Tishby et al. 1999). Furthermore, it was also demonstrated that informational efficiency can be used to make sense of an agent's sensor input (Olsson et al. 2005; Lungarella et al. 2005). The general trend observed in these works seems to be that nature optimizes the information processing in organisms in terms of efficiency (Polani 2009). Empowerment is, in this context, another of these efficiency principles.

Empowerment also relies heavily on the notion that cognition has to be understood as an immediate relationship of a situated and embodied agent with its surroundings. This goes back to the "Umwelt" principle by (von Uexküll 1909), which also provides us with an early depiction of what is now commonly referred to as the perception action loop. This idea was also at the center of a paradigm shift in artificial intelligence towards *enactivism* (Varela et al. 1992; Almeida e Costa and Rocha 2005), which postulates that the human mind organizes itself by interacting with its environment. Embodied robotics (Pfeifer et al. 2007) is an approach trying to replicate these processes "in silico".

4.2.1 Intrinsic Motivation

Central to this body of work is the desire to understand how an organism makes sense of the world and decides to act from its internal perspective. Ultimately all behaviour is connected to an organism's survival, but most natural organisms do not have the cognitive capacity to determine this connection themselves. So, if an animal gets burned by fire, it will not consider the fire's negative effect on its health and potential risk of death and then move away. Instead, it will feel pain via its sensors and react accordingly. The ability to feel pain and act upon it is an adaptation that acts as a proxy criterion for survival, while it still offers a certain level of abstraction from concrete hard-wired reactions. We could say the animal is motivated to avoid pain. Having an abstract motivation allows an agent a certain amount of adaptability; instead of acting like a stimulus-response look-up table the agent can evaluate actions in different situations according to how rewarding they are regarding its motivations.

Examining nature also reveals that not all motivations are based on external rewards, e.g. a well-fed and pain-free agent might be driven by an urge to explore or learn. In the following we discuss related work covering different approaches to specify and quantify such *intrinsic motivations*. The purpose of these models is both to better understand nature, as well as to replicate the ability of natural organism to react to a wide range of stimuli in models for artifical systems.

An evolution-based view of intrinsic motivations uses assumptions about preexisting saliency sensors to generate intrinsic motivations (Singh et al. 2005, 2010). However, where one does not want to assume such pre-evolved saliency sensors, one needs to identify other criteria that can operate with unspecialized generic sensors.

One such family of intrinsic motivation mechanisms focusses on evaluating the learning process. *Artificial curiosity* (Schmidhuber 2002, 1991) is one of the earlier models, where an agent receives an internal reward depending on how "boring" the environment is which it currently tries to learn. This causes the agent to avoid situations that are at either of the extremes: fully predictable or unpredictably random.

The *autotelic principle* by Steels (2004) tries to formalize the concept of "Flow" (Csikszentmihalyi 2000): an agent tries to maintain a state were learning is challenging, but not overwhelming (see also Gordon and Ahissar 2012). Another approach (Kaplan and Oudeyer 2004) aims to maximise the learning progress of different classical learning approaches by introducing rewards for better predictions of future states. A related idea is behind the *homeokinesis* approach, which can be considered a dynamic version of homeostasis. The basic principle here is to act in a way which can be well predicted by a adaptive model of the world dynamics (Der et al. 1999). There is a tendency of such mechanisms to place the agent in stable, easily predictable environments. For this reason, to retain a significant richness of behaviours additional provisions need to be taken so that notwithstanding the predictability of the future, the current states carry potential for instability.

The ideas of homeokinesis are originally based on dynamical system theory. Further studies have transferred them into the realm of information-theoretical approaches (Ay et al. 2008). The basic idea here is to maximise the *predictive information*, the information the past states of the world have about the future. Here, also, predictability is desired, but predictive information will only be large if the predictions about the future are decoded from a rich past, which captures very similar ideas to the dynamical systems view of homeokinesis.

The empowerment measure which is the main concept under discussion in the present paper, also provides a universal, task-independent motivation dynamics. However, it focusses on a different niche. It is not designed to *explore* the environment, as most of the above measures are, but rather aims at identify preferred states in the environment, once the local dynamics are known; if not much is known about the environment, but empowerment is high, this is perfectly satisfactory for the empowerment model, but not for the earlier curiosity-based methods. Therefore, empowerment is better described as a complement to the aforementioned methods, rather than a direct competitor.

Empowerment has been motivated by a set of biological hypotheses, all related to informational sensorimotor efficiency, the ability to react to the environment and similar. However, it would be interesting to identify whether there may be a route stemming from the underlying physical principles which would ultimately lead to such a principle (or a precursor thereof). For some time, the "Maximum Entropy Production Principle" (MEPP) has been postulated as arising from first thermodynamic principles (Dewar 2003, 2005). However, unfortunately, and according to current knowledge, the derivation from first principles still remains elusive and the current attempts at doing so unsuccessful (Grinstein and Linsker 2007). If, however, one should be able to derive the MEPP from first principles, then (Wissner-Gross and Freer 2013) show that this would allow a principle to emerge on the physical (sub-biological) level which acts as a simpler proto-empowerment which shares to some extent several of the self-organizing properties with empowerment, even if in a less specific way and without reference to the "bubble of autonomy" which would accompany a cognitive agent. Nevertheless, if successful, such a line may provide a route to how a full-fledged empowerment principle could emerge from physical principles.

4.3 Empowerment Hypotheses

In this section we want to introduce the main hypotheses which motivated the development of empowerment. Neither the work presented here in this chapter, nor the work on empowerment in general is yet a conclusive argument for either of the three main hypotheses, but they should, nevertheless, be helpful to outline what empowerment can be used for, and to what different domains empowerment can be applied. Furthermore, it should also be noted, that the hypotheses are stated in a generic form which might be unsuitable for experimental testing, but this can be alleviated on a case by case basis by applying a hypothesis to a specific scenario or task.

There are two main motivations for introducing the concept of empowerment: one is, of course, the desire to come up with methods to allow artificial agents to flexibly decide what to do generically, without having a specific task designed into them in every situation. This is closely related to the idea of creating a general AI. The other is to search for candidate proxies of prospective fitness, which could be detected and driven towards during the lifetime of an organism to improve its future reproductive success.

From these two starting points, several implicit and explicit claims have been made about empowerment and how it would relate to phenomena in biology. In the following section we structure these claims into three main hypotheses which we would consider as driving the "empowerment program". This should make it easier for the reader to understand what the simulations in the later chapters should actually demonstrate.

4.3.1 Behavioural Empowerment Hypothesis

The adaptation brought about by natural evolution produced organisms that in absence of specific goals behave as if they were maximising their empowerment.

Klyubin et al. (2005a,b) argue that the direct feedback provided by selection in evolution is relatively sparse, and therefore it would be infeasible to assume that evolution adapts the behaviour of organisms specifically for every possible situation. Instead they suggest that organisms might be equipped with local, task-independent utility detectors, which allows them to react well to different situations. Such generic adaptation might have arisen as a solution to a specific problem, and then persisted as a solution to other problems, as well. This also illustrates why such a utility function should be universal: namely, because it should be possible to retain the essential structure of the utility model, even if the morphology, sensor or actuators of the organism change through evolution.

This is also based on our understanding of humans and other organisms. We seem to be, at least in part, adapted to learn, explore and reason, rather than to only have hard-coded reactions to specific stimuli. As these abilities also usually generate actions, such a drive is sometimes called *intrinsic motivation*. Different approaches have been proposed (see Sec. 4.2) to formalize motivation that would generate actions that are not caused by an explicit external reward. Empowerment does not consider the learning process or the agent trajectory through the world, but instead operates as a pseudo-utility which assigns a value (its empowerment) to each state in the world[1]. Highly empowered states are preferred, and the core hypothesis states that an agent or organism attempts to reach states with high empowerment. Empowerment measures the ability of the agent to *potentially* change its future (it does not mean that it is actually doing that). The lowest value for empowerment is 0, which means that an agent has no influence on the world that it can perceive. From the empowerment perspective, vanishing empowerment is equivalent to the agent's death, and the empowerment maximization hypothesis provides a natural drive for death aversion.

The *behaviour empowerment hypothesis* now assumes that evolution has come up with a solution that produces similar behaviour. To support this hypothesis, the first step would be to demonstrate that empowerment can produce behaviour which is similar to biological organisms in analogous situations. In turn, it should also be possible to anticipate behaviour of biological organisms by considering how it would affect their empowerment. If we follow this idea further and assume that humans use empowerment-like criteria to inform their introspection, then one would expect that those states identified by humans as preferable would also be more likely to have high empowerment.

For the hypothesis to be plausible, it would also be good to ensure that empowerment is indeed local and can be computed from the information available to the agent. Similarly, it should also be universally applicable to different kinds of organisms; we would expect organisms which have undergone small changes to their sensory-motor set-up to still produce comparable empowerment values, and for organisms that discover new modalities of interaction that this is then reflected in the empowerment landscape.

So far, we have discussed a weak version of the behavioural empowerment hypothesis. A stronger version of the hypothesis[2] would argue that an agent actually computes empowerment. While this can be easily checked for artificial agents, in a biological scenario, it becomes necessary to explain how empowerment could actually be computed by the agent. The weak version of the hypothesis, instead, says that the agents just act "as if" driven by empowerment, or are using a suitable approximation. The hypothesis then states that natural behaviours favour highly empowered behaviour routes.

---

[1] Here we mostly adopt an "objective" perspective in that the objective states of the world are known and their empowerment computed. However, truly subjective versions of empowerment are easily definable and will be discussed in Sec. 4.4.4 as context-dependent empowerment.

[2] We do not actually put forward this stronger version for the biological realm, but mention it for completeness, and because of its relevance for empowerment in artificial agents.

4.3.2 Evolutionary Empowerment Hypothesis

The adaptation brought about by natural evolution increases the empowerment of the resulting organism.

Due to its universality, empowerment can in principle, be used to compare the average empowerment of different organisms. For instance, today, we could look at a digital organism, and then come back later after several generations of simulated adaptation, asking whether the organisms are now more empowered? Did that new sensor (and/or actuator) increase the agents empowerment? The hypothesis put forward, e.g. by Polani (2009), is that the adaptation in nature, on average, increases an agent's empowerment. He argues that (Shannon) information operates as a "currency of life", which imposes an inherent cost onto an organism, and, for that reason, a well-adapted organism should have efficient information processing. On the one hand, there is some relevant information (Polani et al. 2006) that needs to be acquired by an agent to perform at a given level, but any additional information processing would be superfluous, and should be avoided, as it creates unnecessary costs. Taking a look at agent morphologies, this also means that agents should be adapted to efficiently use their sensors and actuators. For example, a fish population living in perpetual darkness does not have a need for highly developed eyes (Jeffery 2005), and it is expected that adaptation will reduce the functionality and cognitive investment (i.e. brain operation) related to vision. On the other hand, in the dark the detection of sound could be useful; this perceptual channel could be made even more effective by actively generating sound that is then reflected from objects and could then be detected by the organism. The core question is: how can such potential advantageous gradients in the space of sensorimotoric endowment be detected?

Empowerment is the channel capacity from an agent's actuators to its sensors, and as such, measures the efficiency of that channel. Having actuators whose effect on the environment cannot be perceived, or sensors which detect no change relevant to the current actions is inefficient, and should be selected against. In short, this adaptation would be attained by an increase of the agent's average empowerment.

A test for this hypothesis would be to evolve agents in regard to other objectives, and then check how their empowerment develops over the course of the simulated evolution, similar to studies about complexity growth under evolutionary pressures (Yaeger 2009). Another salient effect of this hypothesis would be an adaptation of an agent morphology based on empowerment should produce sensor layouts and actions which are to some degree "sensible" and perhaps could also be compared to those found in nature.

4.3.3 AI Empowerment Hypothesis

Empowerment provides a task-independent motivation that generate AI behaviour which is beneficial for a range of goal-oriented behaviour.

In existing work, it was demonstrated that empowerment can address quite a selection of AI problems successfully (see the remaining chapter for a selection);

amongst these are pole balancing, maze centrality and others. However, a clear contraindication exists for its use: if an *externally* desired goal state is not highly empowered, then an empowerment-maximising algorithm is not going to seek it out. Opposed to that, such tasks are the standard domain of operation for traditional AI algorithms.

However, in the realm or robotics there have been developments to design robots that are not driven by specific goals, but motivated by exploring their own morphology or other forms of intrinsic motivation. The idea is to build robots that learn and explore, rather than engineer solutions for specific problems determined externally and in advance. Here, empowerment offers itself as another alternative. While empowerment is not designed to explicitly favour exploration, it has an inbuilt incentive to avoid behaviour that leads to a robot being stuck. Having no options available to an agent is bad for empowerment. Non-robotic AI could also benefit from this approach, but since empowerment is defined over the agent world dynamics, there needs to be a clear interface between an agent and the world over which it can be computed: in this case, there needs to be some kind of substitute for embodiment or situatedness. On the other hand, for the robotics domain it is also important that empowerment can be computed in real time and be applied to continuous variables.

The concrete and relevant question would be under which circumstances empowerment would provide a good solution, both in robotic and non-robotic settings? Furthermore, in what situations would maximising empowerment be helpful for a later to be specified task? To approach this question it is helpful to apply empowerment to a wider range of AI problems and inspect its operation in the different scenarios. The remaining chapter will showcase several such examples and discuss the insights gained from these.

In the robotic domain, one faces additional challenges, most prominently the necessity to handle empowerment in continuous spaces. This is discussed in Sec. 4.6. Note, however, that there is still very little current experience on deploying empowerment on real robots, with exception of a basic proof-of-principle context reconstruction example on an AIBO robot (Klyubin et al. 2008).

4.4 Formalism

Empowerment is formalized using terms from information theory, first introduced by Shannon (Shannon 1948). To define a consistent notation, we begin by introducing several standard notions. Entropy is defined as

$$H(X) = -\sum_{x \in \mathcal{X}} p(x) \log p(x) \tag{4.1}$$

where X is a discrete random variable with values $x \in \mathcal{X}$, and $p(x)$ is the probability mass function such that $p(x) = Pr\{X = x\}$. Throughout this paper base 2 logarithms are used by convention, and therefore the resulting units are in *bits*. Entropy can be understood as a quantification of uncertainty about the outcome of X before it is observed, or as the average surprise at the observation of X. Introducing another random variable Y jointly distributed with X, enables the definition of *conditional entropy* as

$$H(X|Y) = -\sum_{x\in\mathcal{X}} p(y) \sum_{y\in\mathcal{Y}} p(x|y) \log p(x|y). \tag{4.2}$$

This measures the remaining uncertainty about X when Y is known. Since Eq. (4.1) is the general uncertainty of X, and Eq. (4.2), is the remaining uncertainty once Y has been observed, their difference, called *mutual information*, quantifies the average information one can gain about X by observing Y. Mutual information is defined as

$$I(X;Y) = H(Y) - H(Y|X). \tag{4.3}$$

The mutual information is symmetric (see (Cover and Thomas 1991)), and it holds that

$$I(X;Y) = H(Y) - H(Y|X) = H(X) - H(X|Y). \tag{4.4}$$

Finally, a quantity which is used in communication over a noisy channel to determine the maximum information rate that can be reliably transmitted, is given by the *channel capacity*:

$$C = \max_{p(x)} I(X;Y)\ . \tag{4.5}$$

These concepts are fundamental measures in classical information theory.

Now, for the purpose of formalizing empowerment, we will now reinterpret the latter quantity in a causal context, and specialize the channel we are considering to the actuation-perception channel.

4.4.1 The Causal Interpretation of Empowerment

Core to the empowerment formalism is now the potential *causal* influence of one variable (or set of variables: the actuators) on another variable (or set of variables: the sensors). Further below, we will define the framework to define this in full generality; for now, we just state that we need to quantify the potential *causal effect* that one variable has on the other.

When we speak about causal effect, we specifically consider the interventionist notion of causality in the sense of Pearl (2000) and the notion of causal information flow based upon it (Ay and Polani 2008). We sketch this principle very briefly and refer the reader to the original literature for details.

To determine the causal information flow $\Phi(X \to Y)$ one cannot simply consider the observed distribution $p(x,y)$, but has to probe the distribution by actively intervening in X. The change resulting from the intervention in X (which we denote by $\hat{X}$) is then observed in the system and used to construct the interventional conditional $p(y|\hat{x})$. This interventional condition will then be used as the causal channel of interest. While (causal) information flow according to (Ay and Polani 2008) has been defined as the mutual information over that channel for an independent interventional input distribution, empowerment considers the *maximal* potential information flow, i.e. it is not based on the actual distribution of the input variable

X (with or without intervention), but considers the maximal information flow that could possibly be induced by a suitable choice of X. This, however, is nothing other than the channel capacity

$$C(X \to Y) = \max_{p(\hat{x})} I(\hat{X};Y). \tag{4.6}$$

for the channel defined by $p(y|\hat{x})$, where by the hat we indicate that this is a channel where we intervene in X.

There is a well-developed literature on how to determine the conditional probability distribution $p(y|\hat{x})$ necessary to compute empowerment, for some approaches, see (Pearl 2000; Ay and Polani 2008). This interventional conditional probability distribution can then be treated as the channel; and the channel capacity, or empowerment, can be computed with established methods, such as the Blahut-Arimoto algorithm (Blahut 1972; Arimoto 1972).

For the present discussion, it shall suffice to say that empowerment can be computed from the conditional probability distribution of observed actuation/sensing data, as long as we can ensure that the channel is a causal pair, meaning we can rule out any common cause, and any reverse influence from y onto x.

4.4.2 Empowerment in the Perception Action Loop

The basic idea behind empowerment is to measure the influence of an agent on its environment, and how much of this influence can be perceived by the agent. In analogy to control theory, it is essentially a combined measure of controllability (influence on the world) and observability (perception by the agent), but, unlike in the control-theoretic context, where controllability and observability denote the dimensionality of the respective vector spaces or manifolds, empowerment is a fully information-theoretic quantity: This has two consequences: the values it can assume are not confined to integer dimensionalities, but can range over the continuum of non-negative real numbers; and, secondly, it is not limited to linear subspaces or even manifolds, but can, in principle, be used in all spaces for which one can define a probability mass measure.

We formalize the concept of empowerment, as stated earlier, as the channel capacity between an agent's actions at a number of times and its sensoric stimuli at a later time. To understand this in detail, let us first take a step back and see how to model an agent's interaction with the environment as a causal Bayesian network (CBN). In general we are looking at a time-discrete model where an agent interacts with the world. This can be expressed as a perception-action loop, where an agent chooses an action for the next time step based on its sensor input in the current time step. This influences the state of the world (in the next time step), which in turn influences the sensor input of the agent at that time step. The cycle then repeats itself, with the agent choosing another action. Note that this choice of action might also be influenced by some internal state of the agent which carries information about the agent's past. To model this, we define the following four random variables:

A: the agent's actuator[3] which takes values $a \in \mathcal{A}$
S: the agent's sensor which takes values $s \in \mathcal{S}$
M: the agent's internal state (or memory) which takes values $m \in \mathcal{M}$
R: the state of the environment which takes values $r \in \mathcal{R}$

Their relationship can be expressed as a time-unrolled CBN, as seen in Fig. 4.1a.

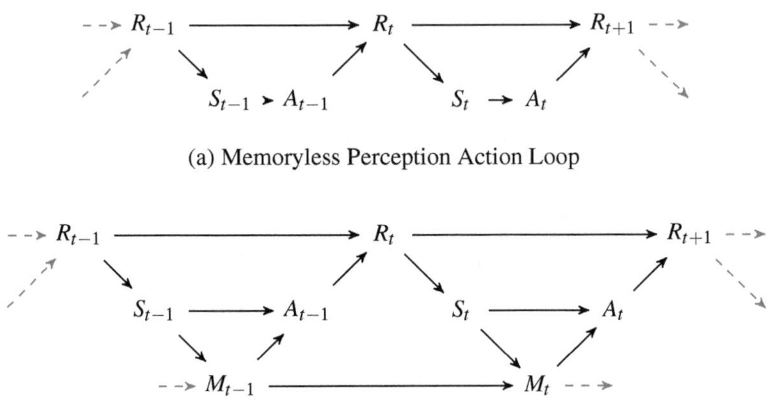

Fig. 4.1 Causal Bayesian network of the perception-action loop, unrolled in time, showing (a) a memoryless model, (b) a model including agent memory. In the memoryless model the agent's actions A_t only depend on its current sensor inputs S_t, while the perception action loop with memory allows for agent models in which the agent can store information from sensor inputs in the past in M, and use this information later for its decision making in A.

Empowerment is then defined as the channel capacity between the agent's actuators A and its own sensors S at a later point in time, here, for simplicity, we assume the next time step:

$$\mathfrak{E} := C(A_t \to S_{t+1}) \equiv \max_{p(a_t)} I(S_{t+1};A_t) \,. \tag{4.7}$$

This is the general empowerment of the agent. In the following text we will use $\mathfrak{E}$ as a shorthand for the causal channel capacity from the sensors to the actuators.

Note that the maximization implies that it is calculated under the assumption that the controller which chooses the action A is free to act, and is not bound by possible behaviour strategy $p(a|s,m)$. Importantly, the distribution $p^*(a)$ that achieves the channel capacity is different from the one that defines the actions of

[3] Saying *actuator* implicitly includes the case of multiple actuators. In fact, it is the most general case. Multiple actuators (which can be independent of each other) can always be written as being incorporated into one single actuator variable.

an empowerment-driven agent. Empowerment considers only the *potential* information flow, so the agent will only calculate how it *could* affect the world, rather than actually carry out its potential.

4.4.3 *n*-Step Empowerment

In Sec. 4.4.2, we considered empowerment as a consequence of a single action taken and the sensor being read out in the subsequent state. However, empowerment, as a measure of the sensorimotor efficiency, may start distinguishing the characteristics of the agent-environment interaction only after several steps. Therefore, a common generalization of the concept is the *n-step* empowerment. In this case we consider not a single action variable, but actually a sequence of action variables for the next n time steps: $(A_{t+1},\dots,A_{t+n})$. We we will sometimes condense these into a single action variable A for notational convenience. The sensor variable is the resulting sensor state in the following timestep S_{t+n+1}, again sometimes denoted by S. Though it is not the most general treatment possible, here we will consider only "open-loop" action sequences, i.e. action sequences which are selected in advance and then carried out without referring to a sensor observation until the final observation S_{t+n+1}. This drastically simplifies both computations and theoretical considerations, as the different possible action sequences A can be treated as if they were separate atomic actions with no inner structure[4].

As mentioned A can typically contain actuator variables from several time steps and can also incorporate several variables per time step. S is typically chosen to contain variables that are strictly temporally downstream from all variables in A, to ensure a clean causal interpretation of the effect of A on S. However, the less studied concept of *interleaved empowerment* has been mentioned in (Klyubin et al. 2008), where S contains sensor variables that lie before some variables in A[5].

4.4.4 Context-Dependent Empowerment

Until now, we have considered empowerment as a generic characterization of the information efficiency of the perception-action loop. Now we go a step further and resolve this informational efficiency in more detail; specifically, we are going to consider empowerment when the system (e.g. agent and environment) is in different states. Assuming that the state of the system is given by r, it will in general affect the effect of the actions on the later sensor states, so that one now considers $p(s|a,r)$ and defines empowerment for the world being in state r as

[4] Future work will investigate the effect of feedback, i.e. closed-loop sequences. However, the current hypothesis is that there will be little qualitative and quantitative difference for most scenarios, with significantly increased computational complexity.

[5] The interpretation of interleaved empowerment is slightly subtle and still subject to study, as in this case S is then capturing rather an aspect of the richness of the action sequences and the corresponding action history, in addition to the state dynamics of the system.

$$\mathfrak{E}(r) = \max_{p(a)} I(S;A|r), \tag{4.8}$$

which is referred to as *state-dependent empowerment*. This also allows us to define the average state-dependent empowerment for an agent that knows what state the world is in as

$$\mathfrak{E}(R) = \sum_{r \in R} p(r)\mathfrak{E}(r) \tag{4.9}$$

Note that this is different from the general empowerment: the general empowerment in Sec. 4.4.2 does not distinguish between different states. If different perception-action loop characteristics $p(s|a)$ are not resolved, the general empowerment can be vanishing, while average state-dependent empowerment is non-zero. In other words, empowerment can depend on being able to resolve states which affect the actuation-sensing channel.

In general, an agent will not be able to resolve all states in the environment, and will operate using a limited sensoric resolution of the world state. When we assume this, the agent might still be able to recognize that the world is in a certain context $k \in K$, based on memory and sensor input. So, an agent might not know its precise state in the world, but may be able to identify some coarse position, e.g. that it might be north or south of some distinct location. Klyubin et al. (2008) demonstrate an example of how such a context can be created from data. Based on this context, it is then possible to define the marginal conditional distribution $p(s|a,k)$, which then allows us to compute the (averaged) contextual empowerment for K as

$$\mathfrak{E}(K) = \sum_{k \in K} p(k)\mathfrak{E}(k) \tag{4.10}$$

In comparison, context free empowerment $\mathfrak{E}_{\text{free}}$ has no assumption about the world and is based on the marginal distribution $p(s|a) = \sum_r p(s|a,r)p(r)$ of all world states. This is the empowerment that an agent would calculate which has no information about the current world state. It can be shown (Capdepuy 2010) with Jensen's Inequality that

$$\mathfrak{E}_{\text{free}} \leq \mathfrak{E}(K) \leq \mathfrak{E}(R) \tag{4.11}$$

This implies (see also Klyubin et al. 2008) that there is a (not necessarily unique) minimal optimal context K_{opt} that best characterizes the world in relation to how the agent's actions affect the world, defined by:

$$K_{\text{opt}} = \underset{\substack{K \\ \mathfrak{E}(K) = \mathfrak{E}(R)}}{\arg\min}\ H(K). \tag{4.12}$$

Such a context K_{opt} is one which leads to the maximal increase in contextual empowerment. Klyubin et al. (2008) argues that such an agent internal measure could be useful to develop internal contexts which are purely intrinsic and based on the agent sensory-motor capacity, and thereby allow developing an understanding of the world based on the way they are able to interact with it.

4.4.5 Open vs. Closed-Loop Empowerment

An important distinction to make is the one between open- and closed-loop empowerment. Open-loop empowerment treats the perception-action loop like a unidirectional communication channel, and assumes that all inputs are chosen ahead of time and without getting any feedback about their source. Closed-loop empowerment is computed under the assumption that some of the later actions in n-step empowerment can change in reaction to the current sensor state.

In most of the existing work, empowerment calculations have been performed with open-loop empowerment only. The framework for this simplest of cases of communication theory is well developed and long known. For the more intricate cases using feedback, Capdepuy (2010) pointed out that *directed information* (Massey 1990) could be used to simplify the computation of closed loop empowerment, and demonstrated for an example how feedback increases empowerment.

4.4.6 Discrete Deterministic Empowerment

A *deterministic* world is one where each action leads to one specific outcome, i.e. for every $a \in \mathcal{A}$ there is exactly one $s_a \in \mathcal{S}$ with the property that

$$p(s|a) = \begin{cases} 1 \text{ if } s = s_a \\ 0 \text{ else} . \end{cases} \tag{4.13}$$

Since here every action only has one outcome, it is clear that the conditional uncertainty of S given A is zero, i.e., $H(S|A) = 0$. From Eq. (4.4) it follows then that

$$\mathfrak{E} = \max_{p(a)}(A;S) = \max_{p(a)} H(S). \tag{4.14}$$

Since the entropy is maximal for a uniform distribution, S can be maximised by choosing any input distribution $p(a)$ which results in a uniform distribution over the set of all reachable states of S, i.e over the set $\mathcal{S}_{\mathcal{A}} = \{s \in \mathcal{S} | \exists a \in \mathcal{A} : p(s|a) \geq 0\}$. As a result, empowerment for the discrete deterministic case reduces to

$$\mathfrak{E} = -\sum_{s \in \mathcal{S}_{\mathcal{A}}} \frac{1}{|\mathcal{S}_{\mathcal{A}}|} \log\left(\frac{1}{|\mathcal{S}_{\mathcal{A}}|}\right) = \log(|\mathcal{S}_{\mathcal{A}}|). \tag{4.15}$$

The bottom line is that in a discrete deterministic world empowerment reduces to the logarithm of the number of sensor states reachable with the available actions. This means empowerment, in the deterministic case, is fully determined by how many distinguishable states the agent can reach.

4.4.7 Non-deterministic Empowerment Calculation

If noise is present in the system, an action sequence a will lead to multiple outcomes s, and thus, we have to consider an actual output distribution $p(s|a)$. In this case, the optimizing distribution needs to be determined using the standard Blahut-Arimoto (BA) algorithm (Blahut 1972; Arimoto 1972) which is an expectation maximization-type algorithm for computing the channel capacity.

BA iterates over distributions $p_k(\mathbf{a})$, where k is the iteration counter, converging towards the distribution that maximises channel capacity, and thereby towards the empowerment value defined in Eq. (4.8). Since the action variable A is discrete and finite we are dealing with a finite number of actions $a_\nu \in \mathcal{A}$, with $\nu = 1,...,n$. Therefore $p_k(\mathbf{a})$ in the k-th iteration can be compactly represented by a vector $p_k(\mathbf{a}) \equiv (p_k^1,...,p_k^n)$, with $p_k^\nu \equiv \Pr(A = a_\nu)$, the probability that the action A attains the value a_ν. Furthermore, let $s \in \mathcal{S}$ be the possible future states of the sensor input as a result of selecting the various actions with respect to which empowerment is being calculated, and $r \in \mathcal{R}$ is the current state of the environment. If we assume that S is continuous we can follow the general outline from (Jung et al. 2011), and define, for notational convenience, the variable $d_{\nu,k}$ as:

$$d_{\nu,k} := \int_{\mathcal{S}} p(s|r,\mathbf{a}_\nu) \log\left[\frac{p(s|r,\mathbf{a}_\nu)}{\sum_{i=1}^{n} p(s|r,\mathbf{a}_i)p_k^i}\right] ds. \tag{4.16}$$

While this is the more general case, this integral is difficult to evaluate for arbitrary distributions of S. We will later discuss, in Sec. 4.6.6, how this integral can be approximated, but even the approximations are very computationally expensive. If we are dealing with discrete and finite S we can simply define $d_{\nu,k}$ with a sum as

$$d_{\nu,k} := \sum_{s \in \mathcal{S}} p(s|r,\mathbf{a}_\nu) \log\left[\frac{p(s|r,\mathbf{a}_\nu)}{\sum_{i=1}^{n} p(s|r,\mathbf{a}_i)p_k^i}\right]. \tag{4.17}$$

The definition of $d_{\nu,k}$ encapsulates the differences between a continuous and discrete S. Therefore, the following parts of the BA algorithm are identical for both cases. The BA begins with initialising $p_0(\mathbf{a})$ to be (e.g.) uniformly distributed[6], by simply setting $p_0^\nu = \frac{1}{n}$ for all actions $\nu = 1,...,n$. At each iteration $k \geq 1$, the new approximation for the probability distribution $p_k(\mathbf{a})$ is obtained from the old one $p_{k-1}(\mathbf{a})$ using

$$p_k^\nu := \frac{1}{z_k} p_{k-1}^\nu \exp(d_{\nu,k-1}) \tag{4.18}$$

where z_k is a normalisation parameter ensuring that the approximation for the probability distribution $p_k(\mathbf{a})$ sum to one for all actions $\nu = 1,...,n$, and is defined as

$$z_k := \sum_{\nu=1}^{n} p_{k-1}^\nu \exp(d_{\nu,k-1}). \tag{4.19}$$

[6] In principle, any distribution can be selected, provided none of the initial probabilities is 0, as the BA-algorithm cannot turn a vanishing probability into a finite one.

Thus $p_k(\mathbf{a})$ is calculated for iteration step k, it can then be used to obtain an estimate $\mathfrak{E}_k(r)$ for the empowerment $\mathfrak{E}(r)$ using

$$\mathfrak{E}_k(r) = \sum_{\nu=1}^{n} p_k^{\nu} \cdot d_{\nu,k}. \tag{4.20}$$

The algorithm can be iterated over a fixed number of times or until the absolute difference $|\mathfrak{E}_k(r) - \mathfrak{E}_{k-1}(r)|$ drops below an arbitrary chosen threshold ε.

4.5 Discrete Examples

4.5.1 Maze

Historically, the first scenario used to illustrate the properties of empowerment was a maze setting introduced in (Klyubin et al. 2005a). Here, the agent is located in a two-dimensional grid world. The agent has five different actions; it can move to the adjacent squares north, east, south and west of it, or do nothing. An outer boundary and internal walls block the agents movement. If an agent chooses the action to move against a wall, it will not move.

The states of the agent's action variable A for n-step empowerment are constituted by all 5^n action sequences that contain n consecutive actions. The resulting sensor value S consists of the location of the agent at time step t_{n+1}, after the last action was executed. Since we are dealing with a discrete and deterministic world, empowerment can be calculated as in Eq. (9.22) in Sec. 4.4.6 by taking the logarithm of all states reachable in n steps.

4.5.2 Average Distance vs. Empowerment

In this maze example, empowerment is directly related to how many states an agent can reach within the next n steps. Now, note that, via the agent's actions, a Finsler metric-like (Wilkens 1995; López and Martínez 2000) structure is implied on the maze, namely the minimum number of action steps necessary to move from one given position in the maze to a target position. Calculating n-step empowerment for the current location in the maze then is simply the logarithm of all states with a distance of n or less to the current state.

Although this n-step horizon provides empowerment with an essentially local "cone of view", Klyubin et al. (2005a) showed in the maze example that empowerment of a location is negatively correlated with the average distance of that location to *all* other locations in the maze. The first is a local, the latter, however, a global property. This indicates that the local property of n-step reachability (essentially n-step empowerment) would relate to a global property, namely that of average distance.

It is a current study objective to which extent this local/global relation might be true, and under which conditions. Wherever it applies, the empowerment of an agent

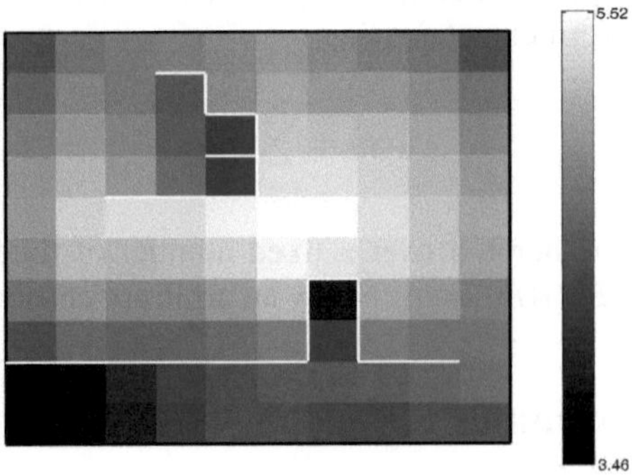

Fig. 4.2 The graph depicts the empowerment values for 5 step action sequences for the different positions in a 10×10 maze. Walls are shown in white, and cells are shaded according to empowerment. As the key suggests empowerment values are in the range [3.46, 5.52] bits. This figure demonstrates that by simply assessing its options (in terms of movement possibilities) reflected in its empowerment, the agent can discover various features of the world. The most empowered cells in the labyrinth are those that can reliable reach the most positions within the next 5 steps. The graph is a reproduction of the results in (Klyubin et al. 2005a).

(which can be determined from knowledge of the local dynamics, i.e. how are my next n-steps going to affect the world) could then be used as a proxy for certain global properties of the world, such as the average distance to all other states. It is clear that this cannot, in general, be true, as outside of the empowerment horizon n, an environment could change its characteristics drastically, unseen to the "cone of view" of the agent's local empowerment. However, many relevant scenarios have some regularity pervading the whole system which has the opportunity to be detected by empowerment.

This motif was further investigated by Anthony et al. (2008), who studied in more detail the relationship between graph centrality and empowerment. The first chosen model was a two-dimensional grid world that contained a pushable box, similar to (Klyubin et al. 2005a). The agent could take five actions; move north, south, west, east, or do nothing. If the agent moves into the location with the box, the box would be pushed into the next square. The state space, the set of possible world configurations, included the position of the agent, and also the position of the box.

The complete system can be modelled as a directed labelled graph, where each node represents a different state of the world and the directed edges, labelled with actions, represent the transitions from one state to another under a specific action. For an agent with 5 possible actions, all nodes have 5 edges leading away from them. This is a generic representation of any discrete and deterministic model. The advantage of this representation is that it provides a core characterization of the system in graph-theoretic language which is abstracted away from a physical representation. As before, the distance from one state to another depends on how many actions an agent needs to move from the first to the second state. In general, this defines a

Finsler metric-like structure (see Sec. 4.5.2), and is not necessarily tied to physical distance.

Anthony et al. (2008) then studied the correlation between closeness centrality and empowerment, both for the previously described box pushing scenario. In addition, he considered a different scenario, namely scale-free random networks as transition graphs. As before, one can consider closeness centrality (which is a global property), and empowerment (which can be calculated from a local subset of the graph). Anthony et al. (2008) find that:

> "these results show a strong indication of certain global aspects of various worlds being 'coded' at a local level, and an appropriate sensory configuration can not only detect this information, but can also use it..."

It is, however, currently unknown how generally and under which circumstances this observation holds. As mentioned before, it is possible to construct counterexamples. A natural example is the one that Anthony et al. note in their discussion, namely that the relationship breaks down for the box pushing example when the agents horizon does not extend to the box; in this case, the agent is too far away for n-step empowerment to be affected by the box. This might indicate that a certain degree of structural homogeneity throughout the world is necessary for this relation to hold, and that the existence of different "pockets" in the state space with different local rules would limit the ability of empowerment to estimate global properties. After all, if there is a part of the world that is radically different from the one the agent is in, and the agent is not able to observe it in the near future, the current situation may not be able to be informative concerning that remote part of the world.

At present, however, it remains an open question how empowerment relates to global properties, such as in the example of graph centrality or average distance. No full or even partial characterization of scenarios where empowerment correlates to global values is currently known.

4.5.3 Sensor and Actuator Selection

An agent's empowerment is not only affected by the state of the world, i.e., the context of the agent, but also depends on what the agent's sensors and actions are. This was illustrated by Klyubin et al. (2005a) by variation of the previously mentioned box-pushing example. In all scenarios we are dealing with a two dimensional grid world where the agent has five different actions. The center of the world contains a box. In Fig. 4.3 we see the 5-step empowerment values for the agent's starting position in four different scenarios. The scenarios differ depending on

1. whether the agent can perceive the box and
2. whether the agent can push the box.

In Fig 4.3.b the agent can push the box but cannot sense it. The box neither influences the agent's outcome, nor is the agent able to perceive it. Basically, this is just like a scenario without a box. Consequently, the empowerment map of the world is flat, i.e., all states have the same empowerment. For empowerment applications

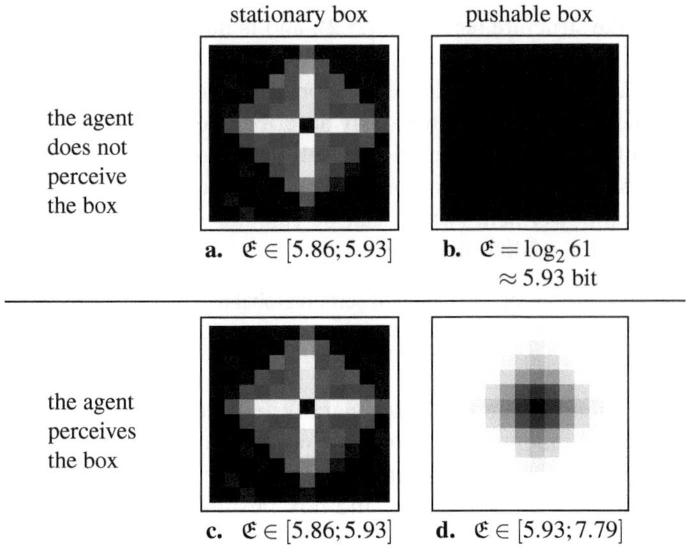

Fig. 4.3 Empowerment maps for 5-step empowerment in a 2 dimensional grid world, containing a box in the center. The scenarios differ by whether the box can be pushed by the agent or not, and whether the agent can perceive the box. Black indicates the highest empowerment. Figure reproduced from (Klyubin et al. 2005a).

this is typically the least interesting case, as it provides no gradient for action selection (see also the comment on the "Tragedy of the Greek Gods" towards the end of Sec. 4.5.4).

Fig. 4.3.d shows the empowerment map for an agent which can perceive the box, the agent's sensor input is both its own position and the position of the box. This different sensor configuration changes the empowerment map of the world. Being close to the box to affect it now allows the agent to "reach" more different outcomes, because different paths that lead to the same final agent location might affect the box differently, thereby resulting in different final states. This results in higher empowerment closer to the box. Note that, comparing this to the previous scenario where the box was not visible, the agent's actions are not suddenly able to create a larger number of resulting world states. Rather, the only change is that the agent is now able to discriminate between different world states that where present all along.

Figures 4.3.a and 4.3.c show the empowerment map for an non-pushable box, so when the agent moves into the box's square, its movement fails. As opposed to the earlier cases, here we see that the empowerment around the box is lowered, because the box is blocking the agents way, thereby reducing the number of states that the agent can reach with its 5-step action sequence. We also see that the empowerment maps in Fig. 4.3.a and 4.3.c are identical, and that it does not matter if the agent can perceive the box or not. This connects back to our earlier arguments that empowerment is about influencing the world one can perceive. As it is not possible for the agent to affect the box's positions, it is also not beneficial or relevant, from an empowerment perspective, to perceive the box position. This also relates back to earlier arguments about sensor and motor co-evolution. Once an agent loses it ability to affect the box, it might just as well lose it ability to sense the box.

One important insight that is demonstrated by this experiment is how different sensor and actuator configurations can lead to significantly different values for the state-dependent empowerment maps. Thus, which state has the highest empowerment might depend on an agent-sensor configuration (and not only on the world dynamics). This can be helpful when using empowerment to define an action policy. If an agent chooses its actions based on expected empowerment gain, then this method is a candidate for causing an agent to change its behaviour by only calculating empowerment for partial sensor input. For example, to drive an agent to focus on changing its location, then selecting a corresponding location sensor might be a good strategy.

4.5.4 Horizon Extension

Extending the horizon, i.e., using a larger n in n-step empowerment, is another way to change the actions under consideration. Since the n-step action sequences can be treated just like atomic actions, lengthening the considered sequences creates more distinct actions to consider, which usually also have a bigger effect on the environment. Returning to the previous maze example, Fig. 4.4 illustrates how the empowerment map changes for action sequences of different length.

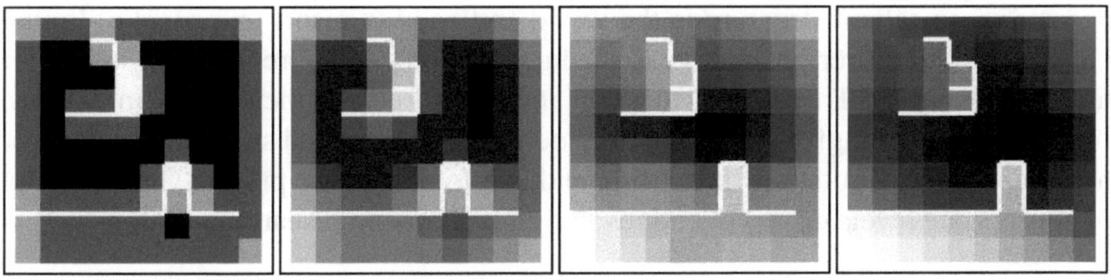

$\mathfrak{E} \in [1;2.32]$	$\mathfrak{E} \in [1.58;3.70]$	$\mathfrak{E} \in [3.46;5.52]$	$\mathfrak{E} \in [4.50;6.41]$
1-step	2-step	5-step	10-step

Fig. 4.4 The n-step empowerment map for the same maze with different horizons. Figure based on (Klyubin et al. 2005a).

The short-term, 1-step empowerment only takes into account its immediate local surroundings. All that matters are if there are walls immediately next to the agent. In general, an agent locked in a room with walls just one step away would have the same empowerment as an agent on an open field. Also, this map only realizes 5 different empowerment values because the world is deterministic, and there can be maximally 5 different outcome states.

With more steps, the empowerment map starts to reflect the immediate surrounding of the agent and measures, as discussed by Anthony et al. (2008), how "central" an agent is located on the graph of possible states. But, as discussed earlier, the world could be shaped in a way that something just beyond the horizon of the agent's empowerment calculation could change this completely. A possible solution would be to further extend the horizon of the agent. One problem, which we will address in the next section is that of computational feasibility.

Another problem is that the agent needs the sensor capacity to adequately reflect an increase in possible actions. Consider the following case: computing, say, 100-step empowerment, then the agent could reach every square from every other square, creating a flat empowerment landscape with an empowerment of $\log(100)$ everywhere. Since the agent itself is very (indeed maximally) powerful now, being able to reach every state of the world, its empowerment landscape is meaningless, as empowerment is incapable of distinguishing states via the number of options they offer. In principle, an analogous phenomenon can be created by massively extending the sensor capacity. Imagine an agent would not only be able to sense it current position, but also sense every action it has taken in the past. Now the agent could differentiate between every possible action sequence, as every one is reflected as a different sensor state. This again leads to a flat empowerment landscape, with empowerment being the logarithm of all possible actions.

So, in short, one has to be careful when the state-space of either actions or sensors is much larger than the other. In this case it is possible that the channel capacity becomes the maximal entropy of the smaller variable for all possible contexts, thereby creating a flat empowerment landscape. This phenomenon can be subsumed under the plastic notion of the "Tragedy of the Greek Gods": all-knowing, all-powerful agents see no salient structure in the world and need to resort to avatars of limited knowledge and power (in analogy to the intervention of the Greek gods with the human fighters in the Trojan War) to attain any structured and meaningful interaction. In short, for meaningful interaction to emerge from a method such as an empowerment landscape, limitations in sensing and acting need to be present. The selection of appropriate levels of power and resolution is a current research question.

4.5.5 Impoverished Empowerment

While seeking the right resolution for actions and sensors can be an issue in worlds of limited complexity, a much more imminent challenge is the fact that as the empowerment horizon grows, the number of action sequences one needs to consider grows exponentially with the horizon. Especially when noise is involved, this becomes quickly infeasible.

To address this dilemma, Anthony et al. (2011) suggest a modified technique that allows for the approximation of empowerment-like quantities for longer action sequences, arguing, among other, that this will bring the empowerment approach in principle closer to what is cognitively plausible.

The basic idea of the *impoverished empowerment* approach is to consider all n-step action sequences (as in the simple empowerment computation), but then to select only a limited amount of sequences from these, namely those which contribute the most to the empowerment at this state. From the endpoint of this "impoverished" action sequence skeleton, this process is then repeated for another n-step sequence, thereby iteratively building up longer action sequences.

In the deterministic case, the selection is done so that the collection of action sequences has the highest possible empowerment. So, if several action sequences would lead to the same end state, only one of them would be chosen.

Interestingly, a small amount of noise is useful for this process, as it favours selecting action sequences which are further apart, because their end states overlap less. If no noise is present, then two action sequences which would end in neighbouring locations would be just as valid as two that lead to completely different locations, but the latter is more desirable as it spans a wider space of potential behaviours.

4.5.6 Sensor and Actuator Evolution

Since empowerment can be influenced by the choice of sensors, it is possible to ask what choice of sensors is maximising an agent's empowerment. Klyubin et al. (2005b, 2008) addressed this question by using a Genetic Algorithm-based optimization for a scenario in which sensors are being evolved to maximize an agent's empowerment. An agent is located in an infinite two-dimensional grid world. On each turn it can take one of five different actions which are to move in one of four directions, or to do nothing. Each location now has a value representing the concentration of a marker substance which is inversely proportional to the distance of the current location to the center at location $(0,0)$.

In this scenario, the agents sensors can change, both in positioning and number. A sensor configuration is defined by where each of the n sensors of the agent is located relative to the agent. The sensor value has n states, and represents which of the n sensors detects the highest concentration value of the marker.

Klyubin et al. (2005b, 2008) then evolved the agents sensor configuration to maximise empowerment for different starting locations with respect to the centre. So, for example, they evolved the sensor configurations to achieve the highest empowerment when the agent starts its movement at location $(0,0)$. To avoid degeneracy, a slight cost factor for the number of sensors was added. In this way the adaptation has to evaluate if the added cost of further sensors are worth the increase in empowerment. The resulting sensor configurations for a 4-step empowerment calculation can be seen in Figure 4.5.

The result was unsurprisingly that different starting positions would lead to different sensor layouts. More interestingly, they realized that the space of possible solutions can be more constrained in some places, so there is only one good solution, while other locations offer several different, nearly equally empowered solutions. More importantly is the observation that empowerment agnostically selects

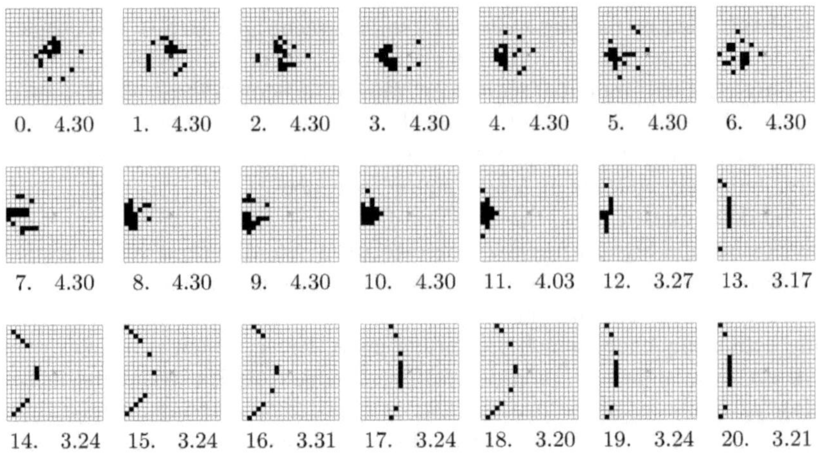

Fig. 4.5 The Figures show what sensor configurations empowerment evolves for different starting positions. The first number indicates how many spaces east of the center the agent starts, and the second number is the resulting empowerment value of the sensor configuration. Figure taken from (Klyubin et al. 2008).

modalities which are most appropriate for the various starting locations. Consider, for instance, Fig. 4.5 which shows how the sensors are placed relative to the agent as the agent moves increasingly away from the center of the world, and to the right of it. The first images show the sensor placement when the agent is at the center of the world. The sensors are placed with more-or-less precision around the center, and there is some indifference as to their exact placement.

In the second row, when the agent has been moved seven and more fields to the right of the centre, a more prominent “blob” is placed at around the location of the centre (the diagram shows the relative placement of the sensors with respect to the agent, so a blob of black dots is covering roughly the location at which the centre of the world will be with respect to the agent.

Finally, as the agent moves further to the right (end of second and last row in Fig. 4.5), a striking effect takes place: the blob sensor, which roughly determines a two-dimensional location of the centre, collapses into a “heading” sensor which is no longer a two-dimensional blob, but rather has 1-dimensional character. This demonstrates that empowerment is able to switch to different modalities (or, in this case, from a 2-dimensional to a 1-dimensional sensor). Because of its information-theoretic nature, empowerment is not explicitly using any assumptions about modality or dimensionality of sensors. The resulting morphologies are purely a result of the selection pressure via empowerment in interaction with the dynamics and structure of the world under consideration.

Another result of the evolutionary scenario involved the evolution of actuators. Without repeating the full details that can be found in (Klyubin et al. 2008), we would like to mention one important result, namely that the placement of actuators via empowerment-driven evolution, unlike the sensors, was extremely unspecific.

Many configurations led to maximum empowerment solutions. The authors suggest that this results as a consequence of the agent being unable to choose what form the 'information' takes, that it has to extract from the environment. Hence, the sensors have to adapt to the information structure available in the environment, leaving the agent free to choose its actions. Therefore many different actuator settings can be used as the agent can utilize each of them to full effect by generation of suitable action sequences. This is an indicator that an agent's action choices should be a more valuable and "concentrated" source of information than the information extracted from the environment, as every action choice is significant, while sensoric information needs to be "scooped" in on a wide front to capture some relevant features. This insight has been taken onboard in later work in form of the the concept of *digested information* (Salge and Polani 2011) where agents observe other agents because their actions are more informationally dense than other aspects of the environment. The core idea of *digested* information is that relevant information (as defined in (Polani et al. 2006)) is often spread out in the environment, but since an agent needs to act upon the information it obtains, the same information is also present in the agent's actions. Because the agent's action state-space is usually much smaller than the state-space of the environment, the agent "concentrates" the relevant information in it actions. From the perspective of another, similar agent this basically means that the agent digests the relevant information and then provides it in a more compact format. It should be noted that all structure in the above example emerges purely from informational considerations; no other cost structure (such as e.g. energy costs) have been taken into account to shape the resulting features.

4.5.7 Multi-agent Empowerment

If two or more agents share an environment, so that their actions all influence the state of the world, then their empowerment becomes intertwined. Capdepuy (2010); Capdepuy et al. (2007, 2012) investigate this phenomenon in detail. Here, due to lack of space, we will limit ourselves to briefly outline his results.

If both agents selfishly optimize their empowerment, then the outcome depends heavily on the scenario. A fully formal categorization is still outstanding, but the qualitative phenomenon can be described in terms similar to different game solution types in game theory. One finds situations that are analogous to zero-sum games where the empowerment of one agent can only be raised to the detriment of the other. In other situations, selfish empowerment maximisation leads to overall high empowerment, and, finally, there are scenarios where agent's strategies converge onto the analoga of intricate equilibria reminiscent of the Nash equilibria in games.

An interesting aspect in relation to biology is Capdepuy's work on the emergence of structure from selfish empowerment maximisation (Capdepuy et al. 2007). The model consists of a two-dimensional grid world where agents are equipped with sensors that measure the density of other agents in the directions around them. In this case, there is a tension between achieving proximity to other agents (to attain any variation in sensor input, as empty space does not provide any) and being sufficiently distant (as to attain sufficient freedom for action and not to be stuck without ability to move); this tension, in turn, provides an incentive to produce non-trivial dynamical structures. Some examples of agent populations evolved for greedy empowerment maximization and some of the better empowered structures resulting from this process can be seen in Fig. 4.6 Capdepuy et al. (2007).

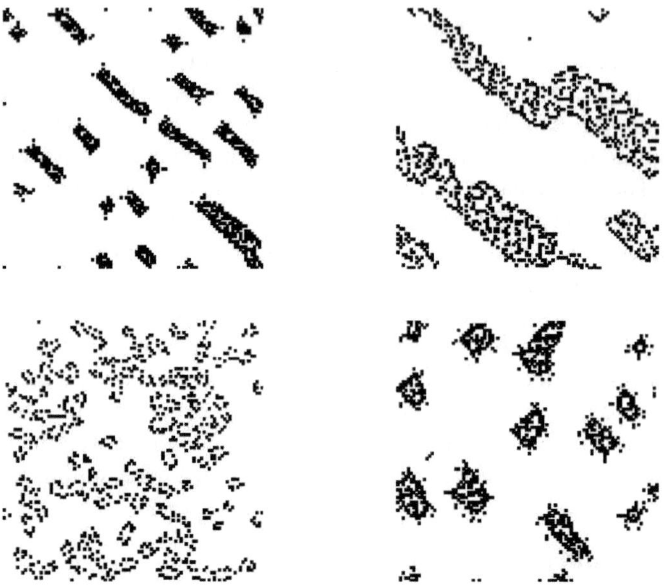

Fig. 4.6 Structures resulting from agent behaviour that was evolved to maximise the agents' individual empowerment. Each black dot in the figure represents an agent in one of the empowerment-maximizing scenarios. Agents are equipped with directional density sensors, measuring the number of other agents present in that particular direction. Creating structures becomes beneficial for the agents, as it gives features to the environment that allow different resulting sensor inputs. The different structures are high empowered solutions of the artificial evolution. Figure taken from (Capdepuy 2010).

4.6 Continuous Empowerment

The empowerment computations that we considered earlier were all operating in discrete spaces. But if we want to apply empowerment to the real world we need to consider that many problems, especially those related to motion or motor control, are continuous in nature. We could apply naive discretizations with finer and finer resolutions, but this will quickly lead to large state and actions spaces, with a forbidding number of options where direct computation of empowerment become very computationally expensive (Klyubin et al. 2008); therefore, different approaches need to be taken to deal with continuous dynamics effectively.

In this section, we will take a closer look at empowerment for continuous actuator and sensor variables. Compared to the discrete case, while channel capacity is still well defined for continuous input/output spaces, there are some important conceptual differences to be considered as compared to the discrete case.

One problem, as we shall illustrate, is that the continuous channel capacity could — in theory — be infinite. The reason for this is as follows: if there is no noise, and arbitrary continuous actions can be selected, these actions now allow to inject continuous, i.e. real-valued quantities (or vectors) into the world state. Reading in their (again) noiseless effect through real-valued sensors means that the full precision of a real number can be used in such a case. As arbitrary amounts of information can be stored in an infinite precision — noiseless — real number, this implies (in nondegenerate cases) an infinite channel capacity. Of course, such a situation is not realistic; in particular, relevant real-world systems always have noise and therefore the channel capacity will be limited.

However, when modeling a deterministic system with floating-point precision in simulation, there is no natural noise level. In a nondegenerate system, empowerment can be made as large as the logarithm of the number of actions (action sequences) available. This is, of course, meaningless. To be meaningful, one needs to endow the system with additional assumptions (such as an appropriate noise level) which are not required in the deterministic case.

But the main problem in the continuous case is that there is at the time of this review no known analytic solution to determine the channel capacity for a general continuous channel. To address this problem, a number of methods to approximate continuous channel capacity have been introduced. We will discuss them and how they can be used to compute empowerment.

We will briefly discuss naive binning, then the Monte Carlo Integration method developed by (Jung et al. 2011), and then focus mostly on the quasi-linear Gaussian approximation, which is fast to compute.

4.6.1 Continuous Information Theory

The analogy to discrete entropy is rigorously defined for continuous random variables as *differential entropy*

$$h(X) = -\int_{\mathcal{X}} p(x)\log(p(x))\,\mathrm{d}x\,, \tag{4.21}$$

where $p(x)$ now denotes not the probability, but the probability density function of X, defined over a support set of $\mathcal{X} \subseteq \mathbb{R}$. Similarly, the *conditional differential entropy* is defined as

$$h(X|Y) = -\int_{\mathcal{Y}} p(y)\int_{\mathcal{X}} p(x|y)\log(p(x|y))\,\mathrm{d}x\mathrm{d}y\,. \tag{4.22}$$

The differential entropies cannot be directly interpreted in the same way as discrete entropies: they can become infinite or even negative. However, without delving too much into their individual interpretation, we will just state here that the difference of two differential entropy terms again can be interpreted as a proper mutual information: $I(X;Y) := h(X) - h(X|Y)$, which shares essentially all characteristics of the discrete mutual information[7]. Thus, consequently, the channel capacity is again defined by maximising the mutual information for the input probability density function

$$\mathfrak{E} = C(A \to S) = \max_{p(a)} I(A;S). \tag{4.23}$$

We will still be dealing with discrete time steps. Just like in the discrete case, we will use the notation A_t and S_t not just for single, but also for compound random variables. So, for each time t, both variables A_t and S_t can consist of vectors of multiple random variables. The variables A and (where relevant) S itself are then again a selection of actuator and sensor variables at different times t, so for example, the actuator input for n-step empowerment might be written compactly as $A = (A_t, ..., A_{t+n-1})$.

4.6.2 *Infinite Channel Capacity*

As mentioned above, in contrast to the discrete case, the continuous channel capacity can be infinite for some $p(s|a)$. Formally, this results from the fact that differential entropy *can* become negative. For instance, it becomes negative infinity for a Dirac $\delta_x(.)$ "distribution". The Dirac "distribution" is a probability measure concentrated on a single point: it can be mathematically defined in a precise fashion, but for the following discussion, the intuition is sufficient that $\delta_x(.)$ is normalized (the integral over this "distribution" is 1), and is 0 everywhere with exception of the one point x at which it is concentrated, where it assumes an infinite value.

To illustrate, imagine that the channel $p(s|a)$ exactly reproduces the real-valued input value of $a \in \mathbb{R}$, i.e. that it implements $s = a$, i.e. $p(s|a) \equiv \delta_a(s)$. Every input a precisely determines the output s, so $h(S|a) = -\infty$. This remains negative infinity when we integrate over all possible inputs, so $h(S|A) = -\infty$. If we now choose for $p(a)$ the uniform input distribution between 0 and 1, which has a differential entropy of 0, we then get the following mutual information[8]

$$I(A;S) = h(S) - h(S|A) = h(A) - (-\infty) = \infty\,. \tag{4.24}$$

It holds $H(S) = H(A)$, because the channel just copies the input distribution to the output. Since this is the largest possible value, this is also the channel capacity.

[7] One exception is that the continuous version of mutual information can become infinite in the continuum. This, however, is perfectly consistent with the ability to store infinite amount of information in continuous variables and does not change anything substantial in the interpretation.

[8] Strictly spoken, we should denote this quantity as *differential* mutual information, but unlike the differential entropy, this term retains the same interpretation in the continuous as in the discrete case, and therefore we will not especially qualify it by terming it "differential".

4.6.3 Continuous Empowerment Approximation

While channel capacity is well defined for any relationship between S and A, it can only be computed for a subset of all possible scenarios. We will here approximate the model of the world with one for which empowerment can be computed. The following section discusses different approaches for doing so.

4.6.4 Binning

The most straightforward and naive approximation for continuous empowerment is to discretize all involved continuous variables and then compute the channel as described in the discrete empowerment section.

However, there are different ways to bin real-valued numbers and, as Olsson et al. (2005) demonstrated, they clearly affect the resulting informational values. *Uniform binning* considers the support of a real-valued random variable (i.e. the set of values of x for which $p(x) > 0$), splits it into equally sized intervals and assigns to each real number the bin it falls into. Of course, this does not necessarily result in the same number of events in each bin and, furthermore, many bins can be left empty or with very few events while others contain many events. This unevenness can mean that significant "information" (in the colloquial sense) in the data is being discarded. The response is to choose the binning in a not necessarily equally spaced way, that ensures that all bins are used, and that the events are well distributed. This is achieved by *Max-Entropy binning* where one adaptively resizes the bins so the resulting distribution has the highest entropy, which usually results in bins containing the approximately same number of events Olsson et al. (2005).

There are two caveats for this case: If adaptive binning is chosen, one needs to take care that the informational values of different measurements are comparable, and that the binning is the same throughout the same context of use. Therefore, it is important to choose the binning in advance, say, adapted only to the overall, context-free channel, and not adapt to each state-dependent channel separately. The second caveat is that, while adaptive binning distributes the events more-or-less evenly over the bins, this can thin out the sampling very considerably and cause the bins to be almost empty or containing very few elements each. This can induce the appearance of nonzero mutual information which, however, is spurious. In this case, it is better to choose a binning that is wide enough to ensure a sufficient number of events per bin. Both approaches require the availability of actual samples, so if the channel in question is only specified as a continuous conditional probability, it is necessary to generate random samples based on $p(s|a)$.

A final note on information estimation: much more robust approaches for mutual information estimation are known, such as the Kraskov-Stögbauer-Grassberger (KSG) estimator (Kraskov et al. 2004). Unfortunately, this method is not suitable for use with empowerment, as it requires the full joint distribution of the variables to be given in advance. When computing empowerment, however, one iteratively selects an input distribution, computes a joint distribution and then applies the information estimator. This means that if one uses the KSG-estimator, it affects the joint distributions and hence its own estimates of mutual information at later iterations of the process, and thus the conditions for correct operation of KSG cease to hold[9].

4.6.5 Evaluation of Binning

One problem with this approach is that it can introduce binning artefacts. Consider the following example: imagine one bins by proper rounding to integers. In this case, outcomes such as, say, 0.6 and 1.4 become the same state, while 1.4 and 1.6 are considered different. If now an agent which moves along the real valued line by an amount of 0.2 at each time step, this binning would make the agent appear to be more empowered at 1.5 then it would be at 1.0, because it could move to two different resulting states from 1.5. If the binning would reflect true sensoric resolution of the agent, this would conform with the empowerment model of being able to resolve the corresponding states; however, in our example, we did not imply anything like that — the underlying continuous structure is completely uniform, and we did not introduce any special sensoric structure. Thus, the difference in empowerment is a pure artefact introduced by the binning itself.

Another problem that emerges with the use of a binning approach is the right choice of granularity. If too few bins are chosen, then, while one has a good number of samples in the bins, interesting structural effect and correlations are lost. If too many bins are chosen, then many (or all bins) contain very few samples, perhaps as few as only one or even none. Such a sparse sampling can significantly overestimate the mutual information of the involved variables. Another problem, specifically in conjunction with empowerment, is that such a sparse sampling is often likely to cause one action to produce exactly one distinguishable sensoric outcome. This means that empowerment reaches its maximum $\log|\mathcal{A}|$ for every context r depriving it of any meaning. However, if the resolution is high enough and sufficiently many samples are collected, binning can produce a quickly implemented (but typically slow to compute) approximation for empowerment. Examples of its application to the simple pendulum can be seen in (Klyubin et al. 2008).

4.6.6 Jung's Monte Carlo Integration

Another approximation to compute empowerment which can still deal with any kind of $p(s|a)$ is Monte Carlo Integration (Jung et al. 2011). It is computed by sampling the outcomes of applying a representative set of available action sequences.

Assume that you have a model, so for a state r you can take actions a_v, with $v = 1,...,n$, and draw N_{MC} samples, which will result in sensor states $s_{v,j}$, with $j = 1,...,N_{MC}$. This method then approximates the term $d_{v,k}$ from Eq. (4.16) in the BA by

$$d_{v,k} \approx \frac{1}{N_{MC}} \sum_{j=1}^{N_{MC}} \log \left[\frac{p(s_{v,j}|r,\mathbf{a}_v)}{\sum_{i=1}^{n} p(s_{v,j}|r,\mathbf{a}_i) p_k^i} \right]. \tag{4.25}$$

[9] The authors thank Tobias Jung for this information (private communication).

To compute this the model needs to provide a way to compute how probable it is that the outcome of one action was produced by another. The necessary noise in the model basically introduces a "distance measure" that indicates how hard it is to distinguish two different actions.

One simple model is to assume that $p(s|r,\mathbf{a}_v)$ is a multivariate Gaussian (dependent on the current state of the world r), or can be reasonably well-approximated by it, i.e.,

$$s|r,\mathbf{a_v} \sim \mathcal{N}(\mu_v,\Sigma_v) \tag{4.26}$$

where $\mu_v = (\mu_{v,1},...,\mu_{v,n})^T$ is the mean of the Gaussian and the covariance matrix is given by $\Sigma_v = diag(\sigma_{v,1}^2,...,\sigma_{v,n}^2)$. The mean and covariance will depend upon the action $\mathbf{a}_v$ and the state r. Samples from the distribution will be denoted $\tilde{s}_v$ and can be generated using standard algorithms.

The following algorithm summarises how to approximate the empowerment $\mathfrak{E}(r)$ given a state $r \in \mathcal{R}$ and transition model $p(s|r,\mathbf{a}_v)$:

1. **Input:**
   a. Specify state r whose empowerment is to be calculated.
   b. For every action a_v with $v = 1,...,n$, define a (Gaussian) state transition model $p(s|r,\mathbf{a}_v)$, which is fully specified by its mean μ_v and covariance Σ_v.
2. **Initialise:**
   a. $p_0(\mathbf{a}_v) := 1/n$ for $v = 1,...,n$.
   b. Draw N_{MC} samples $\tilde{s}_{v,i}$ each, according to distribution density $p(s|r,\mathbf{a}_v) = \mathcal{N}(\mu_v,\Sigma_v)$ for $v = 1,...,n$.
   c. Evaluate $p(\tilde{s}_{v,i}|r,\mathbf{a}_\mu)$ for all $v = 1,...n$; $\mu = 1,...n$; and sample $i = 1,...,N_{MC}$.
3. **Iterate** the following variables for $k = 1,2,...$ until $|\mathfrak{E}_k - \mathfrak{E}_{k-1}| < \varepsilon$ or the maximum number of iterations is reached:
   a. $z_k := 0$, $\mathfrak{E}_{k-1} := 0$
   b. For $v = 1,...,n$
      i. $d_{v,k} := \frac{1}{N_{MC}} \sum_{j=1}^{N_{MC}} \log\left[\frac{p(\tilde{s}_{v,j}|r,\mathbf{a}_v)}{\sum_{i=1}^{n} p(\tilde{s}_{v,j}|r,\mathbf{a}_i)p_k^i}\right]$
      ii. $\mathfrak{E}_k := \mathfrak{E}_{k-1} + p_{k-1}(\mathbf{a}_v) \cdot d_{v,k-1}$
      iii. $p_k := p_{k-1}(\mathbf{a}_v) \cdot \exp(d_{v,k-1})$
      iv. $z_k := z_k + p_k(\mathbf{a}_v)$
   c. For $v = 1,...n$
      i. $p_k(\mathbf{a}_v) := p_k(\mathbf{a}_v) \cdot z_k^{-1}$
4. **Output:**
   a. Empowerment $\mathfrak{E}(r) \approx \mathfrak{E}_{k-1}$ (estimated).
   b. Distribution $p(\mathbf{a})$ achieving the maximum mutual information.

4.6.7 Evaluation of Monte Carlo Integration

Monte Carlo Integration can still deal with the same generic distributions $p(s|a)$ as the binning approach, and it removes the artefacts caused by the arbitrary boundaries of the bins. On the downside, it requires a model with a noise assumption. In the solution suggested by Jung et al. (2011) this lead to the assumption of Gaussian Noise.

The other problem is computability. For good approximations the number of selected representative action sequences should be high, but this also leads to a quick growth of computation time. The several applications showcased in (Jung et al. 2011) all had to be computed off-line, which makes them infeasible for robotic applications.

4.6.8 Quasi-Linear Gaussian Approximation

In the previous section we saw that Jung's Monte Carlo Integration method could deal with the rather general case where the relationship between actuators and sensor can be characterized by $s = f(r,a) + Z$, where f is a deterministic mapping, and Z is some form of added noise. The noise is necessary to limit the channel capacity, and an integral part of the Monte Carlo Integration in Eq. 4.25. While the noise can have different distributions, Jung's example assumed it to be Gaussian.

We will now outline how the assumption of Gaussian noise, together with an assumption regarding the nature of f, will allow us to accelerate the empowerment approximation. Consider now actuation-sensing mappings of the form $s = f(r,a) + \mathcal{N}(0,N_r)$, i.e. which can be described by a deterministic mapping f on which Gaussian noise (which may depend on r) is superimposed[10].

In principle, if the actions A were distributed in an arbitrarily small neighbourhood around 0, one would need f to be differentiable in a with the derivative $D_a f$ depending continuously on r. In practice, that neighbourhood will not be arbitrarily small, so the mapping from a to s needs to be "sufficiently well" approximated at all states r by an affine (or shifted linear) function in $f_r(a)$ for the allowed distributions of actions $p(a)$. To limit the channel capacity there has to be some constraint on the possible action distributions, and the linear approximation has to be sufficiently good for the actions that A can actually attain.[11]

In other words, assuming the channel can be adequately approximated by a linear transformation applied to A with added Gaussian noise, then it is possible to speed up the empowerment calculation significantly by reducing the general problem of continuous channel capacity to parallel Gaussian channels which can be solved with

[10] We will treat this as centred noise, with a mean of 0, but this is not necessary, as any non-zero mean would just shift the resulting distribution, which would leave the differential entropies and mutual information unaffected.

[11] We will not make this notion more precise or derive any error bounds at this point; we just informally assume that the Gaussian action distribution A is concentrated well enough for f_r to appear linear in a.

well-established algorithms. This provides us with the *quasi-linear Gaussian approximation* for empowerment which will now be presented in detail.

Let S be a multi-dimensional, continuous random variable defined over the vector space $\mathbb{R}^n$. Let A be a multidimensional random variable defined over $\mathbb{R}^m$. As in the discrete, A is the action variable, and S the perception variable. According to the quasi-linear Gaussian approximation assumption, we assume that there is a linear transformation $T : \mathbb{R}^m \to \mathbb{R}^n$ that allows us to express the relation between these variables via

$$S = TA + Z. \tag{4.27}$$

Z is a suitable multi-dimensional, Gaussian variable defined over $\mathbb{R}^n$, modelling the combined acting/sensing noise in the system and is assumed to be independent of A and S.

Consider first the simpler white noise case. Here we assume that the noise in each dimension $q \leq n$ of Z is independent of the noise in all other dimensions, and has a normal distribution with $Z_q \sim \mathcal{N}(0, N_q)$ for each dimension (where N_q depends on the dimension). This particular form of noise can be interpreted as having n sensoric channels where each channel q is subject to a source of independent Gaussian noise.

We now further introduce a limit to the *power* P available to the actions A, i.e. we are going to consider only action distributions A with $E(A^2) \leq P$. The reason for that is that without this constraint, the amplitude of A could be made arbitrarily large and this again would render all outcomes distinguishable and thus empowerment infinite[12]. The actual mean of the distributions is irrelevant for our purpose, as a constant shift does not affect the differential entropies. However, we need to ensure that the actuation range considered does not extend the size for which our linearity assumption holds.

It is plausible to consider this limitation as a physical power constraint[13]. Under these constraints, the quantity of interest now becomes

$$\mathfrak{E} = \max_{p(a):E(A^2) \leq P} I(S;A) \tag{4.28}$$

and the maximum being attained for normally distributed A (thus we only need to consider Gaussian distributions for A in the first place).

[12] This specific power limit also implies that the optimal input distributions for the channel capacity results is Gaussian (Cover and Thomas 1991).

[13] This point is subtle: throughout the text, we had made a point that empowerment is determined by the structure of the actuation-perception loop, but otherwise purely informational. In particular, we did not include any further assumptions about the physics of the system. In the quasi-linear Gaussian case, the choice of a "physics-like" quadratic form of power limitation is only owed to the fact that it makes the problem tractable. Other constraints are likely to be more appropriate for a realistic robotic actuator model, but need to be addressed in future work.

4.6.9 MIMO Channel Capacity

Now, assume for a moment that, in addition to our assumption of independent noise, the variance of the noise Z in each dimension has the same value, namely 1, then the problem becomes equivalent to computing the channel capacity for a linear, Multiple-Input/Multiple-Output channel with additive and isotropic Gaussian noise. Though the methods to compute this quantity are well established in the literature, for reasons of self-containedness, we reiterate them here.

The MIMO problem can be solved by standard methods (Telatar 1999), namely by applying a Singular Value Decomposition (SVD) to the transformation matrix T, which decomposes T as

$$T = U\Sigma V^T \tag{4.29}$$

where U and V are unitary matrices and Σ is a diagonal matrix with non-negative real values on the diagonal. This allow us to transform Eq. (4.27) to

$$U^T S = \Sigma V^T A + U^T Z. \tag{4.30}$$

It can be shown that each dimension of the resulting vectorial variables $U^T S$, $\Sigma V^T A$ and $U^T Z$ can be treated as an independent channel (see (Telatar 1999)), and thus reducing the computation of the overall channel capacity to computing the channel capacity for linear, parallel channels with added Gaussian noise, as in (Cover and Thomas 1991),

$$C = \max_{P_i} \sum_i \frac{1}{2} \log\left(1 + \frac{\sigma_i P_i}{E\left[(U^T Z)_i^2\right]}\right) = \max_{P_i} \sum_i \frac{1}{2} \log(1 + \sigma_i P_i) \tag{4.31}$$

where σ_i are the singular values of Σ, and P_i is the average power used in the i-th channel, following the constraint that

$$\sum_i P_i \leq P. \tag{4.32}$$

The simplification in the last step of Eq. (4.31) is based on the assumption of isotropic noise. Because the expected value for the noise is 1.0 and the unitary matrix applied to Z does not scale, but only rotates Z, the noise retains its original value of 1.0.

We remind that the channel capacity achieving distribution for a simple linear channel with added Gaussian Noise is Gaussian (Cover and Thomas 1991). In particular, the optimal input distribution for each subchannel is a Gaussian with a variance of P_i. The optimal power distribution which maximizes Eq. (4.31) can then be found with the *water-filling algorithm* (Cover and Thomas 1991). The basic idea is to first assign power to the channel with the lowest amount of noise. This has an effect that could be described as one of "diminishing returns": once a certain power level is reached, where adding more power to that channel has the same return as adding to the next best channel, additional power is now allocated to the *two* best channels. This is iterated to the next critical level and so on, until all power is allocated. Depending on the available total power, not all channels necessarily get power assigned to them.

We can also see, directly from the formula in Eq. (4.31), that since we divide by the variance of the noise Z, this value needs to be larger than zero. For vanishing noise, the channel capacity becomes infinite. Only the presence of noise induces an "overlap" of outcome states that allows one to obtain meaningful empowerment values. However, this is not a significant limitation in practice, as virtually all applications need to take into account actuator, system and/or sensor noise.

4.6.10 Coloured Noise

In a more general model, the Gaussian noise added to the multi-inputs, multi-output channel might also be coloured, meaning that the noise distributions in the different sensor dimensions are not independent. Let us assume that the noise is given by $Z \sim \mathcal{N}(0, K_s)$, where K_s is the covariance matrix of the noise. As above, we assume that the distribution has a mean of zero, which is without loss of generality since translations are information invariant. The relationship between S and A is again expressed as

$$S = T'A + Z'. \tag{4.33}$$

Conveniently, this can also be reduced to a channel with i.i.d. noise. For this, note that rotation, translation and scaling operators do not affect the mutual information $I(S;A)$. We start by expressing Z' as

$$Z' = U\sqrt{\Sigma}ZV^T, \tag{4.34}$$

where $Z \sim \mathcal{N}(0, I)$ is isotropic noise with a variance of 1, and $U\Sigma V^T = K_s$ is the SVD of K_s. U and T are orthogonal matrices, and Σ contains the singular values. Note that all singular values have to be strictly larger than zero, otherwise there would be a channel in the system without noise, which would allow the empowerment maximizer to inject all power into the zero-noise component of the channel and to achieve infinite channel capacity. $\sqrt{\Sigma}$ is a matrix that contains the square roots of the singular values, which should scale the variance of the isotropic noise to the singular values. The orthogonal matrices then rotate the distributions, so that they resemble Z'.

If we consider $\sqrt{\Sigma}^{-1}$, a diagonal matrix whose entries are the inverse of the square root of the singular values in Σ, this allows us to reformulate:

$$\begin{aligned}
S &= TA + U\sqrt{\Sigma}ZV^T && (4.35)\\
U^T SV &= U^T TAV + \sqrt{\Sigma}Z && (4.36)\\
\sqrt{\Sigma}^{-1}U^T SV &= \sqrt{\Sigma}^{-1}U^T TAV + Z && (4.37)\\
\sqrt{\Sigma}^{-1}U^T S &= \sqrt{\Sigma}^{-1}U^T TA + ZV^T && (4.38)\\
\sqrt{\Sigma}^{-1}U^T S &= \sqrt{\Sigma}^{-1}U^T TA + Z && (4.39)
\end{aligned}$$

The last step follows from the fact that the rotation of isotropic Gaussian noise remains isotropic Gaussian noise. This reduces the whole problem to a MIMO channel with isotropic noise and with the same channel capacity. We simply redefine the transformation matrix T as

$$T = \sqrt{\Sigma}^{-1} U^T T', \tag{4.40}$$

and solve the channel capacity for $S = TA + Z$, as outlined in section 4.6.9.

4.6.11 Evaluation of QLG Empowerment

The advantage of the quasi-linear Gaussian approximation is that it is quick to compute, the computational bottleneck being the calculation of a singular value decomposition that has the same dimensions as the sensors and actuators.

The drawbacks are its introduction of several assumptions. Like Jung's integration, the approximation forces us to assume Gaussian noise. However, a more aggressive assumption than Jung's approximation is that the QLG approximation also needs a locally linear model. So it is not possible to represent locally non-linear relationships between the actions and sensors. In particular the abrupt emergence of novel degrees of freedom which the empowerment formalism is so apt at discovering (see above, e.g. box pushing in Sec. 4.5.3) becomes softened by the Gaussian bell of the agent's actuations.

Finally, the quasi-linear Gaussian approximation also introduces a new free parameter, the power-constraint P which will be discussed in a later example. A more detailed examination of QLG empowerment can be found in (Salge et al. 2012).

4.7 Continuous Examples

We are aware of currently only two publications dealing with continuous empowerment. The first, by Jung et al. (2011), provides a good technical tutorial, and introduces the Monte Carlo Integration technique. Furthermore, it demonstrates that those states generally chosen as goals have high state-dependent empowerment, and that an empowerment-driven controller will tend to drive the system into them, even when initialized from a far away starting point. So, for example, the simple pendulum swings up, and stabilizes in the upright position, even when multiple swing-up phases are required; unlike traditional Reinforcement Learning, there is no value function that needs to be learnt over the whole phase space, but only the transition dynamics, and that needs only to be determined around the actual path taken. In principle, the algorithm does not need to visit any states but those in the neighbourhood of the path taken by the empowerment-driven controller. The empowerment-driven control method can be applied also to other, quite more intricate models, such as bicycle riding or the acrobot scenario (double-pendulum hanging from the top joint and driven by a motor at the middle joint).

The second paper (Salge et al. 2012) discusses the quasi-linear Gaussian method as a faster approximation for empowerment, and focusses on the pendulum; both to compare the QLG method with previous approximations, and to investigate how different parameters affect the empowerment map. In the following section we will use the simple pendulum from the second paper to outline some of the challenges in applying continuous empowerment.

4.7.1 Pendulum

The scenario we will focus on is that of a simple pendulum, because it incarnates many features typical for the continuous empowerment scenarios. First we will produce an empowerment map, which assigns an empowerment value for each state the pendulum can be in. Then we demonstrate empowerment-driven control; an algorithm that generates actions for the pendulum by greedily maximising its expected empowerment in the following step.

We start by observing that the pendulum's current state at the time t is completely characterized by its angle ϕ and its angular velocity $\dot{\phi}$.

For the model we time-discretize the input. So, the actuator variable A contains real values a_t, which represents the external acceleration applied to the pendulum. So, at time t, the motor acceleration is set to a_t, and this acceleration is then applied for the duration Δt. At the end of Δt, we will consider the system to be in time $t + \Delta t$, and the next value is applied.

4.7.2 Action Selection

In general, just having a state-dependent utility function, which assigns a utility to each state (such as empowerment) does not immediately provide a control strategy. One way to address this is to implement a greedy action selection strategy, where each action is chosen based on the immediate expected gain in empowerment. Note that empowerment is not a true value function, i.e. following its maximum local gradient does not necessarily correspond to optimizing some cumulated reward.

For the *discrete and deterministic* case, implementing a greedy control is simple. Since we have local model knowledge, we know what state each action a will lead to. We can then evaluate the empowerment for each action a that can be taken in the current state, and pick that action that leads to the subsequent state with the largest empowerment. This basically provides a gradient ascent approach (modulated by the effect of the action on the dynamics) on the empowerment landscape, with all is benefits and drawbacks.

If we are dealing with a *discrete but noisy* system, one needs to specify in more detail what a "greedy" action selection should look like, since empowerment is not a utility function in the strict sense of utility theory, and the average empowerment over the successor states is not the same as the empowerment of the averaged dynamics. This means that one has different ways of selecting the desired action for the next step.

However, the most straightforward way remains, of course, the selection of the highest average empowerment when a particular action is selected. Assume that, given an action a, and a fixed starting state which we do not denote separately, one has the probability $p(s|a)$ of getting into a subsequent state s[14]. Each of these successor states s has an associated empowerment value $\mathfrak{E}(s)$. Thus, the expected empowerment for carrying out the action a is given by

$$E[\mathfrak{E}(S)|a] = \sum_{s \in \mathcal{S}} \mathfrak{E}(s)p(s|a) \tag{4.41}$$

and one selects the action with the highest expected empowerment.

The necessity of distinction of deterministic and noisy cases becomes even more prominent in the continuous case, where the situation is more complicated. As we have to treat the continuum as a noisy system, there is usually no unique resulting state for an action a, but rather a continuous distribution density of states $p(s|a)$. Ideally, one would integrate the empowerment values over this distribution, similar to Eq. (4.41), but since empowerment cannot be expressed as a simple, integrable function, this is not practicable. One solution is to simply look at the mean of a sampled distribution over the resulting states and average their empowerment. At this point, however, no bounds have been derived on how well this value represents the empowerment values in the distribution of output states.

The continuity of actions creates another problem. Even if we can compute the expected empowerment for a given action, then we still need to select for which actions we want to evaluate their subsequent empowerment. Again, one possible option is to sample several actions a, distributed in a regular fashion; for example, one could look at the resulting states for maximal positive acceleration, for no acceleration at all, and for maximal negative acceleration, and then select the best. This may miss the action a with the highest expected empowerment which might fall somewhere between these sample points. Potential for future work would lie in developing an efficient method to avoid expensive searches for the highest-valued successive expected empowerment.

4.7.3 Resulting Control

In Fig. 4.7 we can see a empowerment map for the pendulum, and the resulting trajectory generated by greedy empowerment control. The controller sampled over 5 possible actuation choices, and chose the one where the resulting state had the highest expected empowerment.

In this specific case, the pendulum swings up and comes to rest in the upright position. This solution, while typical, is not unique. Varying the parameters for time step length and power constraint can produce different behaviour, such as cyclic oscillation and resting in the lower position. We will discuss these cases further below.

One interesting observation to note here is that the empowerment of the pendulum is not strictly increasing over the run, even though the control chooses the action that leads to the most empowered successor state. If one considers the trajectory, it

[14] To simplify the argument, we consider here only fully observed states s.

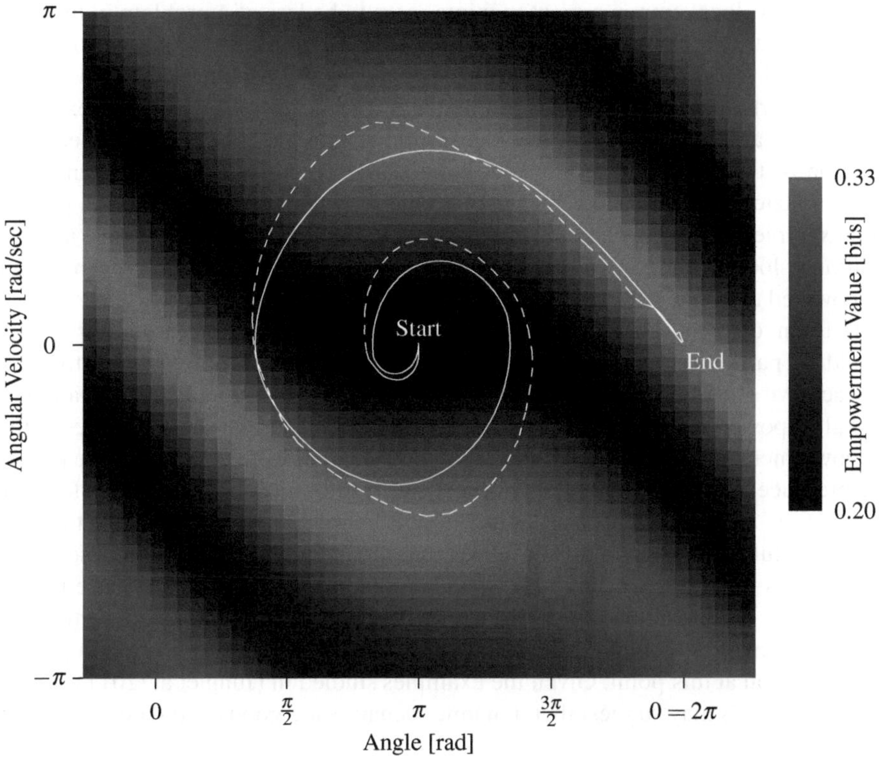

Fig. 4.7 Graph depicting the state space of a pendulum and its associated empowerment values. The solid line shows the trajectory of a pendulum in this state space, controlled by a greedy empowerment maximization algorithm based on the underlying Gaussian quasilinear empowerment landscape (shown in the background). For comparison of control, the dashed line shows the trajectory created by a greedy maximisation based on a Monte Carlo Gaussian empowerment landscape (not depicted here).

is possible to see that the pendulum passes through regions where the empowerment lowers again. This can be seen in Fig. 4.7 where the trajectory passes through the darker regions of lower empowerment after already being in much lighter regions of the empowerment map. This is due to the specific dynamics of the system, in which one can only control the acceleration of the pendulum, but, of course, not its position change, which is mediated by the current velocity. So while the controller chooses the highest empowered future state, all future states have lower empowerment than the current state.

Contrast this with the discrete maze case: in the latter, the agent could maintain any state of the environment, i.e. it position, indefinitely, by doing nothing. Greedy control in the maze therefore moves the agent to increasingly higher empowered states, until it would reach a local optimum, and then remain there.

Strikingly, local empowerment maxima seem to be less of a problem in the pendulum model (which is, in this respect, very similar to the mountain-car problem (Sutton and Barto 1998)). One reason turns out to be that the pendulum cannot maintain certain positions. If the pendulum has a non-zero speed, then its next position will be a different one, because the system cannot maintain both the speed and position of the pendulum at the same time. This sometimes forces the pendulum to enter states that are of lower empowerment than its current state. In the pendulum example this works out well in traversing the low empowered regions; and the continued local optimization of empowerment happens to lead to later, even higher empowered regions.

It is an open question to characterize actuation-perception structures which would be particularly amenable for the local empowerment optimization to actually achieve global empowerment optimization or at least a good approximation of global empowerment optimization. At this point, it is clear that sharp changes in the empowerment landscape (e.g. discovery of new degrees of freedom, e.g. because of the presence of a new manipulable object) need to be inside the local exploration range of the action sequences used to compute empowerment. However, in the case of the pendulum, the maximally empowered point of the upright pendulum seems to "radiate" its basin of attraction into sufficiently far regions of the state space for the local greedy optimization to pick this up. The characterization of the properties that the dynamics of the system needs to have for this to be successful is a completely open question at this point. Given the examples studied in (Jung et al. 2011), a cautious hypothesis may suggest that dynamic scenarios are good candidates for such a phenomenon.

4.7.4 Power Constraint

A closer look at the different underlying empowerment landscapes of the quasi-linear approximation in Fig. 4.8 shows their changes in regard to power constraint P and time step length Δt.

How the change in the time step duration Δt affects the empowerment, and also how it leads to worse approximations is studied in greater detail in (Salge et al. 2012). In general, it is not surprising that empowerment is indeed affected by it, in particular as the time step duration is closely related to the horizon length. The basic insight is, however, that a greater time step length allows a further look-ahead into the future, at the cost of a worsening approximation with the local linear model.

A more interesting effect in regard to the general applicability of the fast QLG method is the varying power constraint P. In general, an increase in power will result in an increase in empowerment, no matter where in the state space the system is. This is not immediately visible in the figures shown, since the colouring of the graphs is normalized, so the black and white correspond to the lowest and highest empowerment value in the respective subgraph.

A more unexpected effect, however, is a potential *inversion* of the empowerment landscape as seen in Fig. 4.8. Inversion means that for two specific points in the state space it might be that for one power level the first has a higher empowerment than the other, but for a different power level this relationship is reversed, and now the second point has a higher empowerment. For example, in Fig. 4.8 we can consider the row of landscapes for a Δt of 0.7. With increasing power there appears a new ridge of local maximal empowerment around the lower rest position of the pendulum.

This slightly counterintuitive effect is a result of how the capacity is distributed on the separate parallel channels. Be reminded, each channel i contributes its own amount to the overall capacity

$$C = \max_{P_i} \sum_i \frac{1}{2} \log(1 + \sigma_i P_i) \tag{4.42}$$

subject to the total power constraint P. Depending on the different values for σ_i, power is first allocated to the channel with the highest amplification value σ_i, up to a point were the return in capacity for the invested power diminishes so much that adding power to a different channel yields more capacity. From that point on the overall system acts as if it was one channel of bigger capacity.

In other words, for low power the factor that determines the channel capacity is the value of the largest σ alone. Once the power increases, the values of both the σ become important. It is therefore possible that for low power, a point with one large σ has comparatively high empowerment, while for a higher power level, another point has a higher empowerment, because the combination of all the σ is better. This is what actually happens in the pendulum example and causes the pendulum to remain in the lower rest position in the examples with higher power.

This indicates that the the empowerment-induced dynamics is sensitive to the given power constraint. One interpretation is that agents with weak actuators need to fine-tune their dynamics to achieve high-empowered states. However, agents with strong actuators can afford to stay in the potential minimum of the system, as their engine is strong enough to reach all relevant points without complicated strategems ("if in doubt, use a bigger hammer"). The inversion phenomenon is a special case for a more generic principle that force may be used to change the landscape in which the agent finds itself.

Another observation emerging from the inversion phenomenon is the general question of whether the Gaussian choice for the input distribution is appropriate. We know that some form of constraint must be applied, otherwise one could just chose input distributions that are spaced so far apart that they would fully compensate for the noise, giving rise to an (unrealistic) infinite channel capacity. Not only is this unhelpful, but also, as realistic actuations will be usually limited. In the current model, inspired by well-established channel capacity applications in communication theory, the power constraints reflects how limited amount of energies are allotted to broadcast a signal. But if we instead look, for example, at the acceleration which a robot could apply to its arm, then for instance an interval constraint would be much more natural to apply. For instance, an action a the robot could chose would lie, for example, between -4.0 and +10.0 m/s^2; a servo-based system may, instead specify a particular location instead, but still constrained by a hard-bounded interval. As consequence, it might be better to have a model where, instead of a general power constraint P, a hard upper and lower limit for each dimension of the actuator input A is imposed. At present, we are not aware of a method to directly compute the channel capacity for a multiple input, multiple output channel with coloured Gaussian noise that uses such a constraint.

4.7.5 Model Acquisition

Before we end this overview we will at least briefly address the problem of model acquisition or model learning. As mentioned, empowerment needs the model $p(s|a,r)$ for its computation. Strictly spoken, the acquisition or adaptation of this model is not part of the empowerment formalism. It is external to it, the model being either given in advance, or being acquired by one of many candidate techniques. However, given that empowerment will be used in scenarios where the model is not known and has to be learnt at the same time as the empowerment gradient is to be followed, model acquisition needs to be treated alongside the empowerment formalism itself.

As mentioned, empowerment only needs a local model of the dynamics from the agent's actuators to the agent's sensor in the current state of the world, but this local model is essential to compute empowerment.

Much of the earlier empowerment work operates under the assumption that the agent in question has somehow obtained or is given a sufficiently accurate model $p(s|a,r)$. Without addressing the "how", this acknowledges the fact that an agent-centric, intrinsic motivation mechanism needs to have this forward model available within the agent.

The earliest work to touch on this (Klyubin et al. 2008) deals with context-dependent empowerment. To model the relationship between an AIBO's discrete actions, and some discrete camera inputs, regular motions of the head are performed to sample the environment. These were then used to construct joint probability distributions and select an appropriate separation of all states of R into different contexts. The choice of context itself was also a decision on how to internally represent the world in an internal model, especially if there is only limited "resources" available to model the world. By grouping together states that behave similarly, the agent gets a good approximation of the world dynamics, and its internal empowerment computation results in high-empowered states. If the agent groups states with different behaviour together, then the resulting contexts have higher levels of uncertainty, and result in comparatively lower empowerment values (from the agent's perspective).

In general, it is clear that the quality of the model will affect the internal evaluation of empowerment. If the dynamics of a state are modelled with a great degree of uncertainty, then this noise will also reflect negatively in the empowerment value for this state. The interesting question here is then how to distinguish between those states that are truly random, and those where the action model is just currently not well known. This also indicates another field of future research. The hypothesis is that, if we would model how exploration or learning would affect our internal

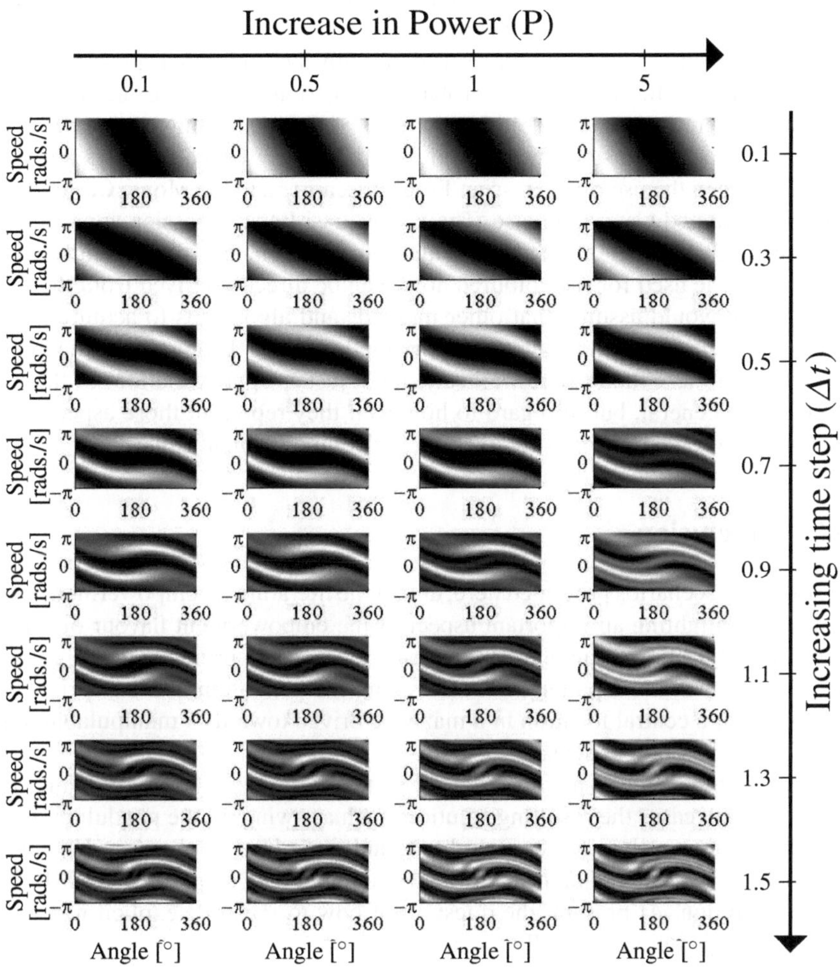

Fig. 4.8 A visualization of the different empowerment landscapes resulting from computation with different parameters for time step length Δt and power constraint P. The graphs plot empowerment for the two dimensional state space (angular speed, angular position) of the pendulum. White areas indicate the highest empowerment, black areas the lowest possible empowerment. The lower rest position is in the middle of the plots, and has low empowerment for less powered scenarios. The upper rest position is high empowered in all cases, it is located in the middle of the right or left edge of the plots. The areas of high empowerment close to the upright angel are those were the angular speed moves the pendulum towards the upper rest position. Figure is taken from (Salge et al. 2012).

model, then the maximisation of (internally computed) empowerment could also lead to exploration and learning behaviour.

In the continuous case, we have to deal with the additional question on how to best represent the conditional probability distributions, since, unlike the discrete case, there is no general and exact way of doing so. Jung et al. (2011) uses Gaussian Processes to store the dynamics of the world. This also offers a good interface between the use of a Gaussian Process Learner and the Monte Carlo integration with assumed Gaussian noise. The faster quasi-linear Gaussian approximation (Salge et al. 2012) also interact well with representation, and, conveniently, the covariance metric used for the coloured noise can be directly derived from the GP. In general, one would assume that other methods and algorithms to acquire a world model could be similarly combined with empowerment. It remains an open question which of these models are well suited, not just as approximations of the world dynamics in general, but in regard to how well they represent those aspects of the world dynamics that are relevant to attain high empowerment values.

4.8 Conclusion

The different scenarios presented here, and in the literature on empowerment in general, are highlighting an important aspect of the empowerment flavour of intrinsic motivation algorithms, namely its *universality*. The same principle that organizes a swarm of agents into a pattern can also swing the pendulum into an upright position, seek out a central location in a maze, be driven towards a manipulable object, or drive the evolution of sensors.

The task-independent nature reflected in this list can be both a blessing and a curse. In many cases the resulting solution, such as swinging the pendulum into the upright position, is the goal implied by default by a human observer. However, if indeed a goal is desired that differs from this default, then empowerment will not be the best solution. At present, the question of how to integrate explicit non-default goals into empowerment is fully open.

Another strong assumption that comes with the use of empowerment is its local character. On the upside, it simplifies the computation and makes the associated model acquisition much cheaper as only a very small part of the state space ever needs to be explored; the assumption of the usefulness of empowerment as a proxy principle for other implicit and less accessible optimization principles depends heavily on how well the local structure of the system dynamics will reflect its global structure. The precise nature of this phenomenon is not fully understood in the successful scenarios, but is believed to have to do with the regularity (e.g. continuity/smoothness) of the system dynamics. Of course, if any qualitative changes in the dynamics happen just outside of the empowerment horizon, the locality of empowerment will prevent them from being seen. This could be due to some disastrous "cliff", or something harmless like the discovery of an object that can be manipulated. Once, however, the change enters the empowerment horizon, and assuming that one can obtain a model of how it will affect the dynamics without losing the agent, empowerment will provide the gradients appropriate to the change.

Another central problem that, in the past, has reappeared across different applications is the computational feasibility. Empowerment quickly becomes infeasible to compute, which is a problem for both the behavioural empowerment hypothesis, and the application of empowerment to real-time AI or robotics problems. Newer methods address both the case for continuous empowerment (such as the QLG), and deeper empowerment horizons (such as the "impoverished" versions of empowerment). They, of course, come with additional assumptions and parameters, and provide only approximate solutions, but maintain the general character of the full solutions, allowing to export empowerment-like characteristics into domains that were hitherto inaccessible.

Let us conclude with a remark regarding the biological empowerment hypotheses in general: the fact that the default behaviours produced by empowerment seem often to match what intuitive expectations concerning default behaviour seem to imply, there is some relevance in investigating whether some of these behaviours are indeed approximating default behaviours observed in nature. A number of arguments in favour of why empowerment maximizing or similar behaviour could be relevant in biology have been made in (Klyubin et al. 2008), of which in this review we mainly highlighted its role as a measure of sensorimotor efficiency and the advantages that an evolutionary process would confer to more informationally efficient perception-action configurations.

Together with other intrinsic motivation measures, empowerment is thus a candidate measure which may help bridge the gap between understanding how organisms may be able to carry out default adaptations into their niche in an effective manner, and methods which would also allow artificial devices to try and copy the success that biological organisms have in doing so.

Acknowledgements. This work was supported by the European Commission via the CORBYS (Cognitive Control Framework for Robotic Systems) project under contract FP7 ICT-270219. The views expressed in this book are those of its editors and authors, and not necessarily those of the consortium.

References

Almeida e Costa, F., Rocha, L.: Introduction to the special issue: Embodied and situated cognition. Artificial Life 11(1-2), 5–12 (2005)

Anthony, T., Polani, D., Nehaniv, C.: On preferred states of agents: how global structure is reflected in local structure. In: Proc. Artificial Life XI, pp. 25–32. MIT Press (2008)

Anthony, T., Polani, D., Nehaniv, C.L.: Impoverished empowerment: 'Meaningful' action sequence generation through bandwidth limitation. In: Kampis, G., Karsai, I., Szathmáry, E. (eds.) ECAL 2009, Part II. LNCS, vol. 5778, pp. 294–301. Springer, Heidelberg (2011)

Arimoto, S.: An algorithm for computing the capacity of arbitrary discrete memoryless channels. IEEE Transactions on Information Theory 18(1), 14–20 (1972)

Atick, J.: Could information theory provide an ecological theory of sensory processing? Network: Computation in Neural Systems 3(2), 213–251 (1992)

Attneave, F.: Some informational aspects of visual perception. Psychological Review 61(3), 183 (1954)

Ay, N., Bertschinger, N., Der, R., Güttler, F., Olbrich, E.: Predictive information and explorative behavior of autonomous robots. The European Physical Journal B-Condensed Matter and Complex Systems 63(3), 329–339 (2008)

Ay, N., Polani, D.: Information flows in causal networks. Advances in Complex Systems 11(1), 17–41 (2008)

Barlow, H.: Sensory mechanisms, the reduction of redundancy, and intelligence. The Mechanisation of Thought Processes, 535–539 (1959)

Blahut, R.: Computation of channel capacity and rate-distortion functions. IEEE Transactions on Information Theory 18(4), 460–473 (1972)

Capdepuy, P.: Informational Principles of Perception-Action Loops and Collective Behaviours. PhD thesis, University of Hertfordshire (2010)

Capdepuy, P., Polani, D., Nehaniv, C.: Maximization of potential information flow as a universal utility for collective behaviour. In: IEEE Symposium on Artificial Life, ALIFE 2007, pp. 207–213. IEEE (2007)

Capdepuy, P., Polani, D., Nehaniv, C.L.: Perception–action loops of multiple agents: informational aspects and the impact of coordination. Theory in Biosciences 131(3), 149–159 (2012)

Cover, T.M., Thomas, J.A.: Elements of Information Theory, 99th edn. Wiley-Interscience (1991)

Csikszentmihalyi, M.: Beyond boredom and anxiety. Jossey-Bass (2000)

Der, R., Steinmetz, U., Pasemann, F.: Homeokinesis: A new principle to back up evolution with learning. Max-Planck-Inst. für Mathematik in den Naturwiss (1999)

Dewar, R.: Information theory explanation of the fluctuation theorem, maximum entropy production and self-organized criticality in non-equilibrium stationary states. Journal of Physics A: Mathematical and General 36(3), 631 (2003)

Dewar, R.C.: Maximum entropy production and the fluctuation theorem. Journal of Physics A: Mathematical and General 38(21), L371 (2005)

Gibson James, J.: The Ecological Approach to Visual Perception. Houghton Mifflin, Boston (1979)

Gordon, G., Ahissar, E.: Hierarchical curiosity loops and active sensing. Neural Networks 32, 119–129 (2012)

Grinstein, G., Linsker, R.: Comments on: a derivation and application of the'maximum entropy production'principle. Journal of physics. A, Mathematical and Theoretical 40(31), 9717–9720 (2007)

Jeffery, W.R.: Adaptive evolution of eye degeneration in the mexican blind cavefish. Journal of Heredity 96(3), 185–196 (2005)

Jung, T., Polani, D., Stone, P.: Empowerment for continuous agent environment systems. Adaptive Behavior 19(1), 16 (2011)

Kaplan, F., Oudeyer, P.-y.: Maximizing learning progress: An internal reward system for development. In: Iida, F., Pfeifer, R., Steels, L., Kuniyoshi, Y. (eds.) Embodied Artificial Intelligence. LNCS (LNAI), vol. 3139, pp. 259–270. Springer, Heidelberg (2004)

Klyubin, A.S., Polani, D., Nehaniv, C.L.: All else being equal be empowered. In: Capcarrère, M.S., Freitas, A.A., Bentley, P.J., Johnson, C.G., Timmis, J. (eds.) ECAL 2005. LNCS (LNAI), vol. 3630, pp. 744–753. Springer, Heidelberg (2005a)

Klyubin, A., Polani, D., Nehaniv, C.: Empowerment: A universal agent-centric measure of control. In: The 2005 IEEE Congress on Evolutionary Computation, vol. 1, pp. 128–135. IEEE (2005b)

Klyubin, A., Polani, D., Nehaniv, C.: Keep your options open: an information-based driving principle for sensorimotor systems. PloS ONE 3(12), e4018 (2008)

Kraskov, A., Stögbauer, H., Grassberger, P.: Estimating mutual information. Phys. Rev. E 69, 066138 (2004)

López, C., Martínez, E.: Sub-finslerian metric associated to an optimal control system. SIAM J. Control Optim. 39(3), 798–811 (2000)

Lungarella, M., Pegors, T., Bulwinkle, D., Sporns, O.: Methods for quantifying the informational structure of sensory and motor data. Neuroinformatics 3(3), 243–262 (2005)

Massey, J.: Causality, feedback and directed information. In: Proc. Int. Symp. Inf. Theory Applic., ISITA 1990, pp. 303–305 (1990)

Olsson, L., Nehaniv, C., Polani, D.: Sensor adaptation and development in robots by entropy maximization of sensory data. In: Proceedings of the 2005 IEEE International Symposium on Computational Intelligence in Robotics and Automation, CIRA 2005, pp. 587–592. IEEE (2005)

Pearl, J.: Causality: Models, Reasoning and Inference. Cambridge University Press (2000)

Pfeifer, R., Bongard, J., Grand, S.: How the body shapes the way we think: a new view of intelligence. The MIT Press (2007)

Polani, D.: Information: Currency of life? HFSP Journal 3(5), 307–316 (2009)

Polani, D., Nehaniv, C., Martinetz, T., Kim, J.T.: Relevant information in optimized persistence vs. progeny strategies. In: Artificial Life X: Proceedings of the Tenth International Conference on the Simulation and Synthesis of Living Systems, pp. 337–343. Citeseer (2006)

Salge, C., Glackin, C., Polani, D.: Approximation of empowerment in the continuous domain. Advances in Complex Systems 16(1&2), 1250079 (2012)

Salge, C., Polani, D.: Digested information as an information theoretic motivation for social interaction. Journal of Artificial Societies and Social Simulation 14(1), 5 (2011)

Schmidhuber, J.: Curious model-building control systems. In: IEEE International Joint Conference on Neural Networks, pp. 1458–1463. IEEE (1991)

Schmidhuber, J.: Exploring the predictable. Advances in Evolutionary Computing 6, 579–612 (2002)

Shannon, C.E.: A mathematical theory of communication. Bell Sys. Tech. Journal 27, 623–656 (1948)

Singh, S., Barto, A.G., Chentanez, N.: Intrinsically motivated reinforcement learning. In: Proceedings of the 18th Annual Conference on Neural Information Processing Systems (NIPS), Vancouver, B.C., Canada (2005)

Singh, S., Lewis, R., Barto, A., Sorg, J.: Intrinsically motivated reinforcement learning: An evolutionary perspective. IEEE Transactions on Autonomous Mental Development 2(2), 70–82 (2010)

Steels, L.: The autotelic principle. In: Iida, F., Pfeifer, R., Steels, L., Kuniyoshi, Y. (eds.) Embodied Artificial Intelligence. LNCS (LNAI), vol. 3139, pp. 231–242. Springer, Heidelberg (2004)

Sutton, R.S., Barto, A.G.: Reinforcement Learning. MIT Press, Cambridge (1998)

Telatar, E.: Capacity of multi-antenna gaussian channels. European Transactions on Telecommunications 10(6), 585–595 (1999)

Tishby, N., Pereira, F., Bialik, W.: The information bottleneck method. In: Proceedings of the 37th Annual Allerton Conference on Communication, Control and Computing, pp. 368–377 (1999)

Varela, F., Thompson, E., Rosch, E.: The Embodied Mind: Cognitive Science and Human Experience. The MIT Press (1992)

von Uexküll, J.: Umwelt und Innenwelt der Tiere. Springer (1909)

Wilkens, G.R.: Finsler geometry in low-dimensional control theory. In: Bao, D.D.-W., Shen Chern, S., Shen, Z. (eds.) Finsler Geometry: Joint Summer Research Conference on Finsler Geometry, Seattle, Washington, July 16-20. Contemporary Mathematics, vol. 196, pp. 245–260. AMS (1995)

Wissner-Gross, A., Freer, C.: Causal entropic forces. Physical Review Letters 110(16), 168702 (2013)

Yaeger, L.S.: How evolution guides complexity. HFSP Journal 3(5), 328–339 (2009)

Chapter 5
A Framework for the Local Information Dynamics of Distributed Computation in Complex Systems

Joseph T. Lizier, Mikhail Prokopenko, and Albert Y. Zomaya

5.1 Introduction

The nature of distributed computation has long been a topic of interest in complex systems science, physics, artificial life and bioinformatics. In particular, emergent complex behavior has often been described from the perspective of computation within the system (Mitchell 1998b,a) and has been postulated to be associated with the capability to support universal computation (Langton 1990; Wolfram 1984c; Casti 1991).

In all of these relevant fields, distributed computation is generally discussed in terms of "memory", "communication", and "processing". Memory refers to the storage of information by some variable to be used in the future of its time-series process. It has been investigated in coordinated motion in modular robots (Prokopenko et al. 2006), in the dynamics of inter-event distribution times (Goh and Barabási 2008), and in synchronization between coupled systems (Morgado et al. 2007). Communication refers to the transfer of information between one variable's time-series process and another; it has been shown to be of relevance in neuroscience (Wibral et al. 2011; Lindner et al. 2011; Marinazzo et al. 2012) and in other biological systems (e.g. dipole-dipole interaction in microtubules (Brown and Tuszynski 1999), and in signal transduction by calcium ions (Pahle et al. 2008)), social animals (e.g. schooling behavior in fish (Couzin et al. 2006)), agent-based systems (e.g. the

Joseph T. Lizier
CSIRO Computational Informatics,
PO Box 76, Epping, NSW 1710, Australia

Mikhail Prokopenko
CSIRO Computational Informatics PO Box 76, Epping, NSW 1710, Australia
& School of Physics The University of Sydney NSW 2006, Australia
& Department of Computing Macquarie University NSW 2109, Australia
e-mail: mikhail.prokopenko@csiro.au

Joseph T. Lizier · Albert Y. Zomaya
School of Information Technologies, The University of Sydney, NSW 2006, Australia

M. Prokopenko (ed.), *Guided Self-Organization: Inception*,
Emergence, Complexity and Computation 9,
DOI: 10.1007/978-3-642-53734-9_5, © Springer-Verlag Berlin Heidelberg 2014

influence of agents over their environments (Klyubin et al. 2005), and in inducing emergent neural structure (Lungarella and Sporns 2006)). Processing refers to the combination of stored and/or transmitted information into a new form; it has been discussed in particular for biological neural networks and models thereof (Kinouchi and Copelli 2006; Atick 1992; Sánchez-Montañés and Corbacho 2002; Yamada and Aihara 1994) (where it has been suggested as a potential biological driver), and also regarding collision-based computing (e.g. (Jakubowski et al. 1997; Adamatzky 2002), and including soliton dynamics and collisions (Edmundson and Enns 1993)).

Significantly, these terms correspond to the component operations of Turing universal computation: **information storage**, **information transfer** (or transmission) and **information modification**. Yet despite the obvious importance of these **information dynamics**, until recently there was no framework for either quantifying them individually or understanding how they interact to give rise to distributed computation.

Here, we review the first complete framework (Lizier et al. 2007, 2008b, 2012c, 2010, 2012b; Lizier and Prokopenko 2010; Lizier 2013) which quantifies each of these information dynamics or component operations of computation within a system, and describes how they inter-relate to produce distributed computation. We refer to the *dynamics* of information for two key reasons here. First, this approach describes the composition of information in the *dynamic state update* for the time-series process of each variable within the system, in terms of how information is stored, transferred and modified. This perspective of state updates brings an important connection between information theory and dynamical systems. Second, the approach focuses on the *dynamics* of these operations on information on a *local scale* in space and time within the system. This focus on the local scale is an important one. Several authors have suggested that a complex system is better characterized by studies of its local dynamics than by averaged or overall measures (Shalizi et al. 2006; Hanson and Crutchfield 1992), and indeed here we believe that quantifying and understanding distributed computation will necessitate studying the information dynamics and their interplay on a local scale in space and time. Additionally, we suggest that the quantification of the individual information dynamics of computation provides three *axes of complexity* within which to investigate and classify complex systems, allowing deeper insights into the variety of computation taking place in different systems.

An important focus for discussions on the nature of distributed computation have been cellular automata (CAs) as model systems offering a range of dynamical behavior, including supporting complex computations and the ability to model complex systems in nature (Mitchell 1998b). We review the application of this framework to CAs here because there is very clear *qualitative* observation of emergent structures representing information storage, transfer and modification therein (Langton 1990; Mitchell 1998b). CAs are a critical proving ground for any theory on the nature of distributed computation: significantly, Von Neumann was known to be a strong believer that "a general theory of computation in 'complex networks of automata' such as cellular automata would be essential both for understanding complex systems in nature and for designing artificial complex systems" (Mitchell (1998b) describing Von Neumann (1966)).

Information theory provides the logical platform for our investigation, and we begin with a summary of the main information-theoretic concepts required. We provide additional background on the qualitative nature of distributed computation in CAs, highlighting the opportunity which existed for our framework to provide quantitative insights. Subsequently, we consider each component operation of universal computation in turn, and describe how to quantify it locally in a spatiotemporal system. As an application, we review the measurement of each of these information dynamics at every point in space-time in several important CAs. We show that our framework provided the first complete quantitative evidence for a well-known set of conjectures on the emergent structures dominating distributed computation in CAs: that blinkers provide information storage, particles provide information transfer, and particle collisions facilitate information modification. Furthermore, we describe the manner in which our results implied that the *coherence* of information may be a defining feature of complex distributed computation. Our findings are significant because these emergent structures of computation in CAs have known analogues in many physical systems (e.g. solitons and biological pattern formation processes, coherent waves of motion in flocks), and as such this work will contribute to our fundamental understanding of the nature of distributed computation and the dynamics of complex systems. We finish by briefly reviewing the subsequent application of the framework to various complex systems, including in analyzing flocking behavior and in a computational neuroscience setting.

5.2 Information-Theoretic Preliminaries

Information theory (Shannon 1948; Cover and Thomas 1991; MacKay 2003) is an obvious tool for quantifying the information dynamics involved in distributed computation. In fact, information theory has already proven to be a useful framework for the design and analysis of complex self-organized systems (e.g. see (Prokopenko et al. 2009)).

We begin by reviewing several necessary information theoretic quantities, including several measures explicitly defined for use with time-series processes. We also describe *local* information-theoretic quantities - i.e. the manner in which information-theoretic measures can be used to describe the information content associated with single observations.

5.2.1 Fundamental Quantities

The fundamental quantity is the Shannon **entropy**, which represents the uncertainty associated with any measurement x of a random variable X (using units in bits):

$$H_X = -\sum_x p(x) \log_2 p(x). \tag{5.1}$$

The **joint entropy** of two (or more) random variables X and Y is a generalization to quantify the uncertainty of the joint distribution of X and Y:

$$H_{X,Y} = -\sum_{x,y} p(x,y) \log_2 p(x,y). \tag{5.2}$$

The **conditional entropy** of X given Y is the average uncertainty that remains about x when y is known:

$$H_{X|Y} = -\sum_{x,y} p(x,y) \log_2 p(x \mid y). \tag{5.3}$$

The **mutual information** (MI) between X and Y measures the average reduction in uncertainty about x that results from learning the value of y, or vice versa:

$$I_{X;Y} = \sum_{x,y} p(x,y) \log_2 \frac{p(x,y)}{p(x)p(y)}. \tag{5.4}$$

$$I_{X;Y} = H_X - H_{X|Y} = H_Y - H_{Y|X}. \tag{5.5}$$

One can also describe the MI as measuring the information contained in X about Y (or vice versa).

The **conditional mutual information** between X and Y given Z is the mutual information between X and Y when Z is known:

$$I_{X;Y|Z} = H_{X|Z} - H_{X|Y,Z} \tag{5.6}$$

$$= H_{Y|Z} - H_{Y|X,Z}. \tag{5.7}$$

Importantly, the conditional MI $I_{X;Y|Z}$ can be larger or smaller than the unconditioned $I_{X;Y}$ (MacKay 2003); it is reduced by redundant information held by Y and Z about X, and increased by synergy between Y and Z about X (e.g. where X is the result of an exclusive-OR or XOR operation between Y and Z).

5.2.2 Measures for Time-Series Processes

Next, we describe several measures which are explicitly defined for time-series processes X.

The **entropy rate** is the limiting value of the rate of change of the joint entropy over k consecutive values of a time-series process X, (i.e. measurements $\mathbf{x}_n^{(k)} = \{x_{n-k+1}, \ldots, x_{n-1}, x_n\}$, up to and including time step n, of the random variable $\mathbf{X}_n^{(k)} = \{X_{n-k+1}, \ldots, X_{n-1}, X_n\}$), as k increases (Cover and Thomas 1991; Crutchfield and Feldman 2003):

$$H_{\mu X} = \lim_{k \to \infty} \frac{H_{\mathbf{X}_n^{(k)}}}{k} = \lim_{k \to \infty} H'_{\mu X}(k), \tag{5.8}$$

$$H'_{\mu X}(k) = \frac{H_{\mathbf{X}_n^{(k)}}}{k}, \tag{5.9}$$

where the limit exists. Note that $\mathbf{X}_n^{(k)}$ is a k-dimensional *embedding vector* of the *state* of X (Takens 1981). A related definition is given by the limiting value of the conditional entropy of the next value of X (i.e. measurements x_{n+1} of the random variable X_{n+1}) given knowledge of the previous k values of X (i.e. measurements $\mathbf{x}_n^{(k)}$ of the random variable $\mathbf{X}_n^{(k)}$):

$$H_{\mu X} = \lim_{k \to \infty} H_{X_{n+1} | \mathbf{X}_n^{(k)}} = \lim_{k \to \infty} H_{\mu X}(k), \tag{5.10}$$

$$H_{\mu X}(k) = H_{\mathbf{X}_{n+1}^{(k+1)}} - H_{\mathbf{X}_n^{(k)}}, \tag{5.11}$$

again, where the limit exists. This can also be viewed as the uncertainty of the next state $\mathbf{x}_{n+1}^{(k)}$ given the previous state $\mathbf{x}_n^{(k)}$, since x_{n+1} is the only non-overlapping quantity in $\mathbf{x}_{n+1}^{(k)}$ which is capable of carrying any conditional entropy. Cover and Thomas (1991) point out that these two quantities correspond to two subtly different notions. These authors go on to demonstrate that for stationary processes X, the limits for the two quantities $H'_\mu(X)$ and $H_\mu(X)$ exist (i.e. the average entropy rate converges) and are equal. For our purposes in considering information dynamics, we are interested in the latter formulation $H_\mu(X)$, since it explicitly describes how one random variable X_{n+1} is related to the previous instances $\mathbf{X}_n^{(k)}$.

Grassberger (1986b) first noticed that a slow approach of the entropy rate to its limiting value was a sign of complexity. Formally, Crutchfield and Feldman (2003) use the conditional entropy form of the entropy rate (5.10)[1] to observe that at a finite block size k, the difference $H_{\mu X}(k) - H_{\mu X}$ represents the information carrying capacity in size k-blocks that is due to correlations. The sum over all k gives the total amount of structure in the system, quantified as the **effective measure complexity** or **excess entropy** (measured in bits):

$$E_X = \sum_{k=0}^{\infty} \left[H_{\mu X}(k) - H_{\mu X} \right]. \tag{5.12}$$

The excess entropy can also be formulated as the mutual information between the semi-infinite past and semi-infinite future of the system:

$$E_X = \lim_{k \to \infty} I_{\mathbf{X}_n^{(k)};\mathbf{X}_{n+1}^{(k^+)}}, \tag{5.13}$$

where $\mathbf{X}_{n+1}^{(k^+)} = \{X_{n+1}, X_{n+2}, \ldots, X_{n+k}\}$ is the random variable (with measurements $\mathbf{x}_{n+1}^{(k^+)} = \{x_{n+1}, x_{n+2}, \ldots, x_{n+k}\}$) referring to the k future values of the process X (from time step $n+1$ onwards). This interpretation is known as the *predictive information* (Bialek et al. 2001), as it highlights that the excess entropy captures the information in a process' past which is relevant to predicting its future.

[1] $H_{\mu X}(k)$ here is equivalent to $h_\mu(k+1)$ in (Crutchfield and Feldman 2003).

5.2.3 Local Information-Theoretic Measures

Finally, we note that the aforementioned information-theoretic quantities are *averages* over all of the observations used to compute the relevant probability distribution functions (PDFs). One can also write down *local* or pointwise measures for each of these quantities, representing their value for one specific observation or configuration of the variables (x,y,z) being observed. The average of a local quantity over all observations is of course the relevant average information-theoretic measure.

Primarily, the **Shannon information content** or **local entropy** of an outcome x of measurement of the variable X is (MacKay 2003):

$$h(x) = -\log_2 p(x). \tag{5.14}$$

Note that by convention we use lower-case symbols to denote local information-theoretic measures throughout this chapter. The quantity $h(x)$ is simply the information content attributed to the specific symbol x, or the information required to predict or uniquely specify that value. Less probable outcomes x have higher information content than more probable outcomes, and we have $h(x) \geq 0$. Specifically, the Shannon information content of a given symbol x is the *code-length* for that symbol in an optimal encoding scheme for the measurements X, i.e. one that produces the minimal expected code length.[2]

Now, note that although the PDF $p(x)$ is *evaluated* for $h(x)$ locally at the given observation x, it is *defined* using all of the available (non-local) observations of the variable X which would go into evaluation of the corresponding $H(X)$. That is to say, we define a certain PDF $p(x)$ from all given measurements of a variable X: we can measure local entropies $h(x)$ by evaluating $p(x)$ for a given observation x, or we can measure average entropies $H(X)$ from the whole function $p(x)$, and indeed we have $H(X) = \langle h(x) \rangle$ when the expectation value is taken over $p(x)$.

Similarly, we have the **local conditional entropy** $h(x \mid y) = -\log_2 p(x \mid y)$ with $H(X \mid Y) = \langle h(x \mid y) \rangle$.

Next, the **local mutual information** (Fano 1961) for a specific observation (x,y) is the information held in common between the specific values x and y:

$$i(x;y) = h(x) - h(x \mid y), \tag{5.15}$$

$$= \log_2 \frac{p(x \mid y)}{p(x)}. \tag{5.16}$$

The local mutual information is the difference in code lengths between coding the value x in isolation (under the optimal encoding scheme for X), or coding the value x given y (under the optimal encoding scheme for X given Y). Similarly, we have the **local conditional mutual information**:

[2] This "optimal code-length" may specify non-integer choices; full discussion of the implications here, practical issues in selecting integer code-lengths, and block-coding optimisations are contained in (Cover and Thomas 1991, Chapter 5).

$$i(x;y \mid z) = h(x \mid z) - h(x \mid y,z), \tag{5.17}$$

$$= \log_2 \frac{p(x \mid y,z)}{p(x \mid z)}. \tag{5.18}$$

Indeed, the form of $i(x;y)$ and $i(x;y \mid z)$ are derived directly from four postulates by Fano (1961, ch. 2): once-differentiability, similar form for conditional MI, additivity (i.e. $i(\{y_n,z_n\};x_n) = i(y_n;x_n) + i(z_n;x_n \mid y_n)$), and separation for independent ensembles. This derivation means that $i(x;y)$ and $i(x;y \mid z)$ are uniquely specified, up to the base of the logarithm. Of course, we have $I(X;Y) = \langle i(x;y) \rangle$ and $I(X;Y \mid Z) = \langle i(x;y \mid z) \rangle$, and like $I(X;Y)$ and $I(X;Y \mid Z)$, the local values are symmetric in x and y.

Importantly, $i(x;y)$ may be positive or negative, meaning that one variable can either positively inform us or actually *misinform* us about the other. An observer is misinformed where, conditioned on the value of y the observed outcome of x was *relatively* unlikely as compared to the unconditioned probability of that outcome (i.e. $p(x \mid y) < p(x)$). Similarly, $i(x;y \mid z)$ can become negative where $p(x \mid y,z) < p(x \mid z)$.

Applied to time-series data, local measures tell us about the *dynamics* of information in the system, since they vary with the specific observations in time, and local values are known to reveal more details about the system than the averages alone (Shalizi 2001; Shalizi et al. 2006).

5.3 Cellular Automata

5.3.1 Introduction to Cellular Automata

Cellular automata (CA) are discrete dynamical systems consisting of an array of cells which each synchronously update their discrete value as a function of the values of a fixed number of spatially neighboring cells using a uniform rule. Although the behavior of each individual cell is very simple, the (non-linear) interactions between all cells can lead to very intricate global behavior, meaning CAs have become a classic example of self-organized complex behavior. Of particular importance, CAs have been used to model real-world spatial dynamical processes, including fluid flow, earthquakes and biological pattern formation (Mitchell 1998b).

The neighborhood of a cell used as inputs to its update rule at each time step is usually some regular configuration. In 1D CAs, this means the same range r of cells on each side and including the current value of the updating cell. One of the simplest variety of CAs – 1D CAs using binary values, deterministic rules and one neighbor on either side ($r = 1$) – are known as the *Elementary CAs*, or *ECAs*. Example evolutions of ECAs from random initial conditions may be seen in Fig. 5.2a and Fig. 5.6a. For more complete definitions of CAs, including the definition of the Wolfram rule number convention for specifying update rules, see Wolfram (2002).

Wolfram (1984c, 2002) sought to classify the asymptotic behavior of CA rules into four classes: I. Homogeneous state; II. Simple stable or periodic structures;

III. Chaotic aperiodic behavior; and IV. Complicated localized structures, some propagating. Much conjecture remains as to whether these classes are quantitatively distinguishable, e.g. see Gray (2003), however they do provide an interesting analogy (for discrete-state and time) to our knowledge of dynamical systems, with classes I and II representing ordered behavior, class III representing chaotic behavior, and class IV representing complex behavior and considered as lying between the ordered and chaotic classes.

More importantly though, the approach seeks to characterize complex behavior in terms of emergent structure in CAs, regarding *gliders*, *particles* and *domains*. Qualitatively, a domain may described as a set of background configurations in a CA, for which any given configuration will update to another such configuration in the set in the absence of any disturbance. Domains are formally defined within the framework of computational mechanics (Hanson and Crutchfield 1992) as spatial process languages in the CA. Particles are qualitatively considered to be moving elements of coherent spatiotemporal structure. Gliders are particles which repeat periodically in time while moving spatially (repetitive non-moving structures are known as *blinkers*). Formally, particles are defined within the framework of computational mechanics as a boundary between two domains (Hanson and Crutchfield 1992); as such, they can also be termed as domain walls, though this is typically used with reference to aperiodic particles.

These emergent structures are more clearly visible when the CA is filtered in some way. Early filtering methods were hand-crafted for specific CAs (relying on the user knowing the pattern of background domains) (Grassberger 1983, 1989), while later methods can be automatically applied to any given CA. These include: ε-machines (Hanson and Crutchfield 1992), input entropy (Wuensche 1999), local information (Helvik et al. 2004), and local statistical complexity (Shalizi et al. 2006). All of these filtering techniques produce a *single* filtered view of the structures in the CA: our measures of local information dynamics will present several filtered views of the distributed computation in a CA, separating each operation on information. The ECA examples analyzed in this chapter are introduced in Section 5.3.3.

5.3.2 Computation in Cellular Automata

CAs can be interpreted as undertaking distributed computation: it is clear that "data represented by initial configurations is processed by time evolution" (Wolfram 1984c). As such, computation in CAs has been a popular topic for study (see Mitchell (1998b)), with a particular focus in observing or constructing (Turing) universal computation in certain CAs. An ability for universal computation is defined to be where "suitable initial configurations can specify arbitrary algorithm procedures" in the computing entity, which is capable of "evaluating any (computable) function" (Wolfram 1984c). Wolfram (1984c,a) conjectured that all class IV complex CAs were capable of universal computation. He went on to state that prediction in systems exhibiting universal computation is limited to explicit simulation of the system, as opposed to the availability of any simple formula or "short-cut", drawing parallels to the halting problem for universal Turing machines (Wolfram 1984c,a) which are echoed by Langton (1990) and Casti (1991). (Casti extended the analogy to undecidable statements in formal systems, i.e. Gödel's Theorem). The capability for universal computation has been proven for several CA rules, through the design of rules generating elements to (or by identifying elements which) specifically provide the component operations required for universal computation: information storage, transmission and modification. Examples here include most notably the Game of Life (Conway 1982) and ECA rule 110 (Cook 2004); also see Lindgren and Nordahl (1990) and discussions by Mitchell (1998b).

The focus on elements providing information storage, transmission and modification pervades discussion of all types of computation in CAs, e.g. (Adamatzky 2002; Jakubowski et al. 2001). Wolfram (1984a) claimed that in class III CAs information propagates over an infinite distance at a (regular) finite speed, while in class IV CAs information propagates at an irregular speed over an infinite range. Langton (1990) hypothesized that complex behavior in CAs exhibited the three component operations required for universal computation. He suggested that the more chaotic a system becomes the more information transmission increases, and the more ordered a system becomes the more information it stores. Complex behavior was said to occur at a phase transition between these extremes requiring an intermediate level of both information storage and transmission: if information propagates too well, *coherent information decays into noise*. Langton elaborates that transmission of information means that the "dynamics must provide for the propagation of information in the form of signals over arbitrarily long distances", and suggests that particles in CAs form the basis of these signals. To complete the qualitative identification of the elements of computation in CAs, he also suggested that blinkers formed the basis of information storage, and collisions between propagating (particles) and static structures (blinkers) "can modify either stored or transmitted information in the support of an overall computation". Rudimentary attempts were made at quantifying the average information transfer (and to some extent information storage), via mutual information (although as discussed later this is a symmetric measure not capturing directional transfer). Recognizing the importance of the emergent structures to computation, several examples exist of attempts to automatically identify CA rules which give rise to particles and gliders, e.g. (Wuensche 1999; Eppstein 2002), suggesting these to be the most interesting and complex CA rules.

Several authors however criticize the aforementioned approaches of attempting to classify CAs in terms of their generic behavior or "bulk statistical properties", suggesting that the wide range of differing dynamics taking place across the CA makes this problematic (Hanson and Crutchfield 1992; Mitchell 1998b). Gray (2003) suggests that there there may indeed be classes of CAs capable of more complex computation than universal computation alone. More importantly, Hanson and Crutchfield (1992) criticize the focus on universal computational ability as drawing away from the ability to identify "generic computational properties", i.e. a lack of ability for universal computation does not mean a CA is not undertaking any computation at all. Alternatively, these studies suggest that analyzing the rich space-time dynamics *within* the CA is a more appropriate focus. As such, these and other studies have analyzed the *local* dynamics of intrinsic or other specific computation, focusing on particles facilitating the transfer of information and collisions facilitating the information processing. Noteworthy examples here include: the method of applying filters from the domain of computational mechanics by Hanson and Crutchfield (1992); and analysis using such computational mechanics filters of CA rules selected via evolutionary computation to perform classification tasks by Mitchell et al. (1994, 1996). Related are studies which deeply investigate the nature of particles and their interactions, e.g. particle types and their interaction products identified for particular CAs (Mitchell et al. 1996; Boccara et al. 1991; Martinez et al. 2006), and rules established for their interaction products by Hordijk et al. (2001).

Despite such interest, until recently there was no complete framework that locally quantifies the individual information dynamics of distributed computation within CAs or other systems. In this review, we describe how the information dynamics can be locally quantified within the spatiotemporal structure of a CA. In particular, we describe the dynamics of how information storage and information transfer interact to give rise to information processing. Our approach is not to quantify computation or overall complexity, nor to identify universal computation or determine what is being computed; it is simply intended to quantify the component operations in space-time.

5.3.3 Examples of Distributed Computation in CAs

In this chapter, we review analysis of the computation carried out by several important ECA rules:

- Class IV complex rules 110 and 54 (Wolfram 2002) (see Fig. 5.4a and Fig. 5.2a), both of which exhibit a number of glider types and collisions. ECA rule 110 is the only proven computationally universal ECA rule (Cook 2004).
- Rules 22 and 30 as representative class III chaotic rules (Wolfram 2002) (see rule 22 in Fig. 5.6a);
- Rules 18 as a class III rule which contains domain walls against a chaotic background domain (Wolfram 1984b; Hanson and Crutchfield 1992).

These CAs each carry out an *intrinsic* computation of the evolution to their ultimate attractor and phase on it (see Wuensche (1999) for a discussion of attractors and state space in finite-sized CAs). That is to say, we view the attractor as the end point of an intrinsic computation by the CA – the dynamics of the transient to the attractor may contain information storage, transfer and modification, while the dynamics on the attractor itself can only contain information storage (since the attractor is either a fixed point or periodic process here). As such, we are generally only interested in studying computation during the transient dynamics here, as non-trivial computation processes.

We also examine a CA carrying out a "human-understandable" computational task. Rule ϕ_{par} is a 1D CA with range $r = 3$ (the 128-bit Wolfram rule number 0xfeedffdec1aaeec0eef000a0e1a020a0) that was evolved by Mitchell et al. (1994,

1996) to classify whether the initial CA configuration had a majority of 1's or 0's by reaching a fixed-point configuration of all 1's for the former or all 0's for the latter. This CA rule achieved a success rate above 70% in its task. An example evolution of this CA can be seen in Fig. 5.5a. The CA appears to carry out this computation using blinkers and domains for information storage, gliders for information transfer and glider collisions for information modification. The CA exhibits an initial emergence of domain regions of all 1's or all 0's storing information about local high densities of either value. Where these domains meet, a checkerboard domain propagates slowly (1 cell per time step) in both directions, transferring information of a *soft* uncertainty in this part of the CA. Some "certainty" is provided where the glider of the leading edge of a checkerboard encounters a blinker boundary between 0 and 1 domains, which stores information about a *hard* uncertainty in that region of the CA. This results in an information modification event where the domain on the opposite side of the blinker to the incoming checkerboard is concluded to represent the higher density state, and is allowed to propagate over the checkerboard. This new information transfer associated with local decision of which is the higher density state has evolved to occur at a faster speed (3 cells per time step) than the checkerboard uncertainty; it can overrun checkerboard regions, and in fact collisions of opposing types of this strong propagation give rise to the (hard uncertainty) blinker boundaries in the first place. The final configuration is therefore the result of this distributed computation.

Quantification of the local information dynamics via these *three axes of complexity* (information storage, transfer and modification) will provide quite detailed insights into the distributed computation carried out in a system. In all of these CAs we expect local measures of information storage to highlight blinkers and domain regions, local measures of information transfer to highlight particles (including gliders and domain walls), and local measures of information modification to highlight particle collisions.

This will provide a deeper understanding of computation than single or generic measures of bulk statistical behavior, from which conflict often arises in attempts to provide classification of complex behavior. In particular, we seek clarification on the long-standing debate regarding the nature of computation in ECA rule 22. Suggestions that rule 22 is complex include the difficulty in estimating the metric entropy (i.e. temporal entropy rate) for rule 22 by Grassberger (1986b), due to "complex long-range effects, similar to a critical phenomenon" (Grassberger 1986a). This effectively corresponds to an implication that rule 22 contains an infinite amount of memory (see Section 5.4.1). Also, from an initial condition of only a single "on" cell, rule 22 forms a pattern known as the "Sierpinski Gasket" (Wolfram 2002) which exhibits clear fractal structure. Furthermore, rule 22 is a 1D mapping of the 2D Game of Life CA (known to have the capability for universal computation (Conway 1982)) and in this sense is referred to as "life in one dimension" (McIntosh 1990), and complex structure in the language generated by iterations of rule 22 has been identified by Badii and Politi (1997). Also, we reported in (Lizier et al. 2012b) that we have investigated the C_1 complexity measure (Lafusa and Bossomaier 2005) (an enhanced version of the variance of the input entropy (Wuensche 1999)) for all

ECAs, and found rule 22 to clearly exhibit the largest value of this measure (0.78 bits to rule 110's 0.085 bits). On the other hand, suggestions that rule 22 is not complex include its high sensitivity to initial conditions leading to Wolfram (2002) classifying it as class III chaotic. Gutowitz and Domain (1997) claim this renders it as chaotic despite the subtle long-range effects it displays, further identifying its fast statistical convergence, and exponentially long and thin transients in state space (see Wuensche (1999)). Importantly, no coherent structure (particles, collisions, etc.) is found for typical profiles of rule 22 using a number of known filters for such structure (e.g. local statistical complexity (Shalizi et al. 2006)): this reflects the paradigm shift to an examination of local dynamics rather than generic, overall or averaged analysis. In our approach, we seek to combine this local viewpoint of the dynamics with a quantitative breakdown of the individual elements of computation, and we will review the application to rule 22 in this light.

5.4 Information Storage

In this section we review the methods to quantify information storage on a local scale in space and time, as presented in (Lizier et al. 2012c). We describe how total information storage used in the future is captured by excess entropy, and introduce active information storage to capture the amount of information storage that is currently in use. We review the application of local profiles of both measures to cellular automata.

5.4.1 Excess Entropy As Total Information Storage

Although discussion of information storage or memory in CAs has often focused on periodic structures (particularly in construction of universal Turing machines), information storage does not necessarily entail periodicity. The excess entropy Eq. (5.12, 5.13) more broadly encompasses all types of structure and memory by capturing correlations across all lengths of time, including non-linear correlations. It is quite clear from the predictive information formulation of the excess entropy Eq. (5.13) – as the information from a process' past that is contained its future – that it is a measure of the total information storage used in the future of a system.[3]

We use the term *univariate excess entropy*[4] to refer to measuring the excess entropy for individual variables X using their one-dimensional time-series process, i.e. $E_X = \lim_{k\to\infty} I_{\mathbf{X}_n^{(k)};\mathbf{X}_{n+1}^{(k+)}}$ from Eq. (5.13). This is a measure of the *average*

---

[3] In (Lizier et al. 2012c) we provide further comment on the relation to the statistical complexity (Crutchfield and Young 1989), which measures *all* information stored by the system which *may be used* in the future, while the *excess entropy* measures that information which *is used* by the system *at some point* in the future. The relation between the two concepts is covered in a more general mathematical context by Shalizi and Crutchfield (2001).

[4] Called "single-agent excess entropy" in (Lizier et al. 2012c).

memory for *each variable* X. Furthermore, we use the term *collective excess entropy* to refer to measuring the temporal excess entropy for a collective of variables $\mathbf{X} = \{X_1, X_2, \ldots, X_m\}$ (e.g. a set of neighboring cells in a CA) using their two-dimensional time-series process. Considered as the mutual information between their joint past and future, i.e. a joint temporal predictive information:

$$E_{\mathbf{X}} = \lim_{k \to \infty} I_{\{\mathbf{X}_{1,n}^{(k)}, \mathbf{X}_{2,n}^{(k)}, \ldots, \mathbf{X}_{m,n}^{(k)}\};\{\mathbf{X}_{1,n+1}^{(k+)}, \mathbf{X}_{2,n+1}^{(k+)}, \ldots, \mathbf{X}_{m,n+1}^{(k+)}\}}, \tag{5.19}$$

this is a measure of the *average* total memory stored in the collective (i.e. stored *collectively* by a set of cells in a CA). Collective excess entropy could be used for example to quantify the "undiscovered *collective memory* that may present in certain fish schools" (Couzin et al. 2006).

Grassberger (1986b,a) studied temporal entropy rate estimates for several ECAs in order to gain insights into their excess entropies. He revealed divergent collective excess entropy for a number of rules, including rule 22, implying a highly complex process. This case has been described by Lindgren and Nordahl (1988) as "a phenomenon which can occur in more complex environments", as with strong long-range correlations a semi-infinite sequence "could store an infinite amount of information about its continuation" (as per the predictive information form of the excess entropy Eq. (5.13)). On the other hand, infinite collective excess entropy can also be achieved for systems that only trivially utilise all of their available memory (e.g. simply copying cell values to the right when started from random initial states). Rule 22 was inferred to have $H_{\mu,N} = 0$ and infinite collective excess entropy, which was interpreted as a process requiring an infinite amount of memory to maintain an aperiodicity (Crutchfield and Feldman 2003).

In attempting to quantify *local* information dynamics of distributed computation here, our focus is on information storage for *single variables or cells* rather than the joint information storage across the collective. Were the univariate excess entropy found to be divergent (this has not been demonstrated), this may be more significant than for the collective case: divergent collective excess entropy implies that the collective is at least trivially utilizing all of its available memory (and even the chaotic rule 30 exhibits this), whereas divergent univariate excess entropy implies that all cells are individually highly utilizing the resources of the collective in a highly complex process. Again though, we emphasize that our focus is on local measures in *time* as well as space, which we present in the next section.

First we note that with respect to CAs, where each cell has only a finite number of values b and takes direct influence from only its single past value and the values of a finite number of neighbors, the meaning of (either average or local) information storage being greater $\log_2 b$ bits (let alone infinite) in the time series process of a single cell is not immediately obvious. Clearly, a cell in an ECA cannot store more than 1 bit of information in isolation. However, the *bidirectional* communication in CAs effectively allows a cell to store extra information in neighbors (even beyond the immediate neighbors), and to subsequently retrieve that information from those neighbors at a later point in time. While measurement of the excess entropy does not explicitly look for such *self-influence* communicated through neighbors,

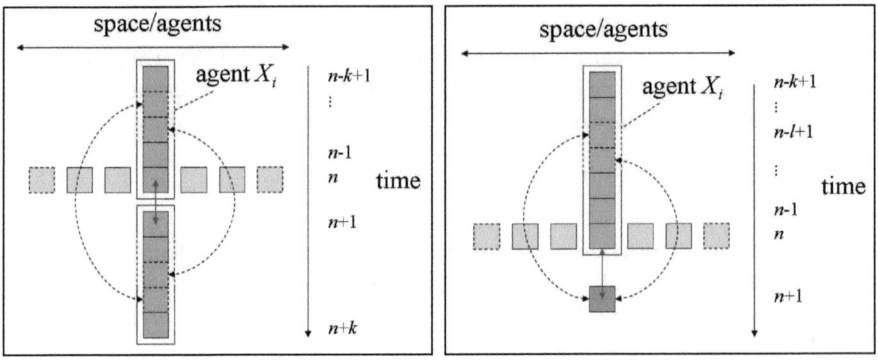

(a) Excess Entropy (b) Active Information Storage

Fig. 5.1 Measures of information storage in the time-series processes of single variables in distributed systems. (a) Excess Entropy: *total* information from the variable's past that is predictive of its future. (b) Active Information storage: the information storage that is *currently in use* in determining the next value of the variable. The stored information can be conveyed directly through the variable itself or via feedback from neighbors. (NB: This figure is reprinted from (Lizier et al. 2012c), Lizier, J. T., Prokopenko, M., and Zomaya, A. Y., Local measures of information storage in complex distributed computation, *Information Sciences*, 208:39–54, Copyright (2012), with permission from Elsevier.)

it is indeed the method by which a significant portion of information is channeled. Considering the predictive information interpretation in Eq. (5.13), it is easy to picture self-influence between semi-infinite past and future blocks being conveyed via neighbors (see Fig. 5.1a). This is akin to the use of stigmergy (indirect communication through the environment, e.g. see Klyubin et al. (2004)) to communicate with oneself.

A measurement of more than $\log_2 b$ bits stored by a cell on average, or indeed an infinite information storage, is then a perfectly valid result: in an infinite CA, each cell has access to an infinite amount of neighbors in which to store information which can later be used to influence its own future. Note however, that since the storage medium is shared by all cells, one should not think about the total memory as the total number of cells multiplied by this average. The total memory would be properly measured by the collective excess entropy, which takes into account the inherent redundancy here.

Following similar reasoning (i.e. that information may be stored and retrieved from one's neighbors), we note that a variable can store information regardless of whether it is causally connected with itself. Also, note that a variable can be perceived to store information simply as a result of how that variable is driven (Obst et al. 2013), i.e. where information is physically stored elsewhere in the system but recurs in the variable at different time steps (e.g. see the description of information storage in feed-forward loop motifs in (Lizier et al. 2012a)).

5.4.2 Local Excess Entropy

We now shift focus to local measures of information storage, which have the potential to provide more detailed insights into information storage structures and their involvement in computation than single ensemble measures.

The *local excess entropy* is a measure of how much information a given variable is storing *at a particular point in time* (Shalizi 2001).[5] The local excess entropy $e_X(n+1)$ of a process is simply the local mutual information Eq. (5.16) of the semi-infinite past and future of the process X at the given time step $n+1$:

$$e_X(n+1) = \lim_{k\to\infty} \log_2 \frac{p(\mathbf{x}_n^{(k)}, \mathbf{x}_{n+1}^{(k^+)})}{p(\mathbf{x}_n^{(k)})p(\mathbf{x}_{n+1}^{(k^+)})}. \tag{5.20}$$

Note that the excess entropy is the average of the local values, $E_X = \langle e_X(n) \rangle$. The limit $k \to \infty$ is an important part of this definition, since correlations at all time scales should be included in the computation of information storage. Since this is not computationally feasible in general, we retain the notation $e_X(n+1,k)$ to denote finite-k estimates of $e_X(n+1)$.

The notation is generalized for lattice systems (such as CAs) with *spatially-ordered* variables to represent the local excess entropy for cell X_i at time $n+1$ as:

$$e(i,n+1) = \lim_{k\to\infty} \log_2 \frac{p(\mathbf{x}_{i,n}^{(k)}, \mathbf{x}_{i,n+1}^{(k^+)})}{p(\mathbf{x}_{i,n}^{(k)})p(\mathbf{x}_{i,n+1}^{(k^+)})}. \tag{5.21}$$

Again, $e(i,n+1,k)$ is used to denote finite-k estimates of $e(i,n+1)$. Local excess entropy is defined for every spatiotemporal point (i,n) in the system. (Alternatively, the collective excess entropy can only be localized in time).

As a local mutual information, the local excess entropy may be positive or negative, meaning the past history of the cell can either positively inform us or actually misinform us about its future. An observer is misinformed where a given semi-infinite past and future are relatively unlikely to be observed together as compared to the product of their marginal probabilities. Another view is that we have misinformative values when $p(\mathbf{x}_{i,n+1}^{(k^+)} \mid \mathbf{x}_{i,n}^{(k)}) < p(\mathbf{x}_{i,n+1}^{(k^+)})$, meaning that taking the past $\mathbf{x}_{i,n}^{(k)}$ into account reduced the probability of the future which was observed $\mathbf{x}_{i,n+1}^{(k^+)}$.

5.4.3 Active Information Storage

The excess entropy measures the total stored information which will be used *at some point* in the future of the time-series process of a variable, possibly but not necessarily at the next time step $n+1$. In examining the local information dynamics of

[5] This is as per the original formulation of the local excess entropy by Shalizi (2001), however this presentation is for a single time-series rather than the light-cone formulation used there.

computation, we are interested in how much of the stored information is actually *in use* at the next time step. As we will see in Section 5.6, this is particularly important in understanding how stored information interacts with information transfer in information processing. As such, the **active information storage** A_X was introduced in (Lizier et al. 2012c) as the average mutual information between the (semi-infinite) past state of the process and its *next value*, as opposed to its whole (semi-infinite) future:

$$A_X = \lim_{k \to \infty} I(\mathbf{X}_n^{(k)}; X_{n+1}). \tag{5.22}$$

The **local active information storage** is then a measure of the amount of information storage in use by the process at a particular time-step $n+1$:

$$a_X(n+1) = \lim_{k \to \infty} \log_2 \frac{p(\mathbf{x}_n^{(k)}, x_{n+1})}{p(\mathbf{x}_n^{(k)})p(x_{n+1})}, \tag{5.23}$$

$$= \lim_{k \to \infty} \log_2 \frac{p(x_{n+1} \mid \mathbf{x}_n^{(k)})}{p(x_{n+1})}, \tag{5.24}$$

and we have $A_X = \langle a_X(n) \rangle$. We retain the notation $a_X(n+1,k)$ and $A_X(k)$ for finite-k estimates. Again, we generalize the measure for variable X_i in a lattice system as:

$$a(i,n+1) = \lim_{k \to \infty} \log_2 \frac{p(\mathbf{x}_{i,n}^{(k)}, x_{i,n+1})}{p(\mathbf{x}_{i,n}^{(k)})p(x_{i,n+1})}, \tag{5.25}$$

and use $a(i,n+1,k)$ to denote finite-k estimates there, noting that the local active information storage is defined for every spatiotemporal point (i,n) in the lattice system.

The average active information storage will always be positive (as for the excess entropy), but is bounded above by $\log_2 b$ bits if the variable takes one of b discrete values. The local active information storage is not bound in this manner however, with values larger than $\log_2 b$ indicating that the particular past of an variable provides strong positive information about its next value. Furthermore, the local active information storage can be negative, where the past history of the variable is actually misinformative about its next value. An observer is misinformed where the past history and observed next value are relatively unlikely to occur together as compared to their separate occurrence.

5.4.4 *Local Information Storage Results*

In this and subsequent results sections, we review the application of these local measures in (Lizier et al. 2007, 2008b, 2012c, 2010, 2012b; Lizier and Prokopenko 2010; Lizier 2013) to sample CA runs. As described earlier, we are interested in studying the non-trivial computation during the transient dynamics before an attractor is reached. Certainly it would be easier to study these information dynamics on attractors – since the dynamics there are cyclo-stationary (because the attractors in finite-length CAs involve only fixed or periodic dynamics) – however as described in Section 5.3.3 the computation there is trivial. To investigate the dynamics of the transient, we estimate the required probability distribution functions (PDFs) from CA runs of 10 000 cells, initialized from random values, in order to generate a large ensemble of transient automata dynamics. We retain only a relatively short 600 time steps for each cell, in order to avoid attractor dynamics and focus on quasi-stationary transient dynamics during that short time period. Alternatively, for ϕ_{par} we used 30 000 cells with 200 time steps retained. Periodic boundary conditions were used. Observations taken at every spatiotemporal point in the CA were used in estimating the required PDFs, since the cells in the CA are homogeneous variables and quais-stationarity is assumed over the relatively short time interval.

The results and the figures displayed here were produced using the open source *Java Information Dynamics Toolkit* (Lizier 2012), which can be used in Matlab/Octave and Python as well as Java. All results can be reproduced using the Matlab/Octave script `GsoChapterDemo2013.m` in the `demos/octave/CellularAutomata` example distributed with this toolkit. We make estimates of the measures with finite values of k, noting that the insights described here could not be attained unless a reasonably large value of k was used in order to capture a large proportion of the correlations. Determination of an appropriate value of k was discussed in (Lizier et al. 2012c), and in (Lizier et al. 2008b) for the related transfer entropy measure presented in Section 9.12. As a rule of thumb, k should at least be larger than the period of any regular background domain in order to capture the information storage underpinning its continuation.

We begin by examining the results for rules 54 and 110, which contain regular gliders against periodic background domains. For the CA runs described above, sample areas of the large CAs are shown in Fig. 5.2a and Fig. 5.4a, while the corresponding local profiles of $e(i,n,k=8)$ generated are displayed in Fig. 5.2b and Fig. 5.4b, and the local profiles of $a(i,n,k=16)$ in Fig. 5.2c and Fig. 5.4c. It is quite clear that positive information storage is concentrated in the vertical gliders or blinkers, and the domain regions. As expected, these results provide quantitative evidence that the **blinkers are the dominant information storage entities**. That the **domain regions contain significant information storage** should not be surprising, since as a periodic sequence its past does indeed store information about its future.

In fact, the local values for each measure form spatially and temporally periodic patterns in the domains, corresponding to the spatial and temporal periodicities exhibited in the underlying raw values. Certainly if the dynamics are only composed of a consistent domain pattern (which is deterministic when viewing single cells' time series), then for $a(i,n,k)$ for example we will always have $p(x_{n+1} \mid \mathbf{x}_n^{(k)}) = 1$ and if $p(x_{n+1})$ is balanced then $a(i,n,k)$ would be constant across the CA. However, the existence of discontinuities in the domain, e.g. gliders, reduces $p(x_{n+1} \mid \mathbf{x}_n^{(k)})$ here, and does so differently for each $\mathbf{x}_n^{(k)}$ configuration in the domain. Imbalances in $p(x_{n+1})$ can also contribute to differences in storage across the domain. These factors leads to the spatiotemporal periodicities of information storage that are observed in the domains.

While the local active information storage indicates a similar amount of stored information in use to compute each space-time point in both the domain and blinker areas, the local excess entropy reveals a larger *total* amount of information is stored in the blinkers. For the blinkers known as α and β in rule 54 (Hordijk et al. 2001) this is because the temporal sequences of the center columns of the blinkers (0-0-0-1, with $e(i,n,k=8)$ in the range 5.01 to 5.32 bits) are more complex than those in the domain (0-0-1-1 and 0-1, with $e(i,n,k=8)$ in the range 1.94 to 3.22 bits), even where they are of the same period. We have $e(i,n,k=8) > 1$ bit here due to the distributed information storage supported by bidirectional communication (as discussed earlier). Such bidirectional communication is also critical to these periodic domain sequences being longer than two time steps – the maximum period that a binary cell could sustain in isolation (e.g. the period-7 domain in rule 110).

Another area of strong information storage appears to be the "wake" of the more complex gliders in rule 110 (see the glider at top right of Fig. 5.4b and Fig. 5.4c). This result aligns well with our observation (Lizier et al. 2008b) that the dynamics following the leading edge of regular gliders consists largely of "non-traveling" information. The presence of the information storage is shown by both measures, although the relative strength of the total information storage is again revealed only by the local excess entropy.

Negative values of $a(i,n,k=16)$ for rules 54 and 110 are also visible in Fig. 5.2c and Fig. 5.4c. Interestingly, negative local components of local active information storage measure are concentrated in the traveling glider areas (e.g. γ^+ and γ^- for rule 54 (Hordijk et al. 2001)), providing a good spatiotemporal filter of these structures. This is because when a traveling glider is encountered at a given cell, the past history of that cell (being part of the background domain) is misinformative about the next value, since the domain sequence was more likely to continue than be interrupted. For example, see the marked positions of the γ gliders in Fig. 5.3. There we have $p(x_{n+1} \mid \mathbf{x}_n^{(k=16)}) = 0.25$ and $p(x_{n+1}) = 0.52$: since the next value occurs relatively infrequently after the given history, we have a misinformative $a(n,k=16) = -1.09$ bits. This is juxtaposed with the points four time steps before those marked "x", which have the same history $\mathbf{x}_n^{(k=16)}$ but are part of the domain, with $p(x_{n+1} \mid \mathbf{x}_n^{(k=16)}) = 0.75$ and $p(x_{n+1}) = 0.48$ giving $a(n,k=16) = 0.66$ bits, quantifying the positive information storage there. Note that the points with misinformative information storage are not necessarily those selected by other filtering techniques as part of the gliders: e.g. the finite state transducers technique (using left to right scanning by convention) by Hanson and Crutchfield (1997) would identify points 3 cells to the right of those marked "x" as part of the γ^+ glider.

The local excess entropy produced some negative values around traveling gliders, though these were far less localized on the gliders themselves and less consistent in occurrence than for the local active information storage. This is because the local excess entropy, as measure of total information storage into the future, is more loosely tied to the dynamics at the given spatiotemporal point. The effect of a glider encounter on $e(i,n,k)$ is smeared out in time, and in fact the dynamics may store more positive information in total than the misinformation encountered at the

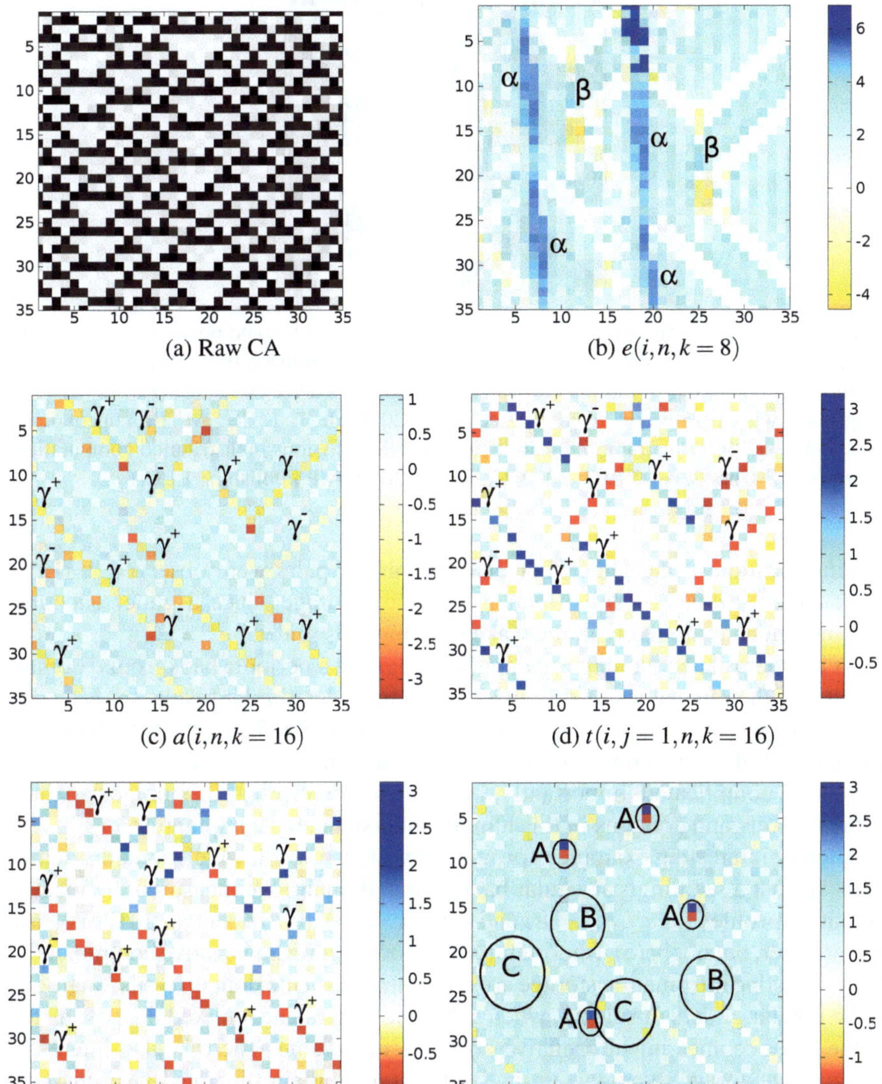

(a) Raw CA (b) $e(i,n,k=8)$

(c) $a(i,n,k=16)$ (d) $t(i,j=1,n,k=16)$

(e) $t(i,j=-1,n,k=16)$ (f) $s(i,n,k=16)$

Fig. 5.2 Local information dynamics in **rule 54** for the raw values in (a) (black for "1", white for "0"). 35 time steps are displayed for 35 cells, and time increases down the page for all CA plots. All units are in bits. (b) Local excess entropy $e(i,n,k=8)$; (c) Local active information storage $a(i,n,k=16)$; Local apparent transfer entropy: (d) one cell to the right $t(i,j=1,n,k=16)$, (e) one cell to the left $t(i,j=-1,n,k=16)$; (f) Local separable information $s(i,n,k=16)$.

Fig. 5.3 Close up of raw values of rule 54. "x" and "+" mark some positions in the γ^+ and γ^- gliders respectively. Note their point of coincidence in collision type "A", with "•" marking the subsequent non-trivial information modification as detected using $s(i,n,k=16)<0$. (Reprinted with permission from (Lizier et al. 2010) J. T. Lizier, M. Prokopenko, and A. Y. Zomaya, "Information modification and particle collisions in distributed computation," *Chaos*, vol. 20, no. 3, p. 037109, 2010. Copyright 2010, AIP Publishing LLC.)

specific location of the glider. For example, glider pairs were observed in (Lizier et al. 2012c) to have positive total information storage, since a glider encounter becomes much more likely in the wake of a previous glider.

As another rule containing regular gliders against a periodic background domain, analysis of the raw values of ϕ_{par} in Fig. 5.5a provides similar results for $e(i,n,k=5)$ (not shown, see (Lizier 2013)) and $a(i,n,k=10)$ in Fig. 5.5b here. One distinction is that the blinker here contains no more stored information than the domain, since it is no more complicated. Importantly, we confirm the information storage capability of the blinkers and domains in this human understandable computation.

Another interesting example is provided by ECA rule 18, which contains domain walls against a seemingly irregular background domain. We measured the local information profiles for $e(i,n,k=8)$ and $a(i,n,k=16)$ in (Lizier et al. 2012c) (shown in that paper, but not here). Importantly, the most significant negative components of the local active information storage are concentrated on the domain walls: analogous to the regular gliders of rule 54, when a domain wall is encountered the past history of the cell becomes misinformative about its next value. There is also interesting information storage dynamics in the background domain for rule 18, discussed in detail in (Lizier et al. 2012c).

Finally, we examine ECA rule 22, suggested to have infinite collective excess entropy (Grassberger 1986b,a) but without any known coherent structural elements (Shalizi et al. 2006). For the raw values of rule 22 displayed in Fig. 5.6a, the calculated local excess entropy profile is shown in Fig. 5.6b, and the local active information storage profile in Fig. 5.6c. While information storage certainly occurs for rule 22, these plots provide evidence that there is no coherent structure to this storage. This is another clear example of the utility of examining local information dynamics over ensemble estimates, given the earlier discussion on collective excess entropy for rule 22.

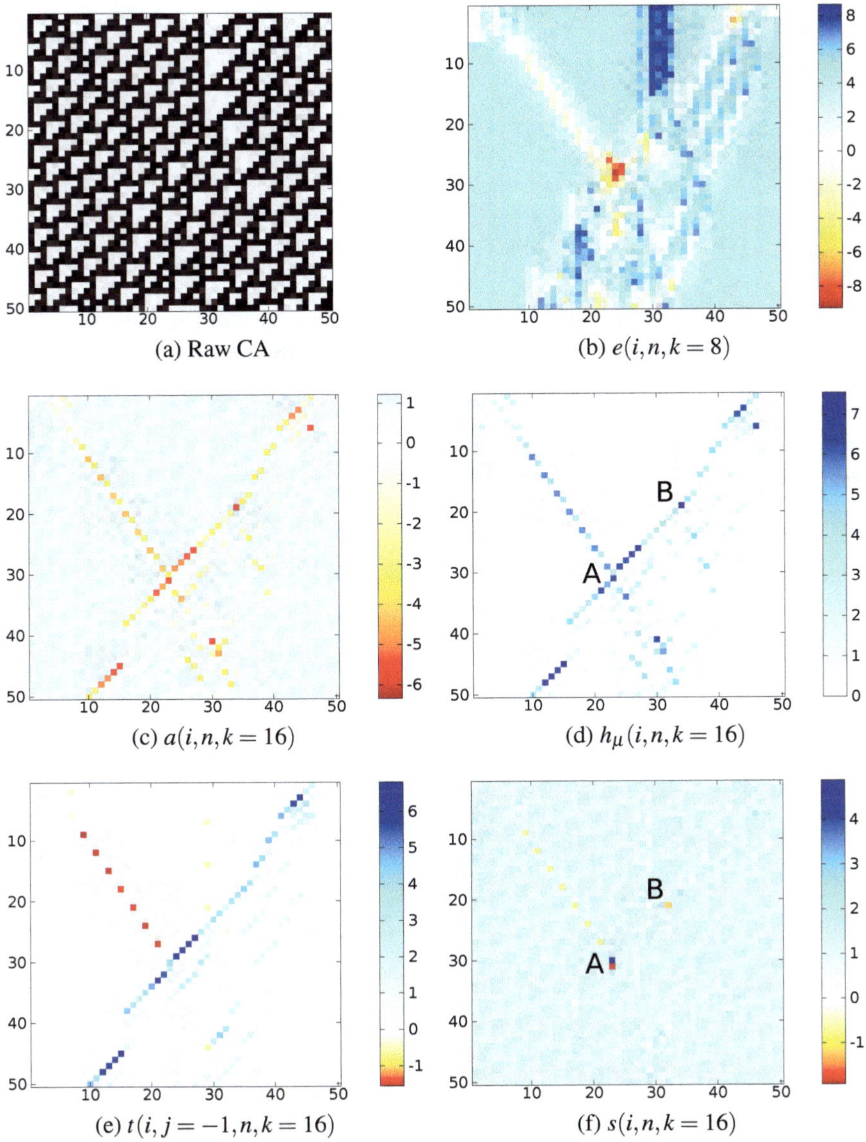

(a) Raw CA

(b) $e(i,n,k=8)$

(c) $a(i,n,k=16)$

(d) $h_\mu(i,n,k=16)$

(e) $t(i,j=-1,n,k=16)$

(f) $s(i,n,k=16)$

Fig. 5.4 Local information dynamics in **rule 110** for the raw values displayed in (a) (black for "1", white for "0"). 50 time steps are displayed for 50 cells, and all units are in bits. (b) Local excess entropy $e(i,n,k=8)$; (c) Local active information storage $a(i,n,k=16)$; (d) Local temporal entropy rate $h_\mu(i,n,k=16)$; (e) Local apparent transfer entropy one cell to the left $t(i,j=-1,n,k=16)$; (f) Local separable information $s(i,n,k=16)$.

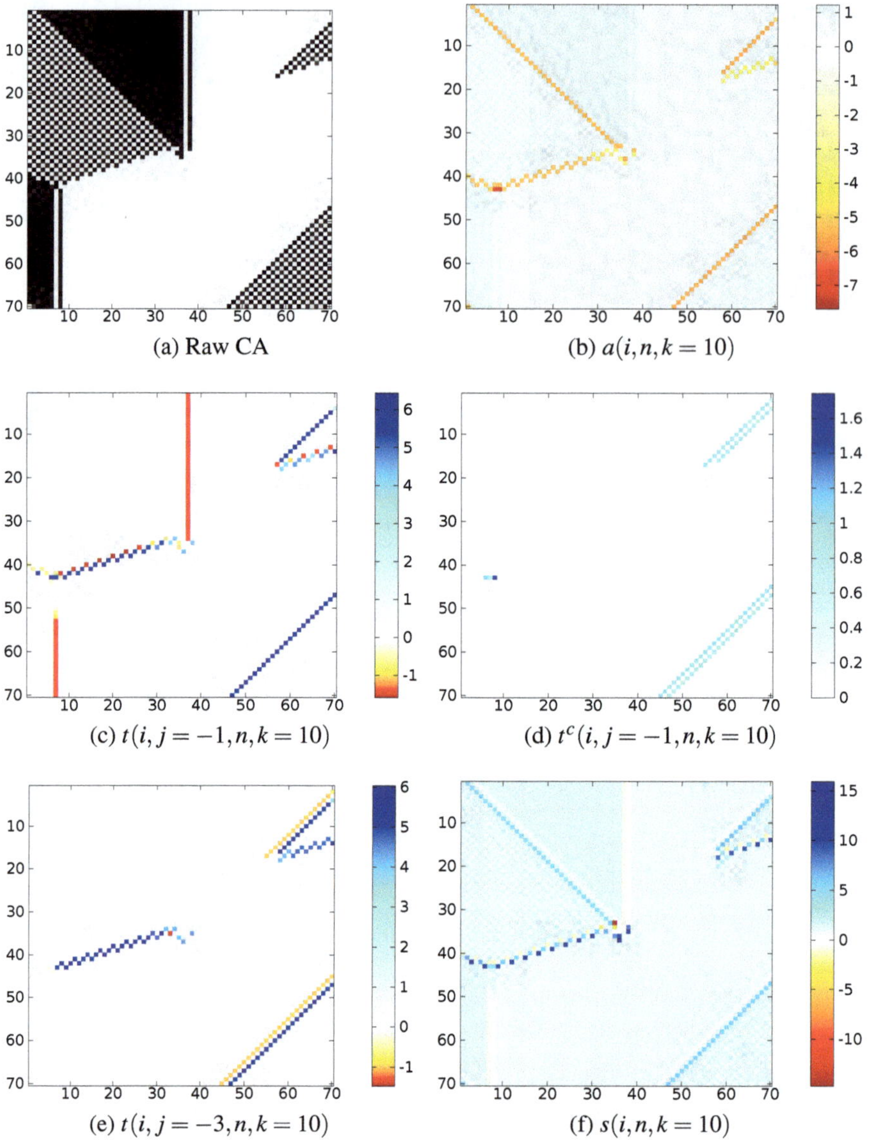

(a) Raw CA

(b) $a(i,n,k=10)$

(c) $t(i,j=-1,n,k=10)$

(d) $t^c(i,j=-1,n,k=10)$

(e) $t(i,j=-3,n,k=10)$

(f) $s(i,n,k=10)$

Fig. 5.5 Local information dynamics in $r=3$ **rule** $\phi_{\mathbf{par}}$ for the raw values displayed in (a) (black for "1", white for "0"). 70 time steps are displayed for 70 cells, and all units are in bits. (b) Local active information storage $a(i,n,k=10)$; Local apparent transfer entropy: (c) one cell to the left $t(i,j=-1,n,k=10)$, and (e) three cells to the left $t(i,j=-3,n,k=10)$; (d) Local complete transfer entropy one cell to the left $t^c(i,j=-1,n,k=10)$; (f) Local separable information $s(i,n,k=10)$.

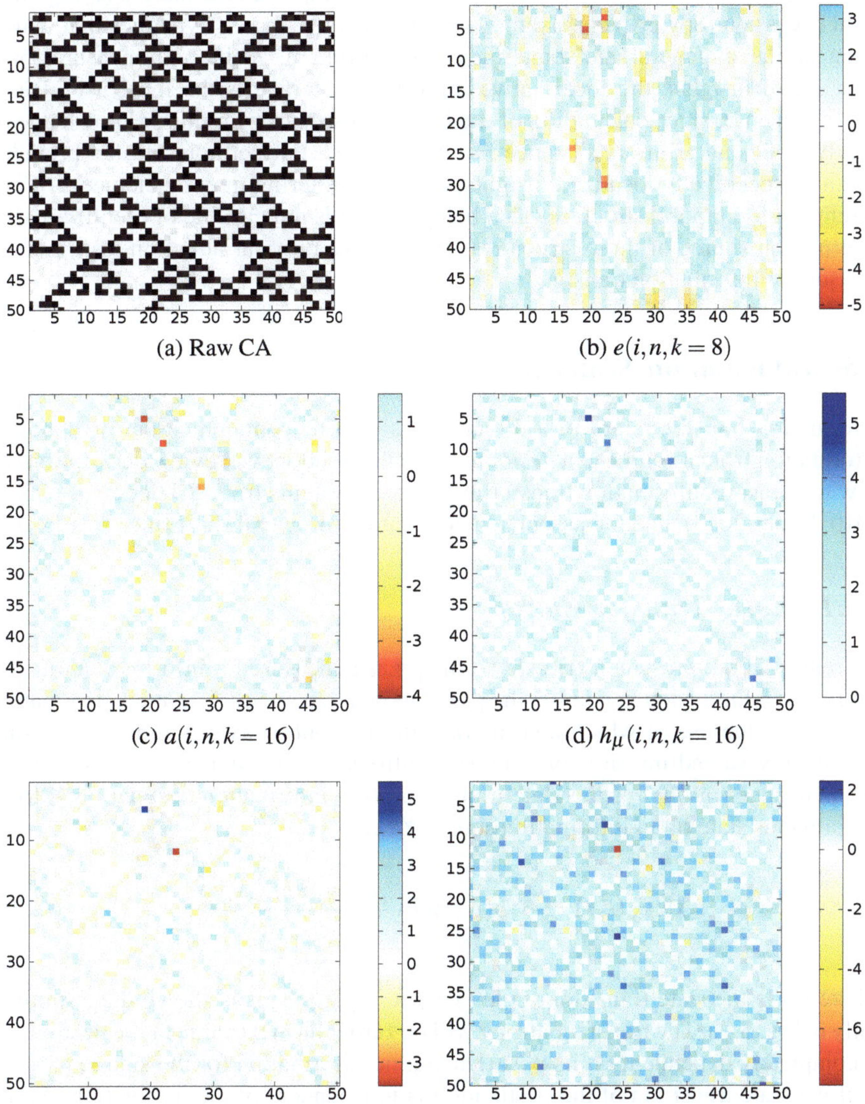

(a) Raw CA (b) $e(i,n,k=8)$

(c) $a(i,n,k=16)$ (d) $h_\mu(i,n,k=16)$

(e) $t(i,j=1,n,k=16)$ (f) $s(i,n,k=16)$

Fig. 5.6 Local information dynamics in **rule 22** for the raw values in (a) (black for "1", white for "0"). 50 time steps displayed for 50 cells, and all units are in bits. (b) Local excess entropy $e(i,n,k=8)$; (c) Local active information storage $a(i,n,k=16)$; (d) Local temporal entropy rate $h_\mu(i,n,k=16)$; (e) Local apparent transfer entropy one cell to the right $t(i,j=1,n,k=16)$; (f) Local separable information $s(i,n,k=16)$.

In summary, we have demonstrated that the local active information storage and local excess entropy provide insights into information storage dynamics that, while often similar in general, are sometimes subtly different. While both measures provide useful insights, the local active information storage is the most useful in a real-time sense, since calculation of the local excess entropy requires knowledge of the dynamics an arbitrary distance into the future.[6] Furthermore, it also provides the most specifically localized insights, including filtering moving elements of coherent spatiotemporal structure. This being said, it is not capable of identifying the information source of these structures; for this, we turn our attention to a specific measure of information transfer.

5.5 Information Transfer

Information transfer refers to a directional signal or communication of dynamic information from a *source* to a *destination*. In this section, we review descriptions of how to measure information transfer in complex systems from (Lizier et al. 2008b, 2010; Lizier 2013), and the associated application to several ECA rules.

5.5.1 *Local Transfer Entropy*

Schreiber (2000) presented **transfer entropy** as a measure for information transfer in order to address deficiencies in the previous de facto measure, mutual information (Eq. (5.4)), the use of which he criticized in this context as a symmetric measure of statically shared information. Transfer entropy is defined as the deviation from independence (in bits) of the state transition of an information destination X from the previous state of an information source Y:

$$T_{Y \to X}(k,l) = \sum_{w_n} p(w_n) \log_2 \frac{p(x_{n+1} \mid \mathbf{x}_n^{(k)}, \mathbf{y}_n^{(l)})}{p(x_{n+1} \mid \mathbf{x}_n^{(k)})}, \tag{5.26}$$

where w_n is the state transition tuple $(x_{n+1}, \mathbf{x}_n^{(k)}, \mathbf{y}_n^{(l)})$. This is shown diagrammatically in Fig. 5.7a. The transfer entropy will be zero if the next value of the destination is completely dependent on its past (leaving no information for the source to add), or if the state transition of the destination is independent of the destination. At the other extreme, it will be maximal if the state transition is completely specified by the source (in the context of the destination's past). As such, the transfer entropy is a *directional*, *dynamic* measure of information transfer. It is a *conditional* mutual information, casting it as the average information in the source about the next

---

[6] Calculation of $e(i,n,k)$ using local block entropies analogous to Eq. (5.12) would also require block entropies to be taken into the future to compute the same local information storage values. Without taking account of the dynamics into the future, we will not measure the information storage that *will* be used in the future of the process, but the information storage that is *likely* to be used in the future.

state of the destination conditioned on the destination's past. We have provided a thermodynamic interpretation of transfer entropy in (Prokopenko et al. 2013).

The role of the past state of the destination $\mathbf{x}_n^{(k)}$ is particularly important here. This past state can indirectly influence the next value via the source or other neighbors: this may be mistaken as an independent flow from the source here (Lizier et al. 2008b). In the context of distributed computation, this is recognizable as the *active information storage*. That is, conditioning on the destination's history $\mathbf{x}_n^{(k)}$ serves to eliminate the active information storage from the transfer entropy measurement. Yet any self-influence transmitted prior to these k values will not be eliminated: in (Lizier et al. 2008b) we suggested that the asymptote $k \to \infty$ is most correct for variables displaying non-Markovian dynamics. Just as the excess entropy and active information storage require $k \to \infty$ to capture all information storage, accurate measurement of the transfer entropy requires $k \to \infty$ to eliminate all information storage from being mistaken as information transfer. Further to these, even if the destination variable does display Markovian dynamics of order k, synergistic interactions between the source and the past of the destination beyond k time steps necessitate the use of a longer destination history to capture the information transfer, again leading us to $k \to \infty$ to capture all transfer. We describe other interpretations of the role of $\mathbf{x}_n^{(k)}$ in (Lizier and Mahoney 2013), including properly capturing the state transition of the destination and capturing the contribution of the source in the context of that state transition; which align with the above. The most generally correct form of the transfer entropy is therefore computed as:

$$T_{Y \to X}(l) = \lim_{k \to \infty} \sum_{w_n} p(w_n) \log_2 \frac{p(x_{n+1} \mid \mathbf{x}_n^{(k)}, \mathbf{y}_n^{(l)})}{p(x_{n+1} \mid \mathbf{x}_n^{(k)})}, \tag{5.27}$$

with $T_{Y \to X}(k,l)$ retained for finite-k estimates.

Also, we note that considering a source *state* $\mathbf{y}_n^{(l)}$ rather than a scalar y_n is most appropriate where the observations y mask a hidden causal process in Y, or where multiple past values of Y in addition to y_n are causal to x_{n+1}. Otherwise, where y_n is directly causal to x_{n+1}, and where it is the only direct causal source in Y (e.g. in CAs), we use only $l = 1$ (Lizier et al. 2008b; Lizier and Prokopenko 2010) and drop it from our notation here. Furthermore, note that one may use source-destination delays other than one time step, and indeed it is most appropriate to match any causal delay from Y to X (Wibral et al. 2013).

Next, we introduced the corresponding **local transfer entropy** at each observation n in (Lizier et al. 2008b):

$$t_{Y \to X}(n+1,l) = \lim_{k \to \infty} t_{Y \to X}(n+1,k,l), \tag{5.28}$$

$$t_{Y \to X}(n+1,k,l) = \log_2 \frac{p(x_{n+1} \mid \mathbf{x}_n^{(k)}, \mathbf{y}_n^{(l)})}{p(x_{n+1} \mid \mathbf{x}_n^{(k)})}. \tag{5.29}$$

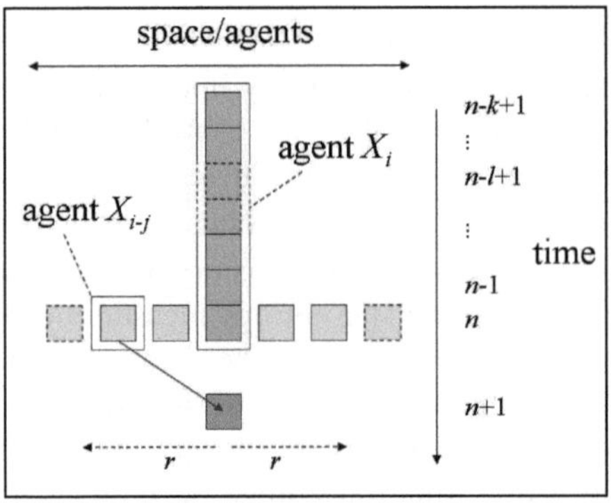

(a) Transfer Entropy

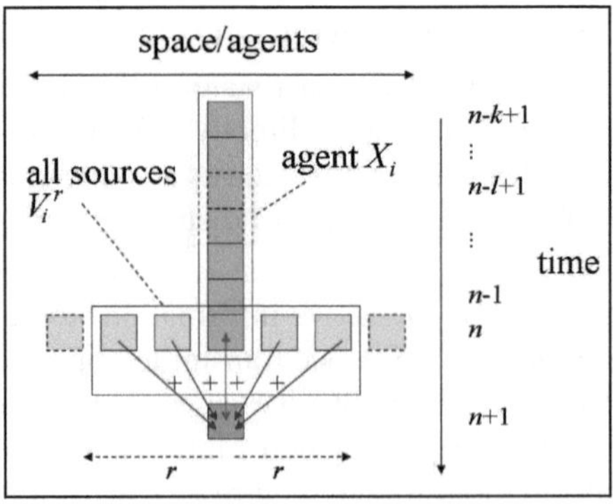

(b) Separable Information

Fig. 5.7 (a) Transfer Entropy $t(i,j,n+1,k)$: information contained in the source cell X_{i-j} about the next value of the destination cell X_i at time $n+1$ in the context of the destination's past. (b) Separable information $s(i,n+1,k)$: information gained about the next value of the destination from separately examining each causal information source in the context of the destination's past. For CAs these causal sources are within the cell range r. (NB: Fig. 5.7a is reprinted with kind permission from Springer Science+Business Media: (Lizier 2013) Lizier, J. T. , *The Local Information Dynamics of Distributed Computation in Complex Systems*, Springer Theses, Springer, Berlin / Heidelberg, Copyright 2013. Fig. 5.7b is reprinted with permission from (Lizier et al. 2010) J. T. Lizier, M. Prokopenko, and A. Y. Zomaya, "Information modification and particle collisions in distributed computation," *Chaos*, vol. 20, no. 3, p. 037109, 2010. Copyright 2010, AIP Publishing LLC.)

The local transfer entropy describes the information added by a specific source state $\mathbf{y}_n^{(l)}$ about x_{n+1} in the context of the past of the destination $\mathbf{x}_n^{(k)}$. Of course, we have $T_{Y\to X}(k,l) = \langle t_{Y\to X}(n+1,k,l)\rangle$.

For lattice systems such as CAs with spatially-ordered variables, the local information transfer to agent X_i from X_{i-j} (across j cells to the right) at time $n+1$ is represented as:

$$t(i,j,n+1,l) = \lim_{k\to\infty} t(i,j,n+1,k,l), \tag{5.30}$$

$$t(i,j,n+1,k,l) = \log_2 \frac{p(x_{i,n+1} \mid \mathbf{x}_{i,n}^{(k)}, \mathbf{x}_{i-j,n}^{(l)})}{p(x_{i,n+1} \mid \mathbf{x}_{i,n}^{(k)})}. \tag{5.31}$$

This information transfer $t(i,j,n+1,k,l)$ to variable X_i from X_{i-j} at time $n+1$ is illustrated in Fig. 5.7a. Then $t(i,j,n,k,l)$ is defined for every spatiotemporal destination (i,n), for every information channel or direction j; sensible values for j correspond to causal information sources, i.e. for CAs, sources within the cell range $|j| \le r$. Again, for homogeneous variables (with stationarity) it is appropriate to estimate the PDFs used in Eq. (5.31) from all spatiotemporal observations, and we write the average across homogeneous variables as $T(j,k) = \langle t(i,j,n,k)\rangle$.

Calculations conditioned on no other information contributors (as in Eq. (5.31))) are labeled as *apparent* transfer entropy (Lizier et al. 2008b). Local apparent transfer entropy $t(i,j,n,k)$ may be either positive or negative, with negative values occurring where (given the destination's history) the source element is actually misleading about the next value of the destination. In deterministic systems, this can only occur where another source is influencing the destination at that time. To counter that effect, the transfer entropy may be conditioned on other possible causal information sources Z, to eliminate their influence from being attributed to the source in question Y (Schreiber 2000). We call this the **conditional transfer entropy** (Lizier et al. 2010), given (as a finite-k estimate) along with the **local conditional transfer entropy** as follows:

$$T_{Y\to X|Z}(k,l) = \langle t_{Y\to X|Z}(n+1,k,l)\rangle, \tag{5.32}$$

$$t_{Y\to X|Z}(n+1,k,l) = \log_2 \frac{p(x_{n+1} \mid \mathbf{x}_n^{(k)}, \mathbf{y}_n^{(l)}, z_n)}{p(x_{n+1} \mid \mathbf{x}_n^{(k)}, z_n)}. \tag{5.33}$$

Z may of course be multivariate, or be an embedded state vector $\mathbf{z}_n^{(m)}$ itself. Indeed, a special case involves conditioning on *all* sources jointly in the set of causal information contributors $\mathbf{V}_X$ to X, except for the source Y, i.e. $\mathbf{V}_X \setminus Y$. This gives the **complete transfer entropy** $T^c_{Y\to X}(k,l) = T_{Y\to X|\mathbf{V}_X\setminus Y}(k,l)$ (Lizier et al. 2008b). At time step n, this set $\mathbf{V}_X \setminus Y$ has joint state $\mathbf{v}_{x,y,n}$, giving the **local complete transfer entropy** (Lizier et al. 2008b):[7]

$$t^c_{Y\to X}(n+1,k,l) = t_{Y\to X|\mathbf{V}_{X,Y}}(n+1,k,l), \tag{5.34}$$

$$\mathbf{v}_{x,y,n} = \{z_n \mid \forall Z \in \mathbf{V}_X \setminus Y\}. \tag{5.35}$$

For CAs the set of causal information contributors to X_i is the neighborhood $\mathbf{V}_i^r$ of X_i, and for the complete transfer entropy we condition on this set except for the source X_{i-j}: $\mathbf{V}_i^r \setminus X_{i-j}$. At time step n this set has joint value $\mathbf{v}^r_{i,j,n}$, giving the following expression for the local complete transfer entropy in CAs (Lizier et al. 2008b):

$$t^c(i,j,n+1,k) = \log_2 \frac{p\left(x_{i,n+1} \mid \mathbf{x}_{i,n}^{(k)}, x_{i-j,n}, \mathbf{v}^r_{i,j,n}\right)}{p\left(x_{i,n+1} \mid \mathbf{x}_{i,n}^{(k)}, \mathbf{v}^r_{i,j,n}\right)}, \tag{5.36}$$

$$\mathbf{v}^r_{i,j,n} = \left\{x_{i+q,n} \mid \forall q : -r \le q \le +r, q \ne -j, q \ne 0\right\}. \tag{5.37}$$

Again, the most correct form is $t^c(i,j,n+1)$ in the limit $k\to\infty$. In deterministic systems (e.g. CAs), complete conditioning renders $t^c(i,j,n) \ge 0$ because the source can only add information about the outcome of the destination.

[7] Note that if past values of X are causal sources to the next value x_{n+1}, they can be included in $\mathbf{v}_{x,y,n}$, but this is irrelevant for complete TE since they are already conditioned on in $\mathbf{x}_n^{(k)}$.

5.5.2 Total Information, Entropy Rate and Collective Information Transfer

The total information required to predict the next value of any process X is the *local entropy* $h_X(n+1)$ Eq. (5.14). Similarly, the *local temporal entropy rate* $h_{\mu X}(n+1,k) = -\log_2 p(x_{n+1} \mid x_n^{(k)})$ is the information to predict the next value of that process given that its past, and the entropy rate is the average of these local values: $H_{\mu X}(k) = \langle h_{\mu X}(n+1,k) \rangle$. For lattice systems we have $h_\mu(i,n+1,k)$. Now, the entropy can be considered as the sum of the active information storage and temporal entropy rate (Lizier et al. 2010, 2012c):

$$H_{X_{n+1}} = I_{X_{n+1};\mathbf{X}_n^{(k)}} + H_{X_{n+1}|\mathbf{X}_n^{(k)}}, \tag{5.38}$$

$$h_X(n+1) = a_X(n+1,k) + h_{\mu X}(n+1,k). \tag{5.39}$$

For deterministic systems (e.g. CAs) there is no intrinsic uncertainty, so the local temporal entropy rate is equal to the **local collective transfer entropy** (Lizier and Prokopenko 2010) and represents a collective information transfer: the information about the next value of the destination *jointly* added by the causal information sources in the context of the past of the destination. This suggested that the local collective transfer entropy (or simply the local temporal entropy rate $h_\mu(i,n,k)$ for deterministic systems) is likely to be a meaningful measure and filter for incoming information.

Also, we showed that the information in a destination variable can be expressed as a sum of incrementally conditioned mutual information terms, considering each of the sources iteratively (Lizier et al. 2010; Lizier and Prokopenko 2010). For ECAs, these expressions become:

$$h(i,n+1) = a(i,n+1,k) + t(i,j=-1,n+1,k) + t^c(i,j=1,n+1,k), \tag{5.40}$$

(and vice-versa in $j = 1,-1$). Clearly, this total information is not simply a simple sum of the active information storage and the apparent transfer entropy from each source, nor the sum of the active information storage and the complete transfer entropy from each source.

5.5.3 Local Information Transfer Results

In this section, we review the application of the local apparent and complete transfer entropies, as well as the local entropy rate, to several ECA rules (Lizier et al. 2007, 2008b, 2010; Lizier 2013). We focus in particular here on the local apparent transfer entropy, whose profiles $t(i,j=1,n,k=16)$ (measuring transfer across one unit to the right per time step) are plotted for rules 54 (Fig. 5.2d) and 22 (Fig. 5.6e), with $t(i,j=-1,n,k=16)$ (transfer across one unit to the left per time step) plotted for rules 54 (Fig. 5.2e), 110 (Fig. 5.4e) and ϕ_{par} (Fig. 5.5c).

Both the local apparent and complete transfer entropy highlight **particles as strong positive information transfer against background domains**. This is true for both regular gliders as well as domain walls in rule 18 (not shown here, see (Lizier et al. 2008b)). Importantly, the particles are measured as information transfer in their direction of macroscopic motion, as expected. As such, local transfer entropy provided the first quantitative evidence for the long-held conjecture that particles are the *dominant* information transfer agents in CAs. For example, at the "x" marks in Fig. 5.3 which denote parts of the right-moving γ^+ gliders, we have $p(x_{i,n+1} \mid \mathbf{x}_{i,n}^{(k=16)}, x_{i-1,n}) = 1.00$ and $p(x_{i,n+1} \mid \mathbf{x}_{i,n}^{(k=16)}) = 0.25$: there is a strong information transfer of $t(i, j = 1, n, k = 16) = 2.02$ bits here because the source (in the glider) added a significant amount of information to the destination about the continuation of the glider.

For ϕ_{par} we confirm the role of the gliders as information transfer agents in the human understandable computation, and demonstrate information transfer across multiple units of space per unit time step for fast-moving gliders in Fig. 5.5e. Interestingly, we also see in Fig. 5.5c ($j = -1$) and Fig. 5.5e ($j = -3$) that the apparent transfer entropy can attribute information transfer to several information sources, whereas the complete transfer entropy (see Fig. 5.5d) is more likely to attribute the transfer to the single causal source. We emphasize though that information transfer and causality are distinct concepts, as discussed in detail in (Lizier and Prokopenko 2010). This result also underlines that **the apparent and complete transfer entropies have a similar nature but are complementary** in together determining the next state of the destination (as in Eq. (5.40)). Neither measure is more correct than the other though – both are required to understand the dynamics fully. A more detailed example contrasting the two is studied for rule 18 in (Lizier et al. 2008b), showing that the complete TE detects transfer to X due to synergies between the source Y and conditioned variable Z, whereas the apparent TE does not.

We also examine the profiles of the local temporal entropy rate $h_\mu(i, n+1, k)$ (which is equal to the local collective transfer entropy in these deterministic systems) here in Fig. 5.4d for rule 110 and Fig. 5.6d for rule 22. As expected, the local temporal entropy rate profiles $h_\mu(i, n+1, k)$ highlight particles moving in *each* relevant channel and are a useful *single* spatiotemporal filter for moving emergent structure. In fact, these profiles are quite similar to the profiles of the negative values of local active information storage. This is not surprising given they are counterparts in Eq. (5.39): where $h_\mu(i, n+1, k)$ is strongly positive (i.e. greater than 1 bit), it is likely that $a(i, n+1, k)$ is negative since the local single cell entropy will average close to 1 bit for these examples. Unlike $a(i, n+1, k)$ however, the local temporal entropy rate $h_\mu(i, n+1, k)$ is never negative.

Note that while achieving the limit $k \to \infty$ is not computationally feasible, a large enough k was required to achieve a reasonable estimates of the transfer entropy; without this, as discussed earlier the active information storage was not eliminated from the transfer entropy measurements in the domains, and the measure did not distinguish the particles from the domains (Lizier et al. 2008b).

We also demonstrated (Lizier et al. 2008b) that while there is zero information transfer in an infinite periodic domain (since the dynamics there only involve information storage), there is a small non-zero information transfer in domains acting as a background to gliders, effectively indicating the *absence* of gliders. These small non-zero information transfers are stronger in the wake of a glider, indicating the absence of (relatively common) following gliders. Similarly, we note here that the local temporal entropy rate profiles $h_\mu(i,n+1,k)$ contain small but non-zero values in these periodic domains. Furthermore, there is interesting structure to the information transfer in the domain of rule 18, described in detail in (Lizier et al. 2008b). As such, while particles are the dominant information transfer agents in CAs, they are not the only transfer entities.

The highlighting of structure by local transfer entropy is similar to results from other methods of filtering for structure in CAs (Shalizi et al. 2006; Wuensche 1999; Hanson and Crutchfield 1992; Helvik et al. 2004), but subtly different in revealing the leading edges of gliders as the major transfer elements in the glider structures, and providing multiple profiles (one for each direction or channel of information transfer).

Also, a particularly relevant result for our purposes is the finding of negative values of transfer entropy for some space-time points in particles moving orthogonal to the direction of measurement in space-time. This is displayed for $t(i,j=1,n,k=16)$ in rule 54 (Fig. 5.2d), and $t(i,j=-1,n,k=16)$ for rule 110 (Fig. 5.4e), and also occurs for rule 18 (see (Lizier et al. 2008b; Lizier 2013)). In general this is because the source, as part of the domain, suggests that this same domain found in the past of the destination is likely to continue; however since the next value of the destination forms part of the particle, this suggestion proves to be misinformative. For example, consider the "x" marks in Fig. 5.3 which denote parts of the right-moving γ^+ gliders. If we now examine the source at the right (still in the domain), we have $p(x_{i,n+1} \mid \mathbf{x}_{i,n}^{(k=16)}, x_{i+1,n}) = 0.13$, with $p(x_{i,n+1} \mid \mathbf{x}_{i,n}^{(k=16)}) = 0.25$ as before, giving $t(i,j=-1,n,k=16) = -0.90$ bits: this is negative because the source (still in the domain) was misinformative about the destination.

Regarding the local information transfer structure of rule 22, we note similar results as for local information storage. There is much information transfer here (in fact the average value $T(j=1,k=16) = 0.19$ bits is greater than for rule 110 at 0.07 bits), although there is no coherent structure to this transfer. Again, this demonstrates the utility of local information measures in providing more detailed insights into system dynamics than their global averages.

In this section, we have described how the local transfer entropy quantifies the information transfer at space-time points within a system, and provides evidence that particles are the dominant information transfer agents in CAs. We also described the collective transfer entropy, which quantifies the joint information contribution from all causal information contributors, and in deterministic systems is equal to the temporal entropy rate. However, we have not yet separately identified collision events in CAs: to complete our exploration of the information dynamics of computation, we now consider the nature of information modification.

5.6 Information Modification

Langton (1990) interpreted information modification as interactions between transmitted and/or stored information which resulted in a modification of one or the other. CAs provide an illustrative example, where the term *interactions* is generally interpreted to mean collisions of particles (including blinkers as information storage), with the resulting dynamics involving something other than the incoming particles continuing unperturbed. The resulting dynamics could involve zero or more particles (with an annihilation leaving only a background domain), and perhaps even some of the incoming particles. Given the focus on perturbations in the definition here, it is logical to associate a collision event with the modification of transmitted and/or stored information, and to see it as an information processing or decision event. Indeed, as an information processing event the important role of collisions in determining the dynamics of the system is widely acknowledged (Hordijk et al. 2001), e.g. in the ϕ_{par} density classification.

Attempts have previously been made to quantify information modification or processing in a system (Sánchez-Montañés and Corbacho 2002; Yamada and Aihara 1994; Kinouchi and Copelli 2006). However, these have either been too specific to allow portability across system types (e.g. by focusing on the capability of a system to solve a known problem, or measuring properties related to the particular type of system being examined), focus on general processing as movement or interpretation of information rather than specifically the modification of information, or are not amenable to measuring information modification at *local* space-time points *within* a distributed system.

In this section, we review the separable information (Lizier et al. 2010) as a tool to detect non-trivial information modification events, and demonstrate it as the first measure which filtered collisions in CAs as such. At the end of the section however, we describe criticisms of the separable information, and describe current efforts to develop new measures of information modification.

5.6.1 Local Separable Information

We begin by considering what it means for a particle to be *modified*. For the simple case of a glider, a modification is simply an alteration to the predictable periodic pattern of the glider's dynamics. At such points, an observer would be surprised or misinformed about the next value of the glider, having not taken account of the entity about to perturb it. The intuition behind the separable information (Lizier et al. 2010) is that this interpretation is reminiscent of the earlier findings that local apparent transfer entropy $t(i,j,n)$ and local active information storage $a(i,n)$ were negative where the respective information sources were *misinformative* about the next value of the information destination (in the context of the destination's past for transfer entropy). Local active information storage was misinformative at gliders, and local apparent transfer entropy was misinformative at gliders traveling in the orthogonal direction to the measurement in space-time. This being said, one expects that the local apparent transfer entropy measured in the direction of glider motion will be *more informative* about its evolution than any misinformation conveyed from other sources. However, where the glider is modified by a collision with another glider, we would no longer expect the local apparent transfer entropy in its macroscopic direction of motion to remain informative about the dynamics. Assuming that the incident glider is also perturbed, the local apparent transfer entropy in its macroscopic direction of motion will also not be informative about the dynamics at this collision point. We expect the same argument to be true for irregular particles, or domain walls.

As such, we made the hypothesis that at the spatiotemporal location of a local information modification event or collision, *separate* inspection of each information source will *misinform* an observer overall about the next value of the modified information destination. More specifically, the information sources referred to here are the past history of the destination (via the local active information storage) and each other causal information contributor (examined in the context of the past history of the destination, via their local apparent transfer entropies).

We quantified the independent sum of information gained from separate observation of the information storage and information transfer contributors $Y \in V$ to a process X as the **local separable information** $s_X(n)$ (Lizier et al. 2010):

$$s_X(n) = a_X(n) + \sum_{Y \in V, Y \neq X} t_{Y \to X}(n). \tag{5.41}$$

$s_X(n,K)$ is used for finite-k estimates. For CAs, where the causal information contributors are homogeneously within the neighborhood r, we write the local separable information in lattice notation as:

$$s(i,n) = a(i,n) + \sum_{j=-r, j \neq 0}^{+r} t(i,j,n). \tag{5.42}$$

We use $s(i,n,k)$ to represent finite-k estimates, and show $s(i,n,k)$ in Fig. 5.7b.

As inferred earlier, we expected the local separable information to be *positive* or *highly separable* where separate observations of the information contributors are informative overall regarding the next value of the destination. This was be interpreted as a trivial information modification, because information storage and transfer are not interacting in any significant manner. More importantly, we expected the local separable information to be *negative* at spatiotemporal points where an information modification event or collision takes place. Here, separate observations are misleading overall because a *non-trivial information modification* is taking place (i.e. the information storage and transfer are interacting).

Importantly, this formulation of non-trivial information modification aligns with the descriptions of complex systems as consisting of (a large number of) elements interacting in a *non-trivial* fashion (Prokopenko et al. 2009), and of emergence as where *"the whole is greater than the sum of its parts"*. "The whole" meant to refer to examining all information sources together; the whole is greater where all information sources must be examined together in order to receive positive information on the next value of the examined entity. The thinking behind the separable information was in the direction of measuring synergies between information storage and transfer sources, prior to the development of a proper framework for examining such synergies (Williams and Beer 2010), as discussed in Section 5.6.3.

5.6.2 Local Separable Information Results

Next, we review the application of the separable information to several ECA rules from (Lizier et al. 2010). The simple gliders in ECA rule 54 give rise to relatively simple collisions which we focus on in our discussion here. Notice that the positive values of $s(i,n,k=16)$ for rule 54 (displayed in Fig. 5.2f) are concentrated in the domain regions and at the stationary gliders (α and β). As expected, these regions are undertaking trivial computations only. The negative values of $s(i,n,k=16)$ are also displayed in Fig. 5.2f, with their positions marked. The dominant negative values are clearly concentrated around the areas of collisions between the gliders, including collisions between the traveling gliders only (marked by "A") and between the traveling gliders and the stationary gliders (marked by "B" and "C").

Collision "A" involves the γ^+ and γ^- particles interacting to produce a β particle ($\gamma^+ + \gamma^- \rightarrow \beta$ (Hordijk et al. 2001)). The only information modification point highlighted is one time step below (or delayed from) that at which the gliders initially appear to collide (see close-up of raw values in Fig. 5.3). The periodic pattern in the past of the destination breaks there, however the neighboring sources are still able to support separate prediction of the value (i.e. $a(i,n,k=16) = -1.09$ bits, $t(i,j=1,n,k=16) = 2.02$ bits and $t(i,j=-1,n,k=16) = 2.02$ bits, giving $s(i,n,k=16) = 2.95$ bits). This is no longer the case however where our measure has successfully identified the modification point; there we have $a(i,n,k=16) = -3.00$ bits, $t(i,j=1,n,k=16) = 0.91$ bits and $t(i,j=-1,n,k=16) = 0.90$ bits, with $s(i,n,k=16) = -1.19$ bits suggesting a non-trivial information modification. A delay is also observed before the identified information modification points of collision types "B" and "C"; possibly these delays represent a time-lag of information processing. Not surprisingly, the results for these other collision types imply that the information modification points are associated with the creation of new behavior: in "B" and "C" these occur along the newly created γ gliders, and for "C" in the new α blinkers.

Importantly, weaker non-trivial information modification points continue to be identified at every second point along all the γ^+ and γ^- particles after the initial collisions. These can also be seen for a similar (right-moving) glider in rule 110 in Fig. 5.4f). This was unexpected from our earlier hypothesis. However, these events can be understood as non-trivial computations of the continuation of the glider in the *absence* of a collision; in effect they are virtual collisions between the real glider and the absence of an incident glider. Interestingly, this finding is analogous to the small but non-zero information transfer in periodic domains indicating the absence of gliders.

We also note that measurements of local separable information must be performed with a reasonably large value of k. Here, using $k < 4$ could not distinguish any information modification points clearly from the domains and particles, and even $k < 8$ could not distinguish *all* the modification points (results not shown). Correct quantification of information modification requires satisfactory estimates of information storage and transfer, and accurate distinction between the two.

We observe similar results in $s(i,n,k=10)$ for ϕ_{par} (see Fig. 5.5f). Note that the collisions at the left and right of the figure do in fact contain significant negative values of $s(i,n,k=10)$ – around 1 to 2 bits – however these are difficult to see in comparison to the much larger negative value at the collision in the centre of the diagram. These results confirm the particle collisions here as non-trivial information modification events, and this therefore completes the evidence for all of the conjectures about this human understandable computation.

The results for $s(i,n,k=16)$ for ECA rule 110 (see Fig. 5.4f) are also similar to those for rule 54. Here, we have collisions "A" and "B" which show non-trivial information modification points slightly delayed from the collision in a similar fashion to those for rule 54. We note that collisions between some of the more complex glider structures in rule 110 (not shown) exhibit non-trivial information modification points which are more difficult to interpret, and which are even more delayed from the initiation of the collision. The larger delay is perhaps this is a reflection of the more complex gliders requiring more time steps for the processing to take place. An interesting result not seen for rule 54 is a collision where an incident glider is absorbed by a blinker, without any modification to the absorbing glider (not shown here, see (Lizier et al. 2010)). No information modification is detected for this absorption event by $s(i,n,k=16)$: this is as expected because the information storage for the absorbing blinker is sufficient to predict the dynamics at this interaction.

As a further test of the measure, we examined collisions between the domain walls of rule 18; see (Lizier et al. 2010). We found that collisions between the domain walls were quite clearly highlighted as the dominant information modification events for this rule - importantly, this result provides evidence that collision of *irregular* particles are information modification events, as expected. The reader is referred to (Lizier et al. 2010) for further discussion of the information modification dynamics of rule 18.

We also apply $s(i,n,k=16)$ to ECA rule 22, as displayed in Fig. 5.6f. As could be expected from our earlier results, there are many points of both positive and negative local separable information here. The presence of negative values implies the occurrence of non-trivial information modification, yet there does not appear to be any structure to these profiles. Again, this aligns well with the lack of coherent structure found using the other measures in this framework and from the local statistical complexity profile of rule 22 (Shalizi et al. 2006).

5.6.3 Outlook for Information Modification

Here, we have reviewed the local separable information, which attempts to quantify information modification at each spatiotemporal point in a complex system. The separable information suggests that information modification events occur where the separable information is negative, indicating that separate or independent inspection of the causal information sources (in the context of the destination's past) is misleading because of non-trivial interaction between these sources. The local separable information was demonstrated to provide the first quantitative evidence that **particle collisions in CAs are the dominant information modification events** therein, and is capable of identifying events involving both creation and destruction.

With that said however, it has been shown that the separable information double-counts parts of the information in the next state of the destination (Flecker et al. 2011). This is clear, and so it is a heuristic more than a measure. Efforts to properly quantitatively define information modification, by combining information dynamics with the partial information decomposition approach (Williams and Beer 2010) to properly measure synergies between information storage and transfer, are ongoing and described in (Lizier et al. 2013). While the separable information is not a proper information-theoretic measure, it remains the only technique which has uniquely filtered particle collision events.

5.7 Importance of Coherent Computation

Our framework has proven successful in locally identifying the component operations of distributed computation. We then considered in (Lizier et al. 2012b) whether this framework can provide any insights into the overall complexity of computation. In other words, what can our results say about the difference in the complex computations of rules 110 and 54 as compared to rule 22 and others? We review those considerations in this section.

We observed that the *coherence* of local computational structure appears to be the most significant differentiator here. "Coherence" implies a property of sticking together or a logical relationship (Oxford English Dictionary 2008): in this context we use the term to describe a logical spatiotemporal relationship between values in local information dynamics profiles. For example, the manner in which particles give rise to similar values of local transfer entropy amongst spatiotemporal neighbors is coherent. From the spatiotemporal profiles presented here, we note that rules 54 and 110 exhibit the largest amount of coherent computational structure, with rule 18 containing a smaller amount of less coherent structure. Rules 22 and 30 (results for rule 30 not shown, see (Lizier et al. 2012b)) certainly exhibit all of the elementary functions of computation, but do not appear to contain any coherent structure to their computations. This aligns well with similar explorations of local information structure for these rules, e.g. by Shalizi et al. (2006). Using language reminiscent of Langton's analysis (Langton 1990), we suggested that complex systems exhibit very *highly-structured coherent* computation in comparison to ordered systems (which exhibit coherence but minimal structure in a computation dominated by information storage) and chaotic systems (whose computations are dominated by rampant information transfer eroding any coherence).

Coherence may also be interpreted as a logical relationship *between* profiles of the individual local information dynamics (as three axes of complexity) rather than

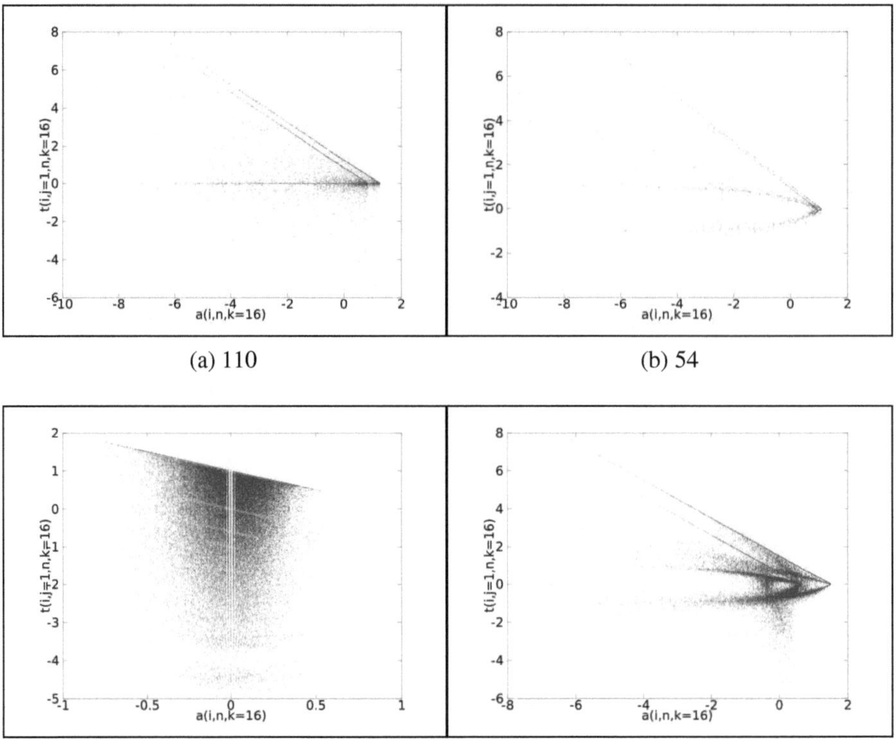

(a) 110 (b) 54

(c) 30 (d) 22

Fig. 5.8 State space diagrams of local transfer entropy (one step to the right) $t(i, j = 1, n, k = 16)$ versus local active information $a(i, n, k = 16)$ (both in bits) at the same space-time point (i, n) for several ECA rules: (a) 110, (b) 54, (c) 30 and (d) 22 (after (Lizier et al. 2012b)).

only within them. To investigate this possibility, Fig. 5.8 plots state-space diagrams of the local apparent transfer entropy for $j = 1$ versus local active information storage (after (Lizier et al. 2012b)). Each point in these diagrams represents the local values of each measure at one spatiotemporal point, thereby generating a complete state-space for the CA. Such state-space diagrams are known to provide insights into structure that are not visible when examining either measure in isolation; for example, in examining structure in classes of systems (such as logistic maps), Feldman et al. (2008) demonstrate that plotting average excess entropy versus entropy rate (while changing a system parameter) reveals loci of the two which are not clear from observing either in isolation. Here however we are looking at structure *within* a single system rather than across a class of systems.

The state-space diagram for rule 110 (Fig. 5.8a) exhibits interesting structure, with significant clustering around certain areas and lines in the state space, reflecting its status as a complex rule. (The two diagonal lines are upper limits representing the boundary condition $t^c(i, j = -1, n, k = 16) \geq 0$ for both destination states "0" and "1"). Rule 54 (Fig. 5.8b) exhibits similar structure in its state-space diagram. On the other hand, the example state space diagram for rule 30 (Fig. 5.8c) exhibits minimal structure (apart from the mathematical upper limit), with a smooth spread of points across the space reflecting its underlying chaotic nature. From the apparent absence of coherent structure in its space-time information profiles, one may expect state-space diagrams for rule 22 to exhibit a similar absence of structure to rule 30. As shown by Fig. 5.8d however this is not the case: the state-space diagram for rule 22 exhibits significant structure, with similar clustering to that of rules 110 and 54.

Importantly, the apparent information structure in the state-space diagrams lends some credence to the claims of complex behavior for rule 22 discussed in Section 5.3.3. However it is a very subtle type of structure, not complex enough to be revealed in the individual local information profiles shown here or by other authors (e.g. by Shalizi et al. (2006)). The structure does not appear to be coherent in these individual profiles, though the state space diagrams indicate a coherent relationship between the local information dynamics which may underpin coherent computation at other scales.

There are certain clues as to the type of coherence which may be displayed by rule 22. Fig. 5.6e does appear to have some traces of coherent transfer entities moving diagonally in space-time; however these seem to be distributed through the CA, seemingly without structure or interactions. More concretely, Grassberger (1983) observed that for rule 22, "there are (at least) four different sets of ordered states, corresponding to $S_i(t) = 0$ for all even/odd i and all even/odd t" – i.e. *rule 22 does have a domain pattern which self-replicates* (with four possible configurations, just offset from each other in space and time). Indeed, ε-machines have been generated to recognize these domains (Crutchfield et al. 2013). Grassberger (1983) goes on to note that "In contrast to the ordered states of rule 18, these states however are unstable: after implanting a kink in an otherwise ordered state, the kink widens without limit, leaving behind it a seemingly disordered state." That is to say, *this domain pattern does not self-organise and it is not robust to perturbations*. This means that, despite the existence of such domains, they are highly unlikely to be found "in the wild" (i.e. when rule 22 is started from random initial states, as we have done for Fig. 5.6). One could also view this as inferring that "life in one dimension" (the perspective that rule 22 is a 1D projection of the 2D Game of Life (McIntosh 1990)) is less stable than "life in two dimensions".

Coming back to Fig. 5.8d – it is possible that these domain patterns, or small versions of them, are what is detected as a signature of coherent information structure by our methods above. Furthermore, emerging evidence suggests that rule 22 can be set up in certain initial states which sustain such domains for a longer period, with certain stable domain walls (Crutchfield 2009), and that these domain walls are detected as information transfer by our methods. Our investigations in this area remain ongoing.

Given the subtlety of structure in the bounds of our analysis, and using our mutual information heuristics, at this stage we conclude that the behavior of this rule is less complex than that exhibited by rules 110 and 54. As such, we suggested that coherent information structure is a defining feature of complex computation, and explored a technique for inferring this property using local information dynamics. These state-space diagrams for local information dynamics produced useful visual results and were shown to provide interesting insight into the nature of computation in rule 22.

5.8 Conclusion

In this chapter, we have reviewed our complete quantitative framework for the information dynamics of distributed computation in complex systems. Our framework quantifies the information dynamics in terms of the component operations of universal computation: information storage, information transfer and information modification. Our framework places particular importance on examining computation on a local scale in space and time. While averaged or system-wide measures have their place in providing summarized results, this focus on the local scale is vital for understanding the information dynamics of computation and provides many insights that averaged measures cannot.

We reviewed the application of the framework to cellular automata, an important example because of the weight of previous studies on the nature of distributed computation in these systems. Significantly, our framework provided the first quantitative evidence for the widely accepted conjectures that blinkers provide information storage in CAs, particles are the dominant information transfer agents, and particle collisions are the dominant information modification events. In particular, this was demonstrated for the human-understandable density classification computation carried out by the rule ϕ_{par}. This is a fundamental contribution to our understanding of the nature of distributed computation, and provides impetus for the framework to be used for the analysis and design of other complex systems.

The application to CAs aligned well with other methods of filtering for complex structure in CAs. However, our work is distinct in that it provides several different views of the system corresponding to each type of computational structure. In particular, the results align well with the insights of computational mechanics, underlining the strong connection between these approaches.

From our results, we also observed that coherent local information structure is a defining feature of complex distributed computation, and used local information state-spaces to study coherent complex computation. Here, our framework provides further insight into the nature of computation in rule 22 with respect to the accepted complex rules 54 and 110. Certainly rule 22 exhibits all of the elementary functions of computation, yet (in line with Shalizi et al. (2006)) there is no apparent coherent structure to the profiles of its local information dynamics ("in the wild" at least). On the other hand, state space views of the interplay between these local information dynamics reveal otherwise hidden structure. Our framework is unique in its ability to resolve both of these aspects. We conclude that rule 22 exhibits more structure than chaotic rules, yet the subtlety of this structure prevents it from being considered as complex as rules 110 and 54.

The major thrust of our work since the presentation of this framework was to apply it to other systems, because the information-theoretic basis of this framework makes it readily applicable as such. For example, we have used the measures in this framework to: quantitatively demonstrate coherent waves of motion in flocks and swarms as information cascades (Wang et al. 2012); evolve a modular robot for maximal information transfer between components, observing the emergence of glider-like information cascades (Lizier et al. 2008a); and to study interactions in robotic football and the relation of information measures to success on the field (Cliff et al. 2013). We have also inferred information structure supporting cognitive tasks using fMRI brain imaging data (Lizier et al. 2012), and studied how the computational capabilities of artificial neural networks relate to underlying parameters and ability to solve particular tasks (Boedecker et al. 2012). We have also made more specific investigations of the relationship between underlying network structure and computational capabilities, including: revealing that intrinsic information storage and transfer capabilities are maximized near the phase transition in dynamics for random Boolean networks (Lizier et al. 2008a); showing that regular networks are generally associated with information storage, random networks with information transfer, and small-world networks exhibit a balance of the two (Lizier et al. 2011b); revealing that feedback and feedforward loop motifs determine information storage capability (Lizier et al. 2012a); and exploring how these information measures relate to synchronization capability of network structures (Ceguerra et al. 2011). We have also explored the relationship of the framework to the context of the observer (Lizier and Mahoney 2013), and provided thermodynamic interpretations of transfer entropy (Prokopenko et al. 2013) and related information-theoreic quantities (Prokopenko et al. 2011). And finally, we have begun reformulating our approach to information modification in seeking a proper measure rather than a heuristic (Lizier et al. 2013). Further developments in all of these directions are expected in the future, due to the utility of the framework.

Acknowledgements. The authors thank Melanie Mitchell for helpful comments and suggestions regarding an early version of this manuscript.

References

Adamatzky, A. (ed.): Collision-Based Computing. Springer, Berlin (2002)

Atick, J.J.: Could information theory provide an ecological theory of sensory processing? Network: Computation in Neural Systems 3(2), 213 (1992)

Badii, R., Politi, A.: Thermodynamics and Complexity of Cellular Automata. Physical Review Letters 78(3), 444 (1997)

Bialek, W., Nemenman, I., Tishby, N.: Complexity through nonextensivity. Physica A: Statistical Mechanics and its Applications 302(1-4), 89–99 (2001)

Boccara, N., Nasser, J., Roger, M.: Particlelike structures and their interactions in spatiotemporal patterns generated by one-dimensional deterministic cellular-Automaton rules. Physical Review A 44(2), 866–875 (1991)

Boedecker, J., Obst, O., Lizier, J.T., Mayer, N.M., Asada, M.: Information processing in echo state networks at the edge of chaos. Theory in Biosciences 131(3), 205–213 (2012)

Brown, J.A., Tuszynski, J.A.: A review of the ferroelectric model of microtubules. Ferroelectrics 220, 141–156 (1999)

Casti, J.L.: Chaos, Gödel and truth. In: Casti, J.L., Karlqvist, A. (eds.) Beyond Belief: Randomness, Prediction and Explanation in Science, pp. 280–327. CRC Press, Boca Raton (1991)

Ceguerra, R.V., Lizier, J.T., Zomaya, A.Y.: Information storage and transfer in the synchronization process in locally-connected networks. In: Proceedings of the 2011 IEEE Symposium on Artificial Life (ALIFE), pp. 54–61. IEEE (2011)

Cliff, O.M., Lizier, J.T., Wang, X.R., Wang, P., Obst, O., Prokopenko, M.: Towards quantifying interaction networks in a football match. In: Proceedings of the RoboCup 2013 Symposium (to be published, 2013)

Conway, J.H.: What is Life? In: Berlekamp, E., Conway, J.H., Guy, R. (eds.) Winning Ways for Your Mathematical Plays, vol. 2, ch. 25, pp. 927–962. Academic Press, New York (1982)

Cook, M.: Universality in Elementary Cellular Automata. Complex Systems 15(1), 1–40 (2004)

Couzin, I.D., James, R., Croft, D.P., Krause, J.: Social Organization and Information Transfer in Schooling Fishes. In: Brown, C., Laland, K.N., Krause, J. (eds.) Fish Cognition and Behavior, Fish and Aquatic Resources, pp. 166–185. Blackwell Publishing (2006)

Cover, T.M., Thomas, J.A.: Elements of Information Theory. Wiley-Interscience, New York (1991)

Crutchfield, J.P.: Personal communication (2009)

Crutchfield, J.P., Ellison, C.J., Riechers, P.M.: Exact complexity: The spectral decomposition of intrinsic computation. arXiv:1309.3792 (2013)

Crutchfield, J.P., Feldman, D.P.: Regularities Unseen, Randomness Observed: Levels of Entropy Convergence. Chaos 13(1), 25–54 (2003)

Crutchfield, J.P., Young, K.: Inferring statistical complexity. Physical Review Letters 63(2), 105–108 (1989)

Edmundson, D.E., Enns, R.H.: Fully 3-dimensional collisions of bistable light bullets. Optics Letters 18, 1609–1611 (1993)

Eppstein, D.: Searching for spaceships. In: Nowakowski, R.J. (ed.) More Games of No Chance. MSRI Publications, vol. 42, pp. 433–453. Cambridge Univ. Press (2002)

Fano, R.M.: Transmission of information: a statistical theory of communications. M.I.T. Press, Cambridge (1961)

Feldman, D.P., McTague, C.S., Crutchfield, J.P.: The organization of intrinsic computation: Complexity-entropy diagrams and the diversity of natural information processing. Chaos 18(4), 43106 (2008)

Flecker, B., Alford, W., Beggs, J.M., Williams, P.L., Beer, R.D.: Partial information decomposition as a spatiotemporal filter. Chaos 21(3), 037104+ (2011)

Goh, K.I., Barabási, A.L.: Burstiness and memory in complex systems. Europhysics Letters 81(4), 48002 (2008)

Grassberger, P.: New mechanism for deterministic diffusion. Physical Review A 28(6), 3666 (1983)

Grassberger, P.: Long-range effects in an elementary cellular automaton. Journal of Statistical Physics 45(1-2), 27–39 (1986a)

Grassberger, P.: Toward a quantitative theory of self-generated complexity. International Journal of Theoretical Physics 25(9), 907–938 (1986b)

Grassberger, P.: Information content and predictability of lumped and distributed dynamical systems. Physica Scripta 40(3), 346 (1989)

Gray, L.: A Mathematician Looks at Wolfram's New Kind of Science. Notices of the American Mathematical Society 50(2), 200–211 (2003)

Gutowitz, H., Domain, C.: The Topological Skeleton of Cellular Automaton Dynamics. Physica D 103(1-4), 155–168 (1997)

Hanson, J.E., Crutchfield, J.P.: The Attractor-Basin Portait of a Cellular Automaton. Journal of Statistical Physics 66, 1415–1462 (1992)

Hanson, J.E., Crutchfield, J.P.: Computational mechanics of cellular automata: An example. Physica D 103(1-4), 169–189 (1997)

Helvik, T., Lindgren, K., Nordahl, M.G.: Local information in one-dimensional cellular automata. In: Sloot, P.M.A., Chopard, B., Hoekstra, A.G. (eds.) ACRI 2004. LNCS, vol. 3305, pp. 121–130. Springer, Heidelberg (2004)

Hordijk, W., Shalizi, C.R., Crutchfield, J.P.: Upper bound on the products of particle interactions in cellular automata. Physica D 154(3-4), 240–258 (2001)

Jakubowski, M.H., Steiglitz, K., Squier, R.: Information transfer between solitary waves in the saturable Schrödinger equation. Physical Review E 56(6), 7267 (1997)

Jakubowski, M.H., Steiglitz, K., Squier, R.K.: Computing with solitons: A review and prospectus. Multiple-Valued Logic 6(5-6), 439–462 (2001)

Kinouchi, O., Copelli, M.: Optimal dynamical range of excitable networks at criticality. Nature Physics 2(5), 348–351 (2006)

Klyubin, A.S., Polani, D., Nehaniv, C.L.: Tracking Information Flow through the Environment: Simple Cases of Stigmergy. In: Pollack, J., Bedau, M., Husbands, P., Ikegami, T., Watson, R.A. (eds.) Proceedings of the Ninth International Conference on the Simulation and Synthesis of Living Systems (ALife IX), Boston, USA, pp. 563–568. MIT Press, Cambridge (2004)

Klyubin, A.S., Polani, D., Nehaniv, C.L.: All Else Being Equal Be Empowered. In: Capcarrère, M.S., Freitas, A.A., Bentley, P.J., Johnson, C.G., Timmis, J. (eds.) ECAL 2005. LNCS (LNAI), vol. 3630, pp. 744–753. Springer, Heidelberg (2005)

Lafusa, A., Bossomaier, T.: Hyperplane Localisation of Self-Replicating and Other Complex Cellular Automata Rules. In: Proceedings of the the 2005 IEEE Congress on Evolutionary Computation, Edinburgh, vol. 1, pp. 844–849. IEEE Press (2005)

Langton, C.G.: Computation at the edge of chaos: phase transitions and emergent computation. Physica D 42(1-3), 12–37 (1990)

Lindgren, K., Nordahl, M.G.: Complexity Measures and Cellular Automata. Complex Systems 2(4), 409–440 (1988)

Lindgren, K., Nordahl, M.G.: Universal computation in simple one-dimensional cellular automata. Complex Systems 4, 299–318 (1990)

Lindner, M., Vicente, R., Priesemann, V., Wibral, M.: TRENTOOL: A Matlab open source toolbox to analyse information flow in time series data with transfer entropy. BMC Neuroscience 12(1), 119+ (2011)

Lizier, J.T.: JIDT: An information-theoretic toolkit for studying the dynamics of complex systems (2012), `https://code.google.com/p/information-dynamics-toolkit/`

Lizier, J.T.: The Local Information Dynamics of Distributed Computation in Complex Systems. Springer Theses. Springer, Heidelberg (2013)

Lizier, J.T., Atay, F.M., Jost, J.: Information storage, loop motifs, and clustered structure in complex networks. Physical Review E 86(2), 026110+ (2012a)

Lizier, J.T., Flecker, B., Williams, P.L.: Towards a synergy-based approach to measuring information modification. In: Proceedings of the 2013 IEEE Symposium on Artificial Life (ALIFE), pp. 43–51. IEEE (2013)

Lizier, J.T., Heinzle, J., Horstmann, A., Haynes, J.-D., Prokopenko, M.: Multivariate information-theoretic measures reveal directed information structure and task relevant changes in fMRI connectivity. Journal of Computational Neuroscience 30(1), 85–107 (2011a)

Lizier, J.T., Mahoney, J.R.: Moving frames of reference, relativity and invariance in transfer entropy and information dynamics. Entropy 15(1), 177–197 (2013)

Lizier, J.T., Pritam, S., Prokopenko, M.: Information dynamics in small-world Boolean networks. Artificial Life 17(4), 293–314 (2011b)

Lizier, J.T., Prokopenko, M.: Differentiating information transfer and causal effect. European Physical Journal B 73(4), 605–615 (2010)

Lizier, J.T., Prokopenko, M., Tanev, I., Zomaya, A.Y.: Emergence of Glider-like Structures in a Modular Robotic System. In: Bullock, S., Noble, J., Watson, R., Bedau, M.A. (eds.) Proceedings of the Eleventh International Conference on the Simulation and Synthesis of Living Systems (ALife XI), Winchester, UK, pp. 366–373. MIT Press, Cambridge (2008a)

Lizier, J.T., Prokopenko, M., Zomaya, A.Y.: Detecting Non-trivial Computation in Complex Dynamics. In: Almeida e Costa, F., Rocha, L.M., Costa, E., Harvey, I., Coutinho, A. (eds.) ECAL 2007. LNCS (LNAI), vol. 4648, pp. 895–904. Springer, Heidelberg (2007)

Lizier, J.T., Prokopenko, M., Zomaya, A.Y.: The information dynamics of phase transitions in random Boolean networks. In: Bullock, S., Noble, J., Watson, R., Bedau, M.A. (eds.) Proceedings of the Eleventh International Conference on the Simulation and Synthesis of Living Systems (ALife XI), Winchester, UK, pp. 374–381. MIT Press, Cambridge (2008b)

Lizier, J.T., Prokopenko, M., Zomaya, A.Y.: Local information transfer as a spatiotemporal filter for complex systems. Physical Review E 77(2), 026110+ (2008c)

Lizier, J.T., Prokopenko, M., Zomaya, A.Y.: Information modification and particle collisions in distributed computation. Chaos 20(3), 037109+ (2010)

Lizier, J.T., Prokopenko, M., Zomaya, A.Y.: Coherent information structure in complex computation. Theory in Biosciences 131(3), 193–203 (2012b)

Lizier, J.T., Prokopenko, M., Zomaya, A.Y.: Local measures of information storage in complex distributed computation. Information Sciences 208, 39–54 (2012c)

Lungarella, M., Sporns, O.: Mapping information flow in sensorimotor networks. PLoS Computational Biology 2(10), e144+ (2006)

MacKay, D.J.C.: Information Theory, Inference, and Learning Algorithms. Cambridge University Press, Cambridge (2003)

Marinazzo, D., Wu, G., Pellicoro, M., Angelini, L., Stramaglia, S.: Information flow in networks and the law of diminishing marginal returns: evidence from modeling and human electroencephalographic recordings. PloS ONE 7(9), e45026+ (2012)

Martinez, G.J., Adamatzky, A., McIntosh, H.V.: Phenomenology of glider collisions in cellular automaton Rule 54 and associated logical gates. Chaos, Solitons and Fractals 28(1), 100–111 (2006)

McIntosh, H.V.: Linear Cellular Automata. Universidad Autónoma de Puebla, Puebla, Mexico (1990)

Mitchell, M.: A Complex-Systems Perspective on the "Computation vs. Dynamics" Debate in Cognitive Science. In: Gernsbacher, M.A., Derry, S.J. (eds.) Proceedings of the 20th Annual Conference of the Cognitive Science Society (Cogsci 1998), Madison, Wisconsin, pp. 710–715 (1998a)

Mitchell, M.: Computation in Cellular Automata: A Selected Review. In: Gramss, T., Bornholdt, S., Gross, M., Mitchell, M., Pellizzari, T. (eds.) Non-Standard Computation, pp. 95–140. VCH Verlagsgesellschaft, Weinheim (1998b)

Mitchell, M., Crutchfield, J.P., Das, R.: Evolving Cellular Automata with Genetic Algorithms: A Review of Recent Work. In: Goodman, E.D., Punch, W., Uskov, V. (eds.) Proceedings of the First International Conference on Evolutionary Computation and Its Applications, Moscow, Russia, Russian Academy of Sciences (1996)

Mitchell, M., Crutchfield, J.P., Hraber, P.T.: Evolving Cellular Automata to Perform Computations: Mechanisms and Impediments. Physica D 75, 361–391 (1994)

Morgado, R., Cieśla, M., Longa, L., Oliveira, F.A.: Synchronization in the presence of memory. Europhysics Letters 79(1), 10002 (2007)

Obst, O., Boedecker, J., Schmidt, B., Asada, M.: On active information storage in input-driven systems. arXiv:1303.5526 (2013)

Oxford English Dictionary (2008), `http://www.oed.com/` (accessed August 5, 2008)

Pahle, J., Green, A.K., Dixon, C.J., Kummer, U.: Information transfer in signaling pathways: a study using coupled simulated and experimental data. BMC Bioinformatics 9, 139 (2008)

Prokopenko, M., Boschietti, F., Ryan, A.J.: An Information-Theoretic Primer on Complexity, Self-Organization, and Emergence. Complexity 15(1), 11–28 (2009)

Prokopenko, M., Gerasimov, V., Tanev, I.: Evolving Spatiotemporal Coordination in a Modular Robotic System. In: Nolfi, S., Baldassarre, G., Calabretta, R., Hallam, J.C.T., Marocco, D., Meyer, J.-A., Miglino, O., Parisi, D. (eds.) SAB 2006. LNCS (LNAI), vol. 4095, pp. 558–569. Springer, Heidelberg (2006)

Prokopenko, M., Lizier, J.T., Obst, O., Wang, X.R.: Relating Fisher information to order parameters. Physical Review E 84, 041116+ (2011)

Prokopenko, M., Lizier, J.T., Price, D.C.: On thermodynamic interpretation of transfer entropy. Entropy 15(2), 524–543 (2013)

Sánchez-Montañés, M.A., Corbacho, F.J.: Towards a New Information Processing Measure for Neural Computation. In: Dorronsoro, J.R. (ed.) ICANN 2002. LNCS, vol. 2415, pp. 637–642. Springer, Heidelberg (2002)

Schreiber, T.: Measuring Information Transfer. Physical Review Letters 85(2), 461–464 (2000)

Shalizi, C.R.: Causal Architecture, Complexity and Self-Organization in Time Series and Cellular Automata. PhD thesis, University of Wisconsin-Madison (2001)

Shalizi, C.R., Crutchfield, J.P.: Computational mechanics: Pattern and Prediction, Structure and Simplicity. Journal of Statistical Physics 104, 817–879 (2001)

Shalizi, C.R., Haslinger, R., Rouquier, J.-B., Klinkner, K.L., Moore, C.: Automatic filters for the detection of coherent structure in spatiotemporal systems. Physical Review E 73(3), 036104 (2006)

Shannon, C.E.: A mathematical theory of communication. Bell System Technical Journal 27 (1948)

Takens, F.: Detecting strange attractors in turbulence. In: Rand, D., Young, L.-S. (eds.) Dynamical Systems and Turbulence, Warwick 1980. Lecture Notes in Mathematics, vol. 21, pp. 366–381. Springer, Heidelberg (1981)

Von Neumann, J.: Theory of self-reproducing automata. University of Illinois Press, Urbana (1966)

Wang, X.R., Miller, J.M., Lizier, J.T., Prokopenko, M., Rossi, L.F.: Quantifying and Tracing Information Cascades in Swarms. PLoS ONE 7(7), e40084+ (2012)

Wibral, M., Pampu, N., Priesemann, V., Siebenhühner, F., Seiwert, H., Lindner, M., Lizier, J.T., Vicente, R.: Measuring Information-Transfer delays. PLoS ONE 8(2), e55809+ (2013)

Wibral, M., Rahm, B., Rieder, M., Lindner, M., Vicente, R., Kaiser, J.: Transfer entropy in magnetoencephalographic data: quantifying information flow in cortical and cerebellar networks. Progress in Biophysics and Molecular Biology 105(1-2), 80–97 (2011)

Williams, P.L., Beer, R.D.: Nonnegative Decomposition of Multivariate Information. arXiv:1004.2515 (2010)

Wolfram, S.: Cellular automata as models of complexity. Nature 311(5985), 419–424 (1984a)

Wolfram, S.: Computation theory of cellular automata. Communications in Mathematical Physics 96(1), 15–57 (1984b)

Wolfram, S.: Universality and complexity in cellular automata. Physica D 10(1-2), 1–35 (1984c)

Wolfram, S.: A New Kind of Science. Wolfram Media, Champaign (2002)

Wuensche, A.: Classifying cellular automata automatically: Finding gliders, filtering, and relating space-time patterns, attractor basins, and the Z parameter. Complexity 4(3), 47–66 (1999)

Yamada, T., Aihara, K.: Spatio-temporal complex dynamics and computation in chaotic neural networks. In: Proceedings of the IEEE Symposium on Emerging Technologies and Factory Automation (ETFA 1994), Tokyo, pp. 239–244. IEEE (1994)

Chapter 6
Quantifying Synergistic Mutual Information

Virgil Griffith* and Christof Koch

6.1 Introduction

Synergy is a fundamental concept in complex systems that has received much attention in computational biology (Narayanan et al. 2005; Balduzzi and Tononi 2008). Several papers (Schneidman et al. 2003a; Bell 2003; Nirenberg et al. 2001; Williams and Beer 2010) have proposed measures for quantifying synergy, but there remains no consensus which measure is most valid.

The concept of synergy spans many fields and theoretically could be applied to any non-subadditive function. But within the confines of Shannon information theory, synergy—or more formally, *synergistic information*—is a property of a set of n random variables $\mathbf{X} = \{X_1, X_2, \ldots, X_n\}$ cooperating to predict (reduce the uncertainty of) a single target random variable Y.

One clear application of synergistic information is in computational genetics. It is well understood that most phenotypic traits are influenced not only by single genes but by interactions among genes—for example, human eye-color is cooperatively specified by more than a dozen genes(White and Rabago-Smith 2011). The magnitude of this "cooperative specification" is the synergistic information between the set of genes $\mathbf{X}$ and a phenotypic trait Y. Another application is neuronal firings where potentially thousands of presynaptic neurons influence the firing rate of a single post-synaptic (target) neuron. Yet another application is discovering the "informationally synergistic modules" within a complex system.

Virgil Griffith
Computation and Neural Systems, Caltech, Pasadena, CA 91125
e-mail: virgil@caltech.edu

Christof Koch
Allen Institute for Brain Science, Seattle, WA 98103

* Corresponding author.

M. Prokopenko (ed.), *Guided Self-Organization: Inception*,
Emergence, Complexity and Computation 9,
DOI: 10.1007/978-3-642-53734-9_6, © Springer-Verlag Berlin Heidelberg 2014

The prior literature(Schneidman et al. 2003b; Anastassiou 2007) has termed several distinct concepts as "synergy". This paper defines synergy as how much the whole is greater than (the union of) its atomic elements.[1]

The prior works on Partial Information Decomposition (Williams and Beer 2010; Harder et al. 2013; Bertschinger et al. 2012; Lizier et al. 2013) start with properties that a measure of redundant information, $\mathrm{I}_\cap$ satisfies and builds a measure of synergy from $\mathrm{I}_\cap$. Although this paper deals directly with measures of synergy on "easy" examples, we are immensely sympathetic to this approach. Our proposed measure of synergy does give rise to an $\mathrm{I}_\cap$ measure.

The properties our $\mathrm{I}_\cup$ satisfies are discussed in Appendix C.

For pedagogical purposes all examples are *determinstic*, however, these methods equally apply to non-deterministic systems.

6.1.1 Notation

We use the following notation throughout. Let

- n: The number of predictors $X_1, X_2, \ldots, X_n$. $n \geq 2$.
- $X_{1\ldots n}$: The *joint* random variable (coalition) of all n predictors $X_1 X_2 \ldots X_n$.
- X_i: The i'th predictor random variable (r.v.). $1 \leq i \leq n$.
- $\mathbf{X}$: The *set* of all n predictors $\{X_1, X_2, \ldots, X_n\}$.
- Y: The *target r.v.* to be predicted.
- y: A particular state of the target r.v. Y.

All random variables are discrete, all logarithms are $\log_2$, and all calculations are in *bits*. Entropy and mutual information are as defined by Cover and Thomas (1991), $\mathrm{H}(X) \equiv \sum_{x \in X} \Pr(x) \log \frac{1}{\Pr(x)}$, as well as $\mathrm{I}(X : Y) \equiv \sum_{x,y} \Pr(x,y) \log \frac{\Pr(x,y)}{\Pr(x)\Pr(y)}$.

6.1.2 Understanding PI-Diagrams

Partial information diagrams (PI-diagrams), introduced by Williams and Beer (2010), extend Venn diagrams to properly represent synergy. Their framework has been invaluable to the evolution of our thinking on synergy.

A PI-diagram is composed of nonnegative *partial information regions* (PI-regions). Unlike the standard Venn entropy diagram in which the sum of all regions is the joint entropy $\mathrm{H}(X_{1\ldots n}, Y)$, in PI-diagrams the sum of all regions (i.e. the space of the PI-diagram) is the mutual information $\mathrm{I}(X_{1\ldots n} : Y)$. PI-diagrams are immensely helpful in understanding how the mutual information $\mathrm{I}(X_{1\ldots n} : Y)$ is distributed across the coalitions and singletons of $\mathbf{X}$.[2]

---

[1] The techniques here are unrelated to the information geometry prospective provided by (Amari 1999). The well-known "total correlation" measure(Han 1978), does not satisfy the desired properties for a measure of synergy.

[2] Formally, how the mutual information is distributed across the set of all nonempty antichains on the powerset of $\mathbf{X}$(Weisstein 2011; Comtet 1998).

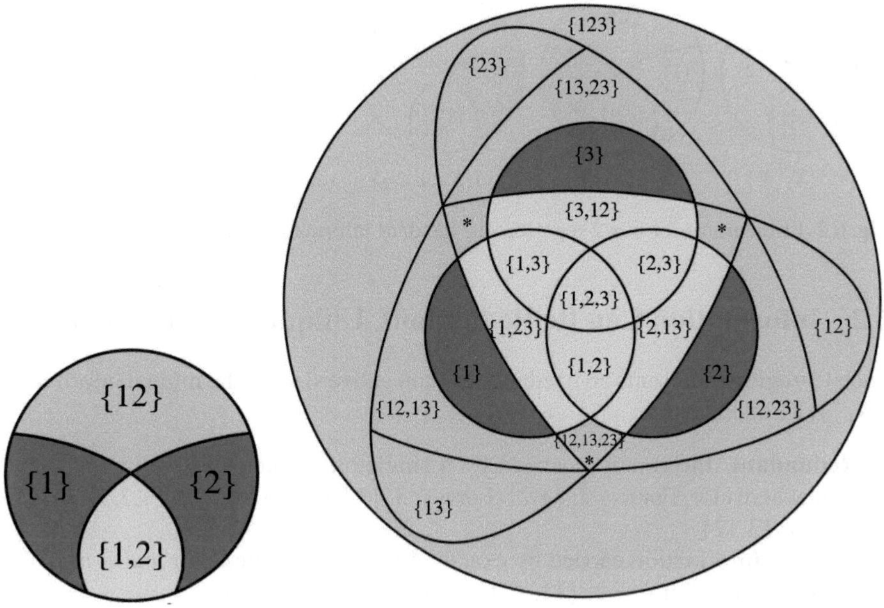

(a) $n = 2$ (b) $n = 3$

Fig. 6.1 PI-diagrams for two and three predictors. Each PI-region represents nonnegative information about Y. A PI-region's color represents whether its information is redundant (yellow), unique (magenta), or synergistic (cyan). To preserve symmetry, the PI-region "$\{12, 13, 23\}$" is displayed as three separate regions each marked with a "*". All three *-regions should be treated as through they are a single region.

How to Read PI-Diagrams. Each PI-region is uniquely identified by its "set notation" where each element is denoted solely by the predictors' indices. For example, in the PI-diagram for $n = 2$ (Fig. 6.1a): $\{1\}$ is the information about Y only X_1 carries (likewise $\{2\}$ is the information only X_2 carries); $\{1,2\}$ is the information about Y that X_1 as well as X_2 carries, while $\{12\}$ is the information about Y that is specified only by the coalition (joint random variable) X_1X_2. For getting used to this way of thinking, common informational quantities are represented by colored regions in Fig. 6.2.

The general structure of a PI-diagram becomes clearer after examining the PI-diagram for $n = 3$ (Fig. 6.1b). All PI-regions from $n = 2$ are again present. Each predictor (X_1, X_2, X_3) can carry unique information (regions labeled $\{1\}$, $\{2\}$, $\{3\}$), carry information redundantly with another predictor ($\{1,2\}$, $\{1,3\}$, $\{2,3\}$), or specify information through a coalition with another predictor ($\{12\}$, $\{13\}$, $\{23\}$). New in $n = 3$ is information carried by all three predictors ($\{1,2,3\}$) as well as information specified through a three-way coalition ($\{123\}$). Intriguingly, for three predictors, information can be provided by a coalition as well as a singleton ($\{1,23\}$, $\{2,13\}$, $\{3,12\}$) or specified by multiple coalitions ($\{12,13\}$, $\{12,23\}$, $\{13,23\}$, $\{12,13,23\}$).

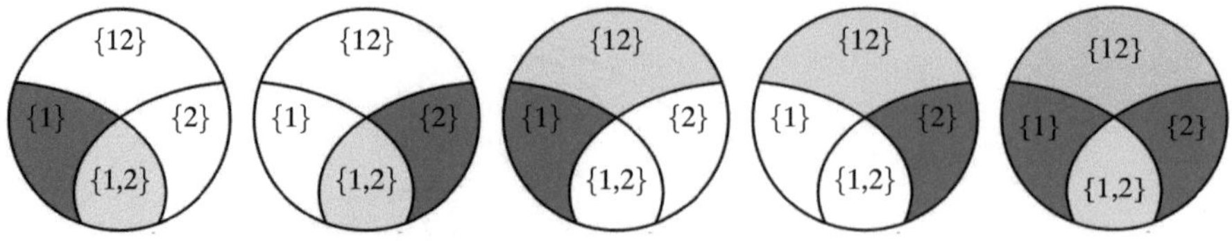

(a) $\mathrm{I}(X_1 : Y)$ (b) $\mathrm{I}(X_2 : Y)$ (c) $\mathrm{I}(X_1 : Y|X_2)$ (d) $\mathrm{I}(X_2 : Y|X_1)$ (e) $\mathrm{I}(X_1 X_2 : Y)$

Fig. 6.2 PI-diagrams for $n = 2$ representing standard informational quantities

6.2 Information Can Be Redundant, Unique, or Synergistic

Each PI-region represents an irreducible nonnegative slice of the mutual information $\mathrm{I}(X_{1\ldots n} : Y)$ that is either:

1. **Redundant.** Information carried by a singleton predictor as well as available somewhere else. For $n=2$: {1,2}. For $n=3$: {1,2}, {1,3}, {2,3}, {1,2,3}, {1,23}, {2,13}, {3,12}.
2. **Unique.** Information carried by exactly one singleton predictor and is available no where else. For $n = 2$: {1}, {2}. For $n = 3$: {1}, {2}, {3}.
3. **Synergistic.** Any and all information in $\mathrm{I}(X_{1\ldots n} : Y)$ that is not carried by a singleton predictor. $n=2$: {12}. For $n=3$: {12}, {13}, {23}, {123}, {12,13}, {12,23}, {13,23}, {12,13,23}.

Although a single PI-region is either redundant, unique, or synergistic, a single state of the target can have any combination of positive PI-regions, i.e. a single state of the target can convey redundant, unique, and synergistic information. This surprising fact is demonstrated in Fig. 6.9.

6.2.1 *Example Rdn: Redundant Information*

If X_1 and X_2 carry some identical[3] information (reduce the same uncertainty) about Y, then we say the set $\mathbf{X} = \{X_1, X_2\}$ has some *redundant information* about Y. Fig. 6.3 illustrates a simple case of redundant information. Y has two equiprobable states: `r` and `R` (`r`/`R` for "redundant bit"). Examining X_1 or X_2 identically specifies one bit of Y, thus we say set $\mathbf{X} = \{X_1, X_2\}$ has one bit of redundant information about Y.

[3] X_1 and X_2 providing identical information about Y is different from providing the same *magnitude* of information about Y, i.e. $\mathrm{I}(X_1 : Y) = \mathrm{I}(X_2 : Y)$. Example UNQ (Fig. 6.4) is an example where $\mathrm{I}(X_1 : Y) = \mathrm{I}(X_2 : Y) = 1$ bit yet X_1 and X_2 specify "different bits" of Y. Providing the same magnitude of information about Y is neither necessary or sufficient for providing some identical information about Y.

6.2.2 *Example Unq: Unique Information*

Predictor X_i carries *unique information* about Y if and only if X_i specifies information about Y that is not specified by anything else (a singleton or coalition of the other $n-1$ predictors). Fig. 6.4 illustrates a simple case of unique information. Y has four equiprobable states: ab, aB, Ab, and AB. X_1 uniquely specifies bit a/A, and X_2 uniquely specifies bit b/B. If we had instead labeled the Y-states: 0, 1, 2, and 3, X_1 and X_2 would still have strictly unique information about Y. The state of X_1 would specify between $\{0,1\}$ and $\{2,3\}$, and the state of X_2 would specify between $\{0,2\}$ and $\{1,3\}$—together fully specifying the state of Y. Accepting the property **(Id)** from (Harder et al. 2013) is sufficient but not necessary for the desired decomposition of example UNQ.

6.2.3 *Example Xor: Synergistic Information*

A set of predictors $\mathbf{X} = \{X_1, \ldots, X_n\}$ has synergistic information about Y if and only if the whole ($X_{1\ldots n}$) specifies information about Y that is not specified by any singleton predictor. The canonical example of synergistic information is the XOR-gate (Fig. 6.5). In this example, the whole X_1X_2 fully specifies Y,

X_1	X_2	Y	
r	r	r	1/2
R	R	R	1/2

(a) $\Pr(x_1, x_2, y)$

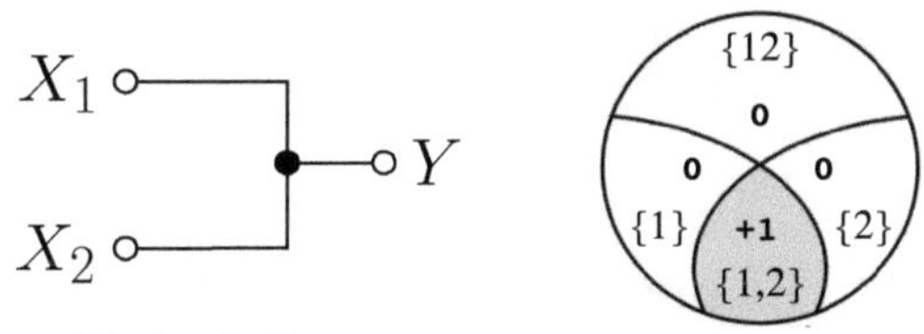

(b) circuit diagram (c) PI-diagram

Fig. 6.3 Example RDN. Fig. 6.3a shows the joint distribution of r.v.'s X_1, X_2, and Y, the joint probability $\Pr(x_1, x_2, y)$ is along the right-hand side of (a), revealing that all three terms are fully correlated. Fig. 6.3b represents the joint distribution as an electrical circuit. Fig. 6.3c is the PI-diagram indicating that set $\{X_1, X_2\}$ has 1 bit of redundant information about Y. $\mathrm{I}(X_1X_2 : Y) = \mathrm{I}(X_1 : Y) = \mathrm{I}(X_2 : Y) = \mathrm{H}(Y) = 1$ bit.

X_1	X_2	Y	
a	b	ab	1/4
a	B	aB	1/4
A	b	Ab	1/4
A	B	AB	1/4

(a) $\Pr(x_1, x_2, y)$

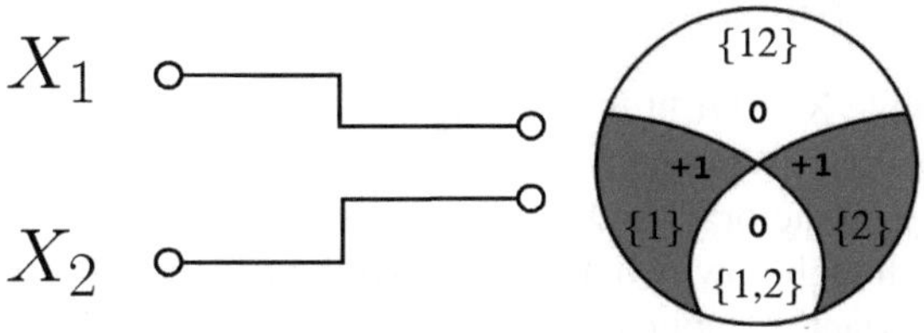

(b) circuit diagram (c) PI-diagram

Fig. 6.4 Example UNQ. X_1 and X_2 each uniquely specify a single bit of Y. $\mathrm{I}(X_1X_2 : Y) = \mathrm{H}(Y) = 2$ bits. The joint probability $\Pr(x_1, x_2, y)$ is along the right-hand side of (a).

$$I(X_1X_2 : Y) = H(Y) = 1 \text{ bit}, \tag{6.1}$$

but the singletons X_1 and X_2 specify *nothing* about Y,

$$I(X_1 : Y) = I(X_2 : Y) = 0 \text{ bits}. \tag{6.2}$$

With both X_1 and X_2 themselves having zero information about Y, we know that there can not be any redundant or unique information about Y—that the three PI-regions $\{1\} = \{2\} = \{1,2\} = 0$ bits. As the information between X_1X_2 and Y must come from somewhere, by elimination we conclude that X_1 and X_2 synergistically specify Y.

X_1	X_2	Y	
0	0	0	1/4
0	1	1	1/4
1	0	1	1/4
1	1	0	1/4

(a) $\Pr(x_1, x_2, y)$

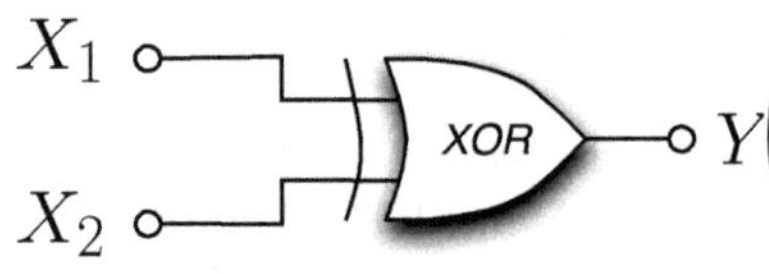

(b) circuit diagram

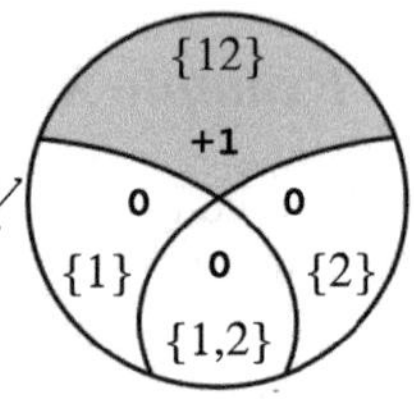

(c) PI-diagram

Fig. 6.5 Example XOR. X_1 and X_2 synergistically specify Y. $I(X_1X_2 : Y) = H(Y) = 1$ bit. The joint probability $\Pr(x_1, x_2, y)$ is along the right-hand side of (a).

6.3 Two Examples Elucidating Properties of Synergy

To help the reader develop intuition for a proper measure of synergy we illustrate two desired properties of synergistic information with pedagogical examples derived from XOR. Readers solely interested in the contrast with prior measures can skip to Section 6.4.

6.3.1 Duplicating a Predictor Does Not Change Synergistic Information

Example XORDUPLICATE (Fig. 6.6) adds a third predictor, X_3, a copy of predictor X_1, to XOR. Whereas in XOR the target Y is specified only by coalition X_1X_2, duplicating predictor X_1 as X_3 makes the target equally specifiable by coalition X_3X_2.

Although now two different coalitions identically specify Y, mutual information is invariant to duplicates, e.g. $I(X_1X_2X_3 : Y) = I(X_1X_2 : Y)$ bit. Likewise for synergistic information to be likewise bounded between zero and the total mutual information $I(X_{1 \ldots n} : Y)$, synergistic information must similarly be invariant to duplicates, e.g. the synergistic information between set $\{X_1, X_2\}$ and Y must be the same as the synergistic information between $\{X_1, X_2, X_3\}$ and Y. This makes sense because if synergistic information is defined as the information in the whole beyond its parts, duplicating a part does not increase the net information provided by the parts. Altogether, we assert that *duplicating a predictor does not change the synergistic information.* Without the property that duplicating a predictor does not change synergistic information, the synergistic mutual information will not be bounded between 0 and $\mathrm{I}(X_{1\ldots n} : Y)$. Synergistic information being invariant to duplicated predictors follows from the equality condition of the monotonicity property (**M**) from Bertschinger et al. (2012).[4]

X_1	X_2	X_3	Y	
0	0	0	0	1/4
0	1	0	1	1/4
1	0	1	1	1/4
1	1	1	0	1/4

(a) $\Pr(x_1, x_2, x_3, y)$

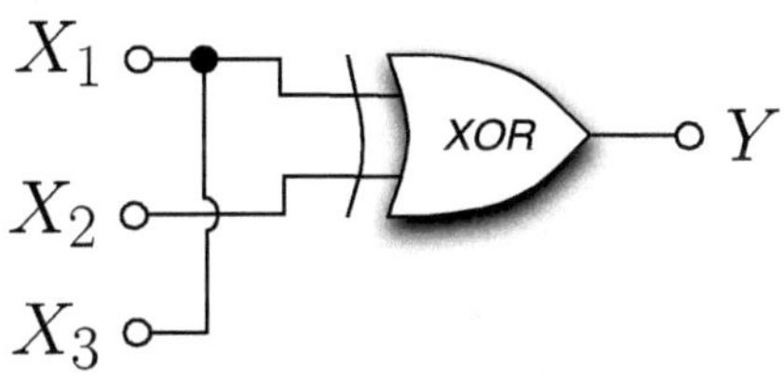

(b) circuit diagram

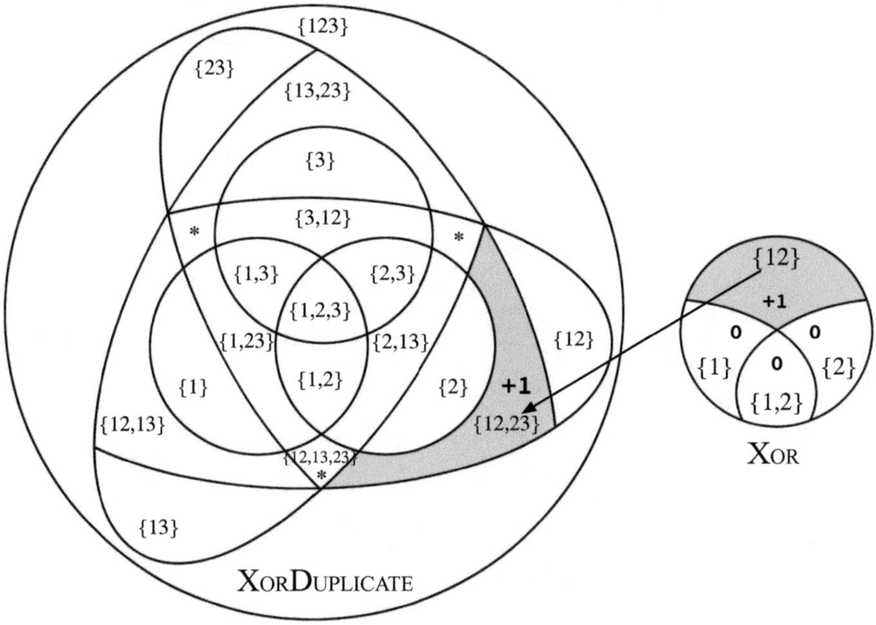

(c) PI-diagram

Fig. 6.6 Example XorDuplicate shows that duplicating predictor X_1 as X_3 turns the single-coalition synergy $\{12\}$ into the multi-coalition synergy $\{12, 23\}$. After duplicating X_1, the coalition X_3X_2 as well as coalition X_1X_2 specifies Y. Synergistic information is unchanged from Xor, $\mathrm{I}(X_3X_2 : Y) = \mathrm{I}(X_1X_2 : Y) = \mathrm{H}(Y) = 1$ bit.

[4] For a proof see Appendix E.

6.3.2 Adding a New Predictor Can Decrease Synergy

Example XORLOSES (Fig. 6.7) adds a third predictor, X_3, to XOR and concretizes the distinction between synergy and "redundant synergy". In XORLOSES the target Y has one bit of uncertainty and just as in example XOR the coalition X_1X_2 fully specifies the target, $\mathrm{I}(X_1X_2:Y) = \mathrm{H}(Y) = 1$ bit. However, XORLOSES has *zero* intuitive synergy because the newly added singleton predictor, X_3, fully specifies Y by itself. This makes the synergy between X_1 and X_2 *completely redundant*—everything the coalition X_1X_2 specifies is now already specified by the singleton X_3.

X_1	X_2	X_3	Y	
0	0	0	0	1/4
0	1	1	1	1/4
1	0	1	1	1/4
1	1	0	0	1/4

(a) $\Pr(x_1, x_2, x_3, y)$

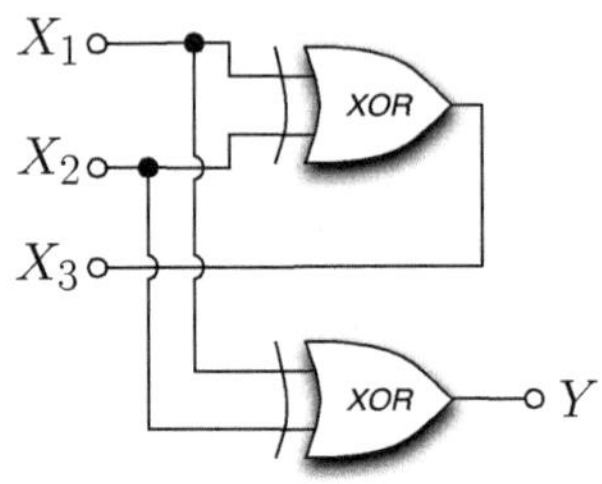

(b) circuit diagram

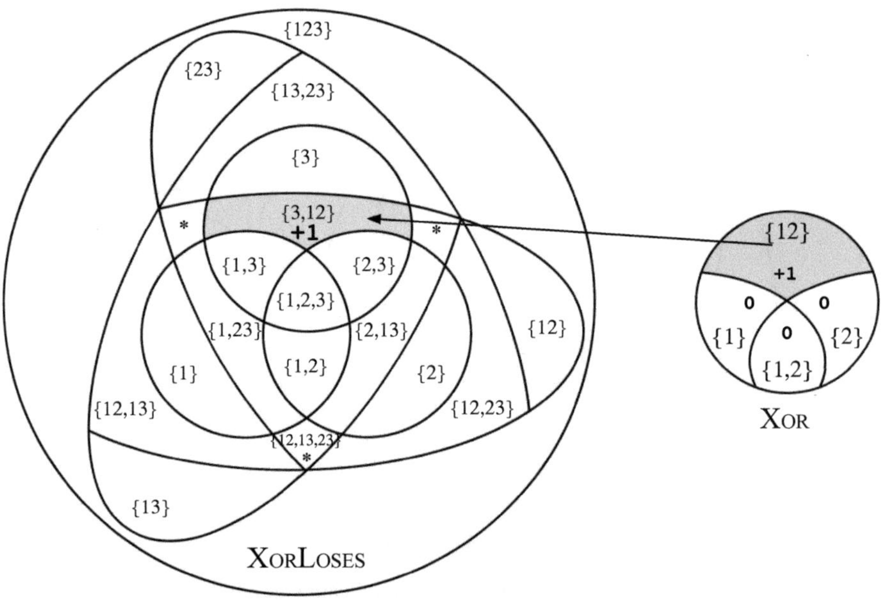

(c) PI-diagram

Fig. 6.7 Example XORLOSES. Target Y is fully specified by the coalition X_1X_2 as well as by the singleton X_3. $\mathrm{I}(X_1X_2:Y) = \mathrm{I}(X_3:Y) = \mathrm{H}(Y) = 1$ bit. Therefore the information synergistically specified by coalition X_1X_2 is a redundant synergy.

6.4 Prior Measures of Synergy

6.4.1 $\mathrm{I}_{\max}$ *synergy:* $\mathcal{S}_{\max}(\mathbf{X} : Y)$

$\mathrm{I}_{\max}$ synergy, denoted $\mathcal{S}_{\max}$, derives from (Williams and Beer 2010). $\mathcal{S}_{\max}$ defines synergy as the whole beyond the state-dependent *maximum* of its parts,

$$\mathcal{S}_{\max}(\mathbf{X} : Y) \equiv \mathrm{I}(X_{1\ldots n} : Y) - \mathrm{I}_{\max}(\{X_1, \ldots, X_n\} : Y) \tag{6.3}$$

$$= \mathrm{I}(X_{1\ldots n} : Y) - \sum_{y \in Y} \Pr(Y = y) \max_i \mathrm{I}(X_i : Y = y) , \tag{6.4}$$

where $\mathrm{I}(X_i : Y = y)$ is (DeWeese and Meister 1999)'s "specific-surprise",

$$\mathrm{I}(X_i : Y = y) \equiv \mathrm{D}_{\mathrm{KL}}[\Pr(X_i|y) \| \Pr(X_i)] \tag{6.5}$$

$$= \sum_{x_i \in X_i} \Pr(x_i|y) \log \frac{\Pr(x_i, y)}{\Pr(x_i)\Pr(y)} . \tag{6.6}$$

There are two major advantages of $\mathcal{S}_{\max}$ synergy. First, $\mathcal{S}_{\max}$ obeys the bounds of $0 \leq \mathcal{S}_{\max}(X_{1\ldots n} : Y) \leq \mathrm{I}(X_{1\ldots n} : Y)$. Second, $\mathcal{S}_{\max}$ is invariant to duplicate predictors. Despite these desired properties, $\mathcal{S}_{\max}$ sometimes miscategorizes merely unique information as synergistic. This can be seen in example Unq (Fig. 6.4). In example Unq the wires in Fig. 6.4b don't even touch, yet $\mathcal{S}_{\max}$ asserts there is one bit of synergy and one bit of redundancy—this is palpably strange.

A more abstract way to understand why $\mathcal{S}_{\max}$ overestimates synergy is to imagine a hypothetical example where there are exactly two bits of unique information for every state $y \in Y$ and no synergy or redundancy. $\mathcal{S}_{\max}$ would be the whole (both unique bits) minus the *maximum* over both predictors—which would be the $\max[1,1] = 1$ bit. The $S_{\max}$ synergy would then be $2 - 1 = 1$ bit of synergy—even though by definition there was no synergy, but merely two bits of unique information.

Altogether, we conclude that $\mathcal{S}_{\max}$ *overestimates* the intuitive synergy by miscategorizing merely unique information as synergistic whenever two or more predictors have unique information about the target.

6.4.2 WholeMinusSum Synergy: $\mathrm{WMS}(\mathbf{X} : Y)$

The earliest known sightings of bivarate WholeMinusSum synergy (WMS) is (Gawne and Richmond 1993; Gat and Tishby 1999) with the general case in Chechik et al. (2002). WholeMinusSum synergy is a signed measure where a positive value signifies synergy and a negative value signifies redundancy. WholeMinusSum synergy is defined by eq. (6.7) and interestingly reduces to eq. (6.9)—the difference of two *total correlations*.[5]

[5] $\mathrm{TC}(X_1; \cdots; X_n) = -\mathrm{H}(X_{1\ldots n}) + \sum_{i=1}^{n} \mathrm{H}(X_i)$ per Han (1978).

$$\mathrm{WMS}(\mathbf{X}:Y) \equiv \mathrm{I}(X_{1\ldots n}:Y) - \sum_{i=1}^{n} \mathrm{I}(X_i:Y) \tag{6.7}$$

$$= \sum_{i=1}^{n} \mathrm{H}(X_i|Y) - \mathrm{H}(X_{1\ldots n}|Y) - \left[\sum_{i=1}^{n} \mathrm{H}(X_i) - \mathrm{H}(X_{1\ldots n})\right] \tag{6.8}$$

$$= \mathrm{TC}(X_1;\cdots;X_n|Y) - \mathrm{TC}(X_1;\cdots;X_n) \tag{6.9}$$

Representing eq. (6.7) for $n=2$ as a PI-diagram (Fig. 6.8a) reveals that WMS is the synergy between X_1 and X_2 *minus* their redundancy. Thus, when there is an equal magnitude of synergy and redundancy between X_1 and X_2 (as in RDNXOR, Fig. 6.9), WholeMinusSum synergy is *zero*—leading one to *erroneously* conclude there is no synergy or redundancy present.[6]

The PI-diagram for $n=3$ (Fig. 6.8b) reaveals that WholeMinusSum double-subtracts PI-regions $\{1,2\}$, $\{1,3\}$, $\{2,3\}$ and triple-subtracts PI-region $\{1,2,3\}$, revealing that for $n>2$ WMS $(\mathbf{X}:Y)$ becomes synergy minus the redundancy *counted multiple times*.

A concrete example demonstrating WholeMinusSum's "synergy minus redundancy" behavior is RDNXOR (Fig. 6.9) which overlays examples RDN and XOR to form a single system. The target Y has two bits of uncertainty, i.e. $\mathrm{H}(Y)=2$. Like RDN, either X_1 or X_2 identically specifies the letter of Y (`r`/`R`), making one bit of redundant information. Like XOR, only the coalition X_1X_2 specifies the digit of Y (`0`/`1`), making one bit of synergistic information. Together this makes one bit of redundancy and one bit of synergy.

Note that in RDNXOR every state $y \in Y$ conveys one bit of redundant information and one bit of synergistic information, e.g. for the state $y = $ `r0` the letter "`r`" is specified redundantly and the digit "`0`" is specified synergistically. Example RDNUNQXOR (Appendix A) extends RDNXOR to demonstrate redundant, unique, and synergistic information for every state $y \in Y$.

In summary, WholeMinusSum *underestimates* synergy for all n with the potential gap increasing with n. Equivalently, we say that WholeMinusSum synergy is a *lowerbound* on the intuitive synergy with the bound becoming looser with n.

6.4.3 *Correlational Importance:* $\Delta\mathrm{I}(\mathbf{X};Y)$

Correlational importance, denoted $\Delta\mathrm{I}$, comes from Panzeri et al. (1999); Nirenberg et al. (2001); Nirenberg and Latham (2003); Pola et al. (2003); Latham and Nirenberg (2005). Correlational importance quantifies the "informational importance of conditional dependence" or the "information lost when ignoring conditional dependence" among the predictors decoding target Y. As conditional dependence is necessary for synergy, $\Delta\mathrm{I}$ seems related to our intuitive conception of synergy. $\Delta\mathrm{I}$ is defined as,

---

[6] This is deeper than Schneidman et al. (2003a)'s point that a mish-mash of synergy and redundancy across different states of $y \in Y$ can average to zero. Fig. 6.9 evaluates to zero for *every state* $y \in Y$.

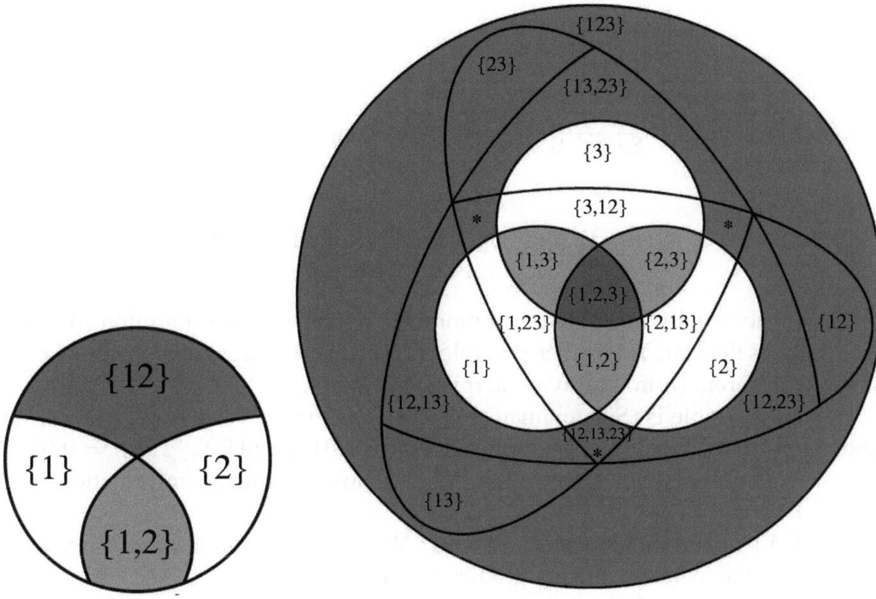

(a) WMS $(\{X_1, X_2\} : Y)$ (b) WMS $(\{X_1, X_2, X_3\} : Y)$

Fig. 6.8 PI-diagrams illustrating WholeMinusSum synergy for $n = 2$ (left) and $n = 3$ (right). For this diagram the colors denote the added and subtracted PI-regions. WMS $(\mathbf{X} : Y)$ is the green PI-region(s), minus the orange PI-region(s), minus two times any red PI-region.

X_1	X_2	Y	
r0	r0	r0	1/8
r0	r1	r1	1/8
r1	r0	r1	1/8
r1	r1	r0	1/8
R0	R0	R0	1/8
R0	R1	R1	1/8
R1	R0	R1	1/8
R1	R1	R0	1/8

(a) $\Pr(x_1, x_2, y)$

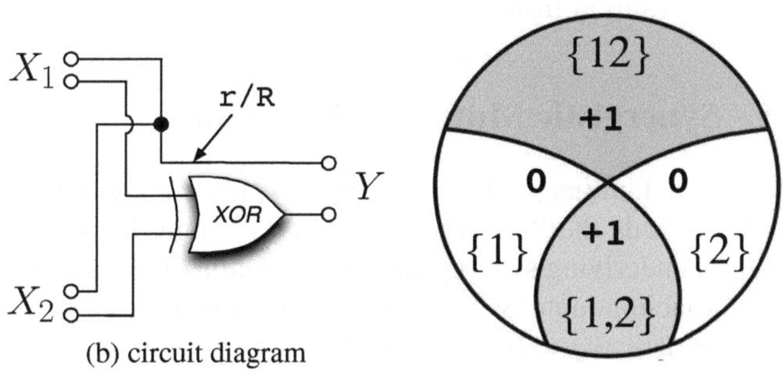

(b) circuit diagram

(c) PI-diagram

Fig. 6.9 Example RDNXOR has one bit of redundancy and one bit of synergy. Yet for this example, $\text{WMS}(\mathbf{X} : Y) = 0$ bits.

$$\Delta\,\mathrm{I}(\mathbf{X};Y) \equiv \mathrm{D_{KL}}[\Pr(Y|X_{1\ldots n})\,||\,\Pr_{\mathrm{ind}}(Y|\mathbf{X})] \quad (6.10)$$

$$= \sum_{y,\mathbf{x}\in Y,\mathbf{X}} \Pr(y,x_{1\ldots n}) \log \frac{\Pr(y|x_{1\ldots n})}{\Pr_{\mathrm{ind}}(y|\mathbf{x})}, \quad (6.11)$$

where $\Pr_{\mathrm{ind}}(y|\mathbf{x}) \equiv \frac{\Pr(y)\prod_{i=1}^{n}\Pr(x_i|y)}{\sum_{y'}\Pr(y')\prod_{i=1}^{n}\Pr(x_i|y')}$. After some algebra[7] eq. (6.11) becomes,

$$\Delta\,\mathrm{I}(\mathbf{X};Y) = \mathrm{TC}(X_1;\cdots;X_n|Y) - \mathrm{D_{KL}}\left[\Pr(X_{1\ldots n}) \middle\| \sum_y \Pr(y)\prod_{i=1}^{n}\Pr(X_i|y)\right] . \quad (6.12)$$

$\Delta\,\mathrm{I}$ is conceptually innovative and moreover agrees with our intuition for all of our examples thus far. Yet further examples reveal that $\Delta\,\mathrm{I}$ measures something ever-so-subtly different from intuitive synergistic information.

The first example is (Schneidman et al. 2003a)'s Figure 4 where $\Delta\,\mathrm{I}$ exceeds the mutual information $\mathrm{I}(X_{1\ldots n}:Y)$ with $\Delta\,\mathrm{I}(\mathbf{X};Y) = 0.0145$ and $\mathrm{I}(X_{1\ldots n}:Y) = 0.0140$. This fact alone prevents interpreting $\Delta\,\mathrm{I}$ as a loss of mutual information from $\mathrm{I}(X_{1\ldots n}:Y)$.[8]

Could $\Delta\,\mathrm{I}$ upperbound synergy instead? We turn to example AND (Fig. 6.10) with $n = 2$ independent binary predictors and target Y is the AND of X_1 and X_2. Although AND's PI-region exact decomposition remains uncertain, we can still bound the synergy. For example AND, the $\mathrm{WMS}(\{X_1,X_2\} : Y) \approx 0.189$ and $\mathcal{S}_{\max}(\{X_1,X_2\} : Y) = 0.5$ bits. So we know the synergy must be between $(0.189, 0.5]$ bits. Despite this, $\Delta\,\mathrm{I}(\mathbf{X};Y) = 0.104$ bits, thus $\Delta\,\mathrm{I}$ does not upperbound synergy.

Finally, in the face of duplicate predictors $\Delta\,\mathrm{I}$ often *decreases*. From example AND to ANDDUPLICATE (Appendix A, Fig. 6.13) $\Delta\,\mathrm{I}$ drops 63% to 0.038 bits.

Taking all three examples together, we conclude $\Delta\,\mathrm{I}$ measures something fundamentally different from synergistic information.

6.5 Synergistic Mutual Information

We are all familiar with the English expression describing synergy as when the whole exceeds the "sum of its parts". Although this informal adage captures the intuition underlying synergy, the formalization of this adage, WholeMinusSum synergy, "double-counts" whenever there is duplication (redundancy) among the parts. A mathematically correct adage should change "sum" to "union"—meaning synergy occurs when the whole exceeds the *union* of its parts. The sum adds duplicate information multiple times, whereas the union adds duplicate information only once. The union of parts never exceeds the sum.

[7] See Appendix F for the steps between eqs. (6.11) and (6.12).

[8] Although $\Delta\,\mathrm{I}$ can not be a loss of mutual information, it could still be a loss of some alternative information such as Wyner's common information (Lei et al. 2010).

X_1	X_2	Y	
0	0	0	1/4
0	1	0	1/4
1	0	0	1/4
1	1	1	1/4

(a) $\Pr(x_1, x_2, y)$

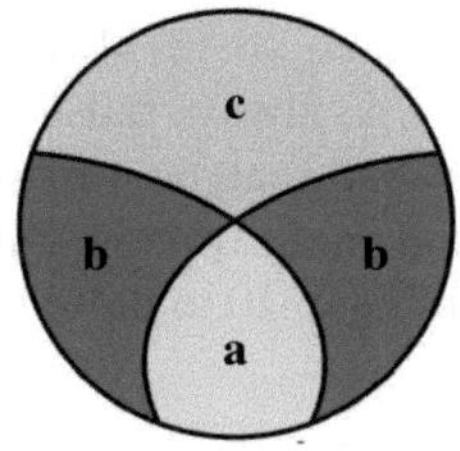

(b) PI-diagram

$$0.189 \le c \le 0.5$$
$$0 \le b \le 0.311$$
$$0 \le a \le 0.311$$

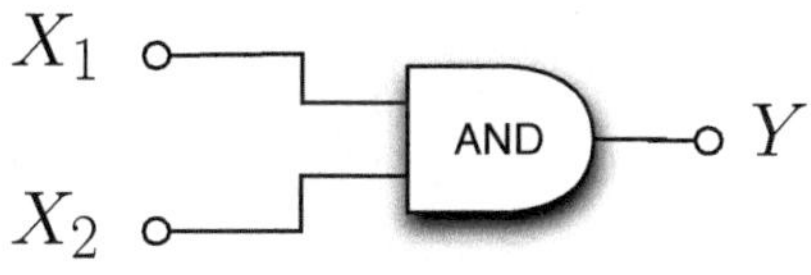

(c) circuit diagram

Fig. 6.10 Example AND. The exact PI-decomposition of an AND-gate remains uncertain. But we can bound a, b, and c using WMS and $\mathcal{S}_{\max}$. In section 6.5 these bounds will be tightened. Most intriguingly, we'll show that $a > 0$ despite $\mathrm{I}(X_1 : X_2) = 0$.

The guiding intuition of "whole minus union" leads us to a novel measure denoted $\mathcal{S}_{\mathrm{VK}}(\{X_1, \ldots, X_n\} : Y)$, or $\mathcal{S}_{\mathrm{VK}}(\mathbf{X} : Y)$, as the mutual information in the whole beyond the union of elements $\{X_1, \ldots, X_n\}$.

Unfortunately, there's no established measure of "union-information" in contemporary information theory. We introduce a novel technique, inspired by Maurer and Wolf (1999), for defining the union information among n predictors. We numerically compute the union information by noisifying the joint distribution $\Pr(X_{1\ldots n}, Y)$ such that only the correlations with singleton predictors are preserved. This is achieved like so,

$$\mathrm{I}_{\mathrm{VK}}(\{X_1, \ldots, X_n\} : Y) \equiv \min_{\Pr^*(X_1, \ldots, X_n, Y)} \mathrm{I}^*(X_{1\ldots n} : Y) \quad \text{subject to: } \Pr^*(X_i, Y) = \Pr(X_i, Y) \ \forall i, \tag{6.13}$$

where $\mathrm{I}^*(X_{1\ldots n} : Y) \equiv \mathrm{D}_{\mathrm{KL}}[\Pr^*(X_{1\ldots n}, Y) \| \Pr^*(X_{1\ldots n}) \Pr^*(Y)]$.

Without any constraint on the distribution $\Pr^*(X_1, \ldots, X_n, Y)$, the minimum of eq. (6.13) is trivially found to be zero bits because simply setting $\Pr^*(X_{1\ldots n})$ to a constant makes $\mathrm{I}^*(X_{1\ldots n} : Y) = 0$ bits. Therefore we must put some constraint on $\Pr^*(X_1, \ldots, X_n, Y)$. As all bits a singleton X_i knows about Y are determined by the joint distribution $\Pr(X_i, Y)$, we simply prevent the minimization from altering these distributions, and presto we arrive at the constraint $\Pr^*(X_i, Y) = \Pr(X_i, Y) \ \forall i$.[9]

[9] We could have instead chosen the *looser* constraint $\mathrm{I}^*(X_i : Y) = \mathrm{I}(X_i : Y) \ \forall i$, but $\Pr^*(X_i, Y) = \Pr(X_i, Y) \ \forall i$ ensures we preserve the "same bits", not just the same magnitude of bits.

Finally, we prove that a minimum of eq. (6.13) always exists because setting $\Pr^*(x_1,\ldots,x_n,y) = \Pr(y)\prod_{i=1}^{n}\Pr(x_i|y)$ always satisfies the constraints.

Unfortunately, we currently have no analytic way to calculate eq. (6.13), however, we do have an analytic upperbound on it. Applying this to AND's PI-decomposition allows us to tighten the bounds in Fig. 6.10 to those in Fig. 6.11.

X_1	X_2	Y	
0	0	0	1/3
0	1	0	1/6
1	0	0	1/6
1	1	0	1/12
1	1	1	1/4

(a) $\Pr^*(x_1,x_2,y)$

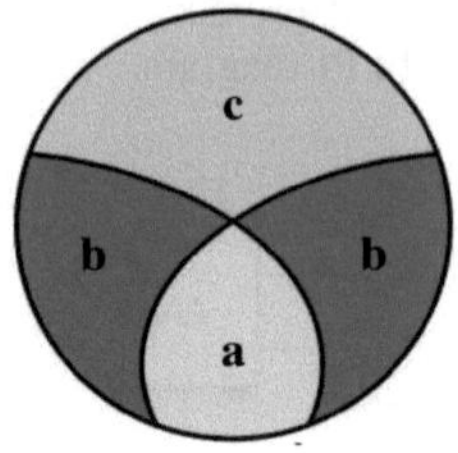

$0.270 \leq c \leq 0.500$

$0 \leq b \leq 0.230$

$0.082 \leq a \leq 0.311$

(b) PI-diagram

Fig. 6.11 Revisiting example AND. Using the analytic upperbound on I_{VK} in Appendix D, we arrive at the $\Pr^*$ distribution in (a). Using this distribution, we tighten the bounds on a, b, and c. Intriguingly, we see that despite $\mathrm{I}(X_1 : X_2) = 0$, that $a > 0$. **Note**: Previous versions (preprints) of this paper *erroneously asserted* independent predictors could not convey redundant information, i.e. that $\mathrm{I}(X_1 : X_2) = 0$ entailed $\mathrm{I}_\cap(\{X_1, X_2\} : Y) = 0$.

Our union-information measure I_{VK} satisfies several desired properties given in (Bertschinger et al. 2012; Harder et al. 2013). Specifically, I_{VK} satisfies: **(GP)**, **(M)**, **(SR)**, $(\mathbf{S_0})$, **(TM)**, $(\mathbf{Id_1})$, and $(\mathbf{LP_0})$. For details see Section 6.6 and Appendix C.

Once the union information is computed, the $\mathcal{S}_{\mathrm{VK}}$ synergy is simply,

$$\mathcal{S}_{\mathrm{VK}}(\{X_1,\ldots,X_n\} : Y) \equiv \mathrm{I}(X_{1\ldots n} : Y) - \mathrm{I}_{\mathrm{VK}}(\{X_1,\ldots,X_n\} \ : \ Y) \ . \tag{6.14}$$

$\mathcal{S}_{\mathrm{VK}}$ synergy quantifies the total "informational work" strictly the coalitions within $X_{1\ldots n}$ perform in reducing the uncertainty of Y. Pleasingly, $\mathcal{S}_{\mathrm{VK}}$ is bounded[10] by the WholeMinusSum synergy (which underestimates the intuitive synergy) and $\mathcal{S}_{\max}$ (which overestimates intuitive synergy),

$$\max\left[0, \mathrm{WMS}\left(\mathbf{X} : Y\right)\right] \leq \mathcal{S}_{\mathrm{VK}}(\mathbf{X} : Y) \leq \mathcal{S}_{\max}\left(\mathbf{X} : Y\right) \leq \mathrm{I}(X_{1\ldots n} : Y) \ . \tag{6.15}$$

6.6 Properties of I_{VK}

Our measure of the union information I_{VK} satisfies a number of properties from the prior literature (for proofs, see Appendix C):

(GP) Global Positivity. $\mathrm{I}_{\mathrm{VK}}(\mathbf{X} : Y) \geq 0$

(SR) Self-Redundancy. The redundant information a single predictor X_1 has about the target Y is equal to the Shannon mutual information between the predictor and the target, i.e. $\mathrm{I}_{\mathrm{VK}}(X_1 : Y) = \mathrm{I}(X_1 : Y)$.

[10] Proven in Appendix E.3.

(S_0) Weak Symmetry. $\mathrm{I}_{\mathrm{VK}}(X_1,\ldots,X_n:Y)$ is invariant under reordering $X_1,\ldots,X_n$.

(M) Monotonicity. $\mathrm{I}_{\mathrm{VK}}(X_1,\ldots,X_n:Y) \leq \mathrm{I}_{\mathrm{VK}}(X_1,\ldots,X_n,W:Y)$ with equality if W is a "subset" (or equivalent to) an $X_i \in \{X_1,\ldots,X_n\}$. W is a subset of a random variable X_i if and only if there exists a function f such that $W = f(X_i)$.

(TM) Target Monotonicity. For all random variables Y and Z, $\mathrm{I}_{\mathrm{VK}}(\mathbf{X}:Y) \leq \mathrm{I}_{\mathrm{VK}}(\mathbf{X}:YZ)$.

(LP_0) Weak Local Positivity. For $n = 2$ predictors, the derived "partial informations" (Williams and Beer 2010) are nonnegative. This is equivalent to,

$$\max\left[\mathrm{I}(X_1:Y),\mathrm{I}(X_2:Y)\right] \leq \mathrm{I}_{\mathrm{VK}}(X_1,X_2:Y) \leq \mathrm{I}(X_1X_2:Y)\ .$$

(Id_1) Strong Identity. $\mathrm{I}_{\mathrm{VK}}(X_1\ldots,X_n:X_{1\ldots n}) = \mathrm{H}(X_{1\ldots n})$.

6.7 Applying the Measures to Our Examples

Table 6.1 summarizes the results of all four measures applied to our examples.

RDN (Fig. 6.3). There is exactly one bit of redundant information and all measures reach their intended answer. For the axiomatically minded, the equality condition of (**M**) is sufficient for the desired answer.

UNQ (Fig. 6.4). $\mathcal{S}_{\max}$'s miscategorization of unique information as synergistic reveals itself. Intuitively, there are two bits of unique information and no synergy. However, $\mathcal{S}_{\max}$ reports one bit of synergistic information. For the axiomatically minded, property (**Id**) is sufficient (but not nessecary) for the desired answer.

XOR (Fig. 6.5). There is exactly one bit of synergistic information. All measures reach the desired answer of 1 bit.

XORDUPLICATE (Fig. 6.6). Target Y is specified by the coalition X_1X_2 as well as by the coalition X_3X_2, thus $\mathrm{I}(X_1X_2:Y) = \mathrm{I}(X_3X_2:Y) = \mathrm{H}(Y) = 1$ bit. All measures reach the expected answer of 1 bit.

XORLOSES (Fig. 6.7). Target Y is specified by the coalition X_1X_2 as well as by the singleton X_3, thus $\mathrm{I}(X_1X_2:Y) = \mathrm{I}(X_3:Y) = \mathrm{H}(Y) = 1$ bit. Together this means there is one bit of redundancy between the coalition X_1X_2 and the singleton X_3 as illustrated by the $+1$ in PI-region $\{3,12\}$. All measures account for this redundancy and reach the desired answer of 0 bits.

RDNXOR (Fig. 6.9). This example has one bit of synergy as well as one bit of redundancy. In accordance with Fig. 6.8a, WholeMinusSum measures *synergy minus redundancy* to calculate $1-1=0$ bits. On the other hand, $\mathcal{S}_{\max}$, $\Delta\,\mathrm{I}$, and $\mathcal{S}_{\mathrm{VK}}$ are not mislead by the co-existance of synergy and redundancy and correctly report 1 bit of synergistic information.

AND (Fig. 6.10). This example is a simple case where correlational importance, $\Delta\,\mathrm{I}(\mathbf{X};Y)$, disagrees with the intuitive value for synergy. The WholeMinusSum synergy—an unambiguous *lowerbound* on the intuitive synergy—is 0.189 bits, yet $\Delta\,\mathrm{I}(\mathbf{X};Y) = 0.104$ bits. We can't perfectly determine $\mathcal{S}_{\mathrm{VK}}$, but we can lowerbound $\mathcal{S}_{\mathrm{VK}}$ using our analytic bound, as well as upperbound it using $\mathcal{S}_{\max}$. This gives $0.270 \leq \mathcal{S}_{\mathrm{VK}} \leq 1/2$.

Table 6.1 Synergy measures for our examples. Answers conflicting with intuitive synergistic information are in red. The $\mathcal{S}_{VK}$ value for AND and ANDDUPLICATE is not conclusively known, but can be bounded.

Example	$\mathcal{S}_{max}$	WMS	ΔI	$\mathcal{S}_{VK}$
RDN	0	−1	0	0
UNQ	1	0	0	0
XOR	1	1	1	1
XORDUPLICATE	1	1	1	1
XORLOSES	0	0	0	0
RDNXOR	1	0	1	1
AND	1/2	0.189	0.104	[0.270,1/2]
RDNUNQXOR	2	0	1	1
ANDDUPLICATE	1/2	−0.123	0.038	[0.270,1/2]
XORMULTICOAL	1	1	1	1

The three supplementary examples in Appendix A: RDNUNQXOR, ANDDUPLICATE, and XORMULTICOAL aren't essential for understanding this paper and are for the intellectual pleasure of advanced readers.

Table 6.1 shows that no prior measure of synergy consistently matches intuition even for $n = 2$. To summarize,

1. I_{max} synergy, $\mathcal{S}_{max}$, overestimates the intuitive synergy when two or more predictors convey unique information about the target (e.g. UNQ).
2. WholeMinusSum synergy, WMS, inadvertently double-subtracts redundancies and thus underestimates the intuitive synergy (e.g. RDNXOR). Duplicating predictors often decreases WholeMinusSum synergy (e.g. ANDDUPLICATE).
3. Correlational importance, ΔI, is not bounded by the Shannon mutual information, underestimates the known lowerbound on synergy (e.g. AND), and duplicating predictors often decreases correlational importance (e.g. ANDDUPLICATE). Altogether, ΔI does not quantify the intuitive synergistic information (nor was it intended to).

6.8 Conclusion

Fundamentally, we assert that synergy quantifies how much the whole exceeds the *union* of its parts. Considering synergy as the whole minus the *sum* of its parts inadvertently "double-subtracts" redundancies, thus *underestimating* synergy. Within information theory, PI-diagrams, a generalization of Venn diagrams, are immensely helpful in improving one's intuition for synergy.

We demonstrated with RDNXOR and RDNUNQXOR that a single state can simultaneously carry redundant, unique, and synergistic information. This fact is under-appreciated, and prior work often implicitly assumed these three types of information could not coexist in a single state.

We introduced a novel measure of synergy, $\mathcal{S}_{\text{VK}}$, (eq. (6.14)). Unfortunately our expression is not easily computable, and until we have an explicit analytic solution to the minimization in I_{VK} the best one can do is numerical optimization using our analytic upperbound (Appendix D) as a starting point.

Along with our examples, we consider our introduction of a candidate for the union information, I_{VK} (eq. (6.13)) and its upperbound our primary contributions to the literature.

Finally, by means of our analytic upperbound on I_{VK} we've shown that, at least for our measure, *independent predictors can convey redundant information about a target*, e.g. Fig. 6.11.

Acknowledgements. We thank Suzannah Fraker, Tracey Ho, Artemy Kolchinsky, Chris Adami, Giulio Tononi, Jim Beck, Nihat Ay, and Paul Williams for extensive discussions. This research was funded by the Paul G. Allen Family Foundation and a DOE CSGF fellowship to VG.

Appendix

A Three Extra Examples

For the reader's intellectual pleasure, we include three more sophisticated examples: RDNUNQXOR, ANDDUPLICATE, and XORMULTICOAL.

Example AndDuplicate

ANDDUPLICATE adds a duplicate predictor to example AND to show how $\Delta\,\text{I}$ responds to a duplicate predictor in a less pristine example than XOR. Unlike XOR, in example AND there's also unique and redundant information. Will this cause the loss of synergy in the spirit of XORLOSES? Taking each one at a time:

- Predictor X_2 is unaltered from example AND. Thus X_2's unique information stays the same. AND's $\{2\} \to$ ANDDUPLICATE's $\{2\}$.
- Predictor X_3 is identical to X_1. Thus all of X_1's unique information in AND becomes redundant information between predictors X_1 and X_3. AND's $\{1\} \to$ ANDDUPLICATE's $\{1,3\}$.
- In AND there is synergy between X_1 and X_2, and this synergy is still present in ANDDUPLICATE. Just as in XORDUPLICATE, the only difference is that now an identical synergy also exists between X_3 and X_2. Thus AND's $\{12\} \to$ ANDDUPLICATE's $\{12,23\}$.

X_1	X_2	Y	
ra0	rb0	rab0	1/32
ra0	rb1	rab1	1/32
ra1	rb0	rab1	1/32
ra1	rb1	rab0	1/32
ra0	rB0	raB0	1/32
ra0	rB1	raB1	1/32
ra1	rB0	raB1	1/32
ra1	rB1	raB0	1/32
rA0	rb0	rAb0	1/32
rA0	rb1	rAb1	1/32
rA1	rb0	rAb1	1/32
rA1	rb1	rAb0	1/32
rA0	rB0	rAB0	1/32
rA0	rB1	rAB1	1/32
rA1	rB0	rAB1	1/32
rA1	rB1	rAB0	1/32

X_1	X_2	Y	
Ra0	Rb0	Rab0	1/32
Ra0	Rb1	Rab1	1/32
Ra1	Rb0	Rab1	1/32
Ra1	Rb1	Rab0	1/32
Ra0	RB0	RaB0	1/32
Ra0	RB1	RaB1	1/32
Ra1	RB0	RaB1	1/32
Ra1	RB1	RaB0	1/32
RA0	Rb0	RAb0	1/32
RA0	Rb1	RAb1	1/32
RA1	Rb0	RAb1	1/32
RA1	Rb1	RAb0	1/32
RA0	RB0	RAB0	1/32
RA0	RB1	RAB1	1/32
RA1	RB0	RAB1	1/32
RA1	RB1	RAB0	1/32

Fig. 6.11 (a) $\Pr(x_1, x_2, y)$

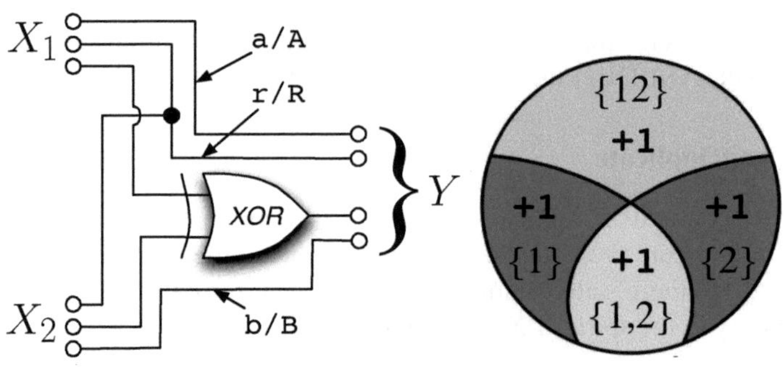

circuit diagram (c) PI-diagram

Fig. 6.12 Example RDNUNQXOR weaves examples RDN, UNQ, and XOR into one. $\mathrm{I}(X_1X_2 : Y) = \mathrm{H}(Y) = 4$ bits. This example is pleasing because it puts exactly one bit in each PI-region.

- Predictor X_3 is identical to X_1. Therefore any information in AND that is specified by both X_1 and X_2 is now specified by X_1, X_2, and X_3. Thus AND's $\{1,2\} \rightarrow$ ANDDUPLICATE's $\{1,2,3\}$.

X_1	X_2	X_3	Y	
0	0	0	0	1/4
0	1	0	0	1/4
1	0	1	0	1/4
1	1	1	1	1/4

(a) $\Pr(x_1, x_2, x_3, y)$

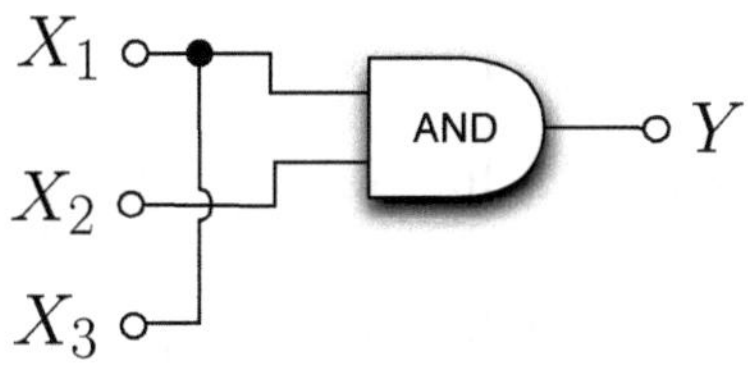

(b) circuit diagram

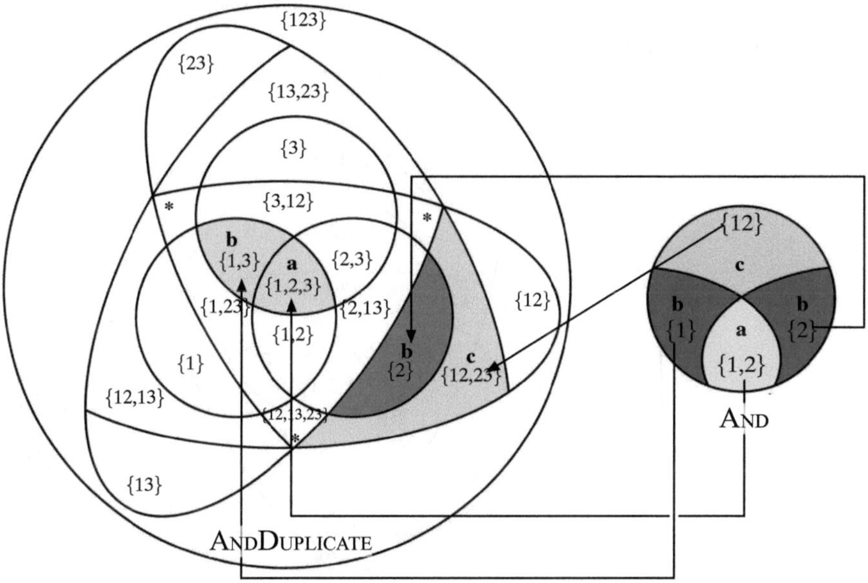

(c) PI-diagram

Fig. 6.13 Example AndDuplicate. The total mutual information is the same as in And, $\mathrm{I}(X_1X_2 : Y) = \mathrm{I}(X_1X_2X_3 : Y) = 0.811$ bits. Every PI-region in example And maps to a PI-region in AndDuplicate. The intuitive synergistic information is unchanged from And. However, correlational importance, $\Delta\,\mathrm{I}$, arrives at 0.104 bits of synergy for And, and 0.038 bits for AndDuplicate. $\Delta\,\mathrm{I}$ is not invariant to duplicate predictors.

X_1	X_2	X_3	Y	
ab	ac	bc	0	1/8
AB	Ac	Bc	0	1/8
Ab	AC	bC	0	1/8
aB	aC	BC	0	1/8
Ab	Ac	bc	1	1/8
aB	ac	Bc	1	1/8
ab	aC	bC	1	1/8
AB	AC	BC	1	1/8

(a) $\Pr(x_1, x_2, x_3, y)$

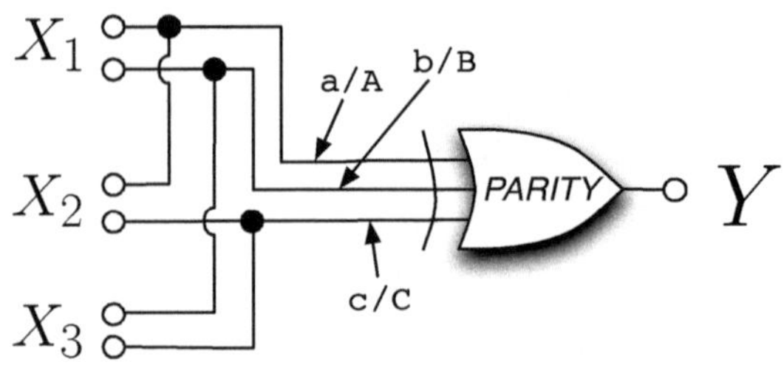

(b) circuit diagram

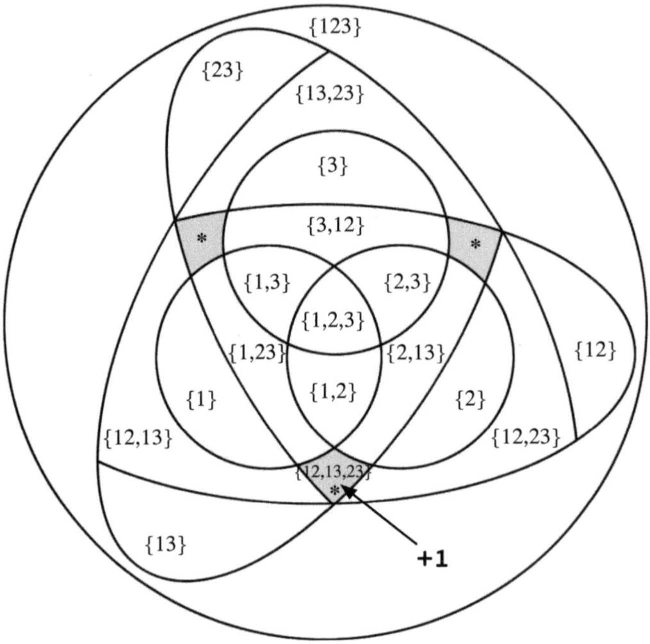

(c) PI-diagram

Fig. 6.14 Example XORMULTICOAL demonstrates how the same information can be specified by multiple coalitions. In XORMULTICOAL the target Y has one bit of uncertainty, $\mathrm{H}(Y) = 1$ bit, and Y is the *parity* of three incoming wires. Just as the output of XOR is specified only after knowing the state of both inputs, the output of XORMULTICOAL is specified only after knowing the state of all three wires. Each predictor is distinct and has access to two of the three incoming wires. For example, predictor X_1 has access to the a/A and b/B wires, X_2 has access to the a/A and c/C wires, and X_3 has access to the b/B and c/C wires. Although no single predictor specifies Y, any coalition of two predictors has access to all three wires and fully specifies Y, $\mathrm{I}(X_1X_2 : Y) = \mathrm{I}(X_1X_3 : Y) = \mathrm{I}(X_2X_3 : Y) = \mathrm{H}(Y) = 1$ bit. In the PI-diagram this puts one bit in PI-region $\{12, 13, 23\}$ and zero everywhere else. All measures reach the expected answer of 1 bit of synergy.

B Connecting Back to $\mathrm{I}_\cap$

Our candidate measure of the union information, I_{VK}, gives rise to a measure of the intersection information denoted $\mathrm{I}_{\mathrm{VK}}^{dual}$. This is done by,

$$\mathrm{I}_{\mathrm{VK}}^{dual}(\mathbf{X}:Y) = \sum_{\mathbf{S}\subseteq\mathbf{X}} (-1)^{|\mathbf{S}|+1}\, \mathrm{I}_{\mathrm{VK}}(\mathbf{S}:Y) \ . \tag{6.16}$$

C Desired Properties of $\mathrm{I}_\cup$

What previously proposed properties does $\mathrm{I}_{\mathrm{VK}}^{dual}$ satisfy? We originally worked on proofs for which properties $\mathrm{I}_{\mathrm{VK}}^{dual}$ satisfies, but for $n > 2$ we were blocked by not having an analytic solution to I_{VK}. So we instead translated the $\mathrm{I}_\cap$ properties into the analogous $\mathrm{I}_\cup$ properties. Although one can't prove the $\mathrm{I}_\cap$ version from the analogous $\mathrm{I}_\cup$ property, it is a start.

In addition to the properties in Section 6.6, we have the properties,

- **($\mathbf{S_1}$)** Strong Symmetry. $\mathrm{I}_\cup(\{X_1,\ldots,X_n\}:Y)$ is invariant under reordering $X_1,\ldots,X_n,Y$.
- **(TC)** Target Chainrule. $\mathrm{I}_\cup(\mathbf{X}:YZ) = \mathrm{I}_\cup(\mathbf{X}:Y) + \mathrm{I}_\cup(\mathbf{X}:Z|Y)$.
- **(UB)** Upperbound. From applying inclusion/exclusion rule to $\mathrm{I}_\cap(\{QX_1,\ldots,QX_n\}:Y) \geq \mathrm{I}(Q:Y)$, we have the following upperbound on the union information,

$$\mathrm{I}_\cup(\{X_1,\ldots,X_n\}:Y) \leq (1-n)\,\mathrm{I}(X_1\wedge\cdots\wedge X_n:Y) + \sum_{i=1}^{n} \mathrm{I}(X_i:Y) \ .$$

- **($\mathbf{LP_1}$)** Strong Local Positivity. For all n, the derived "partial informations" (Williams and Beer 2010) are nonnegative.

We've proven that I_{VK} *does not satisfy* ($\mathbf{S_1}$). And thus far we've been unable to determine whether I_{VK} satisfies **(TC)** and ($\mathbf{LP_1}$).

Proof of (GP)

Proven by the nonnegativity of mutual information.

Proof of (SR)

$$\begin{aligned}\mathrm{I}_{\mathrm{VK}}(X_1:Y) &\equiv \min_{\substack{p^*(x_1,y)\\ p^*(x_1,y)=p(x_1,y)}} \mathrm{I}^*(X_1:Y)\\ &= \mathrm{I}(X_1:Y) \ .\end{aligned}$$

Proof of $(\mathbf{S_0})$

There's only one instance of the terms in $\mathbf{X}$ in the definition of I_{VK}, which is,

$$\mathrm{I}_{\mathrm{VK}}(\mathbf{X}:Y) \equiv \min_{\substack{p^*(X_1,\ldots,X_n,Y) \\ p^*(X_i,Y)=p(X_i,Y) \quad \forall i}} \mathrm{I}^*(X_1\cdots X_n:Y) \ .$$

The term $\mathrm{I}^*(X_1\cdots X_n:Y)$ is invariant to the ordering of $X_1\cdots X_n$. This is due to $\Pr^*(x_1,\ldots,x_n) = \Pr^*(x_n,\ldots,x_1)$. Thus I_{VK} is invariant to the ordering of $\{X_1,\ldots,X_n\}$.

Proof of (M)

We prove the inequality condition of **(M)**, that $\mathrm{I}_{\mathrm{VK}}(X_1,\ldots,X_n:Y) \leq \mathrm{I}_{\mathrm{VK}}(X_1,\ldots,X_n,W:Y)$.

$$\begin{aligned}
\mathrm{I}_{\mathrm{VK}}(X_1,\ldots,X_n:Y) &\equiv \min_{\substack{p^*(X_1,\ldots,X_n,Y) \\ p^*(X_i,Y)=p(X_i,Y) \quad \forall i}} \mathrm{I}^*(X_{1\ldots n}:Y) \\
&= \min_{\substack{p^*(x_1,\ldots,x_n,w,y) \\ p^*(x_i,y)=p(x_i,y)\ \forall i \\ p^*(w,y)=p(w,y)}} \mathrm{I}^*(X_{1\ldots n}:Y) \\
&\leq \min_{\substack{p^*(x_1,\ldots,x_n,w,y) \\ p^*(x_i,y)=p(x_i,y)\ \forall i \\ p^*(w,y)=p(w,y)}} \mathrm{I}^*(X_{1\ldots n}:Y) + \mathrm{I}^*(W:Y|X_{1\ldots n}) \\
&= \min_{\substack{p^*(x_1,\ldots,x_n,w,y) \\ p^*(x_i,y)=p(x_i,y)\ \forall i \\ p^*(w,y)=p(w,y)}} \mathrm{I}^*(X_{1\ldots n}W:Y) \\
&= \mathrm{I}_{\mathrm{VK}}(\{X_1,\ldots,X_n,W\}:Y) \ .
\end{aligned}$$

We prove the equality condition of **(M)**, that,

$$\mathrm{I}_{\mathrm{VK}}(\{X_1,\ldots,X_n,W\} : Y) = \mathrm{I}_{\mathrm{VK}}(\{X_1,\ldots,X_n\} : Y)$$
$$\text{where } \exists i \text{ s.t. } W = f(X_i), \text{ for some function } f \ .$$

Without loss of generality we reorder the predictors so that the X_i above is the last predictor, X_n.

Proof

$$\begin{aligned}
\mathrm{I}_{\mathrm{VK}}(\{X_1,\ldots,X_n,W\} : Y) &= \min_{\substack{p^*(X_1,\ldots,X_n,Y) \\ p^*(X_i,Y)=p(X_i,Y) \quad \forall i \\ p^*(W,Y)=p^*(W,Y)}} \mathrm{I}^*(X_{1\ldots n-1}X_nW : Y) \\
&= \min_{\substack{p^*(X_1,\ldots,X_n,Y) \\ p^*(X_i,Y)=p(X_i,Y) \quad \forall i \\ p^*(W,Y)=p^*(W,Y)}} \mathrm{I}^*(X_{1\ldots n-1}X_n : Y) \ .
\end{aligned}$$

Then, because the constraint $p^*(X_n, Y) = p(X_n, Y)$ wholly encapsulates the constraint $p^*(W, Y) = p(W, Y)$, we can remove the constraint $p^*(W, Y) = p(W, Y)$. This yields,

$$\begin{aligned} \mathrm{I}_{\mathrm{VK}}(\{X_1, \ldots, X_n, W\} : Y) &= \min_{\substack{p^*(X_1,\ldots,X_n,Y) \\ p^*(X_i,Y)=p(X_i,Y) \quad \forall i}} \mathrm{I}^*(X_{1\ldots n-1}X_n : Y) \\ &= \min_{\substack{p^*(X_1,\ldots,X_n,Y) \\ p^*(X_i,Y)=p(X_i,Y) \quad \forall i}} \mathrm{I}^*(X_{1\ldots n} : Y) \\ &= \mathrm{I}_{\mathrm{VK}}(\{X_1, \ldots, X_n\} : Y) . \end{aligned}$$

Proof of (TM)

For notational brevity, we define the following terms,

$$\begin{aligned} \alpha &\equiv \min_{\substack{p^*(x_1,\ldots,x_n,y) \\ p^*(x_i,y)=p(x_i,y) \ \forall i}} \mathrm{I}^*(X_{1\ldots n} : Y) \\ \beta &\equiv \min_{\substack{p^*(x_1,\ldots,x_n,yz) \\ p^*(x_i,yz)=p(x_i,yz) \ \forall i}} \mathrm{I}^*(X_{1\ldots n} : YZ) \\ \gamma &\equiv \min_{\substack{p^*(x_1,\ldots,x_n,y,z) \\ p^*(x_i,y,z)=p(x_i,y,z) \ \forall i}} \mathrm{I}^*(X_{1\ldots n} : Y) \\ \delta &\equiv \min_{\substack{p^*(x_1,\ldots,x_n,yz) \\ p^*(x_i,yz)=p(x_i,yz) \ \forall i}} \mathrm{I}^*(X_{1\ldots n} : Z|Y) . \end{aligned}$$

The proof of **(TM)** is complete by showing $\alpha \leq \beta$. First because no term in γ depends on Z, we can drop γ's constraints on Z leaving $\alpha = \gamma$. Then, by the nonnegativity of mutual information, we know $\alpha, \beta, \gamma, \delta \geq 0$. So thus far we have $\alpha \leq \gamma + \delta$. Next we can prove $\gamma + \delta \leq \beta$ because the sum of two minimums, $\gamma + \delta$, is less than the same minimum over the sum, β.

Taken together,

$$\alpha \leq \gamma + \delta \leq \beta ,$$

and the proof is complete.

Proof of $(\mathbf{LP_0})$

$$\mathrm{I}_{\mathrm{VK}}(\mathbf{X} : Y) \leq \mathrm{I}(X_{1\ldots n} : Y) .$$

This is proven by the condition that $\Pr(X_1, \ldots, X_n, Y)$ satisfies the constraints on the minimizing distribution in I_{VK}. Thus $\mathrm{I}^*(X_{1\ldots n} : Y) \leq \mathrm{I}(X_{1\ldots n} : Y)$.

Disproof of $(\mathbf{S_1})$

We show that, $\mathrm{I}_{\mathrm{VK}}(\{X,Y\}:Z) \neq \mathrm{I}_{\mathrm{VK}}(\{X,Z\}:Y)$ by setting $X = Y$ where $\mathrm{H}(X) > 0$, and Z is a constant, $\mathrm{I}_{\mathrm{VK}}(\{X,Y\}:Z) = 0$ yet $\mathrm{I}_{\mathrm{VK}}(\{X,Z\}:Y) = \mathrm{H}(X)$.

Proof of $(\mathbf{Id_1})$

$$\mathrm{I}_{\mathrm{VK}}(\mathbf{X} : X_{1\ldots n}) \equiv \min_{\substack{p^*(X_1,\ldots,X_n,X_{1\ldots n}) \\ p^*(X_i,X_{1\ldots n})=p(X_i,X_{1\ldots n})\ \forall i}} \mathrm{I}^*(X_{1\ldots n} : X_{1\ldots n}) \tag{6.17}$$

$$= \min_{\substack{p^*(X_1,\ldots,X_n,X_{1\ldots n}) \\ p^*(X_i,X_{1\ldots n})=p(X_i,X_{1\ldots n})\ \forall i}} \mathrm{H}^*(X_{1\ldots n}) , \tag{6.18}$$

Then because $p^*(X_{1\ldots n}) = p(X_{1\ldots n})$,

$$\mathrm{I}_{\mathrm{VK}}(\mathbf{X} : X_{1\ldots n}) = \mathrm{H}(X_{1\ldots n}) . \tag{6.19}$$

D Analytic Upperbound on $\mathrm{I}_{\mathrm{VK}}(\mathbf{X} : Y)$

Our analytic upperbound on I_{VK} starts with the n joint distributions we wish to preserve: $\Pr(X_1,Y),\ldots,\Pr(X_n,Y)$. From one these joint distributions, e.g. $\Pr(X_1,Y)$, we compute the marginal probability distribution $\Pr(Y)$ by summing over the index of $x_1 \in X_1$,

$$\Pr(Y) = \left\{ \sum_{x_1 \in X_1} \Pr(x_1,y) : \forall y \in Y \right\} . \tag{6.20}$$

Then, for every state $y \in Y$ we compute n conditional distributions $\Pr(X_1|y),\ldots,\Pr(X_n|y)$ via,

$$\Pr(X_i|Y=y) = \left\{ \frac{\Pr(x_i,y)}{\Pr(y)} : \forall x_i \in X_i \right\} . \tag{6.21}$$

With the marginal distribution $\Pr(Y)$ and the $|Y| \cdot n$ conditonal distributions, we construct a novel, artificial joint distribution $\Pr^*(X_1,\ldots,X_n,Y)$ defined by,

$$\Pr^*(x_1,\ldots,x_n,y) \equiv \Pr(y) \textstyle\prod_{i=1}^n \Pr(x_i|y) . \tag{6.22}$$

This novel, artificial joint distribution $\Pr^*(X_1,\ldots,X_n,Y)$ satisfies the constraints $\Pr^*(X_i,Y) = \Pr(X_i,Y)\ \forall i$. This is proven by,

$$\Pr^*(x_i,y) = \underbrace{\sum_{x_1\in X_1}\cdots\sum_{x_n\in X_n}}_{\text{All except } x_i\in X_i} \Pr^*(x_1,\ldots,x_n,y) \tag{6.23}$$

$$= \underbrace{\sum_{x_1\in X_1}\cdots\sum_{x_n\in X_n}}_{\text{All except } x_i\in X_i} \Pr(y)\prod_{j=1}^{n}\Pr(x_i|y) \tag{6.24}$$

$$= \underbrace{\sum_{x_1\in X_1}\cdots\sum_{x_n\in X_n}}_{\text{All except } x_i\in X_i} \Pr(x_i,y)\prod_{\substack{j=1\\ j\neq i}}^{n}\Pr(x_j|y) \tag{6.25}$$

$$= \Pr(x_i,y)\underbrace{\underbrace{\sum_{x_1\in X_1}\cdots\sum_{x_n\in X_n}}_{\text{All except } x_i\in X_i}\prod_{\substack{j=1\\ j\neq i}}^{n}\Pr(x_j|y)}_{\text{sums to 1}} \tag{6.26}$$

$$= \Pr(x_i,y)\ . \tag{6.27}$$

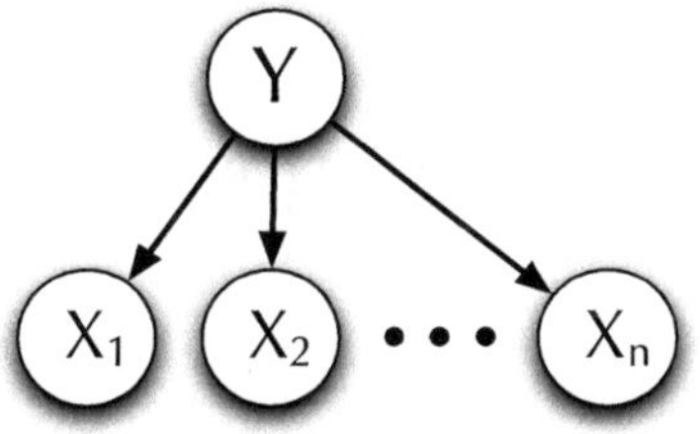

Fig. 6.15 The Directed Acyclic Graph generating the joint distribution $\Pr^*(x_1,\ldots,x_n,y)$. This is a graphical representation of eq. (6.22).

The upperbound on I_{VK} is then the mutual information using this artificial $\Pr^*$ distribution,

$$\mathrm{I}^*(X_1\ldots X_n:Y) = \sum_{x_1\in X_1}\cdots\sum_{x_n\in X_n}\sum_{y\in Y}\Pr^*(x_1,\ldots,x_n,y)\log\frac{\Pr^*(x_1,\ldots,x_n,y)}{\Pr^*(x_1,\ldots,x_n)\Pr^*(y)}\ , \tag{6.28}$$

where the terms $\Pr^*(x_1,\ldots,x_n)$ and $\Pr^*(y)$ are defined by summing over the relevant indices of joint distribution $\Pr^*(X_1,\ldots,X_n,Y)$,

$$\Pr^*(x_1,\ldots,x_n) = \sum_{y'\in Y}\Pr^*(x_1,\ldots,x_n,y') \tag{6.29}$$

$$= \sum_{y'\in Y}\Pr(y')\prod_{i=1}^{n}\Pr(x_i|y')\ ; \tag{6.30}$$

$$
\begin{aligned}
\Pr^*(y) &= \sum_{x_1\in X_1}\cdots\sum_{x_n\in X_n} \Pr^*(x_1,\ldots,x_n,y) && (6.31)\\
&= \sum_{x_1\in X_1}\cdots\sum_{x_n\in X_n} \Pr(y)\prod_{i=1}^{n}\Pr(x_i|y) && (6.32)\\
&= \Pr(y)\underbrace{\sum_{x_1\in X_1}\cdots\sum_{x_n\in X_n}\prod_{i=1}^{n}\Pr(x_i|y)}_{\text{sums to 1}} && (6.33)\\
&= \Pr(y)\ . && (6.34)
\end{aligned}
$$

Putting everything together, our analytic upperbound on I_{VK} is,

$$
\begin{aligned}
\mathrm{I}_{\mathrm{VK}}(\{X_1,\ldots,X_n\}\,:\,Y) &\leq \mathrm{I}^*(X_{1\ldots n}:Y) && (6.35)\\
&= \sum_{x_1}\cdots\sum_{x_n}\sum_{y}\Pr^*(x_1,\ldots,x_n,y)\log\frac{\Pr^*(x_1,\ldots,x_n,y)}{\Pr^*(x_1,\ldots,x_n)\Pr^*(y)} && (6.36)\\
&= \sum_{x_1}\cdots\sum_{x_n}\sum_{y}\Pr^*(x_1,\ldots,x_n,y)\log\frac{\Pr(y)\prod_{i=1}^{n}\Pr(x_i|y)}{\Pr^*(x_1,\ldots,x_n)\Pr(y)} && (6.37)\\
&= \sum_{x_1}\cdots\sum_{x_n}\sum_{y}\Pr^*(x_1,\ldots,x_n,y)\log\frac{\prod_{i=1}^{n}\Pr(x_i|y)}{\Pr^*(x_1,\ldots,x_n)} && (6.38)\\
&= \sum_{y}\Pr(y)\sum_{x_1}\cdots\sum_{x_n}\prod_{i=1}^{n}\Pr(x_i|y)\log\frac{\prod_{i=1}^{n}\Pr(x_i|y)}{\sum_{y'\in Y}\Pr(y')\prod_{i=1}^{n}\Pr(x_i|y')}\ .
\end{aligned}
$$

E Essential Proofs

These proofs underpin essential claims about our introduced measure, synergistic mutual information.

E.1 State-Dependent I_{VK} *and* $\mathcal{S}_{\mathrm{VK}}$

For a single state $y \in Y$, the I_{VK} and $\mathcal{S}_{\mathrm{VK}}$ are defined as,

$$
\begin{aligned}
\mathrm{I}_{\mathrm{VK}}(\mathbf{X}\,:\,Y=y) &\equiv \min_{\substack{p^*(X_1,\ldots,X_n,Y)\\ p^*(X_i,Y)=p(X_i,Y)\quad\forall i}} \mathrm{I}^*(X_{1\ldots n}:Y=y) && (6.39)\\
&= \min_{\substack{p^*(X_1,\ldots,X_n,Y)\\ p^*(X_i,Y)=p(X_i,Y)\quad\forall i}} \mathrm{D_{KL}}\left[\Pr^*(X_{1\ldots n}|y)\,\|\,\Pr^*(X_{1\ldots n})\right]\\
\mathcal{S}_{\mathrm{VK}}(\mathbf{X}:Y=y) &= \mathrm{I}(X_{1\ldots n}:Y=y) - \mathrm{I}_{\mathrm{VK}}(\mathbf{X}\,:\,Y=y) && (6.40)\\
&= \mathrm{D_{KL}}\left[\Pr(X_{1\ldots n}|y)\,\|\,\Pr(X_{1\ldots n})\right] - \mathrm{I}_{\mathrm{VK}}(\mathbf{X}\,:\,Y=y)\ .
\end{aligned}
$$

Naturally, $\mathbb{E}_Y\,\mathrm{I}_{\mathrm{VK}}(\mathbf{X}:y) = \mathrm{I}_{\mathrm{VK}}(\mathbf{X}:Y)$ and $\mathbb{E}_Y\mathcal{S}_{\mathrm{VK}}(\mathbf{X}:y) = \mathcal{S}_{\mathrm{VK}}(\mathbf{X}:Y)$.

E.2 Proof Duplicate Predictors Don't Increase Synergy

We show that synergy being invariant to duplicate predictors follows from the equality condition of **(M)** of the intersection (as well as union) information.

We show that,

$$\mathcal{S}_{\text{VK}}(\mathbf{X}:Y) = \mathcal{S}_{\text{VK}}(\mathbf{X}':Y) \ ,$$

where $\mathbf{X}' \equiv \{X_1, \ldots, X_n, X_1\}$. We show that $\mathcal{S}_{\text{VK}}(\mathbf{X}:Y) - \mathcal{S}_{\text{VK}}(\mathbf{X}':Y) = 0$.

$$\begin{aligned} 0 &= \mathcal{S}_{\text{VK}}(\mathbf{X}:Y) - \mathcal{S}_{\text{VK}}(\mathbf{X}':Y) && (6.41)\\ &= \mathrm{I}(X_{1\ldots n}:Y) - \mathrm{I}_{\text{VK}}(\mathbf{X}:Y) - \mathrm{I}(X_{1\ldots n}X_1:Y) + \mathrm{I}_{\text{VK}}(\mathbf{X}':Y) && (6.42)\\ &= \mathrm{I}_{\text{VK}}(\mathbf{X}':Y) - \mathrm{I}_{\text{VK}}(\mathbf{X}:Y) && (6.43)\\ &= \sum_{\mathbf{T}\subseteq\mathbf{X}'} (-1)^{|\mathbf{T}|+1}\, \mathrm{I}^{dual}_{\text{VK}}(\mathbf{T}:Y) - \sum_{\mathbf{S}\subseteq\mathbf{X}} (-1)^{|\mathbf{S}|+1}\, \mathrm{I}^{dual}_{\text{VK}}(\mathbf{S}:Y) \ . && (6.44) \end{aligned}$$

The terms that **S** enumerates over is a subset of the terms that **T** enumerates. Therefore the $\sum_{\mathbf{S}\subseteq\mathbf{X}}$ completely cancels, leaving,

$$0 = \sum_{\mathbf{T}\subseteq\mathbf{X}} (-1)^{|\mathbf{T}|}\, \mathrm{I}^{dual}_{\text{VK}}\left(\{X_1, T_1, \ldots, T_{|\mathbf{T}|}\}:Y\right) \ . \qquad (6.45)$$

If $\mathrm{I}^{dual}_{\text{VK}}$ obeys **(M)**, then each term of eq. (6.45) s.t. $X_1 \notin \mathbf{T}$ cancels with the same term but with $X_1 \in \mathbf{T}$. This makes eq. (6.45) sum to zero, and completes the proof.

E.3 Proof of Bounds of $\mathcal{S}_{\text{VK}}(\mathbf{X}:Y)$

We show that,

$$\text{WMS}(\mathbf{X}:Y) \leq \mathcal{S}_{\text{VK}}(\mathbf{X}:Y) \leq \mathcal{S}_{\max}(\mathbf{X}:Y) \ . \qquad (6.46)$$

Proof that $\mathcal{S}_{\text{VK}}(\mathbf{X}:Y) \leq \mathcal{S}_{\max}(\mathbf{X}:Y)$

We invoke the standard definitions of $\mathcal{S}_{\text{VK}}$ and $\mathcal{S}_{\max}$,

$$\begin{aligned} \mathcal{S}_{\text{VK}}(\mathbf{X}:Y) &\equiv \mathrm{I}(X_{1\ldots n}:Y) - \mathrm{I}_{\text{VK}}(\mathbf{X}:Y)\\ \mathcal{S}_{\max}(\mathbf{X}:Y) &\equiv \mathrm{I}(X_{1\ldots n}:Y) - \mathrm{I}_{\max}(\mathbf{X}:Y) \ , \end{aligned}$$

where I_{VK} and $\mathrm{I}_{\max}$ are defined as,

$$\begin{aligned} \mathrm{I}_{\text{VK}}(\mathbf{X}:Y) &= \mathbb{E}_Y\, \mathrm{I}_{\text{VK}}(\mathbf{X}:Y=y)\\ &= \mathbb{E}_Y \min_{\substack{p^*(X_1,\ldots,X_n,Y)\\ p^*(X_i,Y)=p(X_i,Y) \quad \forall i}} \mathrm{I}^*(X_{1\ldots n}:Y=y) && (6.47)\\ \mathrm{I}_{\max}(\mathbf{X}:Y) &\equiv \mathbb{E}_Y \max_i \mathrm{I}(X_i:Y=y) \ . && (6.48) \end{aligned}$$

Now we prove $\mathcal{S}_{\text{VK}}(\mathbf{X}:Y) \leq \mathcal{S}_{\max}(\mathbf{X}:Y)$ by showing that $\mathrm{I}_{\text{VK}}(\mathbf{X}:Y) \geq \mathrm{I}_{\max}(\mathbf{X}:Y)$.

Proof

$$\mathbb{E}_Y \, \mathrm{I}_{\mathrm{VK}}(\mathbf{X} : Y = y) \geq \mathbb{E}_Y \, \mathrm{I}_{\max}(\mathbf{X} : Y = y) \tag{6.49}$$

$$\mathbb{E}_Y \left[\mathrm{I}_{\mathrm{VK}}(\mathbf{X} : Y = y) - \mathrm{I}_{\max}(\mathbf{X} : Y = y) \right] \geq 0 \,. \tag{6.50}$$

Now expanding $\mathrm{I}_{\mathrm{VK}}(\mathbf{X} : Y = y)$ and $\mathrm{I}_{\max}(\mathbf{X} : Y = y)$,

$$\mathbb{E}_Y \left[\left(\min_{\substack{p^*(X_1,\ldots,X_n,Y) \\ p^*(X_i,Y) = p(X_i,Y) \quad \forall i}} \mathrm{I}^*(X_{1\ldots n} : Y = y) \right) - \max_i \mathrm{I}(X_i : Y = y) \right] \geq 0 \,. \tag{6.51}$$

We define the index $m \in \{1, \ldots, n\}$ such that $m = \arg\max_i \mathrm{I}(X_i : Y = y)$. The predictor with the most information about state $Y = y$ is thus X_m. This yields,

$$\mathbb{E}_Y \left[\left(\min_{\substack{p^*(X_1,\ldots,X_n,Y) \\ p^*(X_i,Y) = p(X_i,Y) \quad \forall i}} \mathrm{I}^*(X_{1\ldots n} : Y = y) \right) - \mathrm{I}(X_m : Y = y) \right] \geq 0 \,. \tag{6.52}$$

The constraint $p^*(X_i, Y) = p(X_i, Y)$ entails that $\mathrm{I}(X_m : Y = y) = \mathrm{I}^*(X_m : Y = y)$. Therefore we can pull $\mathrm{I}(X_m : Y = y)$ inside the minimization as a constant,

$$\mathbb{E}_Y \left[\min_{\substack{p^*(X_1,\ldots,X_n,Y) \\ p^*(X_i,Y) = p(X_i,Y) \quad \forall i}} \mathrm{I}^*(X_{1\ldots n} : Y = y) - \mathrm{I}^*(X_m : Y = y) \right] \geq 0 \,. \tag{6.53}$$

As X_m is a subset of predictors $X_{1\ldots n}$, we can subtract it yielding,

$$\mathbb{E}_Y \left[\min_{\substack{p^*(X_1,\ldots,X_n,Y) \\ p^*(X_i,Y) = p(X_i,Y) \quad \forall i}} \mathrm{I}^*\left(X_{1\ldots n \setminus m} : Y = y \middle| X_m \right) \right] \geq 0 \,. \tag{6.54}$$

The state-dependent conditional mutual information $\mathrm{I}^*\left(X_{1\ldots n \setminus m} : Y = y \middle| X_m \right)$ is a Kullback-Liebler divergence. As such it is nonnegative. Likewise the minimum of a nonnegative quantity is also nonnegative.

$$\mathbb{E}_Y \left[\underbrace{\min_{\substack{p^*(X_1,\ldots,X_n,Y) \\ p^*(X_i,Y) = p(X_i,Y) \quad \forall i}} \mathrm{I}^*\left(X_{1\ldots n \setminus m} : Y = y \middle| X_m \right)}_{\geq 0} \right] \geq 0 \,. \tag{6.55}$$

Finally, the expected value of a list of nonnegative quantities is nonnegative. And the proof that $\mathcal{S}_{\mathrm{VK}}(\mathbf{X} : Y) \leq \mathcal{S}_{\max}(\mathbf{X} : Y)$ is complete.

Proof that $\text{WMS}(\mathbf{X}:Y) \leq \mathcal{S}_{\text{VK}}(\mathbf{X}:Y)$

We invoke the standard definitions of WMS and $\mathcal{S}_{\text{VK}}$,

$$\text{WMS}(\mathbf{X}:Y) \equiv \text{I}(X_{1\ldots n}:Y) - \sum_{i=1}^{n} \text{I}(X_i:Y) \tag{6.56}$$

$$\mathcal{S}_{\text{VK}}(\mathbf{X}:Y) \equiv \text{I}(X_{1\ldots n}:Y) - \text{I}_{\text{VK}}(X_{1\ldots n} : Y) \tag{6.57}$$

$$= \text{I}(X_{1\ldots n}:Y) - \min_{\substack{p^*(X_1,\ldots,X_n,Y) \\ p^*(X_i,Y)=p(X_i,Y) \quad \forall i}} \text{I}^*(X_{1\ldots n}:Y)\,. \tag{6.58}$$

We prove the conjecture $\text{WMS}(\mathbf{X}:Y) \leq \mathcal{S}_{\text{VK}}(\mathbf{X}:Y)$ by showing,

$$\min_{\substack{p^*(X_1,\ldots,X_n,Y) \\ p^*(X_i,Y)=p(X_i,Y) \quad \forall i}} \text{I}^*(X_{1\ldots n}:Y) \leq \sum_{i=1}^{n} \text{I}(X_i:Y)\,. \tag{6.59}$$

Given:

$$\min_{\substack{p^*(X_1,\ldots,X_n,Y) \\ p^*(X_1,Y)=p(X_1,Y) \\ \vdots \\ p^*(X_n,Y)=p(X_n,Y)}} \text{I}^*(X_{1\ldots n}:Y)\,, \tag{6.60}$$

the individual constraint $p^*(X_1,Y) = p(X_1,Y)$ can add at most $\text{I}(X_1:Y)$ bits to $\text{I}^*(X_{1\ldots n}:Y)$. Therefore we can upperbound eq. (6.60) by dropping the constraint $p^*(X_1,Y) = p(X_1,Y)$ and adding $\text{I}(X_1:Y)$. This yields,

$$\text{I}_{\text{VK}}(\mathbf{X}:Y) \leq \min_{\substack{p^*(X_1,\ldots,X_n,Y) \\ p^*(X_2,Y)=p(X_2,Y) \\ \vdots \\ p^*(X_n,Y)=p(X_n,Y)}} \text{I}^*(X_{1\ldots n}:Y) + \text{I}(X_1:Y)\,. \tag{6.61}$$

Likewise, the righthand-side of eq. (6.61) can be upperbounded by dropping the constraint $p^*(X_2,Y) = p(X_2,Y)$ and adding $\text{I}(X_2:Y)$. This yields,

$$\min_{\substack{p^*(X_2,\ldots,X_n,Y) \\ p^*(X_2,Y)=p(X_2,Y) \\ \vdots \\ p^*(X_n,Y)=p(X_n,Y)}} \text{I}^*(X_{1\ldots n}:Y) \leq \min_{\substack{p^*(X_3,\ldots,X_n,Y) \\ p^*(X_3,Y)=p(X_3,Y) \\ \vdots \\ p^*(X_n,Y)=p(X_n,Y)}} \text{I}^*(X_{1\ldots n}:Y) + \text{I}(X_1:Y) + \text{I}(X_2:Y)\,. \tag{6.62}$$

Repeating this process n times yields,

$$\text{I}_{\text{VK}}(\mathbf{X}:Y) \leq \min_{p^*(X_1,\ldots,X_n,Y)} \text{I}^*(X_{1\ldots n}:Y) + \sum_{i=1}^{n} \text{I}(X_i:Y) \tag{6.63}$$

$$= \sum_{i=1}^{n} \text{I}(X_i:Y)\,. \tag{6.64}$$

F Algebraic Simplification of ΔI

Prior literature (Nirenberg et al. 2001; Nirenberg and Latham 2003; Pola et al. 2003; Latham and Nirenberg 2005) defines $\Delta\,\mathrm{I}\,(\mathbf{X};Y)$ as,

$$\Delta\,\mathrm{I}\,(\mathbf{X};Y) \equiv \mathrm{D_{KL}}\left[\Pr(Y|X_{1\ldots n}) \| \Pr_{\mathrm{ind}}(Y|\mathbf{X})\right] \tag{6.65}$$

$$= \sum_{\mathbf{x},y\in\mathbf{X},Y} \Pr(\mathbf{x},y) \log \frac{\Pr(y|\mathbf{x})}{\Pr_{\mathrm{ind}}(y|\mathbf{x})} \,. \tag{6.66}$$

Where,

$$\Pr_{\mathrm{ind}}(Y=y|\mathbf{X}=\mathbf{x}) \equiv \frac{\Pr(y)\Pr_{\mathrm{ind}}(\mathbf{X}=\mathbf{x}|Y=y)}{\Pr_{\mathrm{ind}}(\mathbf{X}=\mathbf{x})} \tag{6.67}$$

$$= \frac{\Pr(y)\prod_{i=1}^{n}\Pr(x_i|y)}{\Pr_{\mathrm{ind}}(\mathbf{x})} \tag{6.68}$$

$$\Pr_{\mathrm{ind}}(\mathbf{X}=\mathbf{x}) \equiv \sum_{y\in Y}\Pr(Y=y)\prod_{i=1}^{n}\Pr(x_i|y) \tag{6.69}$$

The definition of $\Delta\,\mathrm{I}$, eq. (6.65), reduces to,

$$\Delta\,\mathrm{I}\,(\mathbf{X};Y) = \sum_{\mathbf{x},y\in\mathbf{X},Y} \Pr(\mathbf{x},y) \log \frac{\Pr(y|\mathbf{x})}{\Pr_{\mathrm{ind}}(y|\mathbf{x})} \tag{6.70}$$

$$= \sum_{\mathbf{x},y\in\mathbf{X},Y} \Pr(\mathbf{x},y) \log \frac{\Pr(y|\mathbf{x})\Pr_{\mathrm{ind}}(\mathbf{x})}{\Pr(y)\prod_{i=1}^{n}\Pr(x_i|y)} \tag{6.71}$$

$$= \sum_{\mathbf{x},y\in\mathbf{X},Y} \Pr(\mathbf{x},y) \log \frac{\Pr(\mathbf{x}|y)}{\prod_{i=1}^{n}\Pr(x_i|y)}\frac{\Pr_{\mathrm{ind}}(\mathbf{x})}{\Pr(\mathbf{x})} \tag{6.72}$$

$$= \sum_{\mathbf{x},y\in\mathbf{X},Y} \Pr(\mathbf{x},y) \log \frac{\Pr(\mathbf{x}|y)}{\prod_{i=1}^{n}\Pr(x_i|y)} + \sum_{\mathbf{x},y\in\mathbf{X},Y} \Pr(\mathbf{x},y) \log \frac{\Pr_{\mathrm{ind}}(\mathbf{x})}{\Pr(\mathbf{x})}$$

$$= \sum_{\mathbf{x},y\in\mathbf{X},Y} \Pr(\mathbf{x},y) \log \frac{\Pr(\mathbf{x}|y)}{\prod_{i=1}^{n}\Pr(x_i|y)} - \sum_{\mathbf{x}\in\mathbf{X}} \Pr(\mathbf{x}) \log \frac{\Pr(\mathbf{x})}{\Pr_{\mathrm{ind}}(\mathbf{x})} \tag{6.73}$$

$$= \mathrm{D_{KL}}\left[\Pr(X_{1\ldots n}|Y) \middle\| \prod_{i=1}^{n}\Pr(X_i|Y)\right] - \mathrm{D_{KL}}\left[\Pr(X_{1\ldots n}) \| \Pr_{\mathrm{ind}}(\mathbf{X})\right]$$

$$= \mathrm{TC}\,(X_1;\cdots;X_n|Y) - \mathrm{D_{KL}}\left[\Pr(X_{1\ldots n}) \| \Pr_{\mathrm{ind}}(\mathbf{X})\right] \,. \tag{6.74}$$

where $\mathrm{TC}\,(X_1;\cdots;X_n|Y)$ is the conditional total correlation among the predictors given Y.

References

Amari, S.: Information geometry on hierarchical decomposition of stochastic interactions. IEEE Transaction on Information Theory 47, 1701–1711 (1999)

Anastassiou, D.: Computational analysis of the synergy among multiple interacting genes. Molecular Systems Biology 3, 83 (2007)

Balduzzi, D., Tononi, G.: Integrated information in discrete dynamical systems: motivation and theoretical framework. PLoS Computational Biology 4(6), e1000091 (2008)

Bell, A.J.: The co-information lattice. In: Amari, S., Cichocki, A., Makino, S., Murata, N. (eds.) Fifth International Workshop on Independent Component Analysis and Blind Signal Separation, Springer (2003)

Bertschinger, N., Rauh, J., Olbrich, E., Jost, J.: Shared information – new insights and problems in decomposing information in complex systems. CoRR, abs/1210.5902 (2012)

Chechik, G., Globerson, A., Anderson, M.J., Young, E.D., Nelken, I., Tishby, N.: Group redundancy measures reveal redundancy reduction in the auditory pathway. In: Dietterich, T.G., Becker, S., Ghahramani, Z. (eds.) NIPS 2002, pp. 173–180. MIT Press, Cambridge (2002)

Comtet, L.: Advanced Combinatorics: The Art of Finite and Infinite Expansions. Reidel, Dordrecht (1998)

Cover, T.M., Thomas, J.A.: Elements of Information Theory. John Wiley, New York (1991)

DeWeese, M.R., Meister, M.: How to measure the information gained from one symbol. Network 10, 325–340 (1999)

Gat, I., Tishby, N.: Synergy and redundancy among brain cells of behaving monkeys. In: Advances in Neural Information Proceedings systems, pp. 465–471. MIT Press (1999)

Gawne, T.J., Richmond, B.J.: How independent are the messages carried by adjacent inferior temporal cortical neurons? Journal of Neuroscience 13, 2758–2771 (1993)

Han, T.S.: Nonnegative entropy measures of multivariate symmetric correlations. Information and Control 36(2), 133–156 (1978)

Harder, M., Salge, C., Polani, D.: A bivariate measure of redundant information. Physical Review E 87(1), 012130 (2013)

Latham, P.E., Nirenberg, S.: Synergy, redundancy, and independence in population codes, revisited. Journal of Neuroscience 25(21), 5195–5206 (2005)

Lei, W., Xu, G., Chen, B.: The common information of n dependent random variables. In: Forty-Eighth Annual Allerton Conference on Communication, Control, and Computing (2010), doi:abs/1010.3613:836–843

Lizier, J.T., Flecker, B., Williams, P.L.: Towards a synergy-based approach to measuring information modification. In: IEEE Symposium Series on Computational Intelligence (SSCI 2013) — IEEE Symposium on Artificial Life, Singapore. IEEE Press (April 2013)

Maurer, U.M., Wolf, S.: Unconditionally secure key agreement and the intrinsic conditional information. IEEE Transactions on Information Theory 45(2), 499–514 (1999)

Narayanan, N.S., Kimchi, E.Y., Laubach, M.: Redundancy and synergy of neuronal ensembles in motor cortex. The Journal of Neuroscience 25(17), 4207–4216 (2005)

Nirenberg, S., Carcieri, S.M., Jacobs, A.L., Latham, P.E.: Retinal ganglion cells act largely as independent encoders. Nature 411(6838), 698–701 (2001)

Nirenberg, S., Latham, P.E.: Decoding neuronal spike trains: How important are correlations? Proceedings of the National Academy of Sciences 100(12), 7348–7353 (2003)

Panzeri, S., Treves, A., Schultz, S., Rolls, E.T.: On decoding the responses of a population of neurons from short time windows. Neural Comput. 11(7), 1553–1577 (1999)

Pola, G., Thiele, A., Hoffmann, K.P., Panzeri, S.: An exact method to quantify the information transmitted by different mechanisms of correlational coding. Network 14(1), 35–60 (2003)

Schneidman, E., Bialek, W., Berry II, M.: Synergy, redundancy, and independence in population codes. Journal of Neuroscience 23(37), 11539–11553 (2003a)

Schneidman, E., Still, S., Berry, M.J., Bialek, W.: Network information and connected correlations. Phys. Rev. Lett. 91(23), 238701–238705 (2003b)

Weisstein, E.W.: Antichain (2011), `http://mathworld.wolfram.com/Antichain.html`

White, D., Rabago-Smith, M.: Genotype-phenotype associations and human eye color. Journal of Human Genetics 56(1), 5–7 (2011)

Williams, P.L., Beer, R.D.: Nonnegative decomposition of multivariate information. CoRR, abs/1004.2515 (2010)

Part III
Coordinated Behaviour and Learning within an Embodied Agent

Chapter 7
On the Role of Embodiment for Self-Organizing Robots: Behavior As Broken Symmetry

Ralf Der

7.1 Introduction

Embodiment and SO form two cornerstones of both modern robotics and the understanding of human and animal intelligence. In particular, the role of the embodiment for the behavior of both artificial and natural beings has become of much and increasing interest in recent times. In robotics, there are essentially two attitudes towards the physical embodiment. On the one hand, with rule based systems and/or systems intended to execute a given motion plan, embodiment is more or less considered as a (nasty) problem opposing the execution of the plan. On the other hand, it is well believed and verified by many examples that living beings are taking much advantage from the physico-mechanical properties of their bodies in order to create natural motion patterns. In robotics, this can be of immediate benefit for more robust or energy effective control, The importance of the embodiment for the behavior generation and intelligence in general has been advocated with great impact on the scientific community mainly by the lab of Rolf Pfeifer, see (Pfeifer and Bongard 2006; Pfeifer et al. 2007; Pfeifer and Gomez 1999) for excellent surveys. In recent years this has been further established under the connotation of morphological computation (Pfeifer and Gómez 2009; Hauser et al. 2012; Pfeifer 2012; Hauser et al. 2011). Seminal contributions are also by the lab of Helge Ritter (Ritter et al. 2009; Grossekathofer et al. 2011; Behnisch et al. 2011; Elbrechter et al. 2011; Steffen et al. 2010; Maycock et al. 2010) and others.

Self-Organization (SO) may provide an essential progress in the realization of embodied control. Viewing a robot in its environment as a complex dynamical system, SO can help to let highly coordinated and low dimensional modes emerge in the coupled system of brain, body and environment. In this way, instead of being programmed for solving a specific task, the robot may find out by itself what its bodily

Ralf Der
Max Planck Institute for Mathematics in the Sciences, Inselstraße 22, D-04103, Leipzig, Germany
e-mail: ralfder@mis.mpg.de

M. Prokopenko (ed.), *Guided Self-Organization: Inception,*
Emergence, Complexity and Computation 9,
DOI: 10.1007/978-3-642-53734-9_7, © Springer-Verlag Berlin Heidelberg 2014

affordances are, focusing only in a second step on the exploitation of the emerging motion patterns—by guiding the SO process into the directions of potential benefits.

This paper studies the relation of embodiment to self-organization in autonomous robots. This project faces essentially two challenges. One is how to organize a robotic system in such a way that it starts to self-organize. There are several approaches into that direction based on formulating objective functions (OFs) for SO. In recent years, several such OFs have been proposed, ranging from the maximization of predictive information (Ay et al. 2008, 2012; Martius et al. 2013) or empowerment (Klyubin et al. 2005, 2007; Anthony et al. 2009; Jung et al. 2012), to the minimization of free energy (Friston 2010; Friston et al. 2012; Friston and Stephan 2007; Friston 2012) or the so called time-loop error in the homeokinesis approach (Der 2001; Der and Liebscher 2002; Der and Martius 2012), see also (Prokopenko 2008, 2009; Prokopenko et al. 2009) for more details on how to organize SO. Given an objective function, the optimization process can be translated into a learning rule that is driving the SO process. This paper introduces a new learning rule together with some case studies demonstrating its usefulness. So, there are several pretty satisfying approaches in the SO paradigm.

The second challenge actually is more serious and might be the reason why in the community SO is seen more as a wishful thinking than a systematic approach to autonomous robot development. With genuine SO one must be careful not to plug in (by biasing the system) what one actually wants to get out. The approaches mentioned above all seem to fit that criterion. However, and this seems to be the argument, if nothing is specified from outside, will SO simply make the robot an arbitrary subject that is completely unpredictable in its behaviors and thus rather a thread than a hope. The aim of this paper is to show that this attitude is wrong. Instead, we will develop an understanding of what happens if the system is self-organizing, what the role of the embodiment is and how we can find clues for predicting and shaping the behavior patterns emerging in a genuine SO scenario.

In order to prepare the ground for the role of the embodiment in SO, we give some examples of systems unfolding complex motion patterns with minimalistic control. We start from the famous example of the Braitenberg vehicles (BVs) which we consider as an early case study of machines which develop complicated behaviors on the basis of an extremely simple construction. By these examples, we want to make the reader aware of the phenomenon of spontaneous symmetry breaking that in our opinion is instrumental for understanding how SO can be effective in robotic systems. We think that the robotic community so far has overlooked the importance and substance of that phenomenon. Therefore, this paper will also follow a pedagogical purpose.

The paper is organized as follows. After introducing the original idea of Valentin Braitenberg and variations thereof in Sect. 7.2, we introduce in Sect. 7.3 a much more complex body, our HUMANOID, that develops a complex behavior mode under a minimalistic control. This gives us the opportunity to introduce the concept of fundamental modes, dynamical patterns that are specific for a particular body under homogeneous energy feeding condition. This is also a first example of spontaneous symmetry breaking (SSB). Sect. 7.4 introduces the new learning rules for driving

SO based on a recent result from maximizing predictive information (Martius et al. 2013), followed by a short aside on the principle of homeokinesis in Sect. 7.5 containing a comparison between the two methods. Sect. 7.6 is devoted to a deeper discussion of fundamental modes and the mechanism of SSB. After discussing these phenomena with the example of the autistic vehicles, Sect. 7.7 and Sect. 7.8 study these questions with the HUMANOID and the HEXAPOD robot.

7.2 Vehicles

Let us start with a few ideas on what we want to understand by embodiment and its synergy with the "brain". Let us go back to a very instructive example given by Braitenberg (Braitenberg 1984) in order to better understand the synergy concept.

7.2.1 Braitenbergs Idea

His idea was to use very simple machines in order to demonstrate the emergence of complex behaviors determined by the very physical construction of those machines. In the most simple case, the machine is a two-wheeled robot with two photo cells mounted to the left an right front side of the trunk. The photocells are wired to the motors either diagonally so that the vehicle is approaching the light source and bangs into it, or directly so that light sources are acting as a repeller.

Besides the phenomenon of emerging functionality observed in those machines, there is another fact we want to draw attention to. Both in the basic and also in the more elaborate examples, it seems more or less artificial to make a distinction, as is done in most robotics approaches, between the body and the "brain". Instead, we see a physical system—a mechanical part consisting of the body and the wheels combined with an electrical part consisting of the photo cells, the motors and the wires. With this setting it seems more natural to see the machine as a whole, instead of subdividing it into a body and a controller acting as a (kind of) brain. Of course, which attitude to take is always a matter of taste, preferences shifting more toward a clear partitioning with increasing complexity of the "brain". This paper, keeping control as simple as possible, is more in favor of the holistic attitude: considering the machine—body, sensors, and wiring—as one physical system. In the Braitenberg case, this system may function even completely autonomously since it receives all the energy it needs from its sensors, the photocells.

7.2.2 Autistic Vehicles

The properties of the most simple vehicles are self-explaining given that the behavior is driven by the energy provided by the photo cells together with the specific wiring. If translated to humans or higher animals, this corresponds to a system driven by vision alone. However, motion control is largely depending also on the signals they are receiving from proprioception, like the joint angles and the muscle tensions. In order to mimic the role of those sensors, we are going now to restructure the BVs by including proprioceptive sensors. In particular, we introduce wheel counters, measuring the rotational velocity of each of the wheels. Wiring can be done in the most simple case by connecting the right (left) wheel sensor to the right (left) motor and vice versa. However, we do now need active wires which feed energy into the system (there is no perpetual motion machine of the first kind).

Such an active, nonlinear wire can be realized by a simple neuron, that translates the input (wheel velocity) into an output— the target wheel velocity that is subsequently realized by the motor using electrical energy supplied from outside. In detail, we use a neuron as

$$y = \tanh(cx + h) \tag{7.1}$$

where c is the coupling strength and h a threshold to be put equal to zero for the moment. This setting has been studied before in different contexts. The behavior of the wheel can be seen if we consider the full sensorimotor loop (SML). What we need for that is the connection between the target motor values y and the corresponding wheel velocity x as measured by the wheel counter. Assuming a linear relation between the two, the SML can be modeled by the following dynamical system.

$$x' = ay + \xi = a\tanh(cx + h) + \xi \tag{7.2}$$

where x' is the new vector of sensor values, ξ is to contain all the deviations from the linear law, and a is a hardware constant, with $a = 1$ if the wheel counter is scaled appropriately. In many applications, ξ can be treated as pure noise.

The behavior of the vehicle can be analyzed best by asking for the fixed points (FPs) of the dynamical system which are obtained from $x = ay$ or (with $a = 1$)

$$x = \tanh(cx) \tag{7.3}$$

There are a number of different regimes. Let us consider first the case $0 < c < 1$ with the single FP $x = 0$ corresponding to wheel velocity equal to zero. With noise, each wheel velocity is fluctuating around zero so that the vehicle executes a random walk. With $c > 1$ the system is bistable. If $c = 1 + \delta$, with $\delta > 0$ small, the FP is given approximately as $x = \pm\sqrt{\delta}$, see (Der and Martius 2012) for details. In this case, the system is in one of the FPs but can be switched by the noise. So, the wheel is rotating for some time into one direction with occasional changes of direction. This effect generates an irregular motion of the vehicle due to the noise. In particular, upon colliding with an obstacle, the velocity may switch so that the vehicle is kind of reflected by the obstacles. This may be seen as a basic survival strategy emerging from this minimalistic control. In this sense, the autistic vehicle still does show certain reactions to the environment. However, these are restricted to noise or physical encounters and the perturbations generated thereby.

With $c < 0$, the output of the neuron (motor command) is always opposite in sign to its input so that the motor always gets contradictory commands which might eventually lead to the destruction of the motor. Therefore, this regime was called the regime of self-destruction. However, we will see in Sect. 7.3 below that, based on the specific physical embodiment, this regime may also deploy highly interesting modes.

Altogether there are not much interesting properties of that autistic vehicle. A first interesting result is observed if several such two-wheeled vehicles are coupled passively together. In that case, the physical cross-talk—the coupling forces produced by different wheel velocities of neighboring vehicles—may switch the wheel velocities so that eventually a collective motion of the system emerges. This has been studied extensively earlier (Der et al. 2008; Zahedi et al. 2010; Der and Martius 2012) so that we do not go into details here. Instead we consider other more complex systems where a collective motion is emerging from decentralized control in a surprising way, see Sect. 7.3 below.

7.2.3 Symmetries

Symmetries and their breaking play a central role in this approach. Let us start here with the standard scenario of inducing the breaking of a given symmetry by manually driving certain parameters of the system over a bifurcation point. The self-induced breaking that is of actual interest for this paper will be discussed in Sect. 7.6 below. By way of example, let us consider the autistic vehicle introduced above with a single wheel running on a rail. Assuming there is a perfect forward-backward symmetry of the morphology, the system is invariant against inversion of the x-axis, i. e. it obeys the symmetry $S : x \to -x$.

Given that the controller is invariant against S and assuming that the noise also is, we find that there is no clue for the controlled system to prefer a specific direction in space[1]. In the regime with only a single FP ($0<c<1$), this means that the trajectory is fluctuating around the FP $x = 0$. In the sense of a classical bifurcation analysis, we can now increase the coupling strength c. When crossing the bifurcation point, the system is entering its bistable regime with two stable attractors corresponding to the wheel rotating forward or backward. Which one is chosen is determined by the noise so that we have here a trivial case of noise-induced symmetry breaking.

7.3 The Braitenberg Man—Fundamental Modes

Let us now extend Braitenberg's idea to a more complicated machine—our HUMANOID with a specific wiring.

7.3.1 The HUMANOID

We consider a humanoid robot with 17 active degrees of freedom. Each joint is driven by a simulated servo motor, the motor values $y \in \mathbb{R}^{17}$ sent by the controller

[1] With noise the symmetry is to be understood in the stochastic sense, meaning that each trajectory finds its counterpart—a trajectory that is realized with the same probability—by applying the symmetry operation.

being the target angles of the joints and the sensor values $x \in \mathbb{R}^{17}$ are the true, i. e. observed angles. This is the only knowledge the robot has about its physical state. The physics of the robot is simulated realistically in the LPZROBOTS simulator. Let us introduce a specific wiring in analogy to the autistic vehicle: each joint i is controlled by a single neuron generating the control $y_i = \tanh(cx_i)$ for motor i, with c a constant coupling strength, the same for all joints. In the terminology of Braitenberg the neuron can be considered as an active, nonlinear wire. By this wiring, like in the case of the coupled vehicles mentioned above, we have a system with fully decentralized control (called split control in (Zahedi et al. 2010)).

7.3.2 A Fundamental Mode of the HUMANOID

The interesting behavior emerges if we give the robot much freedom to develop its specific, body-inspired motions. In the experiment, we suspend the HUMANOID like a bungee jumper. After starting it in a position slightly above the equilibrium point of the robot-spring system, the robot is seen to fall down a short distance and then oscillates vertically for a while due to the elastic forces of the spring, without developing internal motions of its own. Its further fate depends essentially on the coupling strength c. We do not observe much of interest for positive coupling strengths. Instead, there is a critical negative value c_{crit} so that, with $c < c_{crit}$, the robot is seen in the experiment—after the oscillatory phase—pretty soon to go into a regular, periodic motion[2]. This is caused by the self-amplification of initial, very small perturbations, see the video S1 (in the supplementary material on `http://robot.informatik.uni-leipzig.de/research/supplementary/GSO2012/`). The interesting and on first sight surprising observation is that this motion is very similar to a running behavior. In particular, despite of the completely decentralized control, arms, legs and hip motions are seen to be synchronized with fixed phase relations like with a running human.

Why is that? How do the individual controller know of each other? In order to discuss this point we must remember that the coupling constant is strongly negative so that actually we are in the regime of self-destruction. However, in the physical system, motor forces are confined to a low range so that—given the inertia due to the masses of the segments—the joint angles can change only slowly. Thus, as long as a joint angle is largely negative, a strong positive motor command is generated that engenders a motion towards the region of positive joint angles. Overshooting (by both inertia and the time averaging of the sensor values) the region of small joint angles, the scenario repeats with an impact towards negative joint angles. Altogether this produces an oscillatory motion of the joint with frequency and amplitude depending on the motor forces and masses of the moved segments in a complicated way.

This effect explains the latent oscillatory mode of each individual joint. But why the (running like) coordination between the segments? Well, the motions are seen to be quite rapid so that there are heavy intra-robot physical forces: by the inertia

[2] We have to smooth the sensor values in a short time horizon to obtain that result.

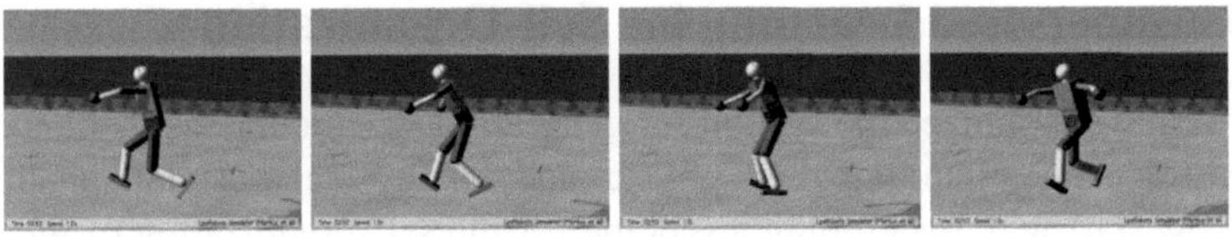

Fig. 7.1 One half-period in the running motion pattern of the Braitenberg man. In the simulation we use a time average over 15 steps of the sensor values.

effects, any motor activity changing a joint angle will generate a heavy impact on its neighbors. So, like with the chain of wheeled robots, there is a strong physical cross-talk that is the communication basis of the intra-robot coordination phenomenon. Spontaneous symmetry breaking (SSB) comes into play here. By its geometry, the specific wiring, and the identity of all the coupling constants, the system is highly symmetric, any emerging motion pattern corresponding to a specific breaking of that symmetry. Due to the overcritical negative feed-back strength given by $c < c_{crit}$, there is a strong tendency of self-amplification of the physical cross talk. Given the high degree of symmetry at the start, this self-amplification induces a symmetry breaking with a specific motion pattern emerging. Choosing the perturbations by hand, many different motion patterns are possible but in the experiments, the running motion pattern (RMP) is dominating, i. e. it seems to be the most stable one—once the pattern is established one may perturb it by external forces but, in most cases, the pattern is recovered after a short time.

The above mechanism also explains why we want to call the emerging dynamics a fundamental mode of the system. We borrowed that term from physics where fundamental modes (often called normal modes) are common phenomena. For instance, in coupled systems of linear or nonlinear oscillators, low frequency modes are emerging where all oscillators are synchronized with a fixed phase relation. In neuroscience such modes are also known for networks of integrate-and-fire neurons, e. g. Synchrony and fixed phase relations are also featuring in the RMP—the individual constituents of the body are moving in a highly coordinated manner. Moreover, the mode may be called fundamental since it is a dynamical pattern that is specific for this particular body under the homogeneous energy feeding condition: the intake of energy is regulated for each motor in the same way by an individual feed-back loop with the same coupling constant. Yet, despite this structural homogeneity we get a highly structured dynamics.

The existence of the fundamental mode heavily relies on the specific conditions, in particular the freedom of movement as realized in the case of the bungee jumper. When lying on the ground, the robot will not develop any interesting activity. With most values of $|c| < c_{crit}$, the robot will reach a fixed point behavior, c_{crit} corresponding to the overall critical feed-back strength. With super-critical values, the motions are confined by the contact with the ground. These restrictions are overcome with a convenient method for self-organization of the behavior as we will develop it in the following.

7.4 Unsupervised Learning for Self-Organization

So far, we have organized the system for generating specific modes by hand, guided by an intuitive understanding of the physics and the physical cross-talk mechanisms. Let us now start with developing a scheme for the self-organization of the system, hoping that the system itself discovers such interesting phenomena like the existence of fundamental modes.

In recent work, the so called predictive information (PI) was introduced as a general objective function for SO (Ay et al. 2008, 2012; Zahedi et al. 2010; Martius et al. 2013). By maximizing the PI, a general learning rule for the synaptic strengths of a neural controller network was derived on the basis of such an Infomax principle. Different from Infomax principles derived so far, that method solves the task of relating the general principle, formulated at the level of behavior, to the internal world of the robot, in to its brain so to say.

The controller of the robot is given in our case by an artificial neural network (ANN) transforming sensor values $x \in \mathbb{R}^n$ into motor values $y \in \mathbb{R}^m$ like

$$y = K(x, ..., w) \tag{7.4}$$

where w are the parameters (synaptic strengths) and ... means some other, internal variables which we will drop for the moment. The translation between the external and the internal world can be done if there is a forward model predicting future sensor values on the basis of the current sensor and motor values. In particular we need only the relation between two time steps so that

$$x_{t+1} = \phi(x_t, y_t) + \xi_{t+1}$$

where ξ is the prediction error and the parametrized function $\phi : \mathbb{R}^n \times \mathbb{R}^m \to \mathbb{R}^n$ is the predictor.

7.4.1 *Learning Rules for Self-model and Control*

We consider here only the result for a specific controller—a one layer feed-forward neural network realized as

$$K(x) = g(Cx + h) \tag{7.5}$$

so that the set of parameters $w = (C, h)$ is given by the synaptic values (matrix C) and the vector h of threshold values for the neurons. In the concrete applications done in this paper, we specifically use $g_i(z) = \tanh(z_i)$ (so that $g : \mathbb{R}^m \to \mathbb{R}^m$ is to be understood as a vector function).

Moreover, the forward model ϕ is given by a layer of linear neurons, so that

$$\phi(x, y) = Ay + Sx + b\,. \tag{7.6}$$

The matrices A and S, and the vector b represent the parametrization of the forward model that can be adapted on-line by a supervised gradient procedure as

$$\Delta A = \eta \xi y^\top, \ \Delta S = \eta \xi x^\top, \ \Delta b = \eta \xi \tag{7.7}$$

with $\xi_t = x_t - \psi(x_{t-1})$. In the applications, the learning rate η may be large such that the low complexity of the model is compensated by a fast adaptation process.

The learning rule for the controller, given in (Martius et al. 2013), was derived from maximization of the predictive information. Based on that, we postulate a new unsupervised learning rule (ULR) which will not be derived here but considered as intuitively grounded by the discussion in Sect. 7.4.2 below. The rule is written as (all quantities are at time t)

$$\frac{1}{\varepsilon} \Delta C_{ij} = \delta y_i \delta x_j - \gamma_i y_i x_j \tag{7.8}$$

$$\frac{1}{\varepsilon} \Delta h_i = -\gamma_i y_i \tag{7.9}$$

where δx_t originally was the prediction error but in the new rule we are free to consider δx as any perturbation of the sensor dynamics. In the experiments described below we used the change of the sensor values in one time step, i. e. $\delta x_t = x_t - x_{t-1}$. δy_t is obtained by backpropagating δx_{t+1} through the world model as

$$\delta y_t = J^T \delta x_{t+1} \tag{7.10}$$

where

$$J = \frac{\partial \phi(x, y)}{\partial y}$$

is the Jacobian matrix of the model relating its output to the input y. In our linear model, we simply have $J = AG'(z)$ where $G'(z) = diag[g'_1 \ldots g'_m]$. Moreover, γ_i is a neuron specific learning rate defined as

$$\gamma_i = 2\beta \delta y_i \delta z_i \tag{7.11}$$

where β is the so called sensitivity, an empirical quantity with $\beta > 1$, and

$$\delta z = C \delta x$$

is the perturbation of the post-synaptic potential due to δx_t.

7.4.2 Anti-Hebbian and Differential Hebbian Learning: A Productive Competition

The specific form of the learning rule allows for a very basic interpretation. Let us start with the last term $-\gamma_i y_i x_j$ contributing to ΔC_{ij} which is easily recognized as a Hebbian like term since it is the product of the input x_j into the synapse C_{ij} times the activation y_i of the neuron. In standard situations, the prefactor γ_i is positive, so that overall the term represents an anti-Hebbian mechanism. As such it would weaken all tracks in the SM loop for which there is a strong output of the motor neuron combined with a strong response from the outside world as reported by the sensor value x_i. In general, this would weaken the activation of any neuron in such a loop, preventing it from saturation.

The first contribution given by $\delta y_i \delta x_j$ is formulated not in the excitations itself but in their time derivatives. This is a differential Hebbian mechanism as it has found much interest recently, see (Kolodziejski et al. 2008, 2009; Kulvicius et al. 2010). Given the relation between δy_t and δx_{t+1}, see Eq. (7.10), C_{ij} is strengthened if there is a strong correlation between[3] δx_j^t and those components of δx_{t+1} which are fed by δy_i^t. Roughly speaking, the first term in the learning rule tries to increase the dynamical correlations across time, driving the system towards activity, while the second term keeps the neurons in their sensitive regions, away from saturation.

This has an important impact on the symmetry breaking scenarios. In the bifurcation scenario discussed in Sect. 7.2.3, symmetry breaking was induced by changing the controller parameter from outside: when crossing the bifurcation point, the system became bistable, the system state jumping into one of the two emerging alternatives. What we have now is a self-referential system, a dynamical system that changes its parameters by itself (Der and Martius 2012). The decisive point in this scenario is the fact that (i) the learning rule does not introduce **explicitly** any violations of symmetries of the physical system it is applied to, but that (ii) the learning is driving the physical dynamics towards activity, eventually causing a **spontaneous** breaking of existing symmetries. This paper will present many examples of behaviors arising from this mechanism of spontaneous symmetry breaking.

7.4.3 Relation to Infomax Principles

As mentioned already in (Martius et al. 2013), there is a close relationship of the learning rules to the so called Infomax principles. Maximizing the mutual information between input and output of a unit is widely known as InfoMax. Applied to a layer of neurons the principle yields an explicit parameter dynamics (Bell and Sejnowski 1995) structurally similar to the one presented in (Martius et al. 2013) and the one given here. Also, similar rules have been obtained in (Triesch 2005) where the entropy of the output of a neuron was maximized under the condition of a fixed average output firing-rate. The resulting dynamics is called intrinsic plasticity as it acts on the membrane instead of on the synaptic level and it was shown to result in the emergence of complex dynamical phenomena (Butko and Triesch 2005; Lazar et al. 2006; Triesch 2007; Lazar et al. 2011). In (Markovic and Gros 2010, 2012) a related dynamics is obtained at the synaptic level of a feedback circuit realized by an autaptic (self) connection. In a recurrent network of such neurons it was shown that any finite update rate (ε in our case) destroys all attractors, leading to intermittently bursting behavior and self-organized chaos.

Our work differs in two aspects. On the one hand, while the information theoretical principle—the starting point for our more intuitively based learning rule—was formulated at the level of behaviors of the whole system, explicit learning rules were

[3] We write the time as a superscript if components of a vector are considered.

derived at the neuronal level by rooting the behavioral level (outside world) back to the level of the neurons (internal world). On the other hand, as a direct consequence of that approach, there is no need to specify the average output activity of the neurons as in (Triesch 2005). Instead, the latter is self-regulating by the closed loop setting. Independent of the specific setting, the general message is that these self-regulating neurons realize a specific working regime where they are both active and sensitive to influences of their environment. Instead of studying those neurons in internal (inside the "brain") recurrences, we embed them into a feedback loop with complex physical systems where these self-active and highly responsive neurons produce surprising phenomena at the level of behaviors—in the outside world.

7.5 Homeokinesis: Body Inspired Behavior

The unsupervised learning rule (ULR) given above shares some common features with the corresponding rule forming the basis of homeokinesis. In the Chapter by Martius, Der, and Hermann, the latter approach will be considered in some detail, demonstrating the potential of that approach for guided SO. This section will give a very short account of the principle of homeokinesis in order to discuss the parallels and differences of the two related approaches. Reader not interested in this comparison may skip this section and go directly to the applications.

Homeokinesis (HK), a general principle originally proposed in (Der 2001; Der and Liebscher 2002), stands for the dynamical symbiosis between brain, body, and environment. It was shown by many works to drive robots to a self-determined, individual development in a playful and obviously embodiment-related way, see (Der and Martius 2012) for a detailed consideration. HK is not only the description of a goal, namely the dynamical symbiosis between brain, body, and environment but comes with a concrete realization in the form of a universal, unsupervised learning rule.

In order to discuss similarities and differences with the ULR of this work, we directly consider the HK rule obtained by the minimization of the so-called time loop error (TLE)[4]

$$E_{TLE} = v^T v = \zeta^T \frac{1}{CC^T} \zeta \,. \tag{7.12}$$

where ξ is the prediction error[5], $\zeta = (AG')^{-1} \xi$, $\mu = \left(CC^T\right)^{-1} \zeta$, and $v = C\mu$. Gradient descending the TLE leads to the following unsupervised learning rules for the parameters of the controller

[4] The matrix inverses have to be understood as pseudo inverses if the normal inverse does not exist.

[5] The minimization of the prediction error is also at the basis of more recent approaches on minimizing the free energy of the sensor process (Friston 2010; Friston et al. 2012; Friston and Stephan 2007; Friston 2012). In Friston's approach, the tendency of stasis is overcome by assuming additional priors that drive the system to activities. This is not necessary both in homeokinesis and in the present approach.

$$\Delta C_{ij} = \varepsilon_c \mu_i v_j - \gamma_i y_i x_j \,,$$
$$\Delta h_i = -\gamma_i y_i \tag{7.13}$$

displaying a noteworthy similarity with Eqs. (7.8, 7.9). Indices are running over all sensors and motors as before, i. e. $i = 1,2,\ldots,n$ and $j = 1,2,\ldots,m$ and we introduced the channel specific learning rate

$$\gamma_i = 2\varepsilon_c \mu_i \zeta_i \,.$$

These relatively simple update rules define the parameter dynamics of the controller, the learning of the self model being given by Eq. (7.7). The rules need some numerical precautions due to the matrix inverses which are discussed in detail in (Der and Martius 2012). Both in HK and in the present work, learning is not to be understood as the convergence towards a specific goal. Instead, the learning rates usually are chosen such that the parameter and system dynamics run on comparable time scales. In the neural network interpretation we have a fast synaptic dynamics which is constitutive for the behavior of the system.

7.5.1 Principles of Action

The peculiarities of the general learning rules has been demonstrated in various robot examples using both real and simulated robots, see (Der and Martius 2012). At a general level, essential features of a learning rule are revealed by considering the landscape of the objective over the parameter space. The time loop error, Eq. (7.12) is characterized by singularities acting as repeller (infinitely repulsive regions) for the gradient flow. This is a direct result of learning the system backward in time. These repulsive regions are easily identified. A first class of singularities is given by the zeros of g' (featuring in ζ), i. e. in the saturation regions of the neurons. The effect of this singularity, which leads to the anti-Hebbian term $-2\varepsilon_i y_i x_j$ in Eq. (7.13), is to keep the neurons away from the saturation regime. This is very reasonable since in that region neurons are not sensitive to their inputs.

Typical for the landscape based on the TLE is a further repulsive region resulting from the inverted matrix CC^T. The role of this singularity is most immediately seen if the robot is initialized in the situation $C = 0$, $h = 0$, (the least biased one in the sense explained in Sect. 7.6.1) so that the controller does not react to its sensor values at all. The slightest perturbation will quickly drive the parameters of the controller away from this unstable fixed point of the combined dynamics. In this way, feedback in the sensorimotor loop is generated and the robot is quickly driven away from this "do nothing" region in behavior space. This is a definite advantage if it is important to get away from that singularity as fast as possible. In our new rule, the "do nothing" region is not a repeller but an unstable fixed point. This may be a disadvantage in the bootstrapping process. However, with the new rule, one may start directly with the least biased initialization. This is beneficial if one wants to make maximum use of the symmetry breaking phenomenon. This paper presents several examples which were not possible in the HK approach.

Another singularity is produced by the inverse of the self-model matrix A featuring in ζ. That one can also be present in the new approach if δy is calculated not by backpropagation but by the so called backprojection method. We consider this as a more or less technical issue so that we will not discuss it here, see (Der and Martius 2012) for details.

Behaviors generated by either the HK or our new rule are inherently contingent (there is no influence from outside) but by far not arbitrary since the whole bootstrapping process is driven by the specific reactions of the embodied robot to the controller signals. Thus, it is the body itself which plays the most active part in the emerging control process so that this phenomenon has also been called body-inspired behavior.

7.6 Vehicles: Behavior As Broken Symmetry

Let us now apply the new learning rule to some specific examples chosen such that the characteristic properties of the self-organization process are illustrated. This section will focus on the TwoWheeled robot.

7.6.1 Least Biased Initialization

In applications, a first point is about the choice of the initial parameters of the networks and the initial configuration of the robot. With our specific choice of the controller network, the initialization with $C = 0$ seems most natural because this corresponds to a controller that is completely numb, i. e. deprived of any functionality. Putting additionally $h = 0$, we find that all motor neurons send the command $y_i = 0$ to the controller, independently of any inputs.

Choosing the initialization in the described way has different effects on the initial pose the robot is taking. For example, in the wheeled robots case this means that all wheels are held at rest. In robots consisting of several segments tied together by joints, $y = 0$ means that all joints are driven towards their center position. We will follow up this point in the detailed examples of the Hexapod and the Humanoid further below.

The combined system, comprising the physical and the synaptic dynamics, is fully deterministic, in the virtual case at least. If starting in the least biased initialization the combined system may be in an unstable fixed point so that we have to add, for a short time interval, a little noise to the sensors. After switching the noise off, the actual initial condition is fixed. From this time on, the further time evolution of the entire system is deterministic, being fully determined by the values of the sensor vector, the current pose of the robot and the values of all parameters of the controller (now already different from zero). This set of values may be called the behavior code of the robot.

7.6.2 Symmetry Breaking—A Rule of Thumb

Before going to present the experiments, let us formulate a simple rule of thumb on the development of the robot when starting from the least biased condition: in typical experiments we observe that the behavior of the robot can be described as being active (caused by the differential Hebbian term in the learning rule) while conserving as much of the original symmetries of the system as possible. This was also formulated as the economy or parsimony of symmetry breaking. Note that symmetries involve not only the geometry of the robots body but also all the symmetries of the physical dynamics. In the two-wheel robot case the body geometry is described by the invariance against both left-right and forward-backward transformations. The physical symmetries are based on the robot being an object in space and time, the physics being invariant against both translations and rotations of the frame of reference, taking however account of the physical boundaries (objects, the ground the robot is on). Moreover, there is the invariance against the inversion of time combined with the inversion of all velocities.

Let us also emphasize that symmetry breaking observed in this scenario is emerging as a phenomenon "from inside" the deterministic system itself so that we may speak of a spontaneous symmetry breaking (SSB). We would also like to stress that fundamental modes are singled out by a maximum degree of symmetry. This may give the concept of fundamental modes a more solid founding but this still needs some work to be done. As an additional feature, the breaking of the symmetries can largely be influenced by external impacts (physical forces in the sense of a desired mode) and/or by choosing specific sensor combinations that help to organize the symmetry breaking scenario. We will give an example with the bungee jumper below.

7.6.3 The Autistic Vehicle: Fundamental Modes

We have introduced above the vehicle with the wheel counters as the only sensors. We have called it autistic because its only contact with the external world is by proprioception. Moreover, the learning starts in the least biased way, so that the symmetry breaking should follow the principle of parsimony mentioned above. In particular, the physical system (realized in the simulator by N differential equations, where $N = 8$ with the vehicle), is invariant against spatial transformations, i. e. translations or rotations of the spatial frame of reference. With the constraints given by the (elastic) surface, the remaining symmetry operations are rotations around the z axis and translations in the $x - y$ plane. Also, as discussed, the learning rule gives no clue of how symmetries are to be broken.

We consider now the TwoWheeled with a two-neuron controller receiving sensor vector $x \in \mathbb{R}^2$ representing the wheel velocities and sending its output $y \in \mathbb{R}^2$ to the motors. Starting with the least biased initialization, and believing in our rule of

Fig. 7.2 The TwoWHEELED as a 3D physical object. The ground is elastic so that the wheels are sinking in, depending on their load. Left, wheelsize = 1: with the given elasticity, the robot is lying more or less flat on the ground when driving straight. There is a strong slip effect when accelerating. Moreover, when moving in a curve, there is an inclination due to the physical forces making the effective radius of the wheels different, see the video S2. This is even more pronounced with larger wheel sizes (middle and left, size = 1.2). These 3D effects make both odometry and the execution of motion plans very difficult as they involve the full physics of the robot.

thumb, we expect the robot to start moving after some time[6] while trying to conserve as much of the original symmetries as possible.

However, when using a learning rate ε not too small, the robot is seen to engage in a sequence of left and right turns combined with motions back and forth along curved lines, without any regularity to be seen, see Fig. 7.3, top left panel. Still, we must note that these trajectories are fully deterministic. Restarting the simulation with exactly the same parameters C, h and in the same physical state reproduces exactly the same trajectory however irregular it is. Nevertheless, our rule of thumb obviously is not valid in this regime as there is no visible footprint of the underlying symmetries—the invariance against rotations and translations of the physical space.

The situation changes drastically when using smaller learning rates so that the interplay between learning and physical state dynamics is given time to unfold. Fig. 7.3 is demonstrating a typical behavior of the robot. As the top right panel shows, after starting, the robot is running through a kind of metastable patterns, converging after some time toward a large scale CP. The bottom left panel shows the fine structure of that limit cycle behavior.

In order to discuss this phenomenon, let us take at first the perspective of an external observer in the sense of Braitenberg. From this external point of view, the robot is drawing a complex geometrical pattern with a surprisingly high accuracy. In order to do that, the robot would have to have a detailed plan of the pattern to be drawn. In order to execute that plan, the robot would need a very accurate system of self-localization for finding its exact local position and heading direction. This could be done by either using a compass and a GPS system (which is not available) or by odometry, based on the measured wheel velocities. However, this is very inaccurate, given that the robot in the simulation is moving on an elastic

[6] When using low learning rates, this time can be very long so that we often start the robot with an initialization close to the bifurcation point, choosing $C = c\mathbb{I}$ with c close to 1. Contrary to the HUMANOID and HEXAPOD treated below, in the TwoWHEELED case, no substantial differences in the behaviors were observed.

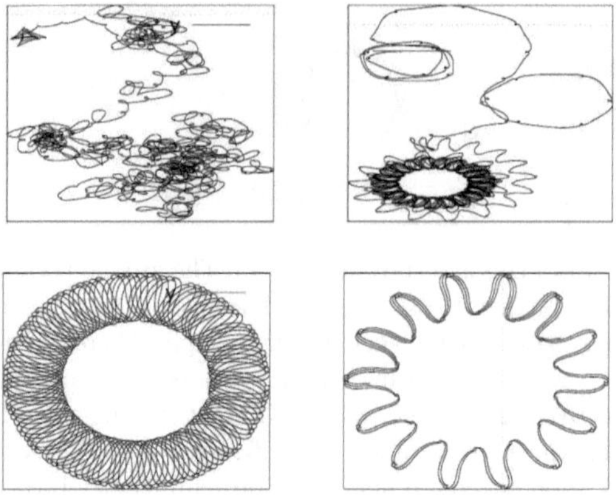

Fig. 7.3 Deterministic trajectories of the robot in the 2D plane emerging with different learning rates ε. If learning is fast ($\varepsilon > 0.01$), irregular trajectories are the rule. With lower rates (here $\varepsilon = 0.001$), after a transient phase of irregular motion through a kind of metastable attractors, the dynamics is converging toward a limit cycle behavior (top right and bottom left panel). A first explanation for the emergence of a circular pattern (CP) is given by using an SO(2) matrix for controlling the robot (bottom right). This leads to a CP but does not explain the fine structure of the CPs observed under learning.

ground so that, when accelerating or moving in a curve, there is an inclination due to the inertia and/or centrifugal forces making the effective radius of the wheels different, see Fig. 7.2. On top of that, the odometry would be very demanding in the computational resources given the required accuracy.

7.6.4 Synergy of Learning and Physical State Dynamics

On the other hand, when looking inside the system, none of the above ingredients for drawing a prescribed pattern with the observed accuracy can be found. Instead, all we see are the two poor neurons with their universal, low-complexity synaptic dynamics. Metaphorically, from this point of view, nothing like a concept of space and/or a plan for drawing a particular pattern can be uncovered. Even more so, it is even impossible of finding the plan inside the robot. In fact, given the wiring, the emerging patterns are specific with the concrete embodiment as they emerge by the interplay with the physics of the system only.

Still, an understanding of the phenomenon is in reach when looking at the structure of the controller in one of the limit cycle patterns. When taking a snapshot of the controller matrix C, we discover a nearly perfect $SO(2)$ structure of the C-matrix, see Fig. 7.4, with tiny modulations in synchrony with the motions of the robot. Any such matrix can be written as a rotation of a given vector v by an angle α and a stretching factor ρ. When using such a matrix in Eq. (7.5), the robot will either draw a (nearly) perfect cycle or, as shown in Fig. 7.3 (bottom right), will generate a circular pattern similar to the co-learning case but with a very different fine structure. Obviously, the fine structure of the patterns is produced by the coupling between physical and learning dynamics so that the rotation matrix does give only a first orientation on the emerging structures. Instead, as we will see below, both the very nature and the fine structure depend additionally on the embodiment in an intricate way.

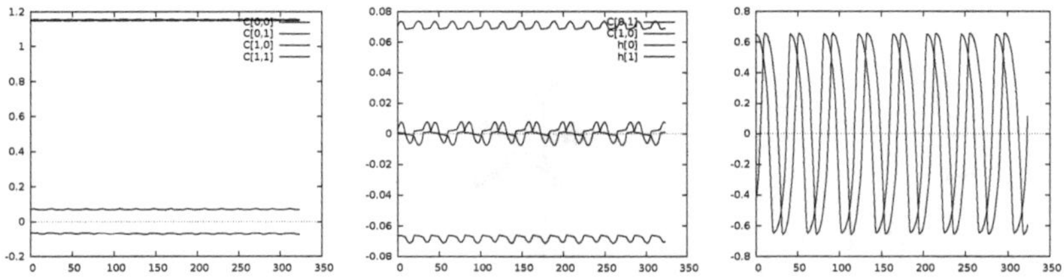

Fig. 7.4 The circular pattern formation is hidden in the dynamics of the controller parameters as driven by the general learning rule. The C-matrix (left and middle panel) is seen to be of a nearly perfect $SO(2)$ structure. This reveals another aspect of the general learning rules—the emerging dimension reduction in the parameter space. Instead of 4 parameters, the $SO(2)$ matrix is described by only one parameter (the rotation angle). The middle panel displays the interplay with the h dynamics which is seen to be periodic with a slight bias. Right panel: the two sensor values (wheel velocities).

So, from looking inside, there is a coarse explanation of how the robot does it, once it has "understood" so to say, what a rotation matrix is and how it can be used. However, this does not explain why the learning discovers these special matrices in the four dimensional matrix space and moreover how to use them in the fine interplay between the parameter and the (physical) state dynamics.

On a general level, an understanding is given by or rule of thumb: a pattern in space can only emerge from breaking the spatial symmetries inherent in the physics of the robot as mentioned above. When trying to make this symmetry breaking as parsimonious as possible, a circle is nearly perfect: while it has broken the translational symmetry (the center is a fixed point in space), rotation symmetry (around that center) is fully conserved. Yet, because of its fine structure, the real patterns emerging in our learning scenario are not circles, but they are still invariant against rotations about a definite angle, see in particular the patterns of Fig. 7.5. This may be seen as a noteworthy parallel to the hexagonal patterns known from many phenomena in nature. So, the observed patterns apparently are the ones with a high degree of preserving the spatial symmetries of the physical system.

There is a more naive argument for the predominance of such regular patterns. We have seen above that, once the initial condition is fixed (see Sect. 7.6.1), the trajectory of the robot is exactly determined for all future by the values of that initialization—its behavior code. Our point is that a completely irregular trajectory needs infinitely much information stored in its code. To the contrary, codes of regular trajectories need much less information so that there are whole sets of values generating one and the same regular trajectory (in a finite but large time interval) so that the latter are realized with much higher probability.

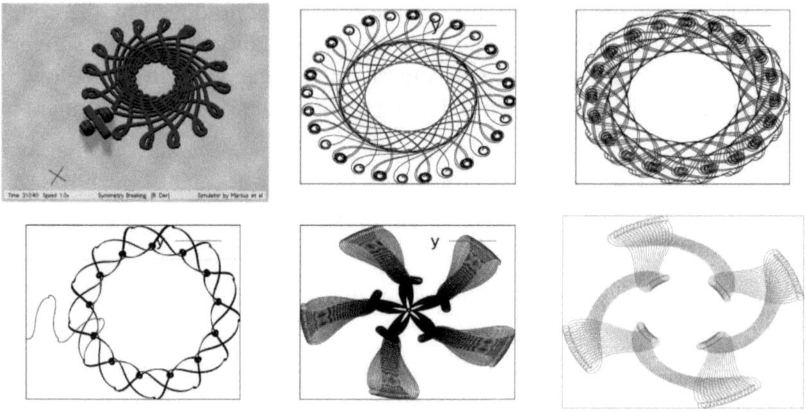

Fig. 7.5 Pattern spin-off effect in the learning on-off scenario. When in a limit cycle (circular pattern) and temporarily switching learning off, a new stable structure is emerging. After switching learning on again, the full dynamics is converging back to the limit cycle. In repetition, this produces a large variety of different structures. Top left: Pattern with robot for size comparison. However, patterns come on very different length scales. See video S3 for an example.

7.6.5 The Pattern Factory

Further insight into the role of the learning dynamics in the pattern formation can be gained in a kind of learning on-off scenario: we start with the full learning dynamics, let a circular pattern, see Fig. 7.3, develop and put $\varepsilon = 0$ for a while. Most astonishingly, the robot starts drawing a new pattern of very high geometric regularity resembling only remotely the circular structure it was spinning off. Upon switching on learning again, the system rapidly returns to the original CP (with a different center in space). This procedure can be repeated with ever new patterns emerging.

In discussing this pattern spin-off effect, we have to remember that once $\varepsilon = 0$, both C and h are fixed so that the spin-off pattern is fully determined by the physical starting state. In a number of experiments we even observed that each parameter set created in the learning on-off scenario is seen to even support several behaviors, each with its own basin of attraction. What we want to emphasize is that all these patterns are off-springs of the CP with behavior codes defined by the snapshots of the parameter dynamics. So, actually, we may say that, once in (or close to) the limit cycle attractor, the learning (parameter dynamics) produces a whole sequence of behavior codes for potential behaviors.

7.6.6 Patterns As Expressions of Embodiment

Let's now have a closer look at the role of the embodiment. As already explained with Fig. 7.2, the TwoWheeled must be considered as a full 3D system moving on

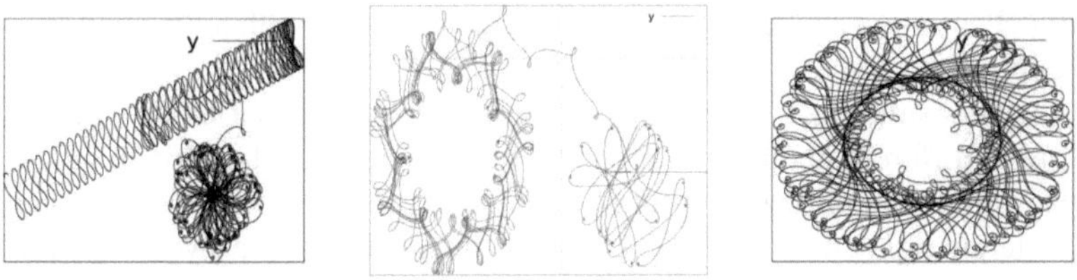

Fig. 7.6 The role of embodiment. Left panel: wheel size 1.1, $\varepsilon = .001$ and cInit=1.2. After a very irregular initial phase, the robot enters an aligned wiggly pattern, running at first to the right and then back toward the left lower corner. Middle: Wheel size 1.15. Right: asymmetry of 1.5:1.75 in the wheel-trunk system with wheel size 1.2. The pattern is less regular.

an elastic ground. Let's consider a few experiments demonstrating how sensitively the emerging behaviors depend on the concrete realization of the robot's body.

As explained with Fig. 7.2, the physics of the robot already changes substantially with changing the wheel size. As demonstrated with the left panel of Fig. 7.6, changing the wheel size from 1 to 1.1 leads to a new pattern—instead of a circular structure we get convergence toward an aligned pattern structure, i. e. a wiggly structure organized along a line. On the general level we may argue that now it is not the rotation symmetry that is partially conserved, but the translational one. In fact, a line is invariant against translations but not rotations in the plane.

As the middle panel of Fig. 7.6 shows, this effect is very sensitive to the choice of the wheel size—increasing the size to 1.15 produces a circular pattern again but with a very different fine structure. The emerging patterns are also very sensitive to the forward-backward asymmetry of the wheel-trunk system, see right panel of Fig. 7.6 for an example. Depending on that degree of asymmetry many different patterns can be observed, mostly circular ones but with widely varying fine structure.

7.6.7 *Modes*

We see that the hidden symmetries of the physical system can in a very intricate way propagate into the combined state-parameter dynamics of the developing system to produce large-scale effects in space and time[7]. When looking at the limit cycle patterns (the circular structures), we see a very complicated periodic or quasi-periodic structure interweaving the physics and the learning dynamics in an irreducible way. This synergy between state and parameter dynamics is breaking down

[7] This effect seems to have been observed already in the homeokinesis (HK) approach (Der and Martius 2012) with the sphere that adapts its rolling motion to the geometry of the basin. Here, we have a more subtle effect, given that there is no formative external geometry, the only link between the internal world (learning dynamics) and outer world (physics) is by the wheel velocities which give information about the behavior in space only in an indirect way and, even in the simulations, with much restricted accuracy. It is only by this window into the outer world by which physical effects, mainly the inertia due to the mass of the body and the wheels, take influence on the behavior generation.

immediately as soon as we freeze the learning dynamics, with a new pattern emerging as explained above.

We may consider each emerging limit-cycle dynamics as a mode of the system. These are of different nature, either with the parameters fixed or driven by the learning dynamics. In the no-learning case, there is an immediate parallel to Braitenberg's idea of interchanging perspectives: looking from inside, we see a vehicle with a fixed wiring of sensors to motors, now also with cross connections and some bias for the neurons (active wires). This extremely simple electro-mechanical system is able of drawing very complex geometrical patterns like the ones shown in Fig. 7.5. Looking from outside, even if the self-localization problem was solved, the system seems to follow a very complex plan for drawing such highly regular geometric patterns. But also in the learning case, the Braitenberg perspectives are still appropriate. If considering the parameter dynamics as the dynamics of some internal state variables, we still have a very simple controller exciting those very complex modes.

Yet another perspective is suggested by the autism metaphor: we could say that, by the learning dynamics, the robot is somehow taking note of its being in space (creating patterns) but this without any internal reflection of that fact. Metaphorically, we may say that the robot does not know what it does but it does this with high accuracy and dedication, discovering by proprioception alone a specific way of being in space and time.

The observed limit cycle attractors, both with and without learning, are examples of the many modes this specific brain-body system is able of developing. We are free to call each of them a fundamental/normal mode but we also may single out the co-learning mode as the most fundamental one as it is the host of all the spin-off modes. This is the attitude taken in this paper.

7.7 The Looping HUMANOID

Let us now continue our considerations with a more complex system, the HUMANOID introduced above. The results in Sect. 7.3 have demonstrated that under certain conditions this system is particularly amenable to fundamental modes. The observed modes may be attributed to a specific form of broken symmetries under minimalistic control. Using the HUMANOID we are now going to look a little deeper into the scenario of spontaneous symmetry breaking in systems of higher complexity than that of the TWOWHEELED.

7.7.1 High Symmetry Motion Patterns

As in the TWOWHEELED case we start with the least biased initialization, wondering what the emerging symmetry breaking scenario will look like. We use the bungee scenario with the same physical setting as in Sect. 7.3. After letting the robot fall, we observe again the vertical oscillations around the equilibrium point but after a while (several minutes real time with $\varepsilon = .001$) the robot develops a specific motion pattern looking like a swimming motion, see Fig. 7.7 and video S4.

On a general level, the observed pattern illustrates very clearly the rule of thumb: the robot is being active but the original symmetries of the physical system are conserved as much as possible. This is obvious for the geometrical left-right symmetry but we may also argue that the original symmetries of the physical system are still dominating the behavior. In the least biased initialization, there is an invariance against inversion of joint angles[8]. Breaking the symmetry means to define for each joint angle where to go by what velocity. Without any external clue, the most parsimonious solution would be to change all the joint angles with the same velocity, generating a phase and frequency locked dynamics of the mechanical system. Actually, this is what we observe at least in the early phases of the behavioral self-organization.

Looking more into the details, we see a qualitative difference to the running motion pattern (RMP) of Sect. 7.3 with its in-phase and anti-phase signature. As all other conditions are equivalent, this must be caused by the different initializations. In the RMP, the strong negative feed-back strength of the system was discussed responsible for the specifics of the RMP, see Sect. 7.3.2. In this setting, we start with $C = 0$ and $h = 0$, the least biased initialization, everything being determined by the learning, starting with the initial perturbations introduced by the physics. In the beginning, the vertical oscillations introduce tiny changes in the joint angles due to inertia effects not fully counteracted by the motors. These are acting on all joints simultaneously as they are caused by the motions of the center of gravity. Once started, these perturbations are self-amplifying which may explain why all angles are changing simultaneously. Later on, there may also be some physical cross-talk as described with the RMP in Sect. 7.3. This is now possible not only in the joints of a single limb but may propagate also between the upper and the lower part of the body and between the arms and legs.

The cross-talk phenomenon shows that the symmetry breaking mechanism is very sensitive to the concrete morphology of the robot. By way of example, we use an experiment where the robot was given an additional back joint, giving the robot more freedom in movements of its lower part. Moreover, we fixed the trunk so that there is no starting perturbation by the vertical oscillations as in the bungee setting. Still, there is a physical starting signal given by the difference of the joint angles in the initial configurations (all equal to zero) and their equilibrium position (different due to gravity). Again, after a transient initial phase, we observe the emergence of a collective mode involving all degrees of freedom of the lower part, see Fig. 7.8. We guess that this collective motion is more a result of the physical cross-talk than the fingerprint of the initial symmetries, but we still have to study this in more detail.

7.7.2 *Exterioception May Guide Self-Organization*

Up to now we were using only proprioceptive sensors so that the orientation of the bungee jumper in space came into play only in a very indirect way, for instance by

[8] In order to make this explicit, think of a linearization of the system around its equilibrium point defined by the initialization. Details will be given elsewhere.

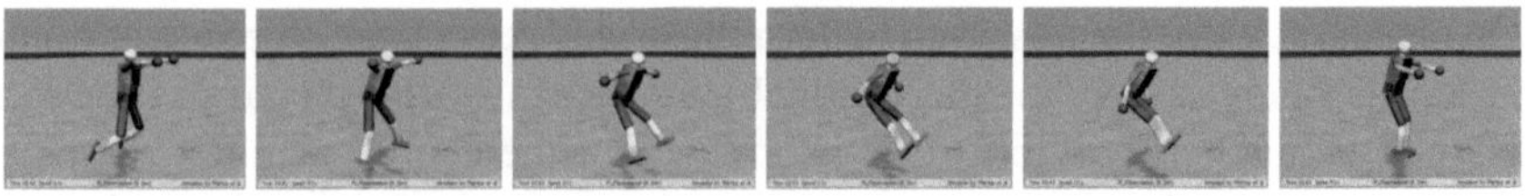

Fig. 7.7 One period of an emerging motion pattern in the bungee scenario. The pattern is an example of minimal symmetry breaking.

Fig. 7.8 One period of an emerging periodic motion pattern with the trunk fixed and an additional back joint giving the robot more freedom of motion. Coherence is now caused not only by the original symmetries but also by the physical cross-talk becoming "louder" with increasing activity of the robot.

the gravity forces providing additional loads on the joint motors. Let us now include a sensor that is orientation sensitive. In a first experiment we use a sensor that measures the rotation velocity of the trunk around its yaw axis (the one running through both the shoulders). This velocity is simply taken as an additional component of the sensor vector x. We let the system run with the learning algorithm of Eq. (7.8) just as in the previous case.

Astonishingly, the emerging behavior is completely different from the previous one. We observe now already after a short time that the robot tries rotational motions around that axis (known to him only by one sensor value without any knowledge what this sensor is measuring). In the video S 5, the amplitude of these rotational motions is seen to increase steadily until the robot is looping, see Fig. 7.9. After that it is in a physically less stable state (by loss of symmetry in its motion pattern) so that it needs some time until the play may begin again.

Again, we can recognize the rule of thumb here. In fact, the motion still—while being of quite some variability—reflects the original symmetries of the physical system in its least biased initialization. In particular, the looping is performed only if the robot is keeping in a motion regime where the left-right symmetry with respect to the yaw axis is nearly perfect, see the video S 5.

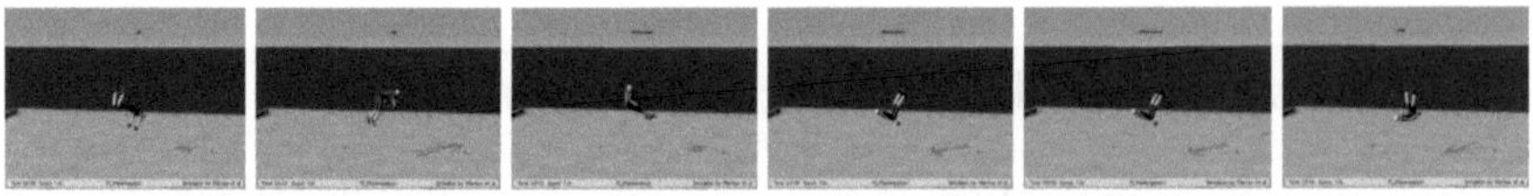

Fig. 7.9 With a sensor measuring the rotation velocity of the trunk around its yaw axis, the robot synchronizes its internal motions with the trunk rotation. After a while, the robot learns rotating its entire body around the trunk axis, eventually executing a loop.

7.7.3 Starting in a Mode

What happens if we start with the above minimalistic controller, i.e. use the initialization $C = c\mathbf{I}$ with $c < -1$? As video S 6 shows, the learning enhances the running motion pattern (RMP) in the course of time more and more by steadily increasing the step length until the whole motion pattern gets unstable after a long time. However, as video S 6 also shows, the stability of the RMP is holding up to a very large step length. This video also shows that the RMP is entirely self-generated. When the bungee force is reduced so that the robot reaches ground, the RMP decays but is resumed rapidly if the conditions of its existence (hanging in air) are reestablished.

7.8 The HEXAPOD

Let us now follow the trace of symmetry breaking a little further down the road with a robot of comparable complexity as the HUMANOID but with a different symmetry and interaction with the environment: the HEXAPOD. We choose this robot since it has not been treated in earlier work on HK (Der and Martius 2012) and because it will be seen to reveal symmetry breaking phenomena in a particular clear way. The robot consists of a trunk with two whiskers (passive joints), and six legs, each one consisting of a shoulder with two joints and a knee with a single joint. Each of the 18 joints is activated by a motor and contains a sensor that is measuring the true joint angle. The effective torques acting on the joint axes are determined by a PID controller so that there is an elastic reaction of the robot to the nominal joint positions $y \in \mathbb{R}^{18}$, similar as in a system controlled by muscles.

In a typical experiment, the HEXAPOD is falling down from a starting position a little above the ground. With the least biased initialization, $y = 0$ in the beginning so that all joints are in their center positions. When hitting the ground, the robot gets into a damped vertical oscillation due to the elasticity of the joint-motor system. This is sufficient for providing an initial perturbation that is further amplified by the bootstrapping mechanism as sketched in Sect. 7.6.

7.8.1 Modes

What can we expect to happen? Depending on the concrete situation (for instance the meta learning parameter, in particular ε) different behaviors may emerge. In most cases the robot starts with a swaying and rolling motion pattern, see Fig. 7.10 and video S 7. We may claim again, that this is in agreement with our rule of thumb since in this motion the joint angles are changing with a pretty high degree of coherence as allowed by the (soft) physical constraints enforced by the ground contacts.

More interesting behaviors emerge after some time. Most exciting is the emergence of a jumping behavior as a stable phenomenon, see Fig. 7.11 and the corresponding video S 8 and S 9. In these jumping patterns the robot is seen to be very active but still with a large coherence in the joint angles as suggested by our rule of thumb. There is still a surprise when looking at the parameters of the controller.

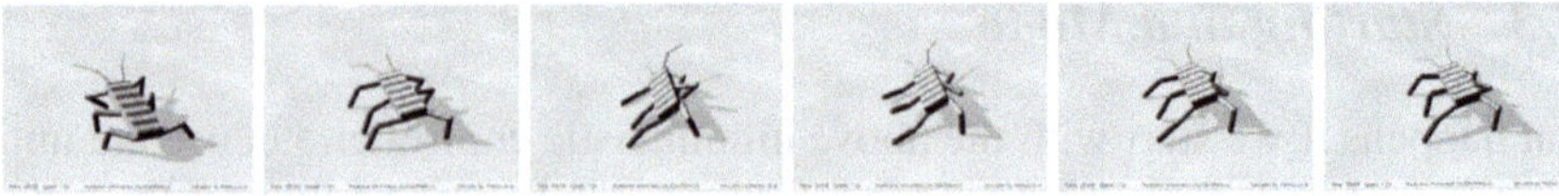

Fig. 7.10 Initially, the robot develops a swaying motion pattern, as if it is very actively trying to move the legs in a coherent way while keeping ground contact

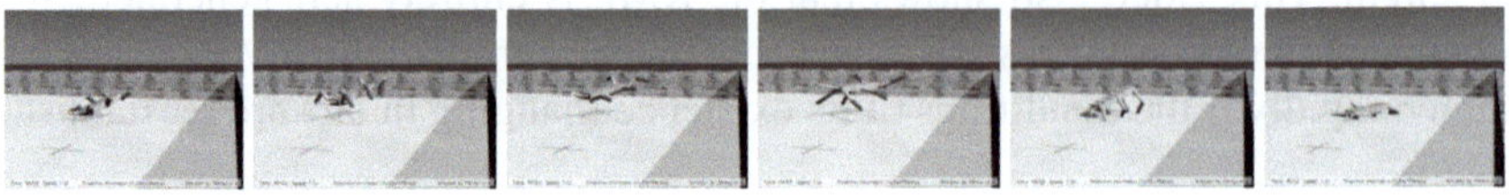

Fig. 7.11 Emerging jumping motion pattern

In the TWOWHEELED case we observed that the C-matrix displays a definite structure, it developed into a $SO(2)$ matrix which is prominent among all $\mathbb{R}^2 \times \mathbb{R}^2$ matrices in forming a group. This is like the symmetries of the physical space leaving a fingerprint since the physical space in $2D$ is invariant against all transformations by the group of $SO(2)$ matrices.

Of course, we can not expect such a clear result in the case of our HEXAPOD because of the much higher dimensionality of the physical space and the fact that the robot is also interacting with the ground. Yet, a look at the C-matrix is the more surprising. As Fig. 7.12 shows, the emerging sensor-to-motor coupling matrix is highly structured, reflecting the original symmetries to an amazingly high degree. Both the shoulders and the knees are seen to follow essentially the same strategy for

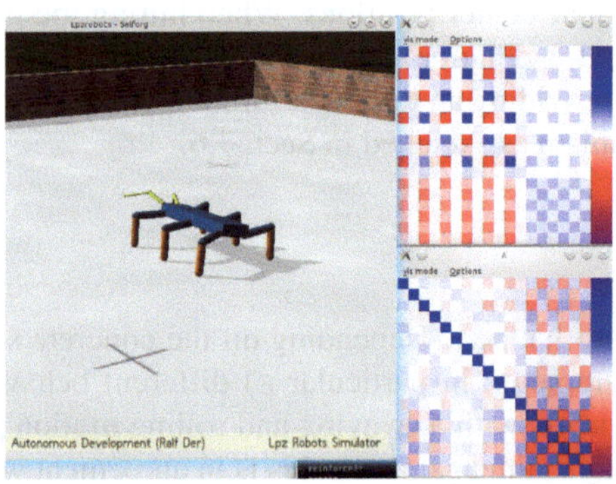

Fig. 7.12 The parameters of the controller (C-matrix, upper right panel) and the model (A-matrix). In the C-matrix, the upper left 12×12 block displays the matrix elements of the shoulder couplings, each row i representing the coupling strength of the 12 angle sensors into the ith motor neuron. Reading each row from the diagonal element to the right and wrapping the reading back to the diagonal, one sees that each shoulder follows essentially the same strategy. A related understanding of the control strategy holds for the knee joints, see the lower right sub-matrix.

moving the body. This is in agreement with our rule of thumb since this collective strategy allows the body to be moving, even to jump, but with a maximum degree of coherence between the individual constituents of the body.

Moreover, we also see from that figure the whole-body nature of the behavior. The periodic jumping pattern are not created by using a central pattern generator producing a master signal that is sent with the necessary phase corrections to the individual motors. Instead, the motion of each body part is generated by combining both excitatory and inhibitory signals from the sensors of all joints in a systematic manner. This is a new control strategy for generating jumping (and, hopefully, also walking like) patterns emerging from simply applying the general learning rule to the physical system.

7.8.2 Perspectives for Guidance and Reinforcement Learning

Important is also the observation that, by including exterioceptive sensors, the development of the modes can be influenced and driven into desired directions. For instance, as our experiments show (to be presented in detail in a later paper), the jumping height can be increased by simply including a sensor measuring the vertical velocity of the robot trunk. The effect can be increased if, like in reinforcement learning, the parameter updates are weighted according to the agreement with the reward. For instance, by using the forward velocity of the robot as a reward, we can guide the SO process towards jumping into the forward direction, see Fig. 7.13 and video S 10 as an example.

There are many more interesting phenomena observed with that kind of robots but we are not going into more details here. We think that the few examples given may serve as an outline for the bunch of phenomena emerging from SSB under the general learning rule, given a convenient initialization and some patience for doing the experiments.

Fig. 7.13 Jumping motion pattern driven by rewarding the forward direction

7.9 Concluding Remarks

This paper tried to answer essentially two questions. The first question is about how to organize self-organization, asking how can we find intrinsic mechanisms that make a system able to self-organize. The answer was given by the unsupervised learning rule (ULR), see Eqs. (7.8, 7.9), that drives systems into self-organization. The ULR stages a weighted competition between a differential Hebbian and an anti-Hebbian learning mechanism. While the former drives the system to activity, the latter acts as a confinement, keeping the system under control. The rule is strictly local—synaptic changes exclusively depend on excitations at the ports of the synapse—yet creates global, whole-body motion patterns of the robotic system. This is demonstrated in applications with (i) a wheeled robot that spontaneously is moving in geometric patterns, and (ii) a hexapod robot with 18 degrees of freedom self-organizing from scratch into several locomotion patterns like jumping and walking.

The second bunch of questions is suggested by exactly that bootstrapping scenario: with nothing specified from outside, what can we expect the learning system to do. What will the emerging behaviors look like and what will the relation to the embodiment of the robot be? How and to what extend are the emerging behaviors determined by the embodiment; and can we find systematic criteria for those behaviors?

Several answers could be given by looking into the role of the underlying symmetries of the system in space and time. The invariance of the physical space against symmetry operations (like translations and rotations) induces corresponding invariances modified by the constraints (like contact with the ground). The point then is that, while driving the system towards instability, the ULR is preserving these symmetries. As a result, the evolution of the system in the learning process is realized by a sequence of spontaneous symmetry breaking following—similar to what we know from nature—a kind of parsimony principle. This lead to our rule of thumb: the emerging behaviors in physical systems (robots) driven by our ULR are qualified by a high activity while preserving as much of the underlying symmetries as possible.

This rule brings the embodiment to the foreground. The symmetries are embodiment specific and, moreover, breaking the symmetries is a process that is related to the very physics of the system. This was demonstrated by a number of examples. The first and probably the most surprising one was given by the TWOWHEELED robot. Controlled by two neurons with a fast synaptic dynamics given by the ULR, the system in many cases was converging towards a limit cycle behavior with the trajectories of the robot forming nearly perfect geometric patterns. The emerging geometric patterns where seen to depend on the embodiment (like the wheel size) in a very intricate and sensitive way. Interestingly, the limit cycle features as a pattern factory: in a learning on-off scenario it was seen to produce a great variety of further patterns, spinning off the limit cycle if the learning is turned down (called the pattern spin-off effect). It would be very interesting to ground this phenomenon in dynamical system theory.

Similar effects of symmetry breaking were obtained by the examples of the HUMANOID and the HEXAPOD. In the latter case, we also observed the excitation of body related, high activity modes with a high degree of coherence between the body parts. In particular, with the HEXAPOD, we observed a jumping mode demonstrating our rule of thumb: emerging behaviors are qualified by high activity while preserving the underlying symmetries of the system as far as possible (the principle of parsimony in spontaneous symmetry breaking). In future work we will be looking for a parallel of the pattern spin-off effect, hoping to thereby uncover a kind of pattern factory for these more complex systems, too.

These results are a step forward as compared to the state of the art. Previous work in self-organizing robot behavior was either restricted to small, easy to analyze systems or produced—like with the principle of homeokinesis—behaviors which looked interesting and were often completely surprising (Der and Martius 2012), as it should be. However, by the same token, it was often not clear what the robot actually is doing. With the concept of behaviors as broken symmetries, this is now (a little) different. The principles and examples given in this paper—in particular the emergence of fundamental modes, the TWOWHEELED as a pattern factory, the looping behavior of the HUMANOID, and last but not least the jumping modes of the HEXAPOD—may help us to better understand and exploit the synergy between embodiment and SO of autonomous robots.

Acknowledgements. The author gratefully acknowledges the great hospitality in the group of Nihat Ay at the Max Planck Institute for Mathematics in the Sciences and many helpful and clarifying discussions with Nihat Ay, Georg Martius, and Keyan Zahedi. I am also grateful to Mikhail Prokopenko (CSIRO Sydney) for very stimulating discussions on self-organization.

References

Anthony, T., Polani, D., Nehaniv, C.L.: Impoverished empowerment: 'Meaningful' action sequence generation through bandwidth limitation. In: Kampis, G., Karsai, I., Szathmáry, E. (eds.) ECAL 2009, Part II. LNCS, vol. 5778, pp. 294–301. Springer, Heidelberg (2011)

Ay, N., Bernigau, H., Der, R., Prokopenko, M.: Information driven self-organization: The dynamical systems approach to autonomous robot behavior. Theory Biosci. (2012)

Ay, N., Bertschinger, N., Der, R., Güttler, F., Olbrich, E.: Predictive information and explorative behavior of autonomous robots. The European Physical Journal B - Condensed Matter and Complex Systems 63(3), 329–339 (2008)

Behnisch, M., Haschke, R., Ritter, H., Gienger, M., Humanoids: Deformable trees - exploiting local obstacle avoidance. In: Humanoids, pp. 658–663 (2011)

Bell, A.J., Sejnowski, T.J.: An information-maximisation approach to blind separation and blind deconvolution. Neural Computation 7, 1129–1159 (1995)

Braitenberg, V.: Vehicles: Experiments in Synthetic Psychology. MIT Press (1984)

Butko, N.J., Triesch, J.: Exploring the role of intrinsic plasticity for the learning of sensory representations. In: ESANN 2006 Proceedings - 14th European Symposium on Artificial Neural Networks Bruges, pp. 467–472. Neurocomputing (2005)

Der, R.: Self-organized acquisition of situated behaviors. Theory in Biosciences 120, 179–187 (2001)

Der, R., Güttler, F., Ay, N.: Predictive information and emergent cooperativity in a chain of mobile robots. In: Artificial Life XI. MIT Press (2008)

Der, R., Liebscher, R.: True autonomy from self-organized adaptivity. In: Proc. of EPSRC/BBSRC Intl. Workshop on Biologically Inspired Robotics. HP Labs Bristol (2002)

Der, R., Martius, G.: The Playful Machine - Theoretical Foundation and Practical Realization of Self-Organizing Robots. Springer (2012)

Elbrechter, C., Haschke, R., Ritter, H.: Bi-manual robotic paper manipulation based on realtime marker tracking and physical modelling. In: IROS, pp. 1427–1432 (2011)

Friston, K.: The free-energy principle: a unified brain theory? Nature Reviews. Neuroscience 11(2), 127–138 (2010)

Friston, K., Adams, R.A., Perrinet, L., Breakspear, M.: Perceptions as Hypotheses: Saccades as Experiments. Frontiers in Psychology, 3 (2012)
Friston, K.J.: A free energy principle for biological systems. Entropy 14(11), 2100–2121 (2012)
Friston, K.J., Stephan, K.E.: Free-energy and the brain. Synthese 159(3), 417–458 (2007)
Grossekathofer, U., Barchunova, A., Haschke, R., Hermann, T., Franzius, M., Ritter, H.: Learning of object manipulation operations from continuous multimodal input. In: Humanoids, pp. 507–512 (2011)
Hauser, H., Ijspeert, A.J., Füchslin, R.M., Pfeifer, R., Maass, W.: Towards a theoretical foundation for morphological computation with compliant bodies. Biological Cybernetics 105(5-6), 355–370 (2011)
Hauser, H., Ijspeert, A.J., Füchslin, R.M., Pfeifer, R., Maass, W.: The role of feedback in morphological computation with compliant bodies. Biological Cybernetics 106(10), 595–613 (2012)
Jung, T., Polani, D., Stone, P.: Empowerment for continuous agent-environment systems. CoRR, abs/1201.6583 (2012)
Klyubin, A.S., Polani, D., Nehaniv, C.L.: Empowerment: a universal agent-centric measure of control. In: Congress on Evolutionary Computation, pp. 128–135 (2005)
Klyubin, A.S., Polani, D., Nehaniv, C.L.: Representations of space and time in the maximization of information flow in the perception-action loop. Neural Computation 19, 2387–2432 (2007)
Kolodziejski, C., Porr, B., Tamosiunaite, M., Wörgötter, F.: On the asymptotic equivalence between differential hebbian and temporal difference learning using a local third factor. In: Advances in Neural Information Processing Systems (2009)
Kolodziejski, C., Porr, B., Wörgötter, F.: Mathematical properties of neuronal td-rules and differential hebbian learning: a comparison. Biological Cybernetics (2008)
Kulvicius, T., Kolodziejski, C., Tamosiunaite, M., Porr, B., Wörgötter, F.: Behavioral analysis of differential hebbian learning in closed-loop systems. Biological Cybernetics (2010)
Lazar, A., Pipa, G., Triesch, J.: The combination of STDP and intrinsic plasticity yields complex dynamics in recurrent spiking networks. In: ESANN, pp. 647–652 (2006)
Lazar, A., Pipa, G., Triesch, J.: Emerging bayesian priors in a self-organizing recurrent network. In: Honkela, T. (ed.) ICANN 2011, Part II. LNCS, vol. 6792, pp. 127–134. Springer, Heidelberg (2011)
Markovic, D., Gros, C.: Self-Organized chaos through polyhomeostatic optimization. Physical Review Letters 105(6), 068702 (2010)
Markovic, D., Gros, C.: Intrinsic adaptation in autonomous recurrent neural networks. Neural Computation 24(2), 523–540 (2012)
Martius, G., Der, R., Ay, N.: Information driven self-organization of complex robotic behavior. PLOS ONE (in press, 2013)
Maycock, J., Dornbusch, D., Elbrechter, C., Haschke, R., Schack, T., Ritter, H.: Approaching manual intelligence. KI 24(4), 287–294 (2010)
Pfeifer, R.: "Morphological computation" - self-organization, embodiment, and biological inspiration. In: IJCCI (2012)
Pfeifer, R., Bongard, J.C.: How the Body Shapes the Way We Think: A New View of Intelligence. MIT Press, Cambridge (2006)
Pfeifer, R., Gomez, G.: Understanding Intelligence (1999)
Pfeifer, R.: Morphological computation – connecting brain, body, and environment. In: Sattar, A., Kang, B.-H. (eds.) AI 2006. LNCS (LNAI), vol. 4304, pp. 3–4. Springer, Heidelberg (2006)

Pfeifer, R., Lungarella, M., Iida, F.: Self-organization, embodiment, and biologically inspired robotics. Science 318, 1088–1093 (2007)
Prokopenko, M. (ed.): Foundations and Formalizations of Self-organization. Springer (2008)
Prokopenko, M.: Information and self-organization: A macroscopic approach to complex systems. Artificial Life 15(3), 377–383 (2009)
Prokopenko, M., Boschetti, F., Ryan, A.J.: An information-theoretic primer on complexity, self-organization, and emergence. Complexity 15(1), 11–28 (2009)
Ritter, H., Haschke, R., Steil, J.J.: Trying to grasp a sketch of a brain for grasping. In: Sendhoff, B., Körner, E., Sporns, O., Ritter, H., Doya, K. (eds.) Creating Brain-Like Intelligence. LNCS, vol. 5436, pp. 84–102. Springer, Heidelberg (2009)
Steffen, J., Elbrechter, C., Haschke, R., Ritter, H.J.: Bio-inspired motion strategies for a bimanual manipulation task. In: Humanoids, pp. 625–630 (2010)
Triesch, J.: A gradient rule for the plasticity of a neuron's intrinsic excitability. In: Duch, W., Kacprzyk, J., Oja, E., Zadrożny, S. (eds.) ICANN 2005. LNCS, vol. 3696, pp. 65–70. Springer, Heidelberg (2005)
Triesch, J.: Synergies between intrinsic and synaptic plasticity mechanisms. Neural Computation 19(4), 885–909 (2007)
Zahedi, K., Ay, N., Der, R.: Higher coordination with less control – A result of information maximization in the sensorimotor loop. Adaptive Behavior 18(3-4), 338–355 (2010)

Chapter 8
Robot Learning by Guided Self-Organization

Georg Martius, Ralf Der, and J. Michael Herrmann

8.1 Introduction

Self-organizing processes are not only crucial for the development of living beings, but can also spur new developments in robotics, e. g. to increase fault tolerance and enhance flexibility, provided that the prescribed goals can be realized at the same time. This combination of an externally specified objective and autonomous exploratory behavior is very interesting for practical applications of robot learning. In this chapter, we will present several forms of guided self-organization in robots based on homeokinesis.

Self-organization in the sense used in natural sciences means the spontaneous creation of patterns in space and/or time in systems consisting of many individual components. This involves the emergence, meaning the spontaneous creation, of structures or functions that are not directly explainable from the interactions between the constituents of the system. Examples are for instance spontaneous magnetization, convection patterns and reaction diffusion systems leading to the wonderful coloring of shells or animal coats. For robotic applications it is important to translate self-organization effects to a single robots considered as complex physical systems consisting of many constituents that are constraining each other in an intensive

Georg Martius · Ralf Der
Max Planck Institute for Mathematics in the Sciences,
Inselstraße 22, D-04103, Leipzig, Germany
e-mail: {martius,ralfder}@mis.mpg.de

J. Michael Herrmann
Bernstein Center for Computational Neuroscience,
Am Faßberg 17, 37077 Göttingen, Germany
Institute for Perception, Action and Behaviour, School of Informatics,
University of Edinburgh, 10 Crichton St, Edinburgh, EH8 9AB, Scotland, U.K.
e-mail: michael.herrmann@ed.ac.uk

M. Prokopenko (ed.), *Guided Self-Organization: Inception*,
Emergence, Complexity and Computation 9,
DOI: 10.1007/978-3-642-53734-9_8, © Springer-Verlag Berlin Heidelberg 2014

manner. This is what homeokinesis (Der 2001; Der and Liebscher 2002; Der and Martius 2012) or information theoretic approaches (Martius et al. 2013; Klyubin et al. 2005) to behavioral self-organization are after.

Homeokinesis or homeokinetic learning is based on a dynamical systems formulation of sensorimotor loops and introduces an objective function, called the time-loop-error. Intuitively it maximizes the sensitivity to sensor inputs while maintaining predictability with respect to an internal adaptive forward model. In practice homeokinetic control enables a robot to self-organize its behavior in a playful interaction with its environment and explores the suitable movement patterns for its particular embodiment. A short introduction to homeokinesis will be given in the following section. Then, we will face the question how goals can be introduced into a self-organizing system. Instead of imposing a goal we will aim at guiding the agents towards the desired behavior using as much of the intrinsic behavior as possible.

For the combination of self-organizing and external drives we coined (Martius et al. 2007) the term *guided self-organization* (GSO), which was before only rarely used e. g. in nano technology (Choi et al. 2005) or swarm robotics (Rodriguez 2007) and gained now a much larger scientific interest (Prokopenko 2009). Goal-oriented methods optimize for a specific task and require a prestructuring of the control problem in high-dimensional systems. Self-organization, on the other side, can generate coherent behavior and structure in the behavior space. Furthermore self-organizing systems show a great flexibility and high tolerance against failures and degrade gracefully rather than catastrophically (Prokopenko 2008, 2009). The perspective of GSO is to obtain a system which unites benefits of both. In the main sections of this chapter we discuss several approaches for guided self-organization with homeokinesis (GSOH). These methods span the range from incorporation of supervised learning signals to reward based methods and to teaching of structural relations.

8.2 Homeokinesis

Homeokinesis (Der 2001; Der and Liebscher 2002; Der and Martius 2012) is about establishing/stabilizing an internally defined dynamic regime of the sensorimotor dynamics and is thus conceptually similar to homeostasis (Cannon 1939; Wikipedia 2013), where a system has a internal set of states that are stabilized against external perturbation. So homeostasis is about keeping things fixed whereas homeokinesis is about keeping things moving. In effect homeokinesis produces a variety of behaviors in dependence on the interaction between control, internal dynamics and environment. Homeokinetic control arises from optimizing the sensorimotor coordination of an embodied agent to stay in a certain dynamical regime of sensitive but well controlled behavior. For that the movements are compared to the predictions by an internal adaptive model, and it works best with a controller and a model of similar complexity. The robot is thus controlled by a quasi-linear controller that receives sensor values and determines the motor values. If the coefficients of the controller are fixed then we have a purely reactive setup which can produce a particular reactive behavior. If, however, the parameters change the robot can produce a variety of behaviors. If done appropriately, e. g. as proposed below, a sequence of behaviors is obtained that are all locally smooth and simple but globally rather complex. The approach consists of adapting the parameters to maximize prediction quality and simultaneously to maximize sensitivity to changes in the sensor values.

Formally, we denote the vector of sensor values at time t by $x_t \in \mathbb{R}^n$. The vector of motor values $y_t \in \mathbb{R}^m$ is generated by a controller function

$$y_t = K(x_t, C, h) = g(Cx_t + h) , \tag{8.1}$$

where $g(\cdot)$ is a componentwise sigmoidal function, e.g. a hyperbolic tangent. The matrix C contains the modifiable parameters of the controller and h is a vector of bias values. The predictive internal model M uses sensor values and motor values to predict the sensory inputs one time step ahead.

$$x_{t+1} = M(x_t, y_t; A, S, b) + \xi_{t+1} , \tag{8.2}$$

$$M(x_t, y_t; A, S, b) = Ay_t + Sx_t + b , \tag{8.3}$$

where ξ is the deviation of the actually observed sensor values from their predictions. The matrices A and S are adapted such as to represent the effect of the actions and the previous sensory values, respectively, onto the new sensor values. The vector b, similar to h above, serves as an offset. Inserting Eq. (8.1) into Eq. (8.2) yields

$$x_{t+1} = M(x_t, K(x_t, C, h); A, S, b) + \xi_{t+1} = \psi(x_t) + \xi_{t+1} , \tag{8.4}$$

which is a stochastic dynamical system describing the temporal evolution of the sensor values. Considering that all the information the robot obtains arrives through its sensors, the dynamics (8.4) describes the behavior of the robot completely. Note, however, in this interpretation, Eq. (8.4) assumes a Markov property with respect to a fixed time step which may not be realizable in real robots in general.

While the controller determines the behavior of the robot and changes its state in the environment, the internal predictive model learns any new arriving sensory inputs by an online adaptation of the parameters A, S, and b via gradient descent. As a consequence, the prediction error $\|\xi\|^2$ (8.4) tends to decrease.

If the parameters C and h of the controller are also adapted by the minimization of the prediction error then the robot dynamics is subject to stabilization. The resulting behavior reflects the complexity of the environment to some extent, but is typically relatively simple or may simply approach a resting state.

Activity in the sensorimotor loop can by achieved by the homeokinetic paradigm, namely by considering instead the reconstruction which is given by

$$v_t = x_t - \psi^{-1}(x_{t+1}) \tag{8.5}$$

between the previous sensory inputs x_t and their reconstructed values obtained by $\psi^{-1}(x_{t+1})$, where it is assumed that ψ is invertible. It can be interpreted as the amount by which the sensor values would have had to be changed in order to

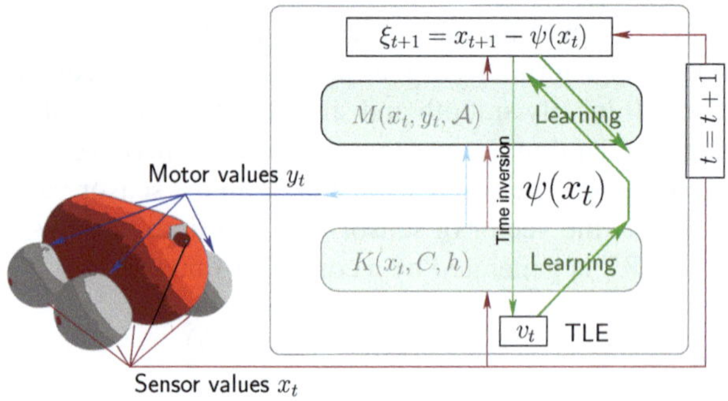

Fig. 8.1 The homeokinetic controller connected to a wheeled robot in a sensorimotor loop. The robot is equipped with wheel counters and a camera. The controller is represented by the function K and the predictor M, both together form the map ψ (Eq. (8.4)). The TLE is obtained by propagating ξ_{t+1} through the inverse of ψ.

preempt any prediction error. The objective function minimizing the reconstruction error v_t is called *time-loop error* (TLE) and it can be approximated using the linearization $v_t = L^{-1}\xi_{t+1}$:

$$E^{\mathrm{TLE}} = \|v_t\|^2 = \xi_{t+1}^\top \left(L_t L_t^\top\right)^{-1} \xi_{t+1} , \tag{8.6}$$

where $(L_t)_{ij} = \frac{\partial \psi(x_t)_i}{\partial (x_t)_j}$ is the Jacobian matrix of ψ at time t. The entire framework is sketched in Fig. 8.1. Note that minimizing this error quantity increases the small eigenvalues of L, i. e. it tends to destabilize the system which is, however, confined by the nonlinearity $g(\cdot)$ (8.1). This eliminates the trivial fixed points (in sensor space) and enables spontaneous symmetry breaking.

The parameters of the controller (C,h) are adapted by a gradient descent on the TLE (8.6). This gives rise to the parameter dynamics

$$\Delta C = -\varepsilon_c \frac{\partial}{\partial C} E = \varepsilon_c \mu v^\top - \varepsilon' y x^\top , \tag{8.7}$$

$$\Delta h = -\varepsilon_c \frac{\partial}{\partial h} E = -\varepsilon' y , \tag{8.8}$$

where ε_c is a global learning rate and ε' is channel-dependent learning rate given by $\varepsilon'_i = 2\varepsilon_c \mu_i \zeta_i$, where $\mu = G' A^\top \left(L^\top\right)^{-1} v$, and $\zeta = Cv$ and G' is the diagonal matrix defined as $G'_{ij} = \delta_{ij} g'_i(Cx+h)$. The derivation of the learning rules can be found in Der and Martius (2012). In our parameterization the Jacobian matrix is given as

$$L = AG'C + S. \tag{8.9}$$

We will generally assume that there are more sensors than motors, which, for $S = 0$, implies that the Jacobian matrix L cannot be inverted such that a pseudo-inverse is being used instead in the above formulas. The parameters A, S and b (8.3) are adapted online in order to minimize the prediction error $||\xi||^2$ (8.4). However, the minimization is ambiguous with respect to A and S because y is a function of x, see (8.1). In order to capture as much as possible of the relationship by the matrix A we introduce a bias by using partly the TLE for learning of the model:

$$\Delta A = \varepsilon_A \xi_{t+1} \left(y_t + \rho_A G' C v\right)^\top , \tag{8.10}$$

$$\Delta S = \varepsilon_S \xi_{t+1} x_t^\top , \quad \Delta b = \varepsilon_b \xi_{t+1} , \tag{8.11}$$

where the parameter $\rho_A = 0.1$ controls the bias. The learning rates ε_S and ε_b are chosen to be smaller than ε_A, but the exact parameter values are not critical; for $\rho_A = 0$ the original delta-rule is restored.

The learning rates are chosen to result in a fast dynamics for the weights. Assuming sensory noise, the TLE is never zero nor has a vanishing gradient such that the rule (8.7) produces an itinerant trajectory in the parameter space, i. e. the robot traverses a sequence of behaviors that are determined by the interaction with the environment. An intuitive idea of the resulting dynamics can be obtained for a robot with just two wheels each equipped with a proprioceptive velocity sensor (see for instance Fig. 8.16(a)). Initially the robot rests, but after a short while it starts to drive autonomously forward and backward or to turn. If the robot arrives at an obstacle, the wheels stop, thus causing a large error because of which the learning dynamics will quickly stop the motors and eventually drive in the free direction. Also high-dimensional systems such as snake- or chain-like robots, quadrupeds, hexapods and wheeled robots can be successfully controlled with the learning dynamics of Eqs. (8.7) to (8.11) (Der and Martius 2012).

8.2.1 Example of Emergent Behavior

To get a more clear idea of what homeokinetic control is about we will present two examples here: the spherical robot and the Cricket robot. The design of the spherical robot is inspired by the artist Julius Popp (2004). It has a ball shaped body and is equipped with three internal masses whose positions are controlled by motors, see Fig. 8.2(a). The motor values define the target positions of the masses along the axes which are realized by simulated servo motors. Collisions of these masses especially at the intersection point are ignored in the simulation.

If we put the spherical robot on level ground and connect the homeokinetic controller initially only small fluctuations due to the sensor noise occur. The learning dynamics increases the feedback strength steadily so that the controller is getting more and more sensitive to the sensor values. Once the critical level is exceeded fluctuations get amplified so that the symmetry of the system is spontaneously broken and the body starts to roll into a decided direction. This is the first moment when the sensor values show a defined response to the actions. The most simple of the natural modes of the robot is realized by rotating around one of the internal axes with the mass on that axis being used for steering and the other ones for shifting the

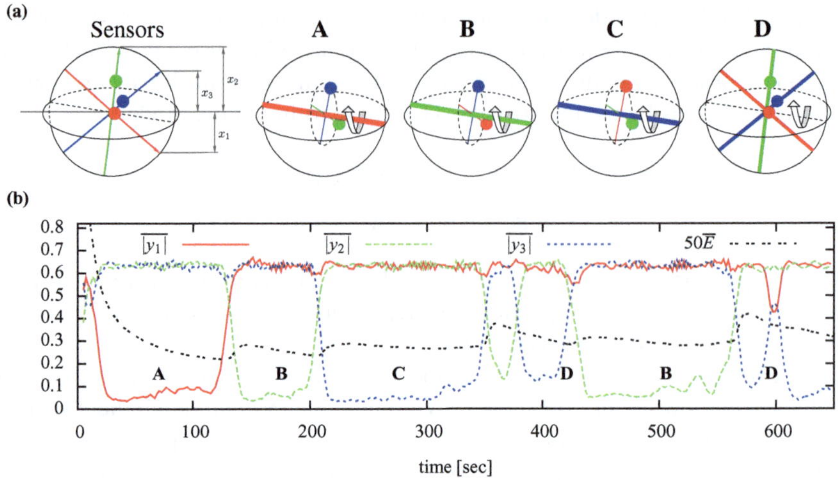

Fig. 8.2 The spherical robot exploring its behavioral capabilities. **(a)** Sensor setup and sketch of four typical behaviors (**A-D**), namely the rolling mode around the three internal axis (**A-C**) and around another axis (**D**). **(b)** Amplitudes of the motor value oscillations ($y_{1\ldots3}$) and the time-loop error E (scaled for visibility) averaged over 10 and 30 sec, respectively. Corresponding behaviors are indicated with letters A-D.

center of gravity. The experiments demonstrate, Fig. 8.2, that the controller picks up such a rolling mode and amplifies it very quickly. The explorative nature of the control algorithm is illustrated in the fact that different rolling modes emerge.

8.2.2 Behavior and Critical Dynamics in High-Dimensional Cricket Robot

In simplified systems the self-organization of the movement parameters of the robots can be studied analytically (Der and Martius 2012), which provides an intuition about the noise amplification and the emergence of behavior in such systems. It is beyond the scope of the present text to represent these results here. Instead we take a phenomenological look at a more realistic system, namely a cricket robot Fig. 8.3.

As before the robot would not move after initialization until the self-amplification of the sensor noise will eventually lead to an initial movement. Because all legs are connected to the trunk their movements are physically coupled, which is automatically extracted by the learning algorithm. The robot starts to sway and becomes more and more active until it starts to lift the feet from the ground. A range of jumping and wobbling motions is emerging that are coming and going.

Theoretically homeokinetic learning should bring the sensorimotor loop into a critical state also termed the edge of chaos. In this state small perturbation in the

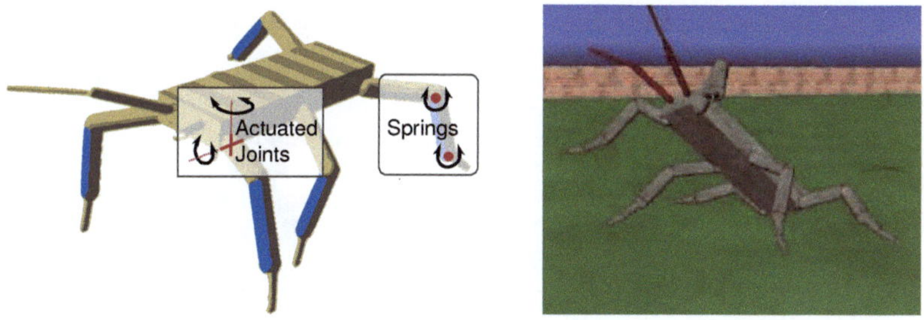

(a) (b)

Fig. 8.3 Cricket robot with realistic leg sizes, ranges and mass distribution, cf. (Cruse et al. 2006). The robot has twelve active degrees of freedom (DoF) and 14 passive DoF (lower legs and antennae). **(a)** Schematic diagram of the robot and actuated joints. **(b)** Screen shot from the computer simulations using LpzRobots (Martius et al. 2012).

sensor values are neither damped nor amplified. An indication for this state can be obtained from the largest eigenvalue of the mapping from current sensor values to future sensor values, which should be 1. That this is indeed the case for the cricket robot shows Fig. 8.4, where the linearization of the map was used. In linear systems, eigenvalues of unity represent an on-going movement, however, the nonlinearity of the controller or of certain interactions with the environment, such as collisions between feet and ground, require a more powerful controller which leads to larger eigenvalues as shown in the figure. So homeokinetic learning works also in dynamically complex systems and leads to an exploration of the behavioral capabilities of the system under control.

8.3 Guided Self-Organization

The homeokinetic learning rule causes a robot to move actively and to react sensitively to its environment. The resulting behaviors are, however, waxing and waning and their time span and transitions are hard to predict. There are only a number of exceptional cases where a robot could directly make use of the above learning scheme. Assume for instance that the robot has a number of options or schemes to follow in specific situations, but when none of these are applicable then a generic search behavior is certainly helpful. Moreover, if the robot has received a prescribed plan it can still explore similar behaviors which may be more smooth or better with respect to an external reward.

In many robotic applications, however, a defined and goal-oriented behavior is desired. With traditional learning methods these may be hard to obtain, especially if the control space is high-dimensional. A promising route, reflecting some properties of biological learning, is to allow the robot to explore its basic behaviors in a playful

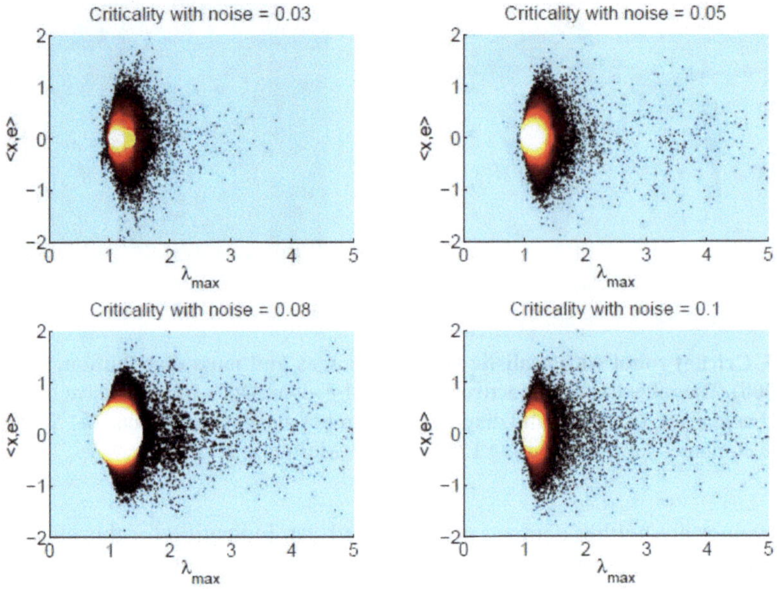

Fig. 8.4 Analysis of the Cricket robot (s. Fig. 8.3). In order to demonstrate the criticality in a complex robot we considered the main eigenvalue of the sensorimotor loop. In accordance with the analytical results for the one-dimensional case (Der and Martius 2012), we observe here eigenvalues with a mean values of approximately 1.2. The y-axes in the plots show the projection of the state of the robot onto the corresponding eigenvector which shows a symmetric distribution. Data was obtained in a single run for each of the plots with a different level of noise in each case. During the run the dominant eigenvector frequently changed its orientation. It is interesting that maximal flexibility is reached at an intermediate noise level as indicated by the heat map. White areas represent a high density of points; in low-density areas individual points are drawn in black.

and self-organized phase and internalize some of the intrinsic properties of the sensorimotor loops. Why should it be more effective? Self-organizing systems tend to scale well to higher dimensions and may exploit the constraints and properties of the embodiment. Also, self-organizing systems show a great flexibility and tolerance against failures and degrade gracefully rather than catastrophically (Prokopenko 2008, 2009). After this or even already during this self-exploratory phase, the robot receives information about the task it is expected to execute. This information can be imposed on the robot in an imperative way, but this is possible only if the exploratory properties cease to have an effect on the robot. Taking it further, we need a continuous balance between external and intrinsic learning: The robot continues to behave exploratory, but will preferentially choose those behavioral patterns that comply best with the external information. This is what *guided self-organization* (GSO) for robot control is about, which we introduced (Martius et al. 2007) for the combination of a desired goal with self-organizing behavior. The term has been used before in contexts such as nanotechnology (Choi et al. 2005), city development (Butera 1998) or swarm robotics (Rodriguez 2007) representing essentially the same idea: exploiting the intrinsic complex dynamics to achieve a goal without much engineering effort or strong interference with the intrinsic dynamics of the system. An illustrative example from nature is again the shell patterns (and animal coat pigmentation). The self-organizing reaction-diffusion systems creating the patterns are guided by comparably simple chemical gradients leading to a specific (species typical) formation. It would have been much more difficult (in terms of e. g. coding length) for evolution to come up with a precise description of pattern in the genome. More importantly this GSO system act as a pattern factory. A new pattern only needs different gradient. However, to engineer a new desired pattern may be difficult, which is part of the challenge of guided self-organization.

By the way, the same general idea also underlies chaos control (Ott et al. 1990) and, more recently, self-motivated learning. GSO is different from active learning, reinforcement learning (Sutton and Barto 1998) and evolutionary learning (Nolfi and Floreano 2001) at least because the exploration is self-organized rather than following a defined scheme or being exhaustive.

To get an intuitive idea how guidance could look like we consider again the emergence of self-organized behavior. In terms of the theory of dynamical systems, the homeokinetically controlled behavior can be considered to consist of series of symmetry breaking events. E. g. a simple robot that is not moving initially, starts to choose to move either forward or backward. If the robot's hardware does not indicate a preference for either direction, the robot chooses a random orientation caused by a possibly tiny fluctuation at the critical moment when the breaking of the symmetry happened. Obviously, the same effect can be achieved if the robot is biased (namely, to move forwards rather than backwards) either by a hardware asymmetry or by any external information. It can be further expected that the external input that the robot receives does not need to be strong. In all cases the robot will continue to self-organize its behavior, but with the difference that the specific decision which was previously due to a noise effect, is now due to an external guidance.

More formally, we will distinguish a number of possibilities for guidance in dependence on the type of information the robot receives. The first one allows for the incorporation of supervised learning signals, e. g. specific nominal motor commands. To make this possible we study the integration of problem-specific error functions into the homeokinetic learning dynamics in the next section. Using a distal learning (Jordan and Rumelhart 1992) setup we also study the use of teaching signals in terms of sensor values and give an example of guidance by visual target stimuli. Interestingly we find a remarkable robustness to sensorimotor disruptions. The second mechanism is discussed in Sect. 8.6 and uses online reward signals to shape the emerging behaviors. The third mechanism for guiding the self-organization can be used to formulate relationships between motors, see Sect. 8.7. This will be proven to be an effective and simple way to introduce constraints into the system and facilitate the unsupervised development of specific behaviors.

8.4 Guidance by Mild Supervision

8.4.1 *Integration of Problem-Specific Error Functions*

The combination of self-organizing processes and additional constraints is not trivial and essentially an instance of the well-known dilemma that arises when both exploration and exploitation is desired at the same time. A problem-specific error function expressed the goal, i.e. a specification what is to be exploited in a given context, while the behavioral self-organization provides an efficient means for exploration. Whether or not the exploration indeed serves the goal in the long run, is a question of the balance between the two which we are going to discuss in this section. A particular goal can be specified in terms of a problem-specific error function (PSEF) that is minimal if the goal is met.

A suggestive way of combining the TLE and a PSEF could be a weighted sum of the two error functions. Performing gradient descent on this sum minimizes then this combination such that we could expect learning to both improve the active engagement with the environment as well as top approach the goal. It is, however, likely that either one of the learning tasks may improve on the cost of the other one.

The optimal balance between the exploration and exploitation depends not only on the specific problem but also on the course of learning and the current state of the system. The reason is that the size of the TLE varies often over several orders of magnitude, whereas the goal-specific terms will usually stay in a smaller range or will not covary with the TLE. Therefore, a fixed weighting in the combined error function cannot be expected to exist in non-trivial problems.

In order to achieve a goal-orienting effect without destroying the self-organization process, we have proposed to scale the gradient on the PSEF in order to be compatible with the TLE (Martius and Herrmann 2010, 2012). This approach was motivated by the natural gradient method (Amari 1998). This method is based on the fact that for an arbitrary Riemann metric of the parameter space the steepest direction is given by the transformed gradient, which is obtained by multiplying with the inverse of the metric. We use a metric which is defined by the matrix $JJ^\top$, where J is the Jacobi matrix of the sensorimotor loop, similar to Eq. (8.9) but now in motor space. We can think of this procedure as map of the error into the action space of the robot.

The PSEF is denoted by E^G and it must be non-trivially dependent on the controller parameters such that the gradient can be effective. So the main formula for guided self-organization with homeokinesis is the new update rule for the controller matrix C as

$$\frac{1}{\varepsilon_c}\Delta C = -(1-\gamma)\frac{\partial E^{\mathrm{TLE}}}{\partial C} - \gamma Q \frac{\partial E^G}{\partial C}\,, \tag{8.12}$$

where $\frac{\partial E^{\mathrm{TLE}}}{\partial C}$ is the homeokinetic learning rule (8.7), $1 \leq \gamma \leq 0$ is the guidance factor defining the weighting between goal following and self-organization, and

$Q = \left(JJ^\top\right)^{-1}$ defines the metric. The latter can also be expressed as $Q = A^\top \left(LL^\top\right)^{-1} A$, see Martius (2013). For $\gamma = 0$ there is no guidance and we obtain the unmodified dynamics, and for $\gamma = 1$ there is no homeokinetic adaptation but only guidance.

The entire update size is still controlled by the learning rate ε_c. For the update of the parameter h we apply an analogous procedure.

Below we will look at a few concrete examples of problem-specific error functions (PSEFs) that implement the guidance by *teaching signals*. In this way a supervised learning procedure is introduced which, however does not imprint its effect on the system but rather have the system explore the learning objective implied by the PSEF.

8.4.2 Direct Motor Teaching

In order use motor-teaching signals we define a PSEF, which penalizes the mismatch

$$\eta_t = y_t^G - y_t \tag{8.13}$$

between motor teaching values y_t^G and actual motor values y_t (output by the homeokinetic controller). Similarly to the prediction error for the forward model we find

$$E^G = \eta_t^\top \eta_t \,. \tag{8.14}$$

Using the gradient descent we get the additional update for the controller matrix C as

$$\frac{\partial E^G}{\partial C} = -G' \eta_t x_t^\top \,, \tag{8.15}$$

where G' is the diagonal matrix given by $G'_{ij} = \delta_{ij} g'_i(Cx + h)$. Similarly, for h we obtain $\frac{\partial E^G}{\partial h} = -G' \eta_t$. These additional terms are integrated into the final learning rule using Eq. (8.12). The guidance factor γ regulates the strength of the additional drive and has to be determined empirically. A small value of γ leads to a small influence of the teaching signal and results in a behavior that is mostly dominated by the original homeokinetic controller. For large values of γ the teaching signals are followed narrowly and few exploratory actions are performed, however, with the increasing danger to break down the self-organization.

8.4.2.1 Experiment

Using a two-wheeled robot, see Fig. 8.16(a), we will show that teaching signals can be used to specify a certain behavior and that the influence of the teaching can be conveniently adjusted using γ. For that let us consider two different motor teaching signals, which are subsequently applied. First the nominal motor values are given by a sine wave and then by a rectangular function with the same value for both motors, i. e.

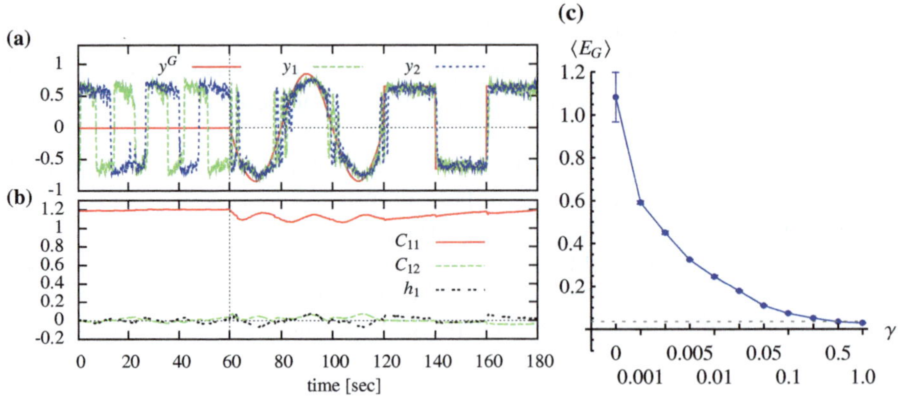

Fig. 8.5 Two-wheeled robot controlled with homeokinetic controller and direct motor teaching signals. **(a)** The teaching signals y^G (identical in both components) are followed partially by the motor values $y_{1,2}$ after teaching was switched on with $\gamma = 0.01$ at 60 sec. **(b)** Time evolution of the controller parameters affecting the first motor is shown to illustrate that only little changes are necessary, however, the adaptations do not vanish. **(c)** Average value of the PSEF E^G (for 5 experiments à 5 min) in dependence of γ (note the logarithmic scale). The noise level (*dotted gray line*) is reached at $\gamma = 1$. Parameters: $\varepsilon_c = \varepsilon_A = 0.1$, $\gamma = 0.01$ (a,b).

$$(y_t^G)_i = \begin{cases} 0.85 \cdot \sin(\omega t) & t < 75 \\ 0.65 \cdot \mathrm{sgn}(\sin(\omega t)) & \text{otherwise}\,, \end{cases} \tag{8.16}$$

with $i = 1,2$ and $\omega = 2\pi/50$. For the choice of the teaching signal we have to consider that the nominal motor values should not be too large because otherwise the controller is driven into the saturation region of the motor neurons. The fixed point of the sensor dynamics in the simplified world condition is at $y \approx \pm 0.65$. This is a good mean teaching signal size, which was also used in Eq. (8.16). As a rule of thumb we recommend confining the motor teaching values to the interval $[-0.85, 0.85]$.

In Fig. 8.5 the produced motor values and the parameter dynamics are displayed for different values of the guidance factor γ. For a low value of γ the desired behavior is only followed by trend, whereas for higher values, e. g. $\gamma = 0.01$, the robot mostly follows the given teaching value with occasional exploratory interruptions. These interruptions cause the robot, for example, to move in curved fashion instead of strictly driving in a straight line as the teaching signals dictate. The exploration around the teaching signals might be useful to find modes which are better predictable or more active. The long performance of a single low-dimensional behavior can lead to the inaccuracy of the adaptive forward model. Thus, the explorative actions can supply the forward model with necessary sensation-actions pairs for complete learning.

The experiment demonstrated that motor teaching signals can be used to achieve a specific behavior. This result is not very surprising, because the system is very simple and the target behavior did not conflict with the homeokinetic principle (sensitive and predictable). However, it served as a proof of principle and showed that the balance between target behavior and remaining self-organized behavior can be adjusted using a single parameter.

8.4.3 *Direct Sensor Teaching and Distal Learning*

In this section we transfer the direct teaching paradigm from motor teaching signals (Sect. 8.4.2) to sensor teaching signals. This is a useful way of teaching because desired sensor values can be more easily obtained than motor values, for instance by moving the robot, or parts of the robot by hand. This kind of teaching is also commonly used when humans learn a new skill, e. g. think of a tennis trainer that teaches a new stroke by moving the arm and the racket of the learner and is a subset of imitation learning (Schaal et al. 2004). Thus, a series of nominal sensations can be acquired that can serve as teaching signals. Setups where the desired outputs are provided in a different domain than the actual controller outputs are called *distal learning* (Jordan and Rumelhart 1992; Stitt and Zheng 1994; Dongyong et al. 2000). Usually a forward model is learned that maps actions to sensations (or more generally to the space of the desired output signals). Then the mismatch between a desired and occurred sensation can be transformed to the required changes of action by inverting the forward model.

The distal learning error is the mismatch between desired sensations x_t^G and actual sensations x_t

$$\xi_t^G = x_t - x_t^G \,. \tag{8.17}$$

The mismatch η_t in motor space can be obtained via the forward model M (8.3) in linear order

$$\eta_t = A^+ \xi_t^G \,, \tag{8.18}$$

where $A = \frac{\partial M(x,y)}{\partial y}$ and the A^+ denotes the pseudoinverse of A. Now the update formulas (8.15) for C and h from the direct motor teaching setup can be used based on the teaching error $E_G = \|\eta_t\|^2$.

8.4.3.1 Experiment

For the two-wheeled robot (Fig. 8.16(a)) the forward model is simply a multiple of the unit matrix. The spherical robot (Fig. 8.6(a)), however, has a non-trivial relation between sensor and motor values and is thus better suited for an illustrative experiment to show that a simple teaching signal in terms of sensor values can be effective in guiding the behavior.

A desired behavior for the spherical robot could be to rotate around the one of its internal axes. For the particular sensor setup we need to assure the the corresponding

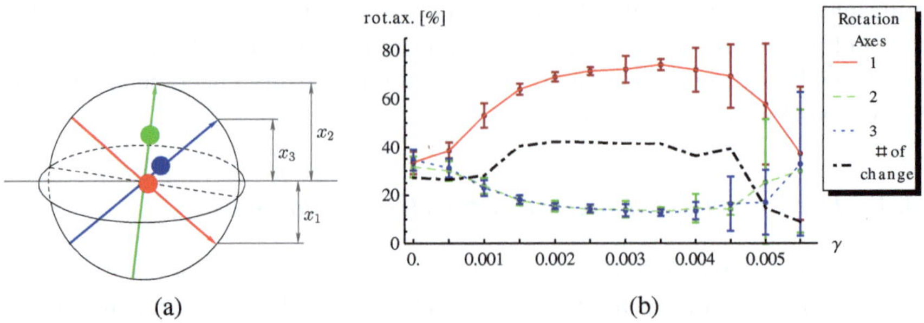

Fig. 8.6 The spherical robot in a homeokinetic plus distal learning setup. (a) Illustration of the robot with its sensor values. **(b)** Behavior with the distal learning signal, Eq. (8.19). The plot shows the percentage of rotation time around each of the internal axes and the number of times the behavior was changed for different values of the guidance factor γ (no teaching for $\gamma = 0$). The rotation around the red (first) axis is clearly preferred for non-zero γ. The mean and standard deviation are plotted for 20 runs each 60 min long, excluding the first 10 min (initial transient, no guidance). For too large values of the guidance factor the self-organization process is too much disturbed such that the robot gets trapped in a random behavior (*dash-dotted line*). Parameters: $\varepsilon_c = \varepsilon_A = 0.1$.

sensor returns consistently a low absolute value. This can be directly specified in the distal learning scheme, here for the first axis:

$$x_t^G = \begin{pmatrix} 0 \\ (x_t)_2 \\ (x_t)_3 \end{pmatrix}. \tag{8.19}$$

Now only the first component of the sensor value produces an error signal. The resulting behavior is characterized in Fig. 8.6.

The distal learning scheme requires a well trained forward model. Therefore pure self-organization was used during the first 10 min of the experiment ($\gamma = 0$). As a descriptive measure of the behavior, we used the index of the internal axis around which the highest rotational velocity was measured at each moment of time. Figure 8.6(b) displays for different values of the guidance factor and for each of the axes the percentage of time it was the major axis of rotation. Without teaching there is no preferred axis of rotation. With distal learning the robot shows a significant preference (up to 75 %) for a rotation around the first axis. For overly strong teaching, a large variance in the performance occurs. This is caused by a destructive influence of the teaching signal on the homeokinetic learning dynamics. Remember that the rolling modes can emerge due to the fine regulation of the sensorimotor loop to the working regime of the homeokinetic controller, which cannot be maintained for large values of γ.

The robot will not stay in the rotational mode about one axis. While the robot is in this rotational mode the teaching signal is negligible. However, the sensitization property of homeokinetic learning increases the impact of the first sensor, such that the mode becomes eventually unstable again. Again this may be considered as an advantage since the temporary breaking out avoids a too narrow specialization of the internal model. Note, moreover, that the learning success in the current setting of controller and forward model could *not* be achieved by the distal learning alone, at least not with a constant learning signal.

To recapitulate, the direct teaching mechanism allows us to specify motor patterns that are more or less closely followed, depending on the strength of integrating the additional drives into the learning dynamics. In this section we considered sensor teaching signals that were transformed into motor teaching signals using the internal forward model. We have shown that the spherical robot with the homeokinetic controller can be guided to locomote mostly around one particular axis, by specifying a constant sensor teaching signal at one of the sensors Martius and Herrmann (2010).

8.5 Self-Organized Interaction with the Environment

Let us know consider a more involved application with direct sensor teaching using a camera sensor.

8.5.1 Integration of Vision into the Sensorimotor Loop

Vision adds a new level of complexity to any robotic system. In particular in most of the applications of the homeokinetic principle, mostly proprioceptive sensors have been used, which helps to generate a sensible control of the body, but may not be sufficient to produce a tight interaction with complex environments. In the following we will discuss the integration of visual information into the framework of self-organizing control, see also Martius (2013).

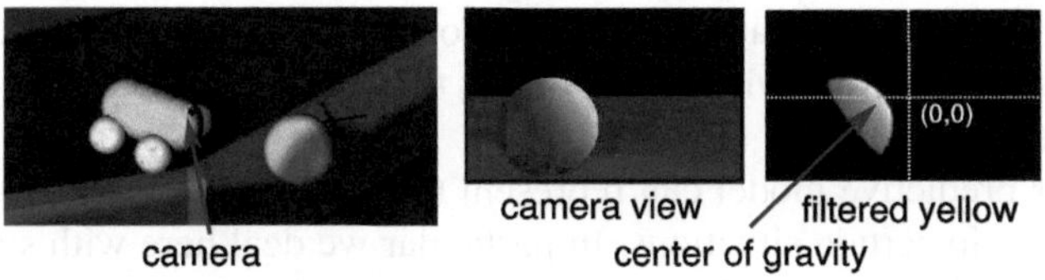

Fig. 8.7 Camera setup, image processing and sensor values

In the following we will describe experiments with a four-wheeled robot (Fig. 8.1). The robot is operated such that the two motors on one side of the robot receive the same target velocity. The two velocity sensors (x_l and x_r) return the average of the actual wheel velocities on one side.

A simplification can be reached by restricting the interacting of the robot to with objects of a certain color, yellow in our case. We start by calculating the center of mass (x_h, x_v) over all pixels of this color based on the assumption that only one yellow object is visible. If this not the case the approach of the robot will help with the disambiguation. Next, we approximate the size (x_s) of the object by the sum of all yellow pixels (normalized to $[0, \sqrt{2}]$). This is prone to light and shadow effects and is again a crude approach but it will turn out to be sufficient for our purposes. In addition we also use the time derivatives of the quantities such that the vector of sensor values reads

$$x = (x_l, x_r, x_h, \dot{x}_h, x_v, \dot{x}_v, x_s, \dot{x}_s)^\top . \tag{8.20}$$

We are often adding a small amount of sensory noise to the simulated sensors which is not only more realistic, but also has the side effect that the TLE does not become zero. The vision sensors are, if any objects are visible at all, rather inaccurate and noisy, e.g. due to illumination, such that additional noise is not required here. Therefore, only the wheel velocities sensors x_l and x_r are subject to Gaussian noise with a small standard deviation.

Exteroceptive sensors in general and our vision sensors in particular may not be active for substantial periods during operation. For instance the position sensor (x_h, x_v) is essentially undefined if no object is in sight. Since the predictive model is to correlate actions with perceptions, the absence of any object nullifies the correlations such that a prediction becomes impossible. A simple solution is to prevent learning of the predictive model on invalid sensor values. We implement this by assuming an undefined sensor value to be zero and set the prediction error $(\xi_i)_t$ to zero as well, if $(x_i)_t = 0$ or $(x_i)_{t-1} = 0$ while it remains unchanged otherwise.

8.5.2 Guiding towards an Object

We will now define a guidance mechanism that drives the robot towards a visible object. In order to fixate the object in the center of the field of vision, the position sensors (x_h, x_v) should approach zero unless a specific target position (p_h, p_v) is given. If the robot should push objects, e. g. a ball, then the value of the size sensor (x_s) should be large. Alternatively if the robot should keep a certain distance, for instance when interacting with other robots, then a smaller value is required. We denote the desired size by s.

The linear predictive model can represent the relation between actions and position/size only in certain situations. In particular we deal here with stationary and moving objects that cause a different sensory response. A new mechanism could make use of the desired value for the derivatives, too. Fortunately, we can use proportional set-point control formula with damping: $\dot{x} = -\alpha(x - x^{\text{desired}}) - \beta\dot{x}$, where α is a rate and β is the damping constant. This differential equation has a fixed point at $x = x^{\text{desired}}$.

The sensor teaching vector x^G is thus given in components as

$$x_l^G = x_l, \qquad x_r^G = x_r, \tag{8.21}$$

$$x_h^G = p_h, \qquad \dot{x}_h^G = -\alpha(x_h - p_h) - \beta\dot{x}_h, \tag{8.22}$$

$$x_v^G = p_v, \qquad \dot{x}_v^G = -\alpha(x_v - p_v) - \beta \dot{x}_v, \tag{8.23}$$

$$x_s^G = s, \qquad \dot{x}_s^G = -\alpha\,(x_s - s) - \beta \dot{x}_s\,, \tag{8.24}$$

where $\beta = 0.1$ and $\alpha = 1$ here. Note, the wheel velocity sensors x_l and x_r produce no teaching signal. For the following experiments we use for the center position $p_h = p_v = 0$ and set the maximal size to $s = \sqrt{2}$.

8.5.3 Emergent Behaviors

The first experiment should test whether the guidance mechanism is able to influence the self-organized behavior to find and push balls. This involves the establishment of the required sensorimotor mappings from scratch in a changing environment (balls can move). All the experiments are performed in virtual reality in our robot simulator (Martius et al. 2012). The formal definition of the goal is specified by the target sensor state x^G Eqs. (8.21–8.24). We place the robot together with five balls into a circular corridor, as displayed in Fig. 8.8(a), such that the robot can possibly push a ball for a long distance without getting stopped by corners. Those parameters of the model (A) connecting to the vision sensors are initialized with zero, such that the guidance has no effect independently of the guidance factor. Recall that the forward model transforms the teaching signal to nominal changes in motor values Eq. (8.18), which will be zero if the model did not learn anything. Once the robot learns to move, the model starts to correlate actions with the visual sensors. In this way the guidance starts to actually influence the behavior, such that the robot sees a ball more often and the model can improve further. Eventually the robot starts to steer at a ball and pushes it along the arena. Note that the robot has a round front shape such that the ball easily drifts away to either side while pushed. From time to time the robot still performs exploratory actions such that the ball gets lost and a ball needs to be found again. A part of a trajectory of the guided robot is shown in Fig. 8.8(b).

Note that there can be more than one ball in the field of view at the same time. However, the sensors cannot distinguish different objects, since the visual sensor (x_h, x_v) provides a position between the objects and the size (x_s) sensor returns a sum of the sizes. Nevertheless, the robot copes with this situation without problems. The robot steers at a group of balls and decides rather spontaneously which one it will touch. The final choice depends on how well the different balls are visible, when they leave the field of view, and other perturbations.

In order to analyze quantitatively the behavior of the robot, we consider the average distance to the closest ball and the cumulative time a ball was in the sight of the robot. This gives a good measure on whether the guidance was followed and the robot is indeed approaching the balls. If the robot is also pushing the balls along the arena, then the traveling distance of the balls raises, which we display together with the other quantities in Fig. 8.9(a). Indeed, for intermediate values of the guidance factor the time a ball was in sight increases from 100 sec to 600 sec. The same holds for the average distance of the robot to the closest ball which decreases from 5 to a

(a)

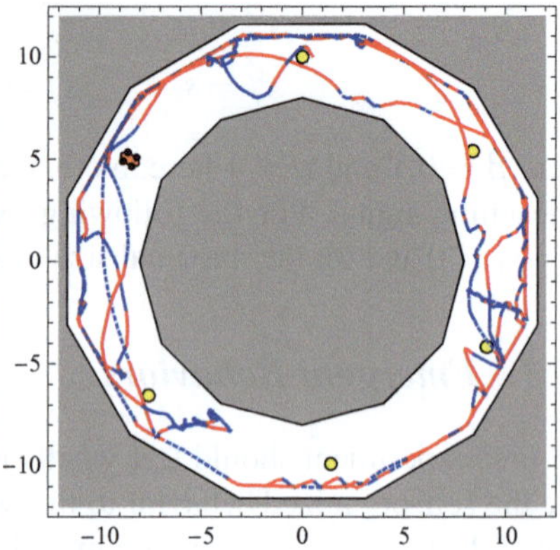

(b) $\gamma = 0.1$

Fig. 8.8 Ball playing scenario. The robot is placed in a circular corridor. **(a)** Screen shot from the simulation. The right inlet shows the camera image and the left displays the color filtered image; **(b)** Part of a sample trajectory of the robot (minutes 5–10) for $\gamma = 0.1$ colored in red (solid) if the robot is close (within two body length) to the ball and it was in sight, and in blue (dashed) otherwise. The yellow disks show the initial positions of the balls.

value of 2. The size of the robot is 1 and the ball has radius of 0.3, resulting in a minimum of 0.8. Why does not the average distance go much below 2? Firstly, the plots include the entire simulation time including the phase where the robot has to acquire basic knowledge about its body. Secondly, it can take a long time and driving distance to find a ball again when it is lost, for instance through an exploratory action. Due to the inner circular walls of the arena the balls are not visible everywhere and finally the distribution of distances is skewed, see below.

The traveling distance of the balls raises from nearly zero to more than 7500 units, which corresponds to about 100 rounds in the arena (in 30 min).

In Fig. 8.9(b) we show that the aspects of the behavior that are not particularly subject to the guidance, namely the covered area of the arena by the robot and its average velocity are not negatively effected by the guidance, at least for moderate guidance strengths. The area coverage and the velocity go up when the task is performed, because the robot drives much more straight and forward than without the guidance.

When the guidance is too strong self-organized adaptation and external pressures become out of balance and the performance drops. Especially visible is this effect at $\gamma = 1$ where no homeokinetic learning takes place (Eq. (8.12)) and the robot fails to move in a coordinated fashion, see Fig. 8.9(b).

Taking a closer look at the distance to the closest ball, we find that the mean is not such an appropriate measure in the guided situation since the distribution of distances is not Gaussian but rather skewed as shown in Fig. 8.10. Without guidance the distribution of distances is almost flat, whereas for weak and intermediate guidance strengths the distribution is skewed with a strong preference for short distances. For

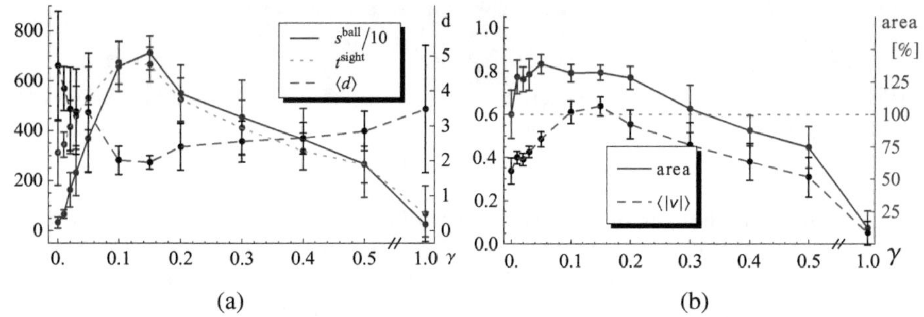

Fig. 8.9 Behavioral quantification of the ball playing scenario. Both panels show the mean and standard deviation of 10 simulations each 30 min long, in dependence of the guidance factor γ. **(a)** Traveling distance of the balls s^{ball} (scaled), cumulative time a ball was in sight t^{sight} (in sec), and average distance to the closest ball $\langle d \rangle$ (right axis, minimum 0.8). **(b)** Average absolute velocity of the robot (left axis) and area coverage (box counting method), given in percent of the case without guidance ($\gamma = 0$) (right axis).

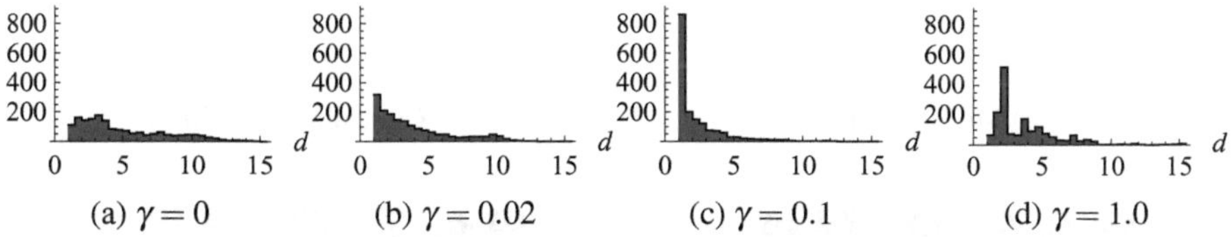

Fig. 8.10 Distribution of distance to the ball in the ball playing scenario. All panels show the histogram (in sec) of the distance d averaged over all simulations for one particular value of the guidance factor γ. **(a)** No guidance; **(b)** weak guidance; **(c)** intermediate guidance; **(d)** overly strong guidance (no self-organization).

overly strong guidance ($\gamma = 1$) the robot gets predominantly stuck at the walls because the sensorimotor coordination is pushed away from its sensitive regime, such that the histogram is rather arbitrary.

8.5.4 Robustness against Structural Changes

In fact we performed quite radical changes to the camera setup, namely to rotate and flip the camera abruptly, see Fig. 8.11. These changes have severe consequences for the sensorimotor dynamics, because some sensor values swap signs or change from being useless to becoming important and vice versa.

We use the same circular arena as in the previous section. In our simulated experiments the camera setup is initially normal and is changed every 10 minutes to the setups shown in Fig. 8.11. Only the backwards view is kept for 20 min. Finally the normal setup is used again, such that an experiment lasted 70 min in total. We conducted 10 experiments with $\gamma = 0.1$ and present the evolution of the relevant model and controller parameters in Fig. 8.12.

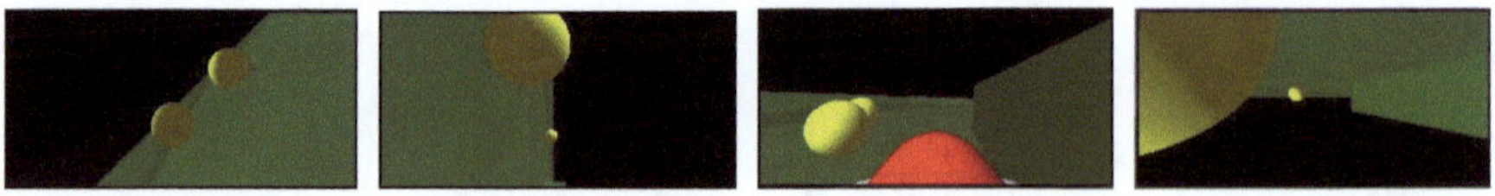

(a) Tilt 45° (b) Tilt −90° (c) Backward view (d) Upside down

Fig. 8.11 Radical changes to the visual perception. In addition to the normal setup of the camera (Fig. 8.8) it is rotated by 45° **(a)**, −90° **(b)**, and 180° **(d)** along the optical axis, and lifted and rotated by 180° **(c)** along the vertical axis yielding a backward view. Note the different perspective and the appearance of the robot's body in the camera view in **(c)**.

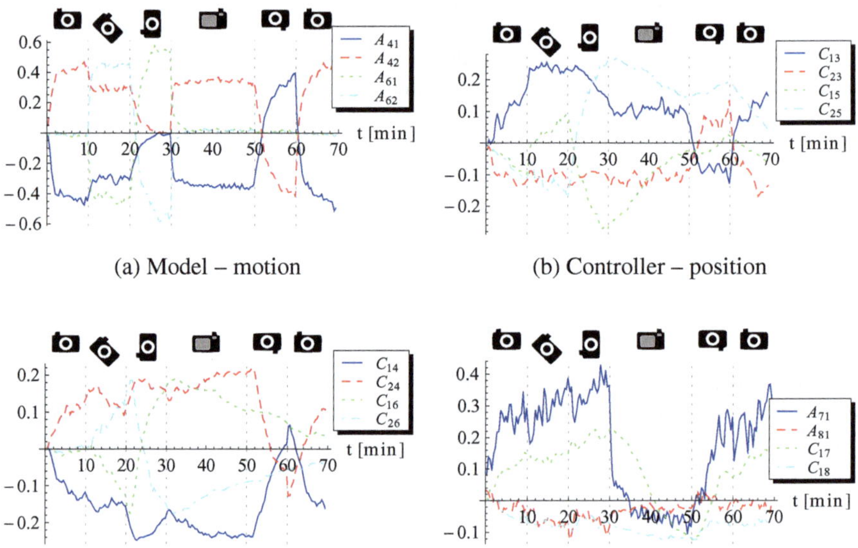

(a) Model – motion (b) Controller – position

(c) Controller – motion (d) Controller & Model – size

Fig. 8.12 Fast relearning: evolution of parameters for a changing camera. The camera is changed every 10 min, illustrated by the vertical lines. Its orientation on the body is shown by the icons. All values are mean values for 10 independent runs. Shown are elements of the model matrix (A) and controller matrix (C). The indexes refer to the sensor and motor value vectors, see Eq. (8.20). **(a)** Model parameters connecting left and right motor command with visual motion input ($\dot{x}_h$,$\dot{x}_v$). **(b)** Controller parameters connecting visual position (x_h,x_v) with left and right motor neuron. **(c)** Controller parameters connecting visual motion ($\dot{x}_h$,$\dot{x}_v$) with left and right motor neuron. **(d)** Controller and Model parameters connecting visual size (x_s,$\dot{x}_s$) and left wheel. The model parameters adapt very quickly to the new camera configurations. The controller utilizes both the position and the motion of the ball, however its adaptation is much slower compared to the model. Parameters: $\gamma = 0.1$.

Especially the model parameters relating motor values with the motion sensors, Fig. 8.12(a), evidently show that the correct correspondence is learned within a few minutes after each switching event. This, however, is only possible if the behavior of the robot is such that a ball remains frequently in the field of vision, which is very hard, if e.g., the positional sensation just swapped sign. In this situation the major strength of the homeokinetic controller shows its fruits, namely its continuous and embodiment related explorative and drive. The controller parameters show that the incorporation of the vision sensors is changed drastically for the different situations, but also that both motion and position information is used. The positional information is required to steer towards the ball and the motion sensor is used avoid overshooting. The parameters C change slower than the model parameters. Note that the behavior is also influenced by the parameters h (not shown). These change more rapidly and help to realize the teaching signals on a shorter timescale until the C matrix captures the correspondence with the sensor values.

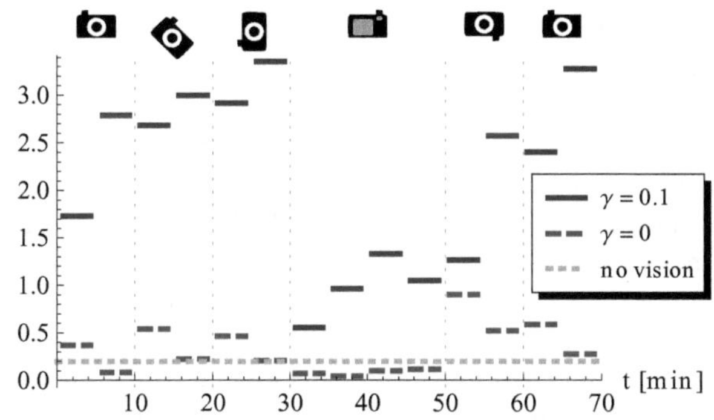

Fig. 8.13 Performance recovery for a changing camera configuration. Depicted is the summed average velocity of all balls within intervals of 5 min corresponding to the simulations in Fig. 8.12. For comparison the case without guidance ($\gamma = 0$) is displayed. The base line (green, dotted) represents the average ball movement of a blind robot.

How is the performance in the task after the structural changes? To answer this question we present in Fig. 8.13 the average ball velocities within 5 minute intervals summed over all balls. Note, that since the balls are subject to rolling friction a constant pushing is required. For comparison the values without guidance and without vision (chance level as a baseline) are displayed. The performance within the first 5 minutes is already far above the baseline and it is doubled from the first to the second 5 minute interval. After each structural disruption the performance drops a bit and is recovered in the second 5 min interval for each setting. Only the setting with camera pointing backward yields worse performance, which is due to the partial obstruction of the visual field by the body. Then the most drastic disruption occurs when the view is switched from backward to forward, but upside down. Here all visual sensor modalities change sign. Nevertheless the performance raises in the second 5 min interval to the performance of before. We can conclude that the performance is rapidly recovered even after severe changes in the sensor modalities.

At the beginning of an experiment the robot learns the behavior from scratch. When the camera is first turned by 45° comparably small adaptations occur, see interval 10-20 min in Fig. 8.12. For instance the sensors for vertical position and motion get slowly integrated, but the remaining structure stays the same and in fact the performance drops only slightly (Fig. 8.13). When the camera is turned to −90° a drastic change occurs. The meaning of the size sensor does not change, but the position and motion sensors require a completely different coupling, which is slowly established (interval 20-30 min). This may be called learning from scratch, but in fact it is worse, it is learning from a wrong configuration. When the switch occurs the controller acts to avoid the ball. To manage this challenge an exploration is required that focuses on the wrong aspects of the model, which is what happens in our approach, where the adaptation speed is actually increased if the prediction errors raise (see Eq. (8.6)). Since the controller does not explicitly know when a structural change occurs it is always adapting in a continuous manner. However, there is no long-term memory such that the controller cannot remember previously experienced configurations.

To summarize, the entire sensorimotor coordination to fulfill the task was learned by the robot within a few minutes. This involves the basic coordination to drive the robot and the integration of the vision sensors such that the balls are approached and balanced while pushed. The task to push the balls is not very complicated and can be achieved with a simple hand-crafted controller. However, to learn it from scratch in a short amount of time is hard. On top of that the orientation of the camera was abruptly changed such that a completely different sensorimotor coordination becomes necessary. We found that guided self-organizing with homeokinesis can cope with a wide range of configuration changes, even those where a complete change in the visual sensation occurs. To our knowledge there is no other system that offers this kind of robustness and the rapid on-line learning.

8.6 Reward-Driven Self-Organization

8.6.1 Reinforcement Learning and Guided Self-Organization

In many applications an explicit objective function is not available, instead a qualitative signal is given that can be interpreted as reward of punishment of a recent state or action of the robot. Reinforcement learning studies the generation of policies under such conditions typically relying on an exploration mechanism that discovers better solutions from present ones. While it is possible to apply a learning rule similar to homeokinesis to shape exploration in reinforcement learning (Smith and Herrmann 2012), we will consider here the usage of the reward signal for the guidance of the homeokinetic exploration.

For example the behavior of the simple robot with one-degree of freedom shows a systematic sweeping through the accessible frequencies of the sensor state reflected by rolling modes with different velocities (Der and Martius 2012). In the case of the spherical robot with its three dimensional motor and sensor space we also observed a sweeping through a large set of possible behaviors. In a setup where the robot can move freely, it will exhibit different slow and fast rolling modes around different axes.

Before introducing the new mechanisms, let us recall that well predictable behaviors persist longer than others. Due to this effect the well predictable behaviors are also quickly found because badly predictable ones are left quickly. Translating this into the case of reward and punishment, we want that rewarded behaviors persist longer than punishment ones and that predictable ones are found quickly. Thus we have to modulate the learning speed according to the online reinforcement signal in a way that in rewarded situations the adaptation speed is reduced and in punished ones the speed in increased. At first glance it seems to be counterintuitive that we have to reduce learning speed in order to keep a behavior, but the self-organized search should be slowed down to find even better behaviors locally. Moreover, the controller is already able to produce the behavior at the time it is exhibited by the robot.

The real-valued reward signal $r(t)$ for each time t is supposed to act as a reward for positive values and as a punishment for negative values. It is incorporated into the error function in the following way

$$E^{\mathrm{r}} = (1 - \tanh(r(t)))E \,, \tag{8.25}$$

where E is the usual TLE (8.6) and $r(t)$ is expected to assume values mainly in the interval between -1 and $+1$. Larger amplitudes are squashed by the hyperbolic tangent such that differences tend to be ignored for high positive or negative rewards. The effect of the factor $(1 - \tanh(r(t)))$ is is the same as a rescaling of the learning rate which is increased for negative and decreased for positive rewards. Therefore, we can expect that rewarded behaviors persist longer and punished behaviors are left quicker. We will demonstrate the effect of the reward-based weighting in shaping the behaviors of the spherical robot (see Fig. 8.5a).

8.6.2 *Modulation of Behavior in a Spherical Robot*

8.6.2.1 Reinforcing Speed

In the following experiment we will use the spherical robot, see Fig. 8.6(a). One of the simplest possible desired behaviors of this robot is fast unidirectional rotation. A reward function for this goal can be constructed from the angular velocity of the robot. For small velocities the reward should be negative, thus causing a stronger change of behavior, whereas larger velocities should result in a positive reward. To achieve that, the reinforcement signal can be expressed as

$$r(t) = \frac{1}{3}\|v_t\| - 1 \,, \tag{8.26}$$

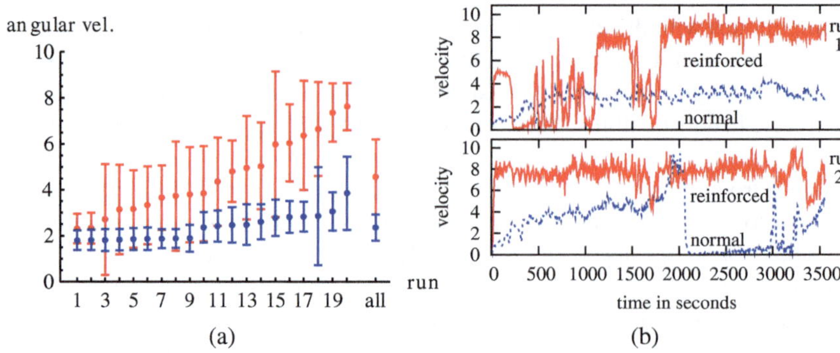

Fig. 8.14 Performance of the spherical robot rewarded for speed. **(a)** Mean and standard deviation of the velocity of the spherical robot for 20 runs each 60 min long with (**red**) and without (*blue*) speed reinforcement, sorted by velocity. The label **'all'** denotes the mean and std. deviation over the means of all runs, which is significantly ($p < 0.001$) higher for the reinforced runs. **(b)** Time course of the robot's velocity for run number 10 and 14, where *blue/dotted* shows the normal case and **red/solid** line shows the reinforced case.

where v_t is the velocity vector of the robot, see Fig. 8.6(a). In order to compare the results with the unguided case the reward is shifted, such that it is zero for the average velocity of normal runs. The scaling is done to keep the reward within the effective range.

We conducted 20 trials with the spherical robot with reinforcement and 20 trials without reinforcement, all with random initial conditions, each for 60 min in simulated real time on a flat surface without obstacles. The robot also experiences rolling friction, so that fast rolling really requires continuous motor activity. In Fig. 8.14 the mean velocity (measured at the center of the robot) for each simulation is plotted and the velocity trace of the robot for two reinforced and two normal runs are displayed as well. The simulations are sorted by performance and plotted pairwise for comparison. As desired, the mean velocities of the reinforced runs are larger than the ones of the normal runs. This is especially evident in the overall mean (mean of means marked by **'all'** in Fig. 8.14(a)), which is significantly different. The null hypothesis that the set of means of the reinforced runs and of the normal runs have an indistinguishable mean was rejected with $p < 0.001$ using the t-test. However, since straight and also fast rolling modes are easily predictable and active they are also exhibited without reinforcement for a long time. It is important to note that the fast rolling modes are also found again, after the robot was moving slower, see Fig. 8.14(b).

The guidance of the homeokinetic controller using a reward for fast motion has shown to increase the average speed of the robot significantly. Although there are also trials where no increased speed was found.

8.6.2.2 Reinforcing Spin

In a different setup we want the robot to follow curves and spin at the spot. We use the angular velocity ω_z around the z-axis of the world coordinates system, which is perpendicular to the ground plane, as depicted in Fig. 8.6(a). The reward function is now given by

$$r(t) = \frac{1}{3}\|\omega_z\| - 1 \,. \tag{8.27}$$

Again the reward is scaled and shifted to be zero for normal runs and to be in an appropriate interval. Positive reward can be obtained by rolling in a curved fashion or by entering a pirouette mode. The latter can be compared to a pirouette done by figure-skaters—with some initial rotation the masses are moved towards the center, so that the robot spins fast in place. The robot also experiences rolling friction, so that fast pirouettes are not persistent.

Again, we conducted 20 trials with reinforcement and 20 trials without reinforcement, each for 60 min simulated real time on a flat surface without obstacles. In Fig. 8.15(a) the mean angular velocity ω_z for each simulation is plotted, again sorted by performance. The time evolution of the angular velocity for two reinforced and two normal runs are displayed in Fig. 8.15(b). In this scenario the differences between the normal runs and the reinforced runs are remarkable. Nearly all reinforced runs show a large mean angular velocity. The reason for this drastic difference is that these spinning modes are less predictable and therefore quickly abandoned in the non-reinforced setup. The traces show that the robot in a normal setup rarely performs spinning motion, whereas the reinforced robot performs, after some time of exploration, very fast spinning motions, which are persistent for several minutes.

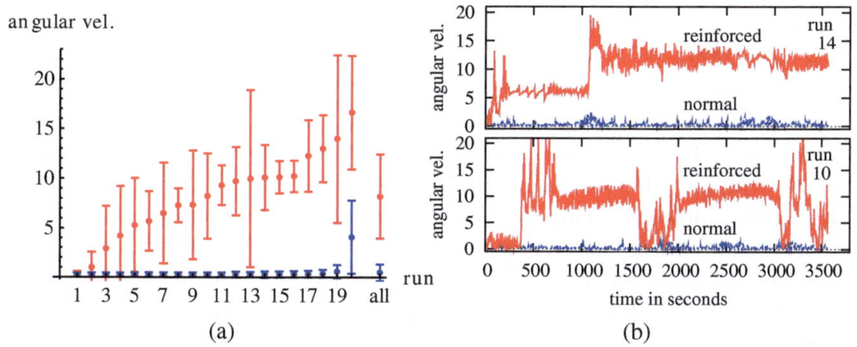

Fig. 8.15 Performance of the spherical robot rewarded for spin. **(a)** Mean and std. deviation of the angular velocity ω_z of the spherical robot for 20 runs each 60 min long with (**red**) and without (*blue*) spin reinforcement, sorted by angular velocity. The label '**all**' denotes the mean and std. deviation over the means of all runs. **(b)** Time course of the velocity for run number 12 and 20, where *blue/dotted* shows the normal case and **red/solid** line shows the reinforced case.

In this setup it can also be seen that the rewarded behaviors are found again after they were lost, see Fig. 8.15(b).

The mechanism to modulate the learning speed by a reward signal showed a strong effect on the behavior of the spherical robot. When controlled by the homeokinetic controller without guidance the robot rarely exhibits narrow curves or spinning behavior. In contrast the guided controller engaged the system into curved motion most of the time. One might wonder how it is possible that this technique is able to reach a behavior that is normally not exhibited. The reason is that when the robot is starting to follow a curve, then the learning rate of the controller goes down, although the forward model is still learning normally. In the unguided case the prediction error rises (because it is a new behavior) and thus the controller will quickly leave this behavior. This actually happens before a fast spinning is reached. In the rewarded case the forward model is able to capture the behavior before it is left (because of the slower drift), which in turn enables the control system to enter modes of more narrow curves.

8.7 Channeling Self-Organization

Periodic behaviors, such as observable in locomotion, are characterized by a particular spatio-temporal structure which can be described in terms of phase relations between the joints. Vice versa, by imposing certain phase relations a bias towards a specific behavior can be conveniently introduced into the dynamical system. For this purpose we will use again soft constraints that break symmetries in a particular way, reduce the effective dimension of the sensorimotor dynamics, and guide thus the self-organizational process towards a subspace of the original control problem. In biological systems similar constraints are known to be effective on a low level of neuronal circuitry, e. g. linking pairs of antagonistic muscles such that the activity of one muscle inhibits activity of the other via inter-neurons in the spinal cord (Pearson and Gordon 2000).

We will apply here an analogous regulation method which refers to motor values of one effector as teaching signals for another one, and will call this scheme *cross-motor teaching*. It will be used to prescribe which motor neuron receives a teaching signal from which other neuron. Note that despite the use of 'teaching signals' the algorithm is completely unsupervised, because the signals are generated internally. The self-organization progress preserves a high amount of symmetries of the physical system. As an example, consider a two-wheeled robot that drives forward and backward and rotates clockwise and counterclockwise equally often. The physical system (morphology of the body and interaction) is essentially symmetric with respect to forward-backward, left-right (lateral), and also straight-rotational behavior. To the contrary, if the robot lacks forward-backward symmetry and, more importantly, also straight-rotational symmetry because of friction and inertia. This is also reflected in behavior in that the robot is more driving straight than rotating.

8.7.1 From Spontaneous to Guided Symmetry Breaking

To achieve symmetry breaking in a predefined way, we will first consider pairwise relations as constraints for the broken-symmetric state. Later we will generalize this by using permutation relations. Let us, e.g. influence the controller to prefer a pairwise *in-phase* or *antiphase* relations in the motor patterns (Martius and Herrmann 2010). For a particular pair of motors (r,s), we place a bidirectional cross-motor connection from r to s, which means that the motor s receives its teaching signal from motor r and vice versa. In this way both motors are guided towards an in-phase activity. The (internal) teaching signal is

$$\left(y_t^G\right)_r = (y_t)_s \qquad \text{and} \qquad \left(y_t^G\right)_s = (y_t)_r\,, \tag{8.28}$$

which is used then in Eqs. (8.12–8.15).

Likewise, an antiphase teaching relation can be expressed by $\left(y_t^G\right)_r = -(y_t)_s$ and vice versa. In this simple setup the cross-motor connections have either a positive or negative sign. For those motors i that are not part of a connected pair we need to set $\left(y_t^G\right)_i = (y_t)_i$, in order to suppress the error signal, see Sect. 8.4.2.

8.7.1.1 Experiment

To illustrate the concept we will consider the above-mentioned two-wheeled robot, cf. Fig. 8.16c. The robot has two motors actuated according to y_1 and y_2 and is subject to the goal of straight driving. This can be obtained by an in-phase relation between both motors following Eq. (8.28), i. e.

$$\left(y_t^G\right)_1 = (y_t)_2 \qquad \text{and} \qquad \left(y_t^G\right)_2 = (y_t)_1\,. \tag{8.29}$$

For experimental evaluation we placed the robot in an environment cluttered with obstacles.

We performed, for different values of the factor γ, five runs of 20 min length. In order to quantify the influence of the cross-motor teaching we recorded the trajectory, the linear velocity, and the angular velocity of the robot. We expect an increase in linear velocity because the robot is to move straight instead of turning. For the same reason the angular velocity should go down. In Fig. 8.16 a sample trajectory and the behavioral quantifications are plotted. Additionally, we plot the relative area coverage which is calculated from the trajectory using a box-counting method. It reflects how much area of the environment was covered by the robot with cross-motor teaching compared to the original homeokinetic controller. As expected, the robot shows a distinct decrease in mean turning velocity and a higher area coverage with increasing values of the guidance factor. Note that the robot is still performing turns and drives both backwards and forwards and does not get stuck at the walls, as seen in the trajectory in Fig. 8.16(c). The properties of the homeokinetic controller, such as sensitivity and exploration, remain.

We have seen that a pairwise cross-motor teaching can be used to guide the self-organizing control to drive mostly straight in the two-wheeled robot. The strength

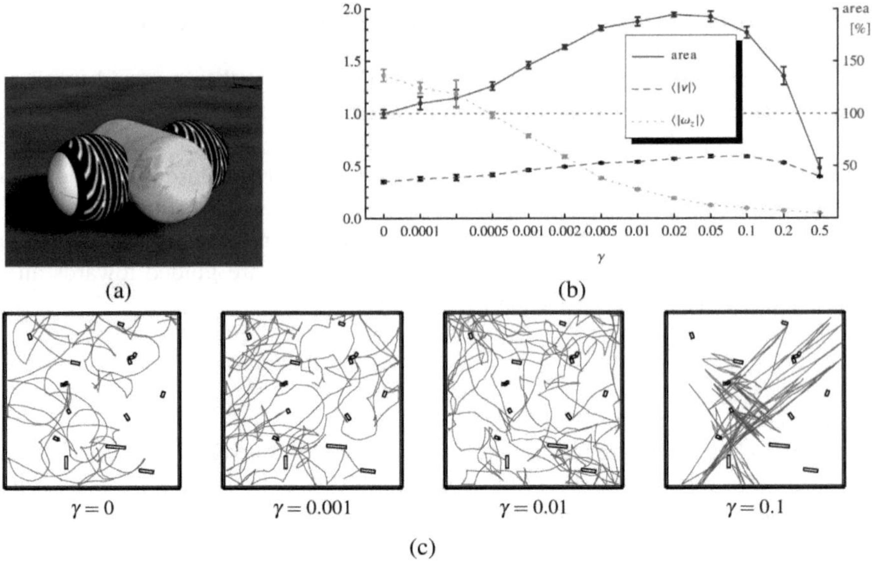

Fig. 8.16 Behavior of a two-wheeled robot (a) guided to move preferably straight. **(b)** Mean and standard deviation (of 5 runs each 20 min) of the area coverage, the average velocity $\langle|v|\rangle$, and the average turning velocity $\langle|\omega_z|\rangle$ for different values of the guidance factor γ. Area coverage (box counting method) is given relative to the the case without influence ($\gamma{=}0$: 100%) (right axis). The robot drives straighter and covers more area for increasing γ, until at large γ the teaching strictly dominates the behavior of the robot. **(c)** Example trajectories for different guidance factors. Parameters: $\varepsilon_c = \varepsilon_A = 0.01$.

of this preference can be adjusted by the guidance factor. The algorithm is self-supervised and the only specific information that is given is the pair of motors to be synchronized.

8.7.2 Multiple Motor Relations

Now we want to consider a more general cross-motor connection setup where each motor has one incoming and one outgoing connection, such that there is still only one teaching signal per motor neuron (Martius and Herrmann 2011). The cross-motor connections can be described by a permutation π_m of the m motor neurons assigning each motor neuron a source of teaching input. The teaching signal is then given by (dropping the time index)

$$y_i^G = y_{\pi_m(i)} \qquad \text{for } i = 1,\ldots,m. \qquad (8.30)$$

Additionally a sign function could be used defines whether the motors are supposed to be in-phase or antiphase, but we do not need it in the following. The pairwise

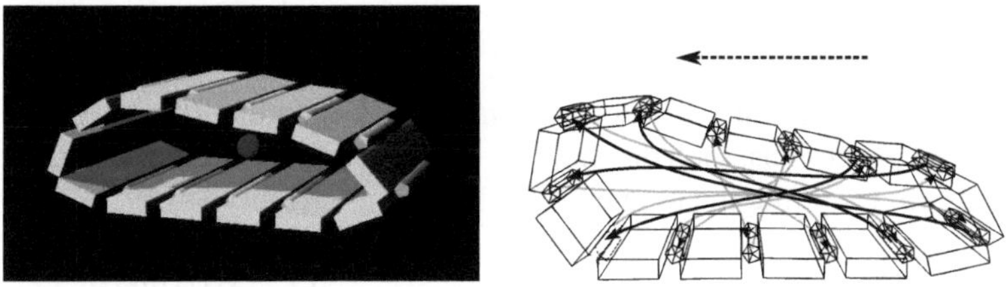

(a) (b)

Fig. 8.17 The armband robot. (a) Screen shots of the simulation. The transparent sphere in the center marks the center of mass of the robot. **(b)** Track-robot armband with cross-motor connections. The arrows indicate unidirectional cross-motor connections, where the head points to the receiving unit. All links are equal, but for visibility reasons only four links are drawn boldly. For this connection setup the robot preferably moves leftwards.

setup (Eq. (8.28)) is of course a special case of this notation. Note, that with a cyclic schema of connections also a group of motors can be synchronized.

8.7.3 Guiding to Directed Locomotion

In order to study a robot with a scalable complexity, we will consider the *armband* robot—a bracelet- or track-like structure. We will see that we can explicitly guide the robot to a directed and fast locomotion by organizing the initially decentralized control into a cooperative mode which can be considered as the emergence of a single controller for the entire robot.

This robot consists of a sequence of m flat segments placed in a ring-like configuration, where subsequent segments are connected by the m hinge joints. The resulting body has the appearance of a bracelet or chain, see Fig. 8.17(a). Each joint is driven by a servo motor and has a joint-angle sensor. The center positions of the joints are such that the robot is in a perfectly circular configuration (deviating by an angle of $2\pi/m$ from a straight positioning). The motor values and sensor values are represented as well as joint angle deviations, see Fig. 8.17(b). The joints are highly coupled through the ring configuration. Therefore, an independent movement of a single joint is not possible. Instead it has to be accompanied by a movement of the neighboring joints and of distant joints.

Since the robot is symmetric there is by construction no preferred direction of motion, meaning that the robot controlled by the homeokinetic controller will equally probable move forward or backward. The robot cannot turn or move sideways, but it can produce a variety of postures and locomotion patterns.

With the method of cross-motor teaching we can help to break different symmetries, such that the robot is more likely to perform a directed motion. One possibility is to connect motors on opposite sides of the robot with a bias in clockwise or counterclockwise direction. For that we define the permutation (used in Eq. (8.30)) as

$$\pi_m(i) = (i + k + \lfloor m/2 \rfloor) \mod m\,, \tag{8.31}$$

where $k \in \{-1, 0, 1\}$ and $\lfloor \cdot \rfloor$ denotes the truncation rounding (floor). We will only use positive connections, such that the sign function is not required. Thus, the teaching signals are (omitting the time index)

$$y_i^G = y_{(i+k+\lfloor m/2 \rfloor) \mod m} \qquad \text{for } i = 1, \ldots, m\,. \tag{8.32}$$

The choice of k depends on the desired direction of motion and on whether the number of joints m is even or odd. If m is even then $k = -1$ and $k = 1$ are used for both directions (forward or backward) and $k = 0$ represents a symmetric connection setup. In the latter case the robot will not prefer a direction of motion and the behavior is similar to the one without guidance. For an odd value of m, which is also used here, $k = 0$ and $k = 1$ need to be used for backward and forward motion.

In the following experiments the robot has $m = 13$ motors. The motor connections for $k = 1$ are illustrated in Fig. 8.17. Each motor connection is displayed by an arrow pointing to the receiving motor. Note that the connections are directed and a motor neuron is not teaching the motor neuron from which it is receiving teaching signals. For $k = 0$ all arrows are inverted, meaning that for each connection the sending and receiving motor neurons swap roles.

To evaluate the performance we conducted, for different values of the guidance factor γ, 5 trials each 30 min long. In a first setting the cross-motor connections were fixed ($k = 1$) for the entire duration of the experiment. Without guidance the robot moves equally to both directions but with comparably low velocity. This can be seen at the mean of the absolute velocity in Fig. 8.18(a). If the value of the guidance factor is chosen conveniently, we observed the formation of a locomotion behavior after a very short time and the robot moves in one direction with varying speed see Fig. 8.18(b) for three velocity traces. Note that this behavior requires all joints of the robot to be highly coordinated. We also observe a peak of high velocity after the first few minutes, which is followed by a dip before a more steady regime is attained. During this time the controller is going from a subcritical regime (at $t = 0$) to a slightly supercritical regime.

The locomotory behavior can also be seen in Fig. 8.19 for a low value of guidance factor ($\gamma_s = 0.001$) and in Fig. 8.20 for a medium value of guidance factor ($\gamma = 0.003$). The average velocity of the robot increased distinctively with rising guidance factors, see Fig. 8.18(a). However, for excessively large values of the guidance factor the velocity goes down again. This occurs for two reasons: First, the cross-motor teaching has a too strong influence on the working regime of the homeokinetic controller and second the actual motor pattern of the locomotion behavior does not perfectly obey the relations between the motor values as specified by Eq. (8.32). In order to satisfy the constraints imposed by Eq. (8.32) all motor values need to be equal, which is of course not the case in the locomotion behavior.

In a second setup we changed the cross-motor connections every 5 min, i. e. k was changed from 0 to 1 and back. A value of $k = 0$ should lead to a negative velocity and a value of $k = 1$ to a positive velocity.

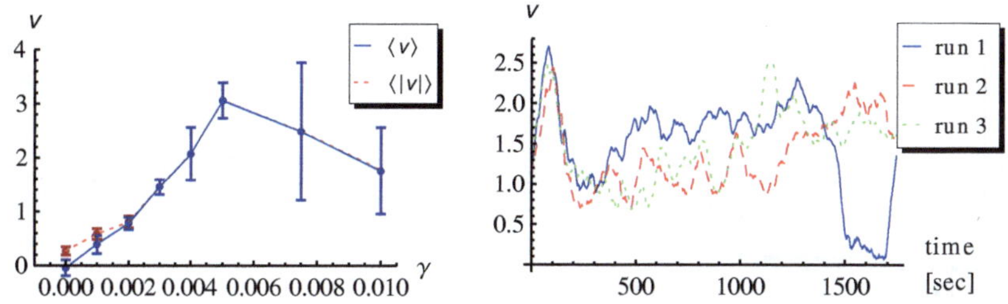

(a) (b)

Fig. 8.18 Performance of the armband robot with constant cross-motor teaching. **(a)** Mean and standard deviation of the average velocity $\langle v \rangle$ and the average absolute velocity $\langle |v| \rangle$ of 5 runs for different value of the guidance factor γ. **(b)** Velocity of the robot $\overline{v}$ (averaged over 1 minute sliding window) for 3 runs at $\gamma = 0.003$. Parameters: $k = 1$, $\varepsilon_c = \varepsilon_A = 0.1$.

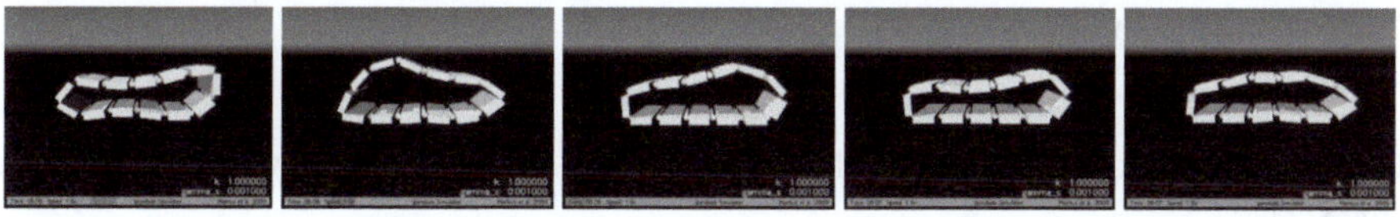

Fig. 8.19 The armband robot learns to locomote by weak guidance. Behavior of the robot with cross-motor teaching and weak guidance ($\gamma = 0.001$). A slow locomotive behavior with different velocities is exhibited. Explorative actions cause the posture of the robot to vary in the course of time.

Fig. 8.20 The armband robot quickly learns to locomote. Behavior of the robot with cross-motor teaching and medium guidance ($\gamma = 0.003$). Comparable fast locomotive behavior emerges quickly and is persistent. Nevertheless the velocity varies. Only small exploratory actions are takes, such that the posture is mainly constant.

To study the dependence on the guidance factor and to measure the performance we use the average absolute velocity $\langle |v| \rangle$ and the correlation of the velocity with the configuration of the coupling $\rho(v,k)$, see Fig. 8.21(a). Without guidance ($\gamma = 0$) there is, as expected, no correlation with the supposed direction of locomotion. For a range of values of the guidance factor we find a high total locomotion speed with a strong correlation to the supposed direction of motion. Note that the size of the correlation depends on the length of the intervals of one connection setting. For long intervals the correlation will approach one. In Fig. 8.21(b) the velocity of the robot is plotted for different runs with the same value of the guidance factor that was used in the previous experiment ($\gamma = 0.003$). We observe that the robot changes the

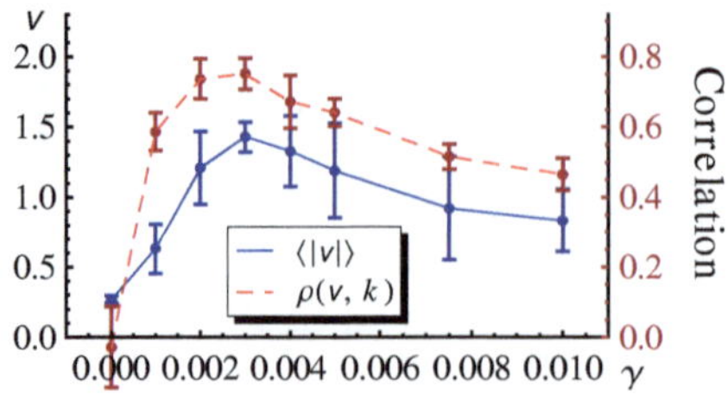

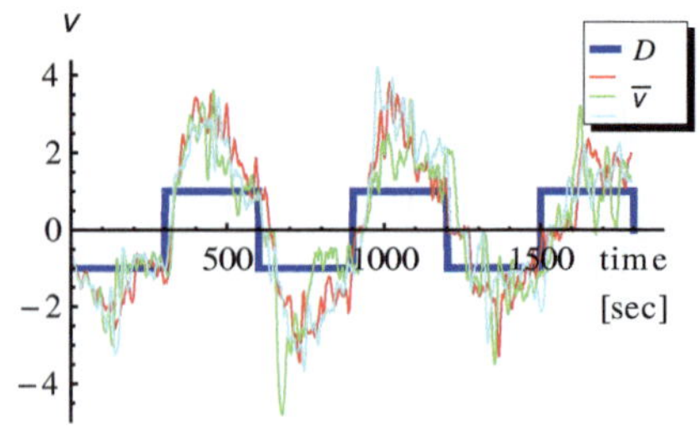

Fig. 8.21 Performance of the armband robot for variable cross-motor teaching. **(a)** Mean and standard deviation of the average absolute velocity $\langle |v| \rangle$ and the correlation $\rho(v,k)$ of the velocity with the configuration of the coupling for 5 runs with different values of the guidance factor γ. **(b)** Velocity (averages over 10 sec sliding windows) of the robot for 3 runs at $\gamma = 0.003$ and the target direction of motion $D = 2k - 1$ for better visibility. Parameters: $\varepsilon_c = \varepsilon_A = 0.1$.

direction of motion shortly after the configuration of the coupling was changed, see Fig. 8.21

8.7.4 Scaling Properties

The locomotion of the robot is essentially influenced by the number of cross-motor connections. To study this we use again the fixed connectivity. In a series of simulations a number $0 \leq l \leq m$ of equally spaced cross-motor connections (Fig. 8.17) are used. With increasing l the robot starts to locomote earlier. Full performance is reached already if 8 out of the 13 connections are used, see Fig. 8.22(a).

In order to study the scaling properties of the learning algorithm we varied the number of segments m of the robot and thus the dimensionality of the control problem. The results are astonishing, see Fig. 8.22(b): The behavior is learned with the same speed also for large number (40) of segments. There is no scaling problem here for the following reason. In the closed loop with an approximate feedback strength (self-regulated by the homeokinetic controller) the robot needs only very little influence to roll. The length of the robot can even help because other behavioral modes (e. g. wobbling) are damped increasingly due to gravitational forces. For the same reason, small robots are slower than medium ones. Large robots are again slower because the available forces at the joints become too weak.

The experiment illustrates that specific behaviors can be achieved in a high-dimensional robot by using cross-motor teaching. Cross-motor connections can break the symmetry between the two directions of motion such that a locomotion behavior is produced quickly. When the connections are switched during runtime, the behavior of the robot changes reliably.

The mechanism proposed here can also be transferred to sensor space using the direct sensor teaching (Sect. 8.4.3) instead of the motor teaching. One obtains a

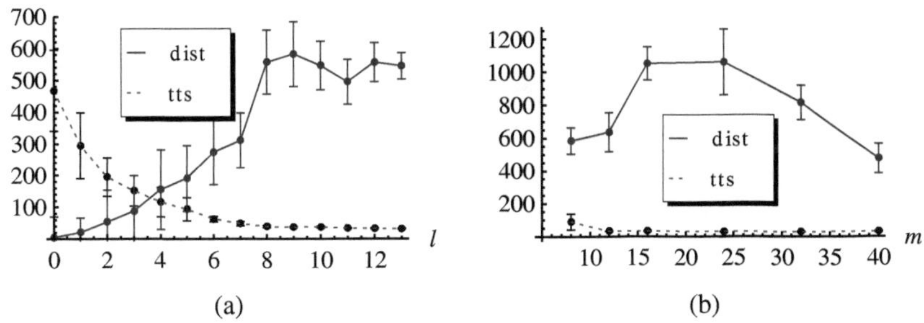

Fig. 8.22 Scaling of learning time and performance for different robot complexity. The plots show mean and standard deviation of the distance traveled by the robot ('dist' in units of 1 segment size) and of the time-to-start ('tts' in seconds) of 20 runs à 10 min ($\gamma = 0.003$). **(a)** Performance as a function of the number of cross-motor connections l (equally spaced around a robot with $m = 13$ joints). **(b)** Performance for different numbers of segments m (DoF) with full cross-motor connectivity ($l = m$).

cross-sensor teaching analogously to the definitions given above. This can become useful, for example if a certain behavior is demonstrated by a human operator by activity moving the robot. In the case of the armband robot, one can imagine pushing the robot along the ground forcing it into a locomotion pattern. Based on the observed sensor readings, the correlations between the sensor channels may be determined and used as a basis for the construction of a specific cross-sensor teaching setup. This highly interesting idea was, however, not yet implemented and remains for future work.

Starting from the guidance by teaching we introduced the concept of cross-motor teaching allowing for the specification of abstract relations between motor channels. There are no external teaching signals required, because the motor values are used mutually as teaching signals. The only specific information put into the system is the cross-motor relation. First we studied simple pairwise relations and shaped the behavior of the two-wheeled robot to drive mostly straight through the coupling between both motors. The couplings introduce soft constraints that guide the self-organization process to a subspace of the entire sensorimotor space and therewith the effective dimension of the search space for behaviors is reduced. This was demonstrated using the high-dimensional armband robot. With a simple cross-motor teaching the robot developed within a short time fast locomotion behaviors from scratch. The direction of motion was altered by a change in the connection setup. Remarkable is also the scaling property with respect to the dimensionality of the control problem.

8.7.5 Coordination of Finger Movements for Grasping

An interesting application of the above method is in neuroprosthetics, where often little information is available in complex control problems. We will consider the control of a prosthetic hand in a simulation. The simulated hand has six controllable

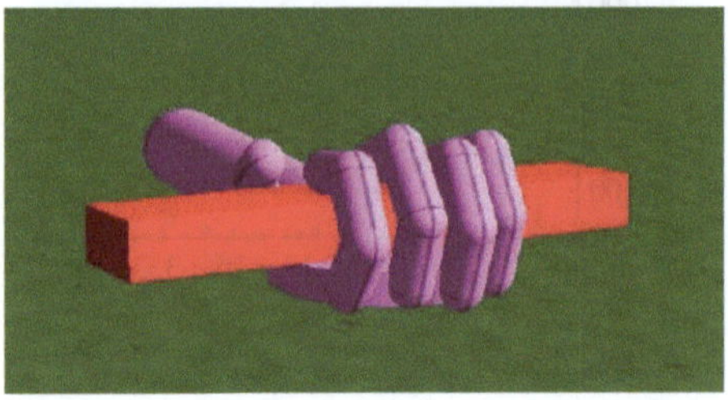

(a)

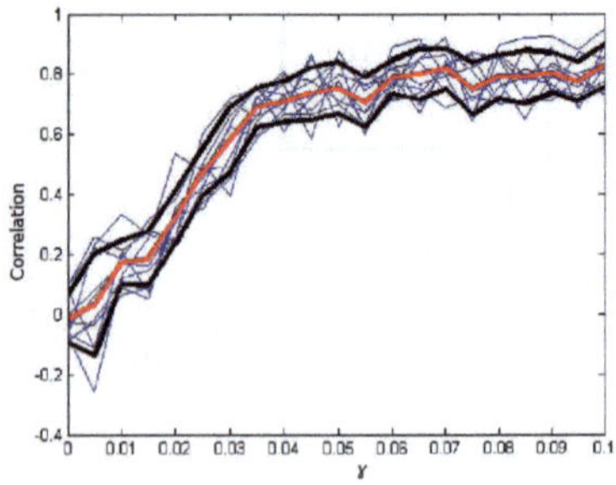

(b)

Fig. 8.23 Guided self-organization with homeokinesis in a simulated hand prosthesis. (**a**) The prostheses has six controllable degrees of freedom. A teaching term based on the similarity of the finger angles keeps the fingers in near synchrony while they are moving independently when controlled by the basic homeokinetic rule. (**b**) For a guidance factor $\gamma \approx 0.03$ correlations between fingers are reached that are similar to observations in healthy humans (Santello and Soechting 2000). The red line shows the mean value over 10 trial and black lines the respective standard deviation.

DoF, two for the thumb and one for the other fingers, i.e. the fingers have a fixed coupling between their three degrees of mobility, see Fig 8.23(a). All fingers are equipped with position sensors of the joints and proximity sensors in the tip. If the fingers are controlled by the homeokinetic controller they develop independent movements because no physical coupling is present in this robotic model. Since in a natural environment the fingers interact mostly because of the manipulation of objects and because of physiological constraints, we can also try learning such correlations by a guidance principle. In this way we will arrive at a measure of the required interaction which then can be compared with observations in healthy humans.

In order to enforce movement synergies between the fingers, we implemented finger correlation by cross motor teaching between all fingers. Here we have multiple teaching signals for each motor neuron, where simply the arithmetic average is used.

The mean correlation of the homeokinetic controller without guidance ($\gamma = 0$) is 0, which means here the fingers move independently. At high values of γ the correlation approaches unity, which indicates that the fingers have lost independence which however would be needed in grasping applications. Considering the correlation value of 0.582 given for human fingers (Santello and Soechting 2000), the optimal value of γ would be around 0.03, see Fig. 8.23(b). Again a very weak guidance is sufficient to influence the behavior in the desired direction.

8.8 Discussion

In this chapter we have presented several mechanisms for guided self-organization of robot behavior based on homeokinetic control (GSOH). Homeokinesis bootstraps the exploration process of embodied systems and leads to self-organization of behavior. Various patterns of behavior emerge depending on the robotic hardware and its environment. With a general framework of problem specific error functions we set the foundation for guidance by teaching signals and guidance by cross-motor teaching. The balance between self-organized behavior and target behavior can be adjusted with a single parameter.

Interestingly, teaching signals can as well be provided in terms of desired sensor values. In this setting, for instance a spherical robot was taught to rotate around one particular axis solely by requesting a zero value of the sensor value corresponding to that axis. In a more elaborate example we show how the task of finding balls and pushing them around in an environment can be achieved by simply providing a desired visual sensor state. The entire sensorimotor coordination to fulfill this goal was learned by the robot within a few minutes. This involves the basic coordination to drive the robot and to integrate the vision sensors such that the balls are approached and balanced while pushed. To probe the robustness of the approach the orientation of the camera was abruptly changed such that a completely different sensorimotor coordination becomes necessary. We found that GSOH can cope with a wide range of configuration changes, even those where a complete change in the visual sensation occurs (signs of all visual sensors swapped).

The teaching mechanisms form the basis for a higher level guiding mechanism, namely cross-motor teaching. It allows to specify relations between motor channels to be in-phase or antiphase activity. This induces soft constraints and therewith reduces the effective dimensionality of the system. This was especially illustrative with the high-dimensional armband robot. A cross-motor teaching with only one connection per joint leads to a fast and coordinated locomotion behavior. Similar to the robustness in the vision experiments, we observe here a rapid and reliable change in the direction of locomotion by an altered connection setup. A particularly promising result is that the performance and speed of learning is almost independent of the dimensionality of the system, at least in the here considered cases of up to 40 DoF. The discrete cross-motor connections offers a good way for higher level control structures to direct the behavior of the robot.

We also presented a simple method to guide the self-organizing behavior using online reward signals, originally published in (Martius et al. 2007). In essence the original time-loop error is multiplied by a strength factor, obtained from the reward signal. The approach was applied to the spherical robot with two goals, fast motion and curved rolling, which was successfully achieved. Notably, the exploratory character of the paradigm still remains intact.

To compare the different methods of GSOH we can ask for the amount and type of information that is required about the behavior and the robotic system. For the direct teaching methods a rather detailed insight into the sensorimotor patterns of the desired behavior is required. In sensor space this is typically easier than in motor space as demonstrated by the examples. For the cross-motor teaching a more high-level knowledge is sufficient, for instance about the symmetries of the body and of the desired motion. In both cases the designer needs to expect a specific behavior, e. g. a locomotion behavior with a certain gait. In the reward based method, on the other hand, the realization is not specified, e. g. the gait would be found autonomously. To be successful, however, the exploration needs to be structured enough to produce short segments of locomotion behavior to be picked and amplified. Here the newest methods for behavioral self-organization using information-theoretic quantities show promising results (Martius et al. 2013; Der and Martius 2013).

Let us briefly compare GSOH with other approaches to learning of autonomous robot behavior, namely evolutionary algorithms (EA) (Nolfi and Floreano 2001) and reinforcement learning (RL) (Sutton and Barto 1998). EA and RL can optimize the parameters of the controller (e. g. a neural network) and can in principle achieve the behaviors demonstrated here. There are many impressive results where systems of similar dynamical complexity have been successfully controlled, see for example (EA) Chemova and Veloso (2004); Bongard et al. (2006); Mazzapioda and Nolfi (2006); de Margerie et al. (2007); Ijspeert et al. (1999) and (RL) Peters and Schaal (2008). In high-dimensional systems, however, identical subcomponents are typically used or the problem is appropriately prestructured by hand. Additionally, long learning times are required (many generations with many individuals or repetitions) which is often prohibitively long for physical robots. Here we see the main strength of our system: The desired behaviors are found very fast even in high-dimensional and dynamically complex systems—we have very fast online-learning. Another difference is that the finally evolved or learned controllers are typically static, such that it only works in the conditions it was evolved/trained in. In contrast we demonstrated the robustness of GSOH to extreme sensor disruption, which is successful due to a continuous self-modeling and exploration.

Of course there is also a downside, namely that the here proposed approaches are rather limited in which behaviors can be achieved and how for instance the reward can be given. Also in GSOH the desired behaviors are only partially followed and no optimality guarantees can be given which is in contrast to RL that was proven to converge to the optimal solution under certain conditions (Sutton and Barto 1998). However, for practical applications these proves are of questionable value because a prohibitive amount of time is required.

To conclude, the GSOH methods offer a fast development of goal-oriented behaviors in high-dimensional continuous-domain robotic systems from scratch, which cannot be achieved with other learning control systems so far. However, the implementation of goals is comparably limited. The reward-based guidance allows any reward signals, but no time delays are tolerated and it is not guaranteed that the reward is maximized. The cross-motor teaching method is suitable to select a subset of behaviors, but cannot be generalized to all behaviors. A combination of both methods is also conceivable, namely using cross-motor teaching to be very effective in high-dimensional systems and additionally using rewards to give a more fine grain control over the behavior. Another line of future research would be the proposed cross-sensor teaching that would allow for the specification of behaviors on the level of sensor relations. We also expect that superior results can be obtained when the here proposed methods are combined with the new algorithms for behavioral self-organization (Der and Martius 2013) as they produce more structure in the emerging behaviors.

Acknowledgment. We thank Raluca Scona and Guillaume de Chambrier for providing us with some of the figures. RD thanks Nihat Ay for the hospitality in his group. The project was supported by BCCN Göttingen grant #01GQ0432, BFNT Göttingen grant #01GQ0811 and DFG (SPP 1527).

References

Amari, S.: Natural gradients work efficiently in learning. Neural Computation 10 (1998)

Bongard, J.C., Zykov, V., Lipson, H.: Resilient machines through continuous self-modeling. Science 314, 1118–1121 (2006)

Butera, F.M.: Urban development as a guided self-organisation process. In: The City and Its Sciences, pp. 225–242. Springer (1998)

Cannon, W.B.: The wisdom of the body. Norton, New York (1939)

Chemova, S., Veloso, M.: An evolutionary approach to gait learning for four-legged robots. In: Proc. IEEE IROS 2004, vol. 3, pp. 2562–2567 (2004)

Choi, J., Wehrspohn, R.B., Gösele, U.: Mechanism of guided self-organization producing quasi-monodomain porous alumina. Electrochimica Acta 50(13), 2591–2595 (2005)

Cruse, H., Dürr, V., Schmitz, J., Schneider, A.: Control of hexapod walking in biological systems. In: Adaptive Motion of Animals and Machines, pp. 17–29. Springer (2006)

de Margerie, E., Mouret, J.-B., Doncieux, S., Meyer, J.-A.: Artificial evolution of the morphology and kinematics in a flapping-wing mini UAV. Bioinspiration and Biomimetics 2, 65–82 (2007)

Der, R.: Self-organized acquisition of situated behaviors. Theory Biosci. 120, 179–187 (2001)

Der, R., Liebscher, R.: True autonomy from self-organized adaptivity. In: Proc. Workshop Biologically Inspired Robotics, Bristol (2002)

Der, R., Martius, G.: The Playful Machine - Theoretical Foundation and Practical Realization of Self-Organizing Robots. Springer (2012)

Der, R., Martius, G.: Behavior as broken symmetry in embodied self-organizing robots. In: Advances in Artificial Life, ECAL 2013 (accepted, 2013)

Dongyong, Y., Jingping, J., Yuzo, Y.: Distal supervised learning control and its application to CSTRsystems. In: SICE 2000. Proc. of the 39th SICE Annual Conference, pp. 209–214 (2000)

Ijspeert, A.J., Hallam, J., Willshaw, D.: Evolving Swimming Controllers for a Simulated Lamprey with Inspiration from Neurobiology. Adaptive Behavior 7(2), 151–172 (1999)

Jordan, M.I., Rumelhart, D.E.: Forward models: Supervised learning with a distal teacher. Cognitive Science 16(3), 307–354 (1992)

Klyubin, A.S., Polani, D., Nehaniv, C.L.: Empowerment: a universal agent-centric measure of control. In: IEEE Congress on Evolutionary Computation, pp. 128–135. IEEE (2005)

Martius, G.: Robustness of guided self-organization against sensorimotor disruptions. Advances in Complex Systems 16(02n03), 1350001 (2013)

Martius, G., Der, R., Ay, N.: Information driven self-organization of complex robotic behaviors. PLoS ONE 8(5), e63400 (2013)

Martius, G., Herrmann, J.M.: Taming the beast: Guided self-organization of behavior in autonomous robots. In: Doncieux, S., Girard, B., Guillot, A., Hallam, J., Meyer, J.-A., Mouret, J.-B. (eds.) SAB 2010. LNCS, vol. 6226, pp. 50–61. Springer, Heidelberg (2010)

Martius, G., Herrmann, J.M.: Tipping the scales: Guidance and intrinsically motivated behavior. In: Advances in Artificial Life, ECAL 2011, pp. 506–513. MIT Press (2011)

Martius, G., Herrmann, J.M.: Variants of guided self-organization for robot control. Theory in Biosci. 131(3), 129–137 (2012)

Martius, G., Herrmann, J.M., Der, R.: Guided self-organisation for autonomous robot development. In: Almeida e Costa, F., Rocha, L.M., Costa, E., Harvey, I., Coutinho, A. (eds.) ECAL 2007. LNCS (LNAI), vol. 4648, pp. 766–775. Springer, Heidelberg (2007)

Martius, G., Hesse, F., Güttler, F., Der, R.: LpzRobots: A free and powerful robot simulator (2012), `http://robot.informatik.uni-leipzig.de/software`

Mazzapioda, M., Nolfi, S.: Synchronization and gait adaptation in evolving hexapod robots. In: Nolfi, S., Baldassarre, G., Calabretta, R., Hallam, J.C.T., Marocco, D., Meyer, J.-A., Miglino, O., Parisi, D. (eds.) SAB 2006. LNCS (LNAI), vol. 4095, pp. 113–125. Springer, Heidelberg (2006)

Nolfi, S., Floreano, D.: Evolutionary Robotics. The Biology, Intelligence, and Technology of Self-organizing Machines. MIT Press, Cambridge (2000) (1st print) (2001) (2nd print)

Ott, E., Grebogi, C., Yorke, J.: Controlling chaos. Phys. Rev. Lett. 64, 1196–1199 (1990)

Pearson, K., Gordon, J.: Spinal reflexes. In: Kandel, E., Schwartz, J.H., Jessell, T.M. (eds.) Principles of Neural Science, 4th edn., pp. 713–736. McGraw-Hill, New York (2000)

Peters, J., Schaal, S.: Natural Actor-Critic. Neurocomputing 71(7-9), 1180–1190 (2008)

Popp, J.: Spherical robots (2004), `http://www.sphericalrobots.com`

Prokopenko, M.: Design vs self-organization. In: Prokopenko, M. (ed.) Advances in Applied Self-organizing Systems, pp. 3–17. Springer (2008)

Prokopenko, M.: Guided self-organization. HFSP Journal 3(5), 287–289 (2009)

Rodriguez, A.: Guided Self-Organizing Particle Systems for Basic Problem Solving. PhD thesis, University of Maryland (College Park, Md., USA) (2007)

Santello, M., Soechting, J.F.: Force synergies for multifingered grasping. Experimental Brain Research 133(4), 457–467 (2000)

Schaal, S., Ijspeert, A., Billard, A.: Computational approaches to motor learning by imitation, vol. 1431, pp. 199–218. Oxford University Press (2004)

Smith, S.C., Herrmann, J.M.: Homeokinetic reinforcement learning. In: Schwenker, F., Trentin, E. (eds.) PSL 2011. LNCS, vol. 7081, pp. 82–91. Springer, Heidelberg (2012)

Stitt, S., Zheng, Y.F.: Distal learning applied to biped robots. In: Proc. of the IEEE Intl. Conf. on Robotics and Automation, pp. 137–142. IEEE Computer Society (1994)

Sutton, R.S.: Reinforcement learning: Past, present and future. In: McKay, B., Yao, X., Newton, C.S., Kim, J.-H., Furuhashi, T. (eds.) SEAL 1998. LNCS (LNAI), vol. 1585, p. 195. Springer, Heidelberg (1999)

Wikipedia (2013). Homeostasis — wikipedia, the free encyclopedia (Online accessed July 23, 2013)

Chapter 9
On the Causal Structure of the Sensorimotor Loop

Nihat Ay and Keyan Zahedi

9.1 Introduction

In recent years, the application of information theory to the field of embodied intelligence has turned out to be extremely fruitful. Here, several measures of information flow through the sensorimotor loop of an agent are of particular interest. There are mainly two ways to apply information theory to the sensorimotor setting.

First, information-theoretic measures can be used within various analysis methods. Sensorimotor interactions of an embodied agent lead to the emergence of redundancy and structure of the agent's intrinsic processes. Understanding the generation of structure in the sensorimotor process and its exploitation is important within the field of embodied intelligence (Pfeifer and Bongard 2006). The quantification and analysis of information flows through an agent's sensorimotor loop from the perspective of an external observer, that is from the perspective of a scientist, proves to be effective in this regard (Lungarella and Sporns 2005, 2006). Here, transfer entropy (Schreiber 2000) has been used in order to quantify the flows of information between various processes of the sensorimotor loop, such as the sensor process on the actuator process. Furthermore, excess entropy, also known as predictive information (Bialek et al. 2001), has been used to analyse the interplay between information-theoretic measures and behavioral patterns of embodied agents (Der et al. 2008).

Second, information-theoretic measures can be used as objective functions for self-organized learning. This is based on the hypothesis that learning in natural intelligent systems is partly governed by an information-theoretic optimisation principle. Corresponding studies aim at the implementation of related principles, so-called

Nihat Ay · Keyan Zahedi
Max Planck Institute for Mathematics in the Sciences,
Inselstraße 22, D-04103, Leipzig, Germany
e-mail: {nay, zahedi}@mis.mpg.de

Nihat Ay
Santa Fe Institute, 1399 Hyde Park Road, Santa Fe, New Mexico 87501, USA

M. Prokopenko (ed.), *Guided Self-Organization: Inception,*
Emergence, Complexity and Computation 9,
DOI: 10.1007/978-3-642-53734-9_9, © Springer-Verlag Berlin Heidelberg 2014

Infomax principles, in artificial systems. Emergent structures at various levels are then analysed in view of corresponding biological structures. In the sensorimotor setting, predictive information maximization has been used as a driving force for self-organised learning (Ay et al. 2008; Zahedi et al. 2010; Ay et al. 2012; Martius et al. 2013). As a result, the emergence of coordinated behavior with distributed control has been shown. The excess entropy has also been applied to similar systems within an evolutionary optimisation context (Prokopenko et al. 2006). Other measures of interest are the notion of relevant information (Polani et al. 2006) and empowerment (Klyubin et al. 2005). In the latter case, the maximization of empowerment determines the behavior of the agent and is not the basis of learning.

Most of the information-theoretic quantities mentioned above have the mutual information of two variables as an important building block. In information theory, this fundamental quantity is used as a measure of the transmission rate of a sender-receiver channel. Therefore, the objective functions for learning that are based on mutual information are usually associated with some kind of information flow. There is one problem with this interpretation, which is not visible in the simple sender-receiver context. Information flows are causal in nature, and related measures should be consistent with this fact. However, the causal aspects are usually not explicitly addressed. In order to do so, it has been proposed to combine information theory with the theory of causal networks (Ay and Polani 2008), based on the causal structure of the sensorimotor loop (Klyubin et al. 2004). This combination allows us to understand how stochastic dependence, and, in particular, the statistical structure of the sensorimotor process, is built up by causal relationships. Various (associational) information flow measures can be formulated in causal terms and lead, in general, to a modification of these measures. Thereby, the repertoire of information-theoretic quantities that can be used within the above-mentioned lines of research is extended. However, currently it is not clear to what extent realistic objective functions for self-organised learning should be causal in nature.

In Section 9.2 we sketch the theory of causal networks and its application to the sensorimotor loop setting. Section 9.3 introduces the notion of a causal effect and the identifiability problem of causal effects. In particular, the identifiability of causal effects from the intrinsic perspective of an agent is discussed. In Section 9.4 basic information-theoretic quantities are introduced. Within the context of temporal processes, such as the sensorimotor process, transfer entropy and predictive information are highlighted as important quantities. The maximization of predictive information is studied in an experimental setup in the final Section 9.5.

9.2 Causal Networks

9.2.1 The Definition of Causal Networks

The formal tool that we use for modelling causality is given in terms of Bayesian networks (Pearl 2000). They have two components, a structural and a functional one. The structural component is given in terms of a network. The network consists

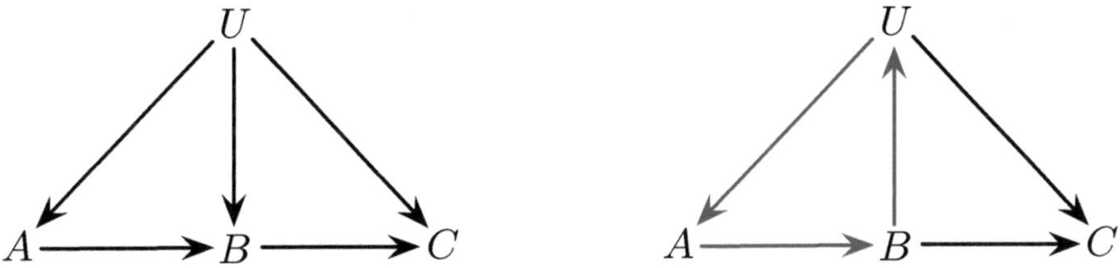

Fig. 9.1 Left: A network without a directed cycle, referred to as directed acyclic network (DAG). Right: A network with a directed cycle.

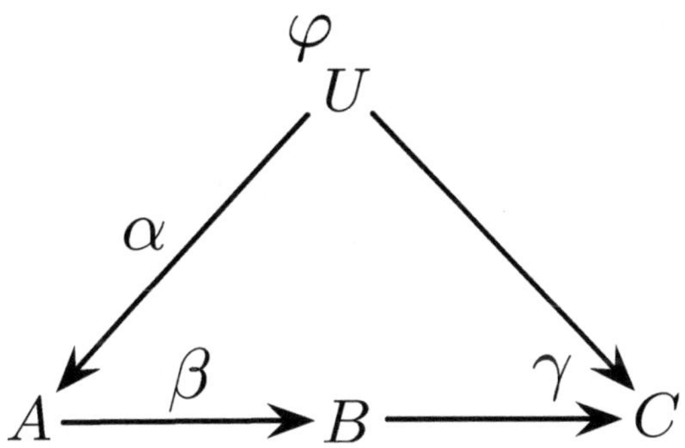

Fig. 9.2 Causal network

of vertices, or nodes, and directed edges. We denote the vertex set by V and the edge set by E which is formally a set of pairs $(v,w) \in V \times V$. Here, the pair (v,w) denotes the edge from v to w. The causal interpretation is that v is a *direct cause* of w and w is a *direct effect* v. The direct causes of a node v are referred to as *parents of* v and denoted by $pa(v)$. Extending this direct cause-effect relation, we say that a node v is a *cause* of a node w, and that w is an *effect* of v, if there is a directed path from v to w, denoted by $v \rightsquigarrow w$ (here, we exclude paths of length 0). According to this causal interpretation of the network we have to postulate that an effect cannot precede its cause. Stated differently, if v is a cause of w then w cannot be a cause of v. This is equivalent to the property that the network does not have any directed cycles as shown in Figure 9.1. A directed network with this property is called a *directed acyclic graph* (*DAG*).

Given a node v with state set $\mathscr{X}_v$, the mechanism of v is formalized in terms of a stochastic map $\kappa(x;x')$, $x \in \mathscr{X}_{pa(v)}$, $x' \in \mathscr{X}_v$, that is $\sum_{x'} \kappa(x;x') = 1$ for all x. We refer to these maps also as (*Markov*) *kernels*. Before we provide a general definition of a Bayesian network, we illustrate its basic concepts in terms of an instructive example. To this end, we consider the DAG with the nodes U,A,B, and C that is shown in Figure 9.2. In addition to the graph, the mechanisms φ, α, β, and γ are given. They describe how the nodes function and are formalized in terms of stochastic maps. For example, $\gamma(u,b;c)$ stands for the probability that node C is in state c given that it has received b and u. Based on these mechanisms, the probability of observing the states u, a, b, and c in the unperturbed system can be computed as the product

$$p(u,a,b,c) = \varphi(u) \cdot \alpha(u;a) \cdot \beta(a;b) \cdot \gamma(u,b;c). \quad (9.1)$$

This equation connects the phenomenological level (left-hand side of equation (9.1)) and the mechanistic level (the individual terms on the right-hand side of equation (9.1)).

Now we come to the general setting of a Bayesian network. Bayesian networks are based on DAGs but have a further structure as model of the involved mechanisms. As in the previous example of Figure 9.2, in a Bayesian network to each node v a mechanism κ^v is assigned. For simplicity of the arguments and derivations, we assume that the nodes v have finitely many states $\mathscr{X}_v$. Each node gets inputs $x_{pa(v)}$ from the parents $pa(v) = \{u \in V : (u,v) \in E\}$ and generates a stochastic output according to the distribution $\kappa^v(x_{pa(v)};\cdot)$. All the mechanisms together generate a distribution of global states. In order to describe this distribution, we choose a numbering of the nodes, that is $v_1, v_2, \ldots, v_n$, which is compatibe with the causal order given by the graph. More precisely, we assume the following: if there is a directed path from v_i to v_j then i is smaller than j. We use this numbering in order to generate the states of the individual nodes.

$$p(x_{v_1}, x_{v_2}, \ldots, x_{v_n}) = \kappa^{v_1}(x_{v_1}) \cdot \kappa^{v_2}(x_{pa(v_2)};x_{v_2}) \cdots \kappa^{v_n}(x_{pa(v_n)};x_{v_n}).$$

This is clearly independent of the particular choice of such an admissible numbering. Therefore, we can write

$$p(x_v : v \in V) = \prod_{v \in V} \kappa^v(x_{pa(v)};x_v). \quad (9.2)$$

On the left-hand side of this equation we have the probability of observing a particular global configuration x_v, $v \in V$. On the right-hand side we have the mechanisms. The equation postulates the transition from the mechanistic level to the phenomenological level. Given a joint distribution describing the phenomenological level, there is always a Bayesian network that generates that distribution. In order to see this, choose an arbitrary ordering $v_1, v_2, \ldots, v_n$ of the nodes. The following equality holds in any case (whenever the conditional probabilities on the right-hand side are defined):

$$p(x_{v_1}, x_{v_2}, \ldots, x_{v_n}) = \prod_{i=1}^{n} p(x_{v_i} \,|\, x_{v_1}, \ldots, x_{v_{i-1}}). \quad (9.3)$$

Consider now the graph in which a pair (v_i, v_j) is an edge if and only if i is smaller than j. With respect to this graph, the parent set of a node v_i is given by $v_1, v_2, \ldots, v_{i-1}$. Defining stochastic maps κ^{v_i} with

$$\kappa^{v_i}(x_{pa(v_i)};x_{v_i}) := p(x_{v_i} \,|\, x_{v_1}, \ldots, x_{v_{i-1}}),$$

whenever $p(x_{v_1},\ldots,x_{v_{i-1}}) > 0$, the equation (9.3) reduces to (9.2). Note that there are many possible mechanistic explanations of a given joint distribution. Having a particular one only means that *one possible* explanation is given, which does not necessarily represent *the* actual mechanisms that underly the joint distribution.

Definition 1 (Causal Markov Property). Given a DAG $G = (V,E)$, we say that a probability measure p on $\times_{v\in V}\mathscr{X}_v$ satisfies the *causal Markov property*, if, with respect to p, each variable is stochastically independent of its non-effects (V minus set of effects), conditional on its direct causes.

This property is also referred to as *local Markov property*. The conditional independence statements of the local Markov property imply also other conditional independence statements that can be deduced from the graph. In order to be more precise, we have to introduce the notion of *d-separation*.

Definition 2. Let $G = (V,E)$ be a DAG, and let S be a (possibly empty) subset of V. We say that a path $(v_1,\ldots,v_k)$ is *blocked* by S if there is a node v_i on the path such that

- either $v_i \in S$, and edges of the path do not meet head-to-head at v_i, or
- v_i and all its descendants are not in S, and edges of the path meet head-to-head at v_i.

Two non-empty and disjoint sets $A,B \subseteq V\setminus S$ are *d-separated* by S if all paths between A and B are blocked by S.

The notion of d-separation is completely structural. The local Markov property provides a way of coupling the structure with the joint probability distribution p. This condition implies a seemingly stronger Markov property: we say that p satisfies the *global Markov property*, if

$$A \text{ and } B \text{ are } d\text{-separated by } S \Longrightarrow A \text{ and } B \text{ are stochastically independent given } S.$$

The following theorem is central in graphical models theory (Lauritzen 1996).

Theorem 1. *Let $G = (V,E)$ be a DAG. For a probability measure p on $\times_{v\in V}\mathscr{X}_v$, the following conditions are equivalent:*

1. *p admits a factorization according to G (a factorization like in formula (9.2)).*
2. *p obeys the global Markov property, relative to G.*
3. *p obeys the causal Markov property, relative to G.*

A simple application of this theorem to the causal graph of Figure 9.2 yields the following conditional independence statements:

$$U \text{ and } B \text{ are stochastically independent given } A. \tag{9.4}$$

$$A \text{ and } C \text{ are stochastically independent given } U \text{ and } B. \tag{9.5}$$

These and similar conditional independence statements will be used in the context of the sensorimotor loop.

9.2.2 The Causal Structure of the Sensorimotor Loop

The following figure (Figure 9.3) illustrates the components of a sensorimotor loop with their respective interactions. In order to apply the theory of causal networks to the sensorimotor loop, we have to consider a causal network that captures the main aspects of this structure. Figure 9.4 shows the general causal network of a sensorimotor loop, where W_t, S_t, C_t, A_t denote the state of the world, the sensor, the controller, and the actuator at some time point t, respectively (Klyubin et al. 2004; Ay and Polani 2008). We denote the corresponding state spaces by $\mathscr{W}, \mathscr{S}, \mathscr{C}$, and $\mathscr{A}$. The stochastic maps α, β, φ, and π describe the mechanisms that are involved in the sensorimotor dynamics:

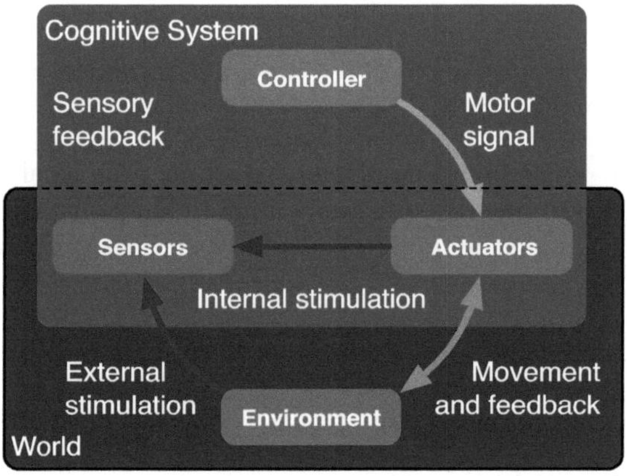

Fig. 9.3 Sensorimotor loop

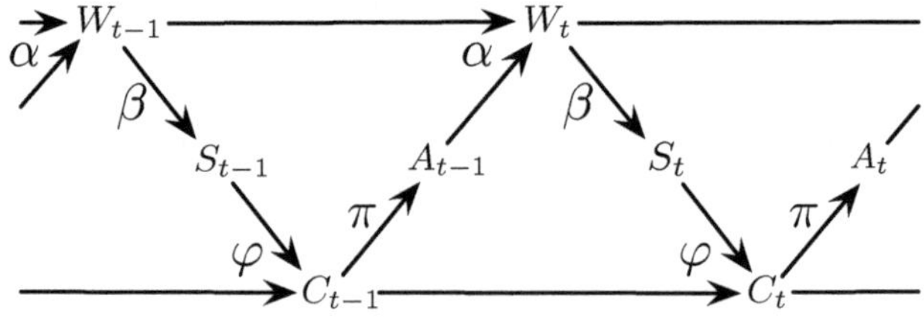

Fig. 9.4 Sensorimotor loop

$$\alpha : \mathscr{W} \times \mathscr{A} \to \mathcal{P}(\mathscr{W}),$$
$$\beta : \mathscr{W} \to \mathcal{P}(\mathscr{S}),$$
$$\varphi : \mathscr{C} \times \mathscr{S} \to \mathcal{P}(\mathscr{C}),$$
$$\pi : \mathscr{C} \to \mathcal{P}(\mathscr{A}).$$

Here, $\mathcal{P}(\mathscr{X})$ denotes the set of probability measures on $\mathscr{X}$. The kernels α and β encode the constraints of the sensorimotor loop due to the agent's morphology and the properties of its environment. The mechanisms α and β are extrinsic and encode the agent's embodiment which sets constraints for the agent's behavior and learning. The kernels φ, π are intrinsic with respect to the agent and are assumed to be modifiable through a learning process (for details, see (Zahedi et al. 2010)).

The process (W_t, S_t, C_t, A_t), $t = 0, 1, 2, \ldots$, is generated by Markov transition kernels. Given the above kernels as models of the mechanisms that constitute the sensorimotor loop, and given an initial distribution μ, we obtain the joint distribution

$$p(w_0, s_0, c_0, a_0, \ldots, w_n, s_n, c_n, a_n)$$
$$= \mu(w_0, s_0, c_0, a_0) \cdot \prod_{t=1}^{n} \alpha(w_{t-1}, a_{t-1}; w_t)\, \beta(w_t; s_t)\, \varphi(c_{t-1}, s_t; c_t)\, \pi(m_t; a_t).$$

As examples, we now consider sensorimotor loops where particular arrows are removed.

Example 1. 1. *Passive observer.* If the agent does not act on the world at all or, stated differently, if it only observes its environment then there is no edge from A_t to W_t and the kernel α does not involve the actuator state. More precisely,

$$\alpha(w, a; w') = \alpha(w; w') \quad \text{for all } w,\, a, \text{ and } w'.$$

Obviously, in this situation we can remove the edges from A to W and from C to A, which leads to the diagram shown in Figure 9.5.

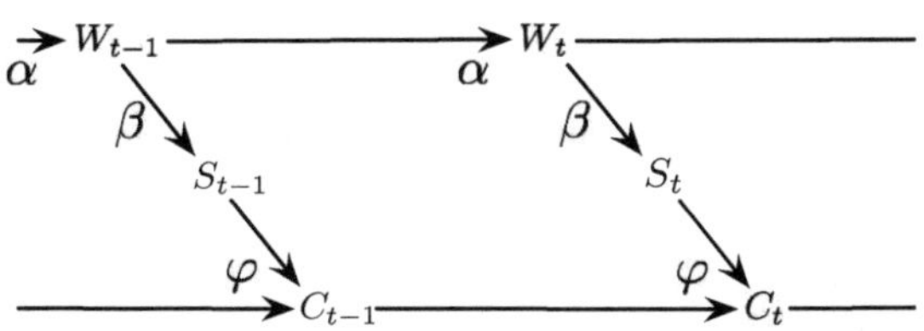

Fig. 9.5 Sensorimotor loop of a passive observer

2. *Open loop controller.* The situation in which the agent does not sense anything in the world is referred to as *open loop control*. Here, the kernel φ does not use the sensor state s, that is

$$\varphi(c,s;c') = \varphi(c;c') \quad \text{for all } c,\, s, \text{ and } c'.$$

Here, the edges from S to C and from W to S can be removed, and we obtain the diagram shown in Figure 9.6.

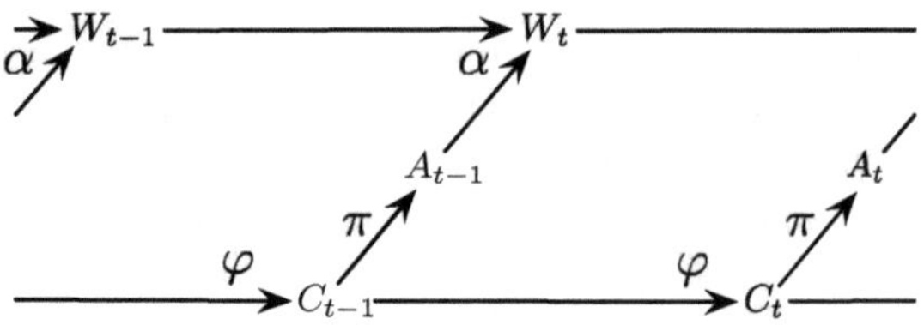

Fig. 9.6 Sensorimotor loop of an open loop controller

3. *Memoryless or reactive controller.* In this example, we assume that there is no edge from the controller state at time $t-1$ to the controller state at time t (see Figure 9.7 (A)). This means

$$\varphi(c,s;c') = \varphi(s;c') \quad \text{for all } c,\, s, \text{ and } c'.$$

If we combine the kernels φ and π to one kernel, which we again denote by π, then we have a representation of a memoryless controller that is often referred to as *reactive controller* (see Figure 9.7 (B)).

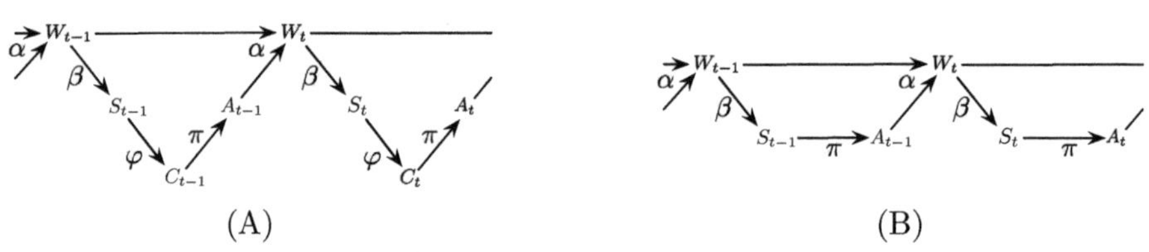

Fig. 9.7 Sensorimotor loop of a memoryless or reactive controller

9.3 Causal Effects

9.3.1 The Definition of Causal Effects

In order to study causal effects, one has to apply an interventional operation, which we also call *clamping*. Clamping the state of a node means a change of the mechanism of that node, and it is formalized by the so-called *do*-operation. In order to explain the main idea behind the *do*-operation, we first consider our example of Figure 9.2. In this example we compute the causal effect of A on C. If we clamp $\hat{a}$, or, in Pearl's terminology (Pearl 2000), do $\hat{a}$, we would have to replace equation (9.1) by

$$p_{\hat{a}}(u,a,b,c) = \varphi(u)\cdot\hat{\alpha}(a)\cdot\beta(a;b)\cdot\gamma(u,b;c), \tag{9.6}$$

where $\hat{\alpha}(a)=1$, if $a=\hat{a}$, and $\hat{\alpha}(a)=0$, if $a\neq\hat{a}$. In particular, after clamping node A the new mechanism $\hat{\alpha}$ is not sensitive to u anymore. In terms of the *do*-formalism, the post-interventional probability measure $p_{\hat{a}}(u,a,b,c)$ is written as $p_{\hat{a}}(u,a,b,c) = p(u,a,b,c\,|\,do(\hat{a}))$. Summation over u,a,b yields the probability of observing c after having clamped $\hat{a}$:

$$p(c\,|\,do(\hat{a})) = \sum_{u,b}\varphi(u)\cdot\beta(\hat{a};b)\cdot\gamma(u,b;c). \tag{9.7}$$

We refer to this post-intervational probability measure as *causal effect* of A on C (see Figure 9.8).

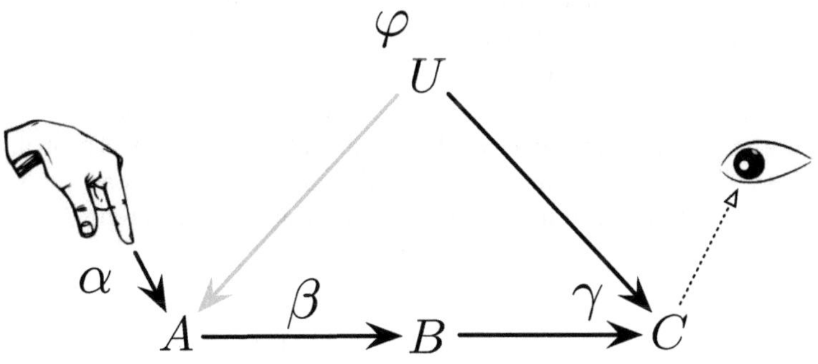

Fig. 9.8 Intervention

Now, we extend this idea of intervention to the general setting of Bayesian networks. Knowing the mechanisms, it is possible to model the consequences of intervention. Consider the equation (9.2), a subset A of V, and assume that the configuration of A is externally set to x_A. What does this mean? It means that the mechanisms of the nodes v in A are replaced by the mechanisms

$$\delta_{x_v}(x'_v) = \begin{cases} 1, & \text{if } x'_v = x_v \\ 0, & \text{otherwise} \end{cases}.$$

This leads to the truncated product

$$p(x_{V\setminus A}\,|\,do(x_A)) = \prod_{v\in V\setminus A}\kappa^v(x_{pa(v)};x_v). \tag{9.8}$$

Although it is not essential for what follows, we present the formula for the usual way of conditioning:

$$p(x_{V\setminus A} \mid x_A) = \frac{p(x_{V\setminus A}, x_A)}{p(x_A)} = \frac{\prod_{v\in V} \kappa^v(x_{pa(v)}; x_v)}{\sum_{x'_V : x'_{V\setminus A} = x_{V\setminus A}} \prod_{v\in V} \kappa^v(x'_{pa(v)}; x'_v)} . \tag{9.9}$$

This shows that the interventional conditioning, given by equation (9.8), is much easier to compute than the standard conditioning of equation (9.9).

In the above derivations we considered a set A where the intervention takes place and observed the complement $V \setminus A$ of A, which corresponds to the causal effect $p(x_{V\setminus A} \mid do(x_A))$ (see Figure 9.9 (A)). In more general situations, the post-interventional distribution is observed in a subset B of $V \setminus A$. In order to define this, we simply have to marginalize out the unobserved nodes:

$$p(x_B \mid do(x_A)) = \sum_{x_{V\setminus(A\cup B)}} p(x_B, x_{V\setminus(A\cup B)} \mid do(x_A)). \tag{9.10}$$

Comparing this interventional conditioning with classical conditioning, we observe one important difference. The latter is only possible when the event that is conditioned on has positive probability. It describes the probability of observing an event F if an event E has already been observed. However, if the probability for E vanishes then, already at an intuitive level, it is not clear with which probability the occurrence of F should be expected. Formally, if $p(x_A) = 0$ then the expression (9.9) is not well defined. The interventional conditioning is different. Already at an

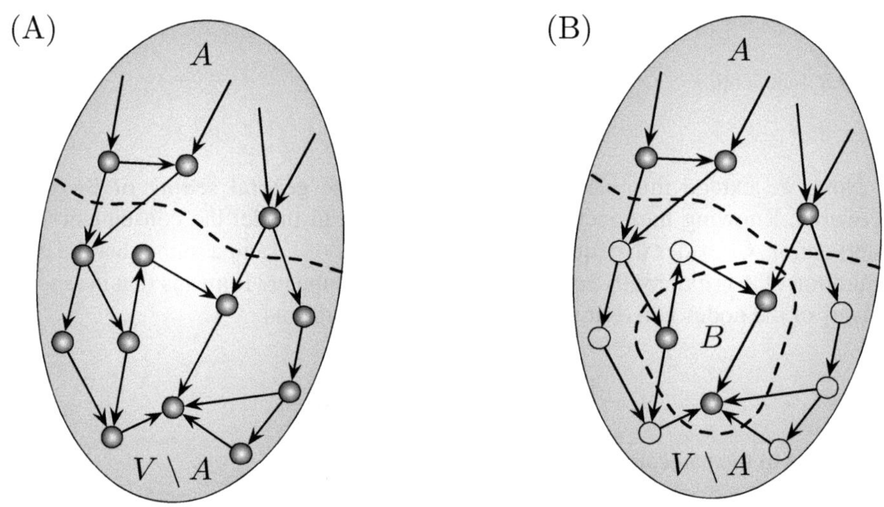

Fig. 9.9 (A) intervention in A and observation in the complement $V \setminus A$ of A in V, (B) intervention in A and observation in B

intuitive level it is clear that any intervention will lead to some reaction of the system. Formally, we see this in equation (9.8). The product on the right-hand side of this equation is always well defined.

Although interventional conditioning differs from observational conditioning, in some cases both operations coincide. In order to be more precise, we note that any observed association of two variables A and B in a Bayesian network has basically three sources: A is a cause of B, B is a cause of A, or there is a common cause of A and B. This is known as the common cause principle (Reichenbach 1956). If we assume that B is not a cause of A and there is no common cause of A and B then there is no explanation for the association of A and B other than A being a cause of B. In that case, all stochastic dependence between A and B is due to the causal effect of A on B. For such a situation we say that the pair (A,B) is a *causal pair*. One can show that for causal pairs interventional and observational conditioning are equivalent.

Proposition 1. *If (A,B) is a causal pair with respect to a DAG G, then for any Bayesian network with graph G, we have $p(x_B \,|\, do(x_A)) = p(x_B \,|\, x_A)$.*

This proposition shows that in some cases the post-interventional distribution, that is $p(x_B \,|\, do(x_A))$, can be obtained by observation only. More precisely, if observations of all joint events (x_A, x_B) allow us to estimate their probabilities $p(x_A, x_B)$, then one can compute the causal effect as

$$p(x_B \,|\, do(x_A)) = \frac{p(x_A, x_B)}{\sum_{x'_B} p(x_A, x'_B)}, \tag{9.11}$$

whenever $p(x_A) = \sum_{x'_B} p(x_A, x'_B) > 0$. If a causal effect can be computed in this way, that is without intervention, we say that it is identifiable. In the next section, this subject is discussed to the extent to which it is used in the context of the sensorimotor loop. We will argue that causal effects that are relevant to an agent should be identifiable with respect to the intrinsic variables of that agent.

9.3.2 *Identification of Causal Effects*

The equation (9.8) provides a formal definition of a causal effect. Such a causal effect can be determined in various ways depending on the available experimental operations. If we can experimentally intervene into the system, then the mechanisms will generate the post-interventional probability measure which can be observed. In many cases, however, experimental intervention is not possible. Then one has to ask the following question: *Is it possible to conclude the consequences of intervention solely on the basis of observation, that is without actual intervention?* At first sight, the answer seems to be clearly *No!* In some sense, this *is* already the whole answer to this question. On the other hand, Proposition 1 proves that in the case of a causal pair, it is indeed possible to compute the causal effect, left-hand side of (9.11), without intervention, from the right-hand side of (9.11). This demonstrates that, in order to identify a causal effect observationally, we require some structural information.

Without any structural information, it is not possible at all to identify causal effects. However, this does not mean that we need to know the complete structure, that is the DAG, in order to identify a causal effect. Partial structural knowledge can be sufficient for the identification of a causal effect, as we will see.

In what follows we further illustrate the problem of causal effect identification using our standard example. We want to compute the causal effect of A on C. Obviously, U is a common cause of A and C, and therefore the pair (A,C) is not a causal pair, which implies $p(c\,|\,do(a)) \neq p(c\,|\,a)$ in general. The following formula shows a different way of obtaining $p(c\,|\,do(a))$ by assuming that we can observe A, B, and C:

$$p(c\,|\,do(a)) = \sum_b p(b\,|\,a) \sum_{a'} p(a') \cdot p(c\,|\,a',b). \tag{9.12}$$

This equality is quite surprising and not obvious at all. Here is the proof:

$$\begin{aligned}
p(c\,|\,do(a)) &= \sum_{u,b} \varphi(u) \cdot \beta(a;b) \cdot \gamma(u,b;c) \qquad \text{(formula (9.7))} \\
&= \sum_{u,b} p(u) \cdot p(b\,|\,a) \cdot p(c\,|\,u,b) \\
&= \sum_{u,b} \left(\sum_{a'} p(a') \cdot p(u\,|\,a') \right) p(b\,|\,a) \cdot p(c\,|\,u,b) \\
&= \sum_b p(b\,|\,a) \sum_{a'} p(a') \sum_u p(u\,|\,a') \cdot p(c\,|\,u,b) \\
&= \sum_b p(b\,|\,a) \sum_{a'} p(a') \sum_u p(u\,|\,a',b) \cdot p(c\,|\,u,b,a') \\
&\quad \text{(conditional inedependence statements (9.4) and (9.5))} \\
&= \sum_b p(b\,|\,a) \sum_{a'} p(a') \cdot p(c\,|\,a',b)
\end{aligned}$$

The structure of this example will be revisited in the context of the sensorimotor loop.

9.3.3 *Causal Effects in the Sensorimotor Loop*

In this section we study the problem of the identification of causal effects in the sensorimotor loop. In this context, there are various causal effects of interest, for instance the effect of actions on sensor inputs. We are mainly interested in causal effects that involve intrinsic variables of the agent, that is the variables S_t, C_t, and A_t, $t = 0,1,2,\ldots$. Other causal effects can not be evaluated by the agent, because extrinsic variables are by definition not directly available to the agent. In the proposition below, we list three causal effects in the sensorimotor loop that are identifiable by the agent without actual intervention and purely based on in situ observations of the agent. In order to be more precise, we have a closer look at the causal diagram of the transition from time $t-1$ to t. Here, as shown in Figure 9.10, we consider the future sensor value of only one time step and summarize the past process by one

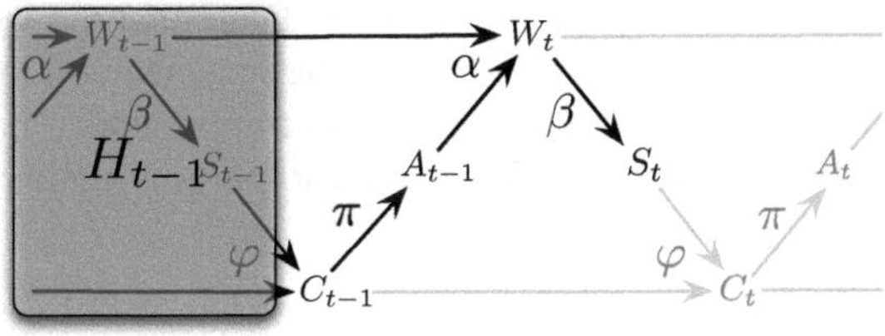

Fig. 9.10 Reduction procedure of the causal diagram

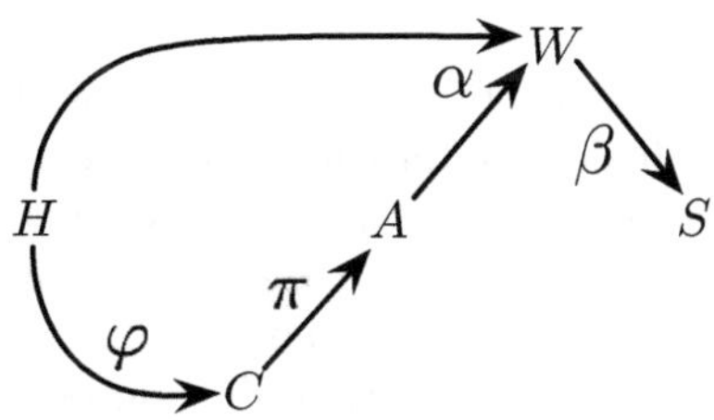

Fig. 9.11 Reduced causal diagram for one time step

variable H_{t-1}. We focus on the resulting causal diagram of Figure 9.11. The joint distribution in the reduced diagram is given as

$$p(h,c,a,w,s) = p(h)\,\varphi(h;c)\,\pi(c;a)\,\alpha(h,a;w)\,\beta(w;s). \tag{9.13}$$

We can now consider, for instance, the causal effect of A on S. In general, we do *not* have $p(s\,|\,do(a)) = p(s\,|\,a)$, which follows from the fact that H is a common cause of A and S. Nevertheless, it turns out that this causal effect, among others, is identifiable with respect to the intrinsic variables of the agent.

Proposition 2. *Let the joint distribution (9.13) be strictly positive. Then the following equations hold:*

$$\begin{aligned}
&(1) \qquad & p(s\,|\,do(a),c) &:= \frac{p(s,c\,|\,do(a))}{p(c\,|\,do(a))} = p(s\,|\,c,a)\\
&(2) \qquad & p(s\,|\,do(a)) &= \sum_c p(s\,|\,c,a)\,p(c)\\
&(3) \qquad & p(s\,|\,do(c)) &= \sum_a p(a\,|\,c) \sum_{c'} p(s\,|\,c',a)\,p(c').
\end{aligned}$$

The proof of Proposition 2 is given in the appendix. In all three causal effects of this proposition, the conditional distribution $p(s\,|\,c,a)$ turns out to be essential as building block for the identification of the causal effects. It describes the expectation of the agent to observe s, given that it is in state c and performs an action a. We refer to this conditional probability distribution as *world model* of the agent. Note that in the strictly positive case, according to Proposition 2 (1), the world model is not dependent on the agent's policy. These results indicate that the world model plays an important role in evaluating causal effects in the sensorimotor loop. Furthermore, it is an essential object within the empowerment approach to behavior (Klyubin et al. 2005). We will see that the world model also plays a fundamental role within learning processes.

9.4 Information Flows

9.4.1 Information-Theoretic Preliminaries

In this section, we introduce a few fundamental quantities known from information theory. First, we consider Shannon information. Assume that $p(x)$ describes the expectation that the outcome of a random experiment is going to be x. The information that we receive by knowing the outcome x is equal to the surprise about that outcome, which is quantified by $-\ln p(x)$. Events x that we expect to occur with very low probability will highly surprise us when they do occur. This is expressed by a large value of $-\ln p(x)$ for a low probability $p(x)$. The expectation value of this function is known as *Shannon entropy* or *Shannon information*:

$$H_p(X) \;:=\; -\sum_{x\in\mathscr{X}} p(x)\ln p(x)\,.$$

The Shannon entropy quantifies the uncertainty about the outcome of the random variable X. Note that for $p(x) = 0$, the term $p(x)\ln p(x)$ is not directly defined. However, the function $t\ln t$ can be continuously extended in $t = 0$ with value equal to 0, which justifies the convention $0\ln 0 = 0$. If we know the outcome of a random experiment beforehand, that is if our expectation p is concentrated around one value x_0, then the Shannon entropy is small. On the other hand, if the distribution is equally spread over all events, then the uncertainty about the outcome of X is maximal. With the uniform distribution u and the cardinality n of $\mathscr{X}$, we have

$$H_u(X) \;=\; -\sum_{x\in\mathscr{X}} \frac{1}{n}\ln\frac{1}{n} \;=\; -\ln\frac{1}{n} \;=\; \ln n\,.$$

Now we consider two variables X and Y. With the expectation $p(x,y)$ we associate the measure $H_p(X,Y)$ of uncertainty. Now assume that we have observed the variable X and have a remaining uncertainty about Y given X. This is given as

$$H(Y|X) \;=\; -\sum_{x\in\mathscr{X}} p(x)\sum_{y\in\mathscr{Y}} p(y|x)\ln p(y|x)\,.$$

Observation of X reduces the uncertainty about the outcome of Y and we define the reduction of uncertainty by

$$I(X;Y) \;:=\; H(Y) - H(Y|X)\,.$$

This quantity is symmetric and has various other representations:

$$\begin{aligned} I(X;Y) &= H(Y) - H(Y|X) \\ &= \sum_{x \in \mathscr{X}} p(x) \sum_{y \in \mathscr{Y}} p(y|x) \ln p(y|x) - \sum_{y \in \mathscr{Y}} p(y) \ln p(y) \\ &= I(Y;X). \end{aligned}$$

This quantity is referred to as *mutual information* and quantifies the stochastic dependence of variables X and Y. We have $I(X;Y) = 0$ if and only if X and Y are stochastically independent, that is $p(x,y) = p(x)\,p(y)$. Introducing a further variable Z, we can consider the conditional mutual information

$$I(Z;Y|X) = H(Z|X) - H(Z|X,Y).$$

This vanishes if and only if Z and Y are conditionally independent given X.

The mutual information of two variables X and Y can be extended to more than two variables. Given variables X_v, $v \in V := \{1,2,\ldots,N\}$, we define the *multi-information* as

$$I(X_1;X_2;\ldots;X_N) := \sum_{v=1}^{N} H(X_v) - H(X_1,\ldots,X_N).$$

This definition is clearly independent of the order of the X_v.

Information-theoretic quantities can be used to characterize conditional independence. They can also be used to quantify the deviation from (conditional) independence as a measure for (conditional) stochastic dependence. Combined with the notion of intervention, this also leads to measures of causal information flows (Ay and Polani 2008).

9.4.2 Transfer Entropy and Causality

In this section, we draw a close connection between the multi-information and another information-theoretic quantity which addresses causal aspects of interacting processes. This quantity has been studied as *transfer entropy* by Schreiber (Schreiber 2000) and a slightly different version of it has been called *directed information* by Massey (Massey 1990). Recently, a thermodynamic interpretation of transfer entropy has been provided (Prokopenko et al. 2013). Using a somewhat implicit terminology, we argue that a more careful consideration of causality is necessary for understanding the sources of stochastic dependence.

In order to simplify the arguments we first consider only a pair X_t, Y_t, $t = 1,2,\ldots$, of stochastic processes which we also denote by X and Y. The extension to more than two processes will be straight-forward. Furthermore, we use the notation X^t for the random vector $(X_1,\ldots,X_t)$ and similarly x^t for the particular outcome $(x_1,\ldots,x_t)$.

Given a time n, consider the following quantity, which we refer to as *mutual information rate*:

$$\mathfrak{I}(X^n;Y^n) := \frac{1}{n} I(X^n;Y^n) \tag{9.14}$$

$$= \frac{1}{n}\Big(H(X^n)+H(Y^n)-H(X^n,Y^n)\Big) \tag{9.15}$$

$$= \frac{1}{n}\sum_{t=1}^{n}\Big\{H(X_t\,|\,X^{t-1})+H(Y_t\,|\,Y^{t-1})-H(X_t,Y_t\,|\,X^{t-1},Y^{t-1})\Big\} \tag{9.16}$$

$$= \frac{1}{n}\sum_{t=1}^{n}\left\{\underbrace{I(X_t;Y_t\,|\,X^{t-1},Y^{t-1})}_{I}+\underbrace{T(Y^{t-1}\to X_t)}_{II}+\underbrace{T(X^{t-1}\to Y_t)}_{III}\right\} \tag{9.17}$$

with the *transfer entropy* terms (Schreiber 2000)

$$T(Y^{t-1}\to X_t) := I(X_t;Y^{t-1}\,|\,X^{t-1})$$
$$T(X^{t-1}\to Y_t) := I(Y_t;X^{t-1}\,|\,Y^{t-1}).$$

Furthermore, using standard arguments we easily see that in the case of stationary processes the following limit exists:

$$\mathfrak{I}(X;Y) := \lim_{n\to\infty}\mathfrak{I}(X^n;Y^n).$$

Note that in the case of independent and identically distributed variables (X_t,Y_t), $t=1,2,\ldots$, the transfer entropy terms *II* and *III* in (9.17) vanish and the only contributions to $\mathfrak{I}(X^n;Y^n)$ then are the mutual informations $I(X_t;Y_t)$. In the case of stationary processes these mutual informations coincide and we have $\mathfrak{I}(X;Y)=I(X_1;Y_1)$. In this sense the quantity $\mathfrak{I}$ extends the mutual information of two variables to a corresponding measure for two processes.

In general, the first term I of (9.17) quantifies the stochastic dependence of X_t and Y_t after "screening off" the causes of X_t and Y_t that are intrinsic to the system, namely X^{t-1} and Y^{t-1}. Assuming the principle of common cause (Reichenbach 1956), which postulates that *all* stochastic dependences are based on causal interactions, we can infer common causes of X_t and Y_t that act from outside on the system, if the first term I is positive. If, on the other hand, the system is closed in the sense that

$$p(x_t,y_t\,|\,x^{t-1},y^{t-1}) = p(x_t\,|\,x^{t-1},y^{t-1})\,p(x_t\,|\,x^{t-1},y^{t-1}), \qquad t=1,2,\ldots,$$

then the first term in (9.17) vanishes and the transfer entropies *II* and *III* are the only contributions to $\mathfrak{I}(X^n;Y^n)$. They refer to causal interactions within the system. As example we consider the term *II*:

$$T(Y^{t-1}\to X_t) = H(X_t\,|\,X^{t-1})-H(X_t\,|\,X^{t-1},Y^{t-1}).$$

It quantifies the reduction of uncertainty about the outcome x_t if, in addition to the knowledge about the previous outcomes $x_1, \ldots, x_{t-1}$ of X, also the previous outcomes $y_1, \ldots, y_{t-1}$ of Y are known. Therefore, the transfer entropy $T(Y^{t-1} \to X_t)$ has been used as a measure for the causal effect of Y^{t-1} on X_t (Schreiber 2000), (Kaiser and Schreiber 2002), which is closely related to the concept of *Granger causality*. On the other hand, a direct interpretation of transfer entropy as a measure for causal effects has some shortcomings which are already mentioned in (Kaiser and Schreiber 2002) and further addressed in (Ay and Polani 2008) within the context of Pearl's (Pearl 2000) causality theory. Below, we will illustrate these shortcomings demonstrating the need for an alternative formalization of causality.

The definitions of the multi-information rate and the transfer entropy naturally generalize to more than two processes.

Definition 3. Let X_v be stochastic processes with state space $\mathscr{X}_v$, $v \in V$. We define the *transfer entropy* and the *multi-information rate* in the following way:

$$\begin{aligned} T({X_{V\setminus v}}^{t-1} \to X_{v,t}) &:= I(X_{v,t}; {X_{V\setminus v}}^{t-1} \,|\, {X_v}^{t-1}) \\ \mathfrak{I}({X_v}^n : v \in V) &:= \frac{1}{n}\left(\sum_{v\in V} H({X_v}^n) - H({X_V}^n)\right) \\ &= \frac{1}{n}\sum_{t=1}^{n}\left\{ I(X_{v,t} : v \in V \,|\, {X_V}^{t-1}) + \sum_{v\in V} T({X_{V\setminus v}}^{t-1} \to X_{v,t})\right\} \end{aligned}$$

and, if stationarity is assumed,

$$\mathfrak{I}(X_v : v \in V) := \lim_{n\to\infty} \mathfrak{I}({X_v}^n : v \in V).$$

In order to illustrate the problem of interpreting transfer entropy in a causal way, consider a finite node set V and a set $E \subseteq V \times V$ of directed edges. This is a way of encoding the structure of the nodes' causal interactions. The mechanisms are described in terms of Markov transition kernels

$$\kappa^v : \mathscr{X}_{pa(v)} \times \mathscr{X}_v \to [0,1], \qquad (x_{pa(v)}, x'_v) \mapsto \kappa^v(x_{pa(v)}; x'_v).$$

They define a "global kernel" as follows:

$$\kappa : \mathscr{X}_V \times \mathscr{X}_V \to [0,1], \qquad (x, x') \mapsto \prod_{v\in V} \kappa^v(x_{pa(v)}; x'_v).$$

With a stationary distribution p_1 of κ, that is $\sum_{x_1} p_1(x_1)\,\kappa(x_1;x_2) = p_1(x_2)$, we consider the process $X_{V,t} = (X_{v,t})_{v\in V}$, $t = 1, 2, \ldots$, that satisfies

$$P\{X_1 = x_1, \ldots, X_n = x_n\} \;=\; p_1(x_1)\,\kappa(x_1;x_2)\cdots\kappa(x_{n-1};x_n).$$

In these definitions, the corresponding multi-information rate simplifies to

$$\mathfrak{I}(X_v{}^n : v \in V) = \frac{1}{n} \sum_{t=1}^{n} \sum_{v \in V} T(X_{V \setminus v}{}^{t-1} \to X_{v,t}).$$

If we assume that the individual kernels κ^v are deterministic then the global kernel κ is also deterministic. This implies that for all nodes v there is a time s such that all transfer entropies $T(X_{V \setminus v}{}^{t-1} \to X_{v,t})$, $t \geq s$, vanish, and thus the multi-information rate converges to zero. This appears counterintuitive because even if we have strong causal interactions within a deterministic system, the dynamics creates redundancy that allows all nodes v to predict their own next states $x_{v,t}$ from their previous states $x_{v,1}, \ldots, x_{v,t-1}$ without the need of additional information from the other nodes $V \setminus v$. This intrinsic prediction is not required to be mechanistically implemented and therefore screens off the actual mechanisms that might involve causal effects from the complement of v. We illustrate this effect by a more specific example.

We consider again two processes X and Y and assume, as illustrated in the following diagram, that the next state of X as well as the next state of Y only depend on the current state of Y.

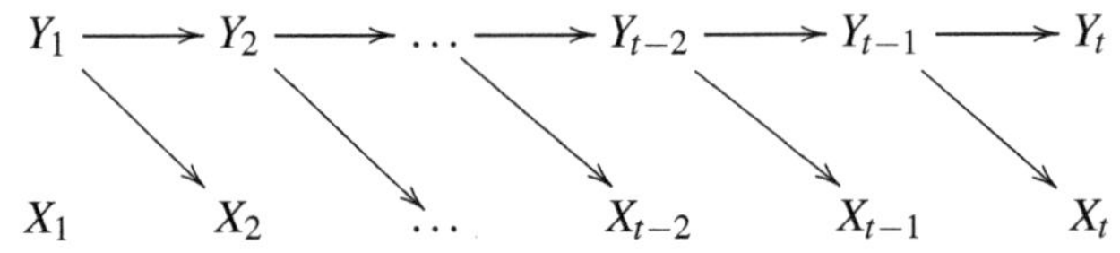

In what follows we define a one-dimensional family of transition kernels. To this end, we first consider two extreme situations. In the first situation, we assume that there is no memory:

$$\kappa^X(x_{t-1}, y_{t-1}; x_t) = \kappa^Y(x_{t-1}, y_{t-1}; y_t) = \frac{1}{2}. \tag{9.18}$$

Obviously, in this situation there is no temporal information flow at all. The other extreme situation is given in the following way: In order to compute the next state, both nodes copy the current state of node Y and invert it.

$$(x, y) \to (-y, -y), \qquad x, y \in \{\pm 1\}. \tag{9.19}$$

In this case, the current state of Y completely determines the next state of X. Therefore, intuitively, one would expect a maximal amount of information flow from Y to X. We now interpolate these two extreme situations of minimal and maximal information flow in order to get a one-parameter family of transition kernels. More precisely, we define

$$\kappa^Y(x_{t-1}, y_{t-1}; y_t) := \frac{1}{1 + e^{2\beta y_t y_{t-1}}}, \qquad \kappa^X(x_{t-1}, y_{t-1}; x_t) := \frac{1}{1 + e^{2\beta x_t y_{t-1}}}.$$

Here, β plays the role of an inverse temperature. In the high-temperature limit ($\beta \to 0$) we recover the completely random transition (9.18), and in the low-temperature limit ($\beta \to \infty$) we recover the maximum information transition (9.19). In order to compute the stationary distribution, we consider the stochastic matrix describing the global dynamics (the rows denote (x_{t-1}, y_{t-1}), the columns (x_t, y_t) and the entries the transition probabilities from a state at time $t-1$ to time t):

	$(-1,-1)$	$(+1,-1)$	$(-1,+1)$	$(+1,+1)$
$(-1,-1)$	a^2	ab	ab	b^2
$(+1,-1)$	a^2	ab	ab	b^2
$(-1,+1)$	b^2	ab	ab	a^2
$(+1,+1)$	b^2	ab	ab	a^2

where

$$a := \frac{1}{1+e^{2\beta}}, \qquad b := \frac{1}{1+e^{-2\beta}}.$$

The stationary distribution is given by

$$p(+1,+1) = p(-1,-1) = \frac{1}{2} - ab, \qquad p(-1,+1) = p(+1,-1) = ab.$$

Obviously, for $\beta = 0$ we have uniform distribution $p(x,y) = \frac{1}{4}$, which implies that there is no correlation between the two nodes. As β increases, we get more and more redundancy, and in the limit $\beta \to \infty$ we get totally correlated nodes with distribution $\frac{1}{2}\left(\delta_{(-1,-1)} + \delta_{(+1,+1)}\right)$. This redundancy increase allows for compensating information about y_{t-1} by x_{t-1} and computing x_k on the basis of this information. More precisely, the two distributions $p(x_t | x_{t-1}, y_{t-1})$ and $p(x_t | x_t)$ come closer to each other. Therefore, the conditional mutual information $I(X_t; Y_{t-1} | X_{t-1})$, which is an upper bound of the transfer entropy $T(Y^{t-1} \to X_t)$, converges to zero. This fact shows that although there is no arrow from X_{t-1} to X_t, the conditional distribution $p(x_t | x_{t-1})$ is effectively dependent on x_{t-1}, which appears causally inconsistent. In order to derive a corresponding causal variant, we consider the conditional mutual information:

$$\begin{aligned} & I(X_t; Y_{t-1} | X_{t-1}) \\ &= \sum_{x_{t-1}} p(x_{t-1}) \sum_{y_{t-1}} p(y_{t-1} | x_{t-1}) \sum_{x_t} p(x_t | x_{t-1}, y_{t-1}) \ln \frac{p(x_t | x_{t-1}, y_{t-1})}{\sum_{y'_{t-1}} p(y'_{t-1} | x_{t-1})\, p(x_t | x_{t-1}, y'_{t-1})} \\ &= \sum_{x_{t-1}} p(x_{t-1}) \sum_{y_{t-1}} p(y_{t-1} | x_{t-1}) \sum_{x_t} p(x_t | y_{t-1}) \ln \frac{p(x_t | y_{t-1})}{\sum_{y'_{t-1}} p(y'_{t-1} | x_{t-1})\, p(x_t | y'_{t-1})} \end{aligned} \qquad (9.20)$$

With an abuse of terminology we refer to this conditional mutual information also as *transfer entropy*, which is plotted in Figure 9.12. Replacing all conditional probabilities in (9.20) by corresponding interventional ones leads to $I(X_t; Y_{t-1})$ as a causal variant of the above measure $I(X_t; Y_{t-1} | X_{t-1})$, which we refer to as *information flow*. More precisely, we have

$$
\begin{aligned}
&\sum_{x_{t-1}} p(x_{t-1}) \sum_{y_{t-1}} p(y_{t-1} \,|\, do(x_{t-1})) \sum_{x_t} p(x_t \,|\, do(y_{t-1})) \times \\
&\qquad\qquad \ln \frac{p(x_t \,|\, do(y_{t-1}))}{\sum_{y'_{t-1}} p(y'_{t-1} \,|\, do(x_{x_{t-1}}))\, p(x_t \,|\, do(y_{t-1}))} \\
&= \sum_{x_{t-1}} p(x_{t-1}) \sum_{y_{t-1}} p(y_{t-1}) \sum_{x_t} p(x_t \,|\, y_{t-1}) \ln \frac{p(x_t \,|\, y_{t-1})}{\sum_{y'_{t-1}} p(y'_{t-1})\, p(x_t \,|\, y'_{t-1})} \\
&= \sum_{y_{t-1}} p(y_{t-1}) \sum_{x_t} p(x_t \,|\, y_{t-1}) \ln \frac{p(x_t \,|\, y_{t-1})}{\sum_{y'_{t-1}} p(y'_{t-1})\, p(x_t \,|\, y'_{t-1})} \\
&= I(X_t; Y_{t-1})
\end{aligned}
$$

Comparing these two quantities, we have

$$
I(X_t;Y_{t-1}) - I(X_t;Y_{t-1} \,|\, X_{t-1}) \;=\; I(X_t;X_{t-1}) \;\geq\; 0.
$$

Intuitively speaking, in this example, the transfer entropy captures only one part of the causal information flow.

The following diagram shows the shape of the conditional mutual information and the information flow as function of β. As we see, the information flow is consistent with the intuition that moving from $\beta = 0$ to $\beta = \infty$ corresponds to an interpolation between a transition with vanishing information flow and a transition with maximal information flow. Near $\beta = 0$ the transfer entropy increases as β becomes larger and is close to the information flow. But for larger β's it starts decreasing and converges to zero for $\beta \to \infty$. The reason for that is simply that the transition for large β generates more redundancy between the two processes X and Y. Therefore, as β grows, an increasing amount of information about Y_{t-1} can be computed from information about X_{t-1}, which lets the transfer entropy decrease towards zero.

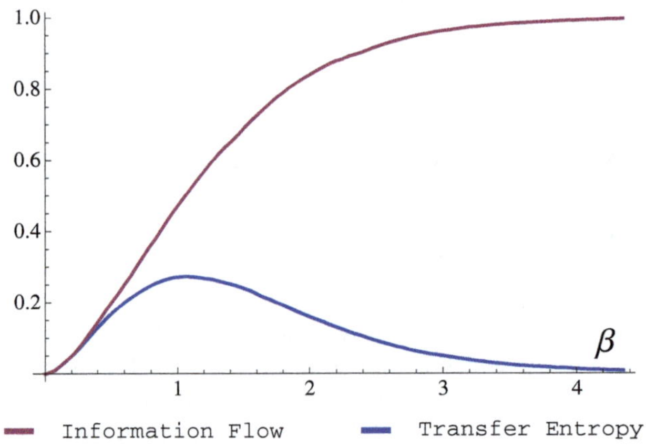

Fig. 9.12 Transfer entropy and information flow

9.4.3 Information Flows in the Sensorimotor Loop

In this section, we apply the notion of transfer entropy to the context of the sensorimotor loop. Sporns and Lungarella used the transfer entropy in order to describe information flows through the sensorimotor loop (Lungarella and Sporns 2006). Here, we approach this subject from a theoretical perspective and point out that one has to be very careful with the interpretation of transfer entropy as a causal measure. Our considerations will be based on the causal diagram shown in Figure 9.13. Here, we have four processes W_t, S_t, C_t, A_t, $t \geq 1$, and the initial node H which stands for "history."

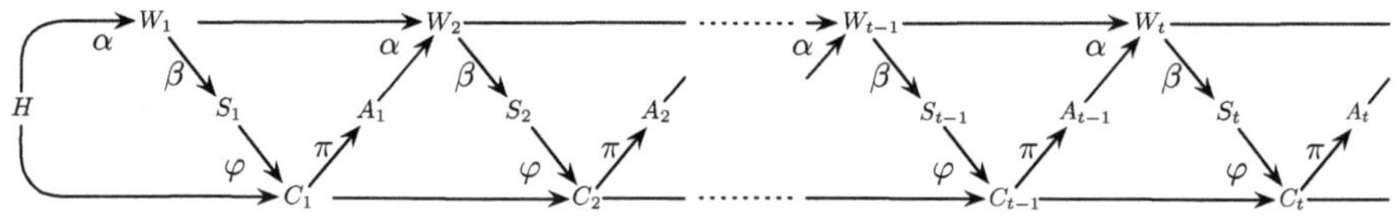

Fig. 9.13 Causal diagram of the sensorimotor loop

Let us study the two simple cases of a passive observer and an open loop controller (see Figures 9.5 and 9.6). In these cases, some of the arrows are missing, and therefore we know from the definition of the respective causal structure that particular causal effects are not present. We can then evaluate the corresponding transfer entropy and thereby test its consistency with the given causal structure. We start with the passive observer, shown in Figure 9.5. Here, the actuator with its arrows is removed, which means that the agent is not able to act on the world. Therefore, we expect $T(C^{t-1} \to W_t) = 0$ and $T(S^{t-1} \to W_t) = 0$. This is confirmed by the fact that C^{t-1} and W_t are d-separated by W^{t-1} (see Figure 9.14). Similarly, S^{t-1} and W_t are d-separated by W^{t-1}.

Now let us consider the open loop controller. Here, the sensor with its arrows is removed, and we expect no causal effect of W on C and also no causal effect of A on C. Therefore, consistency of the transfer entropy with the causal structure would require $T(W^{t-1} \to C_t) = 0$ and $T(A^{t-1} \to C_t) = 0$. This is confirmed in terms of the same d-separation arguments as in the above case of a passive observer (see Figure 9.15).

Summarising the above considerations, we have the following proposition.

Proposition 3. *For the passive observer one has* $T(C^{t-1} \to W_t) = 0$ *and* $T(S^{t-1} \to W_t) = 0$ *for all* $t \geq 1$. *In the context of open loop control we have* $T(W^{t-1} \to C_t) = 0$ *and* $T(A^{t-1} \to C_t) = 0$ *for all* $t \geq 1$.

Proposition 3 confirms our expectation that a passive observer has no causal effect on the world, and that within open loop control there is no causal effect of the world on the agent (note that, with "agent" we mean the internal process C of the agent). However, one has to be careful with this measure. For example, in the situation of a passive observer, there is no directed path from C to S. Therefore, one

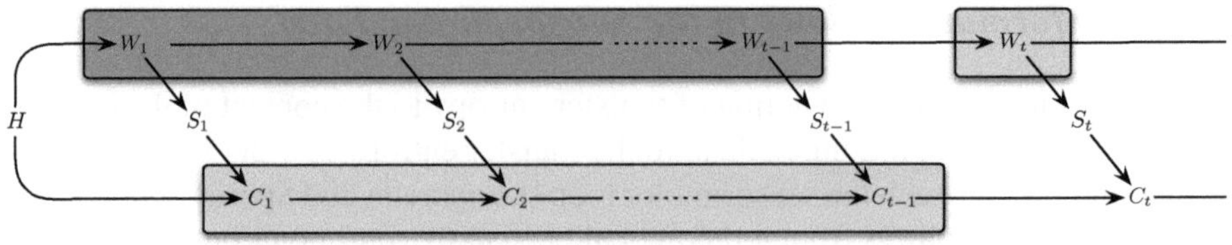

(A)

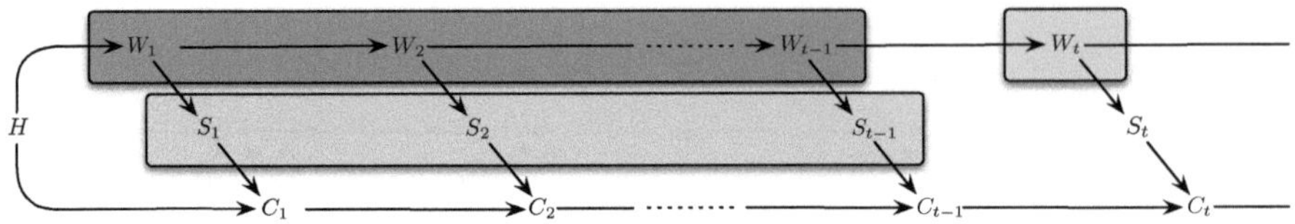

(B)

Fig. 9.14 Sensorimotor loop of a passive observer. The diagram (A) shows that $T(C^{t-1} \to W_t) = 0$, and diagram (B) shows $T(S^{t-1} \to W_t) = 0$.

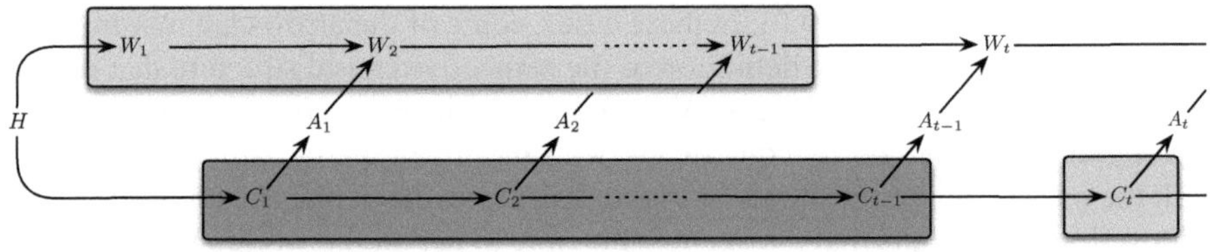

(A)

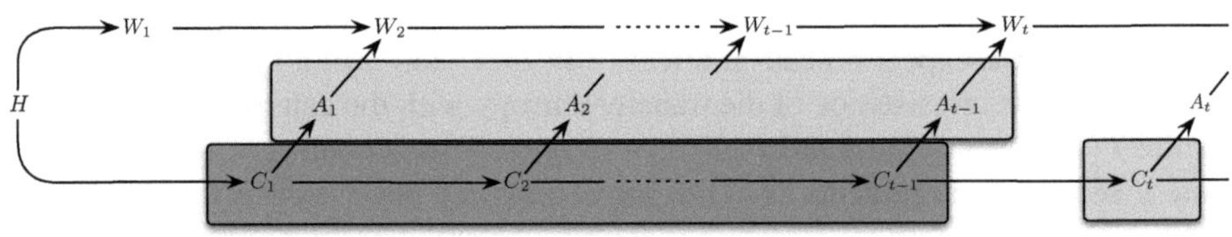

(B)

Fig. 9.15 Sensorimotor loop of an open loop controller. The diagram (A) shows that $T(W^{t-1} \to C_t) = 0$, and diagram (B) shows $T(A^{t-1} \to C_t) = 0$.

would not only expect the absence of causal effects of the agent on the world but also the absence of causal effects on the sensor process, which is obtained from the world process. In other words, it is natural to have not only $T(C^{t-1} \to W_t) = 0$, which is confirmed in Proposition 3, but also $T(C^{t-1} \to S_t) = 0$. From the causal structure

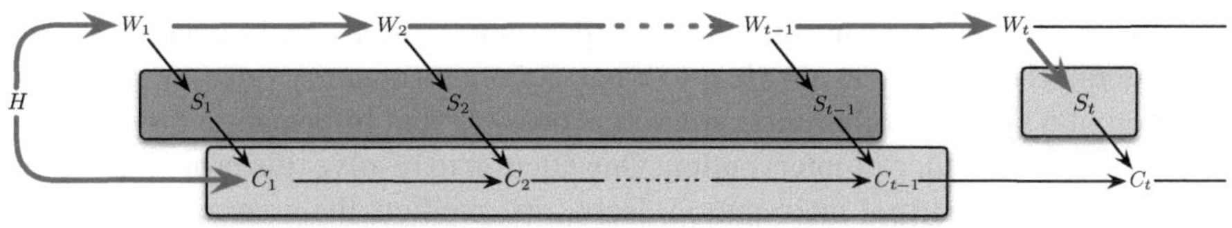

(A)

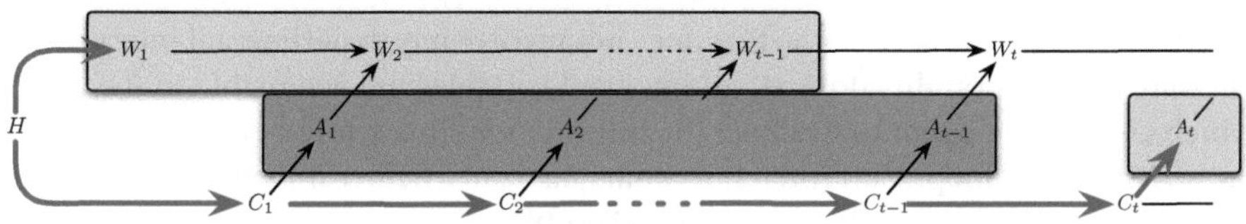

(B)

Fig. 9.16 (A) Sensorimotor loop of a passive observer. Although C^{t-1} is not a cause of S_t, that is $C^{t-1} \not\rightsquigarrow S_t$, it is indeed possible that $T(C^{t-1} \to S_t) > 0$. This is seen by the fact that the sets C^{t-1} and S_t are not d-separated by S^{t-1} (see unblocked path). (B) Sensorimotor loop of an open loop controller. Although W^{t-1} is not a cause of A_t, that is $W^{t-1} \not\rightsquigarrow A_t$, it is indeed possible that $T(W^{t-1} \to A_t) > 0$. This is seen by the fact that the sets W^{t-1} and A_t are not d-separated by A^{t-1} (see unblocked path).

shown in Figure 9.16 (A) this is not confirmed in terms of the d-separation criterion. More precisely, S_t and C^{t-1} are not d-separated by S^{t-1}, and $T(C^{t-1} \to S_t) > 0$ is indeed possible, which appears counterintuitive. We have the same problem with open loop control (see Figure 9.16 (B)). In this case, one would expect that there is no causal effect of the world on the agent's actions, that is $T(W^{t-1} \to A_t) = 0$. However, A_t and W_{t-1} are not d-separated by A^{t-1}, and, again, $T(W^{t-1} \to A_t) > 0$ is possible although W^{t-1} is not a cause of A_t.

A further method to study information flows, which has been proposed in (Ay and Polani 2008), is based on the interventional calculus of conditioning as already applied at the end of Section 9.4.2. Here, we briefly outline how this method can be applied within the context of the sensorimotor loop. In order to do so, we consider the causal effects listed in Proposition 2. The information flow from C to S, for instance, can be quantified by

$$IF(C \to S) := \sum_c p(c) \sum_s p(s \,|\, do(c)) \ln \frac{p(s \,|\, do(c))}{\sum_{c'} p(c')\, p(s \,|\, do(c'))} \tag{9.21}$$

which is a causal variant of the mutual information

$$I(C;S) = \sum_c p(c) \sum_s p(s \,|\, c) \ln \frac{p(s \,|\, c)}{\sum_{c'} p(c')\, p(s \,|\, c')}.$$

This expression (9.21) requires an intervention in C, which will, in general, disturb the system. Therefore, it is not clear to what extent this quantity refers to the in situ situation of a system while functioning. It is possible that information flow patterns change as consequence of intervention. One attempt to resolve this problem is given by the method of virtual intervention. Here, one predicts the consequences of interventions based on observations only, without actually applying an interventional operation. The system remains unperturbed, and one can still evaluate the expression (9.21). To be more precise, we use Proposition 2 and replace in (9.21) all distributions $p(s\,|\,do(c))$ by $\sum_a p(a\,|\,c) \sum_{c'} p(s\,|\,c',a)\, p(c')$. This leads to an expression of the information flow $IF(C \to S)$ that does not involve any experimental intervention but only observation. In addition, all involved variables are accessible to the agent, which enables the agent to evaluate the information flow while being embedded in the sensorimotor loop. Currently, it remains unclear whether this method of virtual intervention resolves the above-mentioned problem.

9.5 Predictive Information and Its Maximization – An Experimental Case Study

In the previous Sections 9.4.2 and 9.4.3 we focussed on information flows between various interacting processes, such as the control process C on the sensor process S in the sensorimotor loop. In this section, we now concentrate on temporal flows within one single process. To this end, consider first a stochastic process X_t, $t \in \mathbb{Z}$. Furthermore, given three time points $t_- < t < t_+$, we consider the mutual information

$$I(X_{t_-},\ldots,X_t;X_{t+1},\ldots,X_{t_+}) \;=\; H(X_{t+1},\ldots,X_{t_+}) - H(X_{t+1},\ldots,X_{t_+}\,|\,X_{t_-},\ldots,X_t).$$

If we assume that the variables $X_{t_-},\ldots,X_t$ represent the past (and present) of observed variables and $X_{t+1},\ldots,X_{t_+}$ the future or unobserved variables with respect to t, then this mutual information quantifies the reduction of uncertainty about the future given the past. In other words, it quantifies the amount of information in the future that can be predicted in terms of past observations. Therefore, the corresponding limit for $t_- \uparrow \infty$ and $t_+ \uparrow \infty$, which always exists but can be infinite, is referred to as *predictive information* (PI) (Bialek et al. 2001). It is also known as *excess entropy* (Crutchfield and Young 1989) and *effective measure complexity* (Grassberger 1986). Note that the predictive information is independent of t if the process is stationary.

In the context of the sensorimotor loop, the predictive information and related quantities serve as objective functions for self-organized learning. Of particular interest is the predictive information of the sensor process $S = (S_t)_{t\in\mathbb{Z}}$, which quantifies the amount of information in the agent's future sensor process that can be predicted by the agent based on the observation of its past (and present) sensor process. For simplicity, we also consider the lower bound $I(S_t;S_{t+1})$ of the predictive information and, with abuse of terminology, we refer to this simplified quantity also as predictive information (PI). Clearly, the predictive information is, on one hand, dependent on the policy and, on the other hand, also dependent on the embodiment of the agent. Policies with high predictive information correspond to niches of predictability within the sensorimotor loop and allow the agent to exploit these niches for task oriented behavior. Learning processes that maximize several variants of predictive information have been proposed and studied in view of their behavioral implications (Ay et al. 2008; Zahedi et al. 2010; Ay et al. 2012; Martius et al. 2013).

Here, we focus on the experimental case study of our previous work (Zahedi et al. 2010). In our experiments, embodied agents maximize the predictive information calculated on their sensor data by modulation of their policies. Different controller types are evaluated and the results are coordinated behaviors of passively coupled, individually controlled robots. From these results, three conclusions will be drawn and discussed. First, the different controller structures lead to a conclusion about optimal design, second, PI maximization leads to the formation of behavioral modes, and third, PI maximization leads to morphological computation (Pfeifer and Bongard 2006).

For the implementation of the learning rule in an embodied system, we chose a discrete-valued representation of the probability distributions for the following reason. At this initial step of evaluating the PI as a self-organised learning principle, we wanted to use as few assumptions as possible about the underlying model. Implementing the learning rule in the continuous domain generally requires more assumptions and restrictions, as the following example demonstrates. An implementation of the policy as a (recurrent) neural network binds the space of possible functions to the structure of the neural network (Pasemann 2002). Changing the weights and biases of the network only permits variation among the functions defined by the structure (and neural models). To avoid such a pruning in the space of policies, we chose stochastic matrices. To reduce the complexity, we first concentrated on reactive control (see Figure 9.7). Given a world model γ and a policy π, an estimate of the PI can be calculated with intrinsically available information by the following set of equations

$$I^{(\gamma)}(S_t;S_{t+1}) = \sum_{s_t,s_{t+1}\in\mathscr{S}} p(s_t,s_{t+1}) \ln \frac{p(s_t,s_{t+1})}{p(s_t)\,p(s_{t+1})} \tag{9.22}$$

$$p(s_t,s_{t+1}) = \sum_{a_t\in\mathscr{A}} p(s_t,a_t,s_{t+1}) = \sum_{a_t\in\mathscr{A}} p(s_t)\,\pi(s_t;a_t)\,\gamma(s_t,a_t;s_{t+1}) \tag{9.23}$$

$$p(s_t) = \sum_{s_{t+1}\in\mathscr{S}} p(s_t,s_{t+1}), \qquad p(s_{t+1}) = \sum_{s_t\in\mathscr{S}} p(s_t,s_{t+1}) \tag{9.24}$$

which can be used to calculate a *natural gradient* iteration (Amari 1998) of the PI with respect to the policy $\pi(s_t;a_t)$ in the following way (Zahedi et al. 2010)

$$\pi^{(0)}(s;a) := \frac{1}{|\mathscr{S}|} \qquad n \in \mathbb{N}\backslash\{0\}$$

$$\pi^{(n)}(s;a) = \pi^{(n-1)}(s;a) + \frac{1}{n+1}\pi^{(n)}(s;a)\left(F(s) - \sum_a \pi^{(n-1)}(s;a)F(s)\right) \tag{9.25}$$

$$F(s) := p^{(n)}(s)\sum_{s'}\gamma^{(n)}(s,a;s')\log_2 \frac{\sum_a \pi^{(n-1)}(s;a)\gamma^{(n)}(s,a;s')}{\sum_{s''} p^{(n)}(s'')\sum_a \pi^{(n-1)}(s'';a)\,\gamma^{(n)}(s'',a;s')}$$

The sensor distribution $p(s_t)$ and the intrinsic world model $\gamma(s_t,a_t;s_{t+1})$ are sampled according to

$$p^{(0)}(s) := \frac{1}{|\mathscr{S}|}$$

$$p^{(n)}(s) := \begin{cases} \frac{n}{n+1}p^{(n-1)}(s) + \frac{1}{n+1} & \text{if } S_{n+1} = s \\ \frac{n}{n+1}p^{(n-1)}(s) & \text{if } S_{n+1} \neq s \end{cases} \tag{9.26}$$

$$\gamma^{(0)}(s,a;s') := \frac{1}{|\mathscr{S}|}$$

$$\gamma^{(n_a^s)}(s,a;s') := \begin{cases} \frac{n_a^s}{n_a^s+1}\gamma^{(n_a^s-1)}(s,a;s') + \frac{1}{n_a^s+1} & \text{if } S_{n_a^s+1} = s',\, S_n = s,\, A_{n_a^s+1} = a \\ \frac{n_a^s}{n_a^s+1}\gamma^{(n_a^s-1)}(s,a;s') & \text{if } S_{n_a^s+1} \neq s',\, S_n = s,\, A_{n_a^s+1} = a \\ \gamma^{(n_a^s-1)}(s,a;s') & \text{if } S_{n_a^s} \neq s \text{ or } A_{n_{s,a}+1} \neq a \end{cases} \tag{9.27}$$

This concludes the brief presentation of the learning rule. See (Zahedi et al. 2010) for a detailed description. The next paragraphs describe the experiments and discuss the results.

All experiments were conducted purely in simulation for the sake of simplicity, speed and analysis. Current simulators, such as YARS (Zahedi et al. 2008), which was chosen for the experiments presented below, are shown to be sufficiently realistic to simulate the relevant physical properties of mobile robots, and designed such that experimental runs can be automated, run at faster than real-time speed, and require minimum effort to set-up.

The experiments are conducted with differential wheeled robots (see Figure 9.17A). The only sensors of each robot are the current wheel velocities S_t and the only actions are the motor commands for the next wheel velocities A_t. Each robot is either controlled by a single controller (see Figure 9.17C) or by two controllers (see Figure 9.17D). We refer to the single controller setting as *combined* control and to the two controller setting as *split* control.

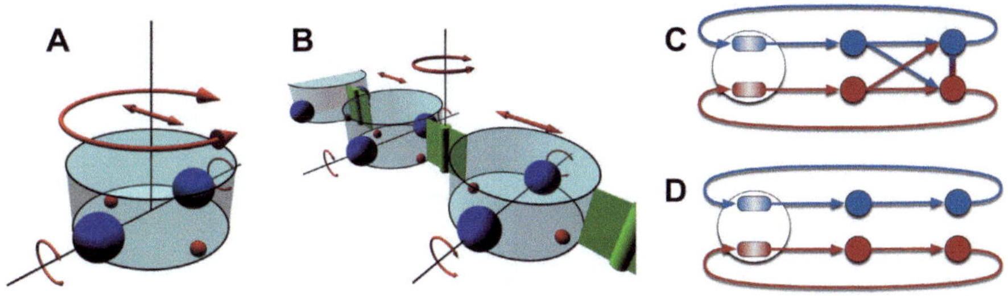

Fig. 9.17 Experimental set-up. A) Single circular differential wheeled robot. The image shows the two driving wheels and the possible movements that they allow for. B) This image shows the passive coupling between the robots. C) This image is a schema of the combined control, where one controller reads the information of both wheels and controls the velocity of both wheels. D) In the split controller configuration, each wheel has a controller that is independent of the other. Hence, each controller only reads the information of one wheel and controls the velocity of only one wheel.

Three and five robots are passively connected to a chain of robots (see Figure 9.17B), which results in the following four experiments:

1. three robots with combined control,
2. three robots with split control,
3. five robots with combined control, and
4. five robots with split control.

We will refer to these settings by the number of robots and the controller type. This means that 3C, 3S, 5C, and 5S refer to three robots with combined control, three robots with split control, etc.

The results of the experiments (see Figure 9.18) show that all systems maximize the PI and that they all perform some sort of exploration behavior. As the maximization of the PI does not specify a behavior of a system that can be evaluated directly, we need to define a measure based on our observations. In a chain of individually controlled robots, the travelled distance over time is a good indication for the quality of the coordination, as only well coordinated robots will be able to travel far. This is why we chose a sliding window coverage entropy (see Figure 9.18 and (Zahedi et al. 2010) for details) to measure the exploration behavior. It must be noted, that we are not interested in the exploration itself, but rather in the quantifiable, qualitative change of the observable behavior of the systems which result from the PI maximization. In this context, two counter-intuitive results are shown in Figure 9.18. First, the robot chains with five robots outperform the robot chains with three robots (compare 5C with 3C and 5S with 3S in Figure 9.18), and second, the split control systems outperform the corresponding combined control systems (compare 5S with 5C and 3S with 3C in Figure 9.18), in both, maximizing the PI and maximizing the coordination of the robots in a chain.

The first result is counter-intuitive, because longer chains means that more robots, and hence, more controllers have to coordinate based on the local information

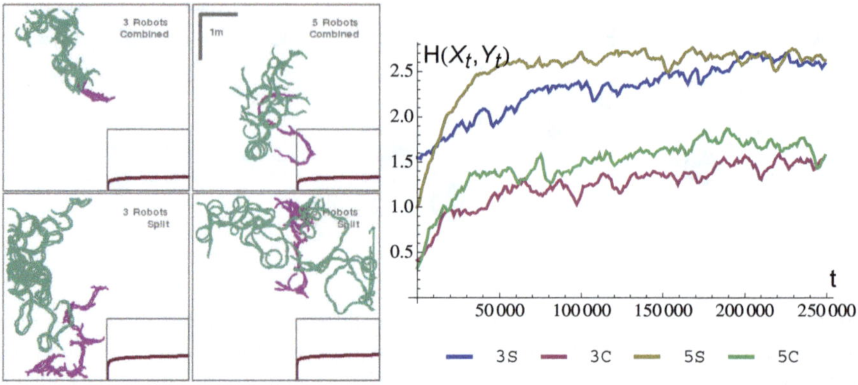

Fig. 9.18 PI maximization results. The graph on the left-hand shows four plots with one sub-plot each. The trajectory plots show the initial ten minutes (purple) and the final ten minutes (green) of the behavior (after 106 policy updates, which is approximately 27 hours). The sub-plots show the progress of the PI averaged over 100 runs for the same amount of time, normalised to the unit interval. The plot on the right-hand side shows how the exploration quality of the controllers progresses over time. For this purpose, the entropy $H(X_t, Y_t)$ over the spatial coordinates X_t, Y_t is calculated for a sliding window (see (Zahedi et al. 2010) for a discussion). Both plots show, indirectly measured by the sliding window coverage entropy, that the chains with five robots show a higher coordination compared to those with three robots, and that split control results in a higher coordination compared to combined control.

available through the wheel velocity sensors only. The second result is counter-intuitive, because the combined controller should, if anything, have additional properties compared to the split control as it combines the sensor information of both wheels. The next paragraph will analyse the behavior of one representative policy of the 5S setting and thereby explain all the results.

The behavior of all robots can be categorised into three modes. The first mode is called *forward* movement and it is characterised by (mainly) positive wheel velocities. The second mode is called *backward* movement and it is analogously characterised by (mainly) negative wheel velocities. The third is a transition between the two previous modes in which there is no clear direction of the robot chain.

To understand how the modes occur, we recorded the data stream of the sensors and actions ($s(t)$ and $a(t)$) of one representative controller of the 5S setting. According to the update rules used for the world model (see Eq. 9.27) we sampled the following four conditional probabilities $p(A_t \geq 0|S_t \geq 0)$, $p(A_t < 0|S_t \geq 0)$, $p(A_t < 0|S_t < 0)$, and $p(A_t \geq 0|S_t < 0)$, where the first two and the latter two sum up to one (see Figure 9.19). The two conditional probabilities $p(A_t \geq 0|S_t \geq 0)$ and $p(A_t < 0|S_t < 0)$ refer to maintaining the current direction of travel, whereas, $p(A_t < 0|S_t \geq 0)$ and $p(A_t \geq 0|S_t < 0)$ refer to a switching of the current direction of travel. At the beginning of the learning process, all conditional probabilities are by definition equal (see Eq. 9.25). After a while, it is seen that the system is more likely to maintain its current movement compared to switching as the sign of the

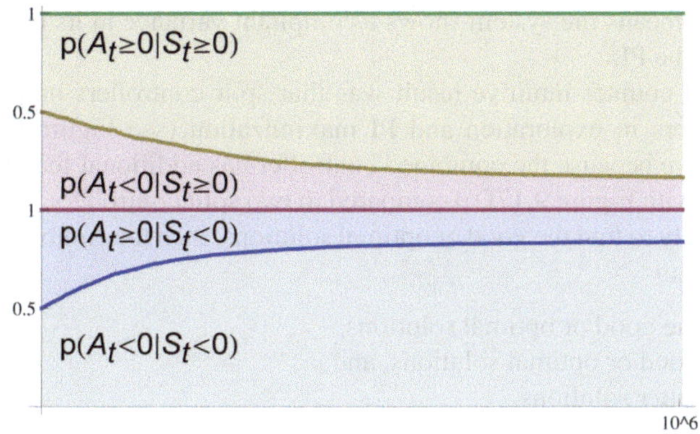

Fig. 9.19 Formation of modes. This graph shows the four conditional probabilities $p(A_t \geq 0|S_t \geq 0)$, $p(A_t < 0|S_t \geq 0)$, $p(A_t < 0|S_t < 0)$, and $p(A_t \geq 0|S_t < 0)$ plotted above each other over time and summed accordingly. The plot reveals that the policies are learned such that they maintain their current mode (forward or backward) with a high probability and that switching between the two modes occurs with the same probability. This fulfils the requirements of diversity and compliance posed by the PI.

action $a(t)$ is more likely chosen to be equal to the sign of the sensed wheel velocity $s(t)$ (see Figure 9.19). At the end of the learning process (10^6 iterations), the conditional probabilities are approximately $p(A_t \geq 0|S_t \geq 0) \approx p(A_t < 0|S_t < 0) \approx 0.85$ and $p(A_t < 0|S_t \geq 0) \approx p(A_t \geq 0|S_t < 0) \approx 0.15$. From these estimations, we can now reconstruct how the modes occur. If a robot chain is currently in the forward mode ($S_t \geq 0$ for all wheels), it requires more than half the controllers to decide on switching the mode for the robot chain to change its direction of movement. For the chain with three robots, it requires at least four controllers, and for the chain with five robots, six controllers to decide to switch directions. Hence, the probability of switching, denoted by p_s, is $p_s \leq 0.15^4$ for the three robot chain and $p_s \leq 0.15^6$ for the five robot configuration, where the probability refers to the controller update frequency (10Hz). For the two chains of robots, this means that the overall probability of maintaining the current behavioral mode in every time step is larger than $1 - 0.15^4 = 99.949\%$ for 3S and larger than $1 - 0.15^6 = 99.999\%$ for 5S. This explains why longer chains outperform shorter chains in terms of exploration as they are more likely to maintain their current direction of movement. The modes of the 5S are more distinctive compare to the 3S due to the larger number of robots in the chain. This can also be considered as a form of morphological computation (Pfeifer and Bongard 2006), which we will address later in this chapter again. The next question to answer is why the modes are beneficial in term of maximizing the PI. From the discussion above it follows that $H(S_{t+1}|S_t)$ is minimized because knowledge of the current wheel velocity reduces the uncertainty of the next wheel velocity significantly due to the formation of the modes. As the switching probabilities are almost equal, all sensor states are equally often perceived, which maximizes the entropy

$H(S_{t+1})$. This means the system shows a compliant variance in its behavior as it is demanded by the PI.

The second counter-intuitive result was that split controllers outperform combined controllers in exploration and PI maximization (see Figure 9.18). This is counter-intuitive because the combined controller has additional features (compare Figure 9.17C with Figure 9.17D)) compared to two split controllers. If the split controllers are likely to find the good or optimal solutions, then the combined controllers should be able to

1. find the same good or optimal solutions,
2. find other good or optimal solutions, and
3. find even better solutions.

The question is, why is this not the case? The space of possible policies spanned by the two split controllers is a subspace of possible controllers spanned by one combined controller. It happens to be that this subset of the split controllers encloses only a few maximizers and that a sufficiently large number of them is optimal with respect to the maximization of the PI. This is more obvious if only one robot is allowed to learn with two split and one combined controller. This is not shown here but discussed in detail in (Zahedi et al. 2010). Due to the low number of parameters defining the subspace of the split controllers, the optimisers are found faster and more reliably. In the superset of the combined controller, we find many suboptimal solutions which are more likely to be found compared to the optimisers. This means, by splitting the controllers we have made a large subspace of the combined controller space inaccessible to learning. The resulting subspace still had all the maximizers of the PI which is why the split controller outperforms the combined controller. Concluding, if one finds a natural way to restrict the policy space (possibly according to the morphology), such that it captures all maximizers of a given function, then this would be called an *optimal control* as the policy is optimally parameterised for learning and control. Finding such a natural method is an ongoing topic in this field of research.

As stated earlier, we want to discuss the results also from the perspective of morphological computation (Pfeifer and Bongard 2006). We already saw that the number of robots in the chain directly influences the exploration behavior. That the maximization of the PI leads to morphological computation is more obvious, if we take the 5S system, and remove the passive joints between the robots (see Figure 9.20). Both plots in the figure show the trajectories of all 10 wheels of the five robots, where the wheels of each robot share one color. The left-hand side shows the exploration behavior that we have already seen (compare with Figure 9.18). The right-hand side of the figure shows that the uncoupled robots loose a lot of their original behavior. They rotate more often on the spot and the trajectories are not as long and smooth as in the coupled system.

As the controllers are identical in both settings, this means that there is a contribution of the world (morphology and environment) to the behavior which cannot be assigned to the controller as they are identical in both systems. This contribution is called *morphological computation* (Pfeifer and Bongard 2006). Various

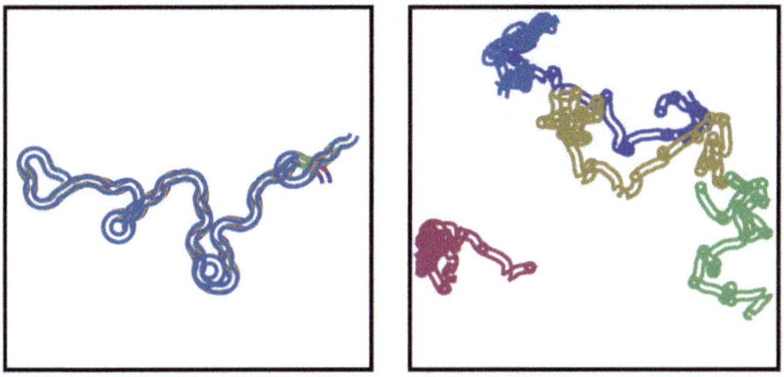

Fig. 9.20 Morphological computation. The two plots show the trajectories of all wheels of all five robots of the split control setting for ten seconds. The wheels of one robot share the same color in the plots. The plot on the left-hand side shows the trajectories of the original, passively coupled robots. The plot on the right-hand side shows the same five robots with the same ten controllers but with the passive connections removed. The comparison of the plots shows that the behavior changes significantly, which leads to two conclusions. First, PI maximization adapts to the world, and second, PI maximization leads to morphological computation, as the behavior is also significantly determined by the morphology of the system.

quantifications of morphological computation are derived and evaluated in experiments in (Zahedi and Ay 2013). They are based on causal and associative information-theoretic measures.

Appendix

Proof of Proposition 2:
(1)

$$p(h,c,w,s\,|\,do(a)) = p(h)\,\varphi(h;c)\,\alpha(h,a;w)\,\beta(w;s).$$

This implies

$$p(s,c\,|\,do(a)) = \sum_{h,w} p(h)\,\varphi(h;c)\,\alpha(h,a;w)\,\beta(w;s)$$

$$\begin{aligned} p(c\,|\,do(a)) &= \sum_{s}\sum_{h,w} p(h)\,\varphi(h;c)\,\alpha(h,a;w)\,\beta(w;s) \\ &= p(c) \end{aligned}$$

$$
\begin{aligned}
p(s\,|\,do(a),c) &= \frac{p(s,c\,|\,do(a))}{p(c\,|\,do(a))} \\
&= \sum_{h,w} \tfrac{p(h)}{p(c)}\, \varphi(h;c)\, \alpha(h,a;w)\, \beta(w;s) \\
&= \sum_{h,w} p(h\,|\,c)\, p(w\,|\,h,a)\, p(s\,|\,w) \\
&= \sum_{h,w} p(h\,|\,c,a)\, p(w\,|\,h,a,c)\, p(s\,|\,w,h,a,c) \\
&\quad \text{(conditional independence, see diagram in Figure 9.11)} \\
&= p(s\,|\,a,c).
\end{aligned}
$$

The second and third equations of the proposition follow from the general theory (see (Pearl 2000), Theorem 3.2.2 (Adjustment for Direct Causes), and Theorem 3.3.4 (Front-Door Adjustment)). For completeness, we prove them directly.

(2)

$$
\begin{aligned}
p(s\,|\,do(a)) &= \sum_{h,c,w} p(h,c,w,s\,|\,do(a)) \\
&= \sum_{h,c,w} p(h)\, \varphi(h;c)\, \pi(c;a)\, \alpha(h,a;w)\, \beta(w;s)\, \frac{1}{p(a|c)} \\
&= \sum_{h,c,w} \frac{p(h,c,a,w,s)}{p(c,a)}\, p(c) \\
&= \sum_{c} p(s|c,a)\, p(c).
\end{aligned}
$$

(3)

$$
\begin{aligned}
p(s\,|\,do(c)) &= \sum_{h,a,w} p(h,a,w,s\,|\,do(c)) \\
&= \sum_{a} \pi(c;a) \sum_{h,w} p(h)\, \alpha(h,a;w)\, \beta(w;s) \\
&= \sum_{a} p(a|c) \sum_{h,w} \left(\sum_{c'} p(c')\, p(h|c') \right) p(w|h,a)\, p(s|w) \\
&= \sum_{a} p(a|c) \sum_{c'} p(c') \sum_{h,w} p(h|c')\, p(w|h,a)\, p(s|w) \\
&= \sum_{a} p(a|c) \sum_{c'} p(c') \sum_{h,w} p(h|c',a)\, p(w|h,a,c')\, p(s|w) \\
&= \sum_{a} p(a|c) \sum_{c'} p(c')\, p(s|c',a).
\end{aligned}
$$

□

Acknowledgements. This work has been supported by the DFG Priority Program 1527, *Autonomous Learning*, and by the Santa Fe Institute.

References

Amari, S.: Natural gradient works efficiently in learning. Neural Computation 10(2), 251–276 (1998)

Ay, N., Bernigau, H., Der, R., Prokopenko, M.: Information driven self-organization: The dynamical system approach to autonomous robot behavior. Theory Biosci. (131), 161–179 (2012)

Ay, N., Bertschinger, N., Der, R., Güttler, F., Olbrich, E.: Predictive information and explorative behavior of autonomous robots. EPJ B 63(3), 329–339 (2008)

Ay, N., Polani, D.: Information flows in causal networks. Advances in Complex Systems 11(1), 17–41 (2008)

Bialek, W., Nemenman, I., Tishby, N.: Predictability, complexity, and learning. Neural Computation 13(11), 2409–2463 (2001)

Crutchfield, J.P., Young, K.: Inferring statistical complexity. Phys. Rev. Lett. 63(2), 105–108 (1989)

Der, R., Güttler, F., Ay, N.: Predictive information and emergent cooperativity in a chain of mobile robots. In: ALife XI. MIT Press (2008)

Grassberger, P.: Toward a quantitative theory of self-generated complexity. International Journal of Theoretical Physics 25(9), 907–938 (1986)

Kaiser, A., Schreiber, T.: Information transfer in continuous processes. Physica D: Nonlinear Phenomena 166(1-2), 43–62 (2002)

Klyubin, A.S., Polani, D., Nehaniv, C.L.: Tracking information flow through the environment: Simple cases of stigmerg. In: Pollack, J. (ed.) Artificial Life IX: Proceedings of the Ninth International Conference on the Simulation and Synthesis of Living Systems, pp. 563–568 (2004)

Klyubin, A.S., Polani, D., Nehaniv, C.L.: Empowerment: A universal agent-centric measure of control. In: Proc. CEC. IEEE (2005)

Lauritzen, S.L.: Graphical Models. Oxford University Press (1996)

Lungarella, M., Sporns, O.: Information self-structuring: Key principle for learning and development. In: IEEE (ed.) Proc. the 4th International Conference on Development and Learning, pp. 25–30. IEEE Press, San Diego (2005)

Lungarella, M., Sporns, O.: Mapping information flow in sensorimotor networks. PLoS Comp. Biol. 2(10), e144 (2006)

Martius, G., Der, R., Ay, N.: Information driven self-organization of complex robotic behaviors. PLOS One 8(5) (2013), doi:10.1371/ journal.pone.0063400

Massey, J.L.: Causality, feedback and directed information. In: Proc. 1990 Intl. Symp. on Info. Th. and its Applications, pp. 27–30 (1990)

Pasemann, F.: Complex dynamics and the structure of small neural networks. Network: Computation in Neural Systems 13(2), 195–216 (2002)

Pearl, J.: Causality: Models, Reasoning and Inference. Cambridge University Press (2000)

Pfeifer, R., Bongard, J.C.: How the Body Shapes the Way We Think: A New View of Intelligence. The MIT Press (Bradford Books) (2006)

Polani, D., Nehaniv, C., Martinetz, T., Kim, J.T.: Relevant Information in Optimized Persistence vs. Progeny Strategies. In: Rocha, L.M., Bedau, M., Floreano, D., Goldstone, R., Vespignani, A., Yaeger, L. (eds.) Proc. Artificial Life X, pp. 337–343. MIT Press, Cambridge (2006)

Prokopenko, M., Gerasimov, V., Tanev, I.: Evolving spatiotemporal coordination in a modular robotic system. In: Nolfi, S., Baldassarre, G., Calabretta, R., Hallam, J.C.T., Marocco, D., Meyer, J.-A., Miglino, O., Parisi, D. (eds.) SAB 2006. LNCS (LNAI), vol. 4095, pp. 558–569. Springer, Heidelberg (2006)

Prokopenko, M., Lizier, J.T., Price, D.C.: On thermodynamic interpretation of transfer entropy. Entropy 15(2), 524–543 (2013)

Reichenbach, H.: The Direction of Time. University of California Press (1956)

Schreiber, T.: Measuring information transfer. Physical Review Letters 85(2) (2000)

Zahedi, K., Ay, N.: Quantifying morphological computation. Entropy 15(5), 1887–1915 (2013)

Zahedi, K., Ay, N., Der, R.: Higher coordination with less control – a result of information maximization in the sensori-motor loop. Adaptive Behavior 18(3-4), 338–355 (2010)

Zahedi, K., von Twickel, A., Pasemann, F.: YARS: A physical 3D simulator for evolving controllers for real robots. In: Carpin, S., Noda, I., Pagello, E., Reggiani, M., von Stryk, O. (eds.) SIMPAR 2008. LNCS (LNAI), vol. 5325, pp. 75–86. Springer, Heidelberg (2008)

Chapter 10
Action Switching in Brain-Body-Environment Systems

Eran Agmon

10.1 Introduction

In recent years, the cognitive sciences have been converging upon an integrated perspective, a perspective that reframes behavior and cognition as a special type of self-organization that arises through the nonlinear, distributed interactions between brain, body and environment (abbreviated BBE). The BBE perspective has been separately developed by multiple lines of research such as the extended mind (Clark and Chalmers 1998), distributed cognition (Hutchins 2000), embodied cognition (Clark 1998), enactive cognition (Noė 2005; Thompson 2007; Varela et al. 1992)), situated cognition (Clancey 1997; Hutchins 1995), and the dynamical approaches to cognition (Beer 1995b; Thelen and Smith 1996; Kelso 1995; Port and van Gelder 1995). These different theories all emphasize different elements of the BBE; either the body, or the environment, or the temporal element. But their different theories are friendly to each other and can be brought together into a broader, integrated perspective. By bringing focus to all of the relevant components and their interactions, cognitive systems are transformed into seemingly self-organizing systems, in which behavior and cognition become a dynamical process that unfolds through distributed interactions (Kelso 1995; Maturana and Varela 1980; Thompson 2007).

We need now proceed with caution; the term self-organization has been a widely used term in scientific fields from physics to human social networks, and has acted as a unifying theme in systems sciences such as cybernetics and complex systems. But when we consider the different definitions surrounding self-organization, there is an abundance of philosophical stances and formal methodologies (Polani 2008). In this chapter we will focus on a single perspective of self-organization, which will be called the "absolute system" perspective, after Ross Ashby's framework for describing what he considers to be adaptive behavior and self-organization (Ashby 1962, 1952). An absolute system takes what some describe as the omniscient perspective (Dupuy 2009), which takes all of a system's relevant variables and puts

Eran Agmon
Indiana University, 107 S. Indiana Avenue, Bloomington, Indiana 47405-7000, United States
e-mail: agmon.eran@gmail.com

M. Prokopenko (ed.), *Guided Self-Organization: Inception*,
Emergence, Complexity and Computation 9,
DOI: 10.1007/978-3-642-53734-9_10, © Springer-Verlag Berlin Heidelberg 2014

them in a model that fully describes the system's dynamics. For Ashby, a system has a finite set of internal states and a transformation rule that maps a state onto itself as it unfolds in time. An absolute system is an autonomous system, in which all relevant variables are accounted for. With such systems an initial state has a regular trajectory that follows, and upon repeated re-initializations to the same initial state there are no divergences in following behaviors. If there are divergences, then some relevant variables must not have been accounted for and the system is not absolute. A characterization of systems in this sense is ideal for science, because it allows for perfect predictability of the system's behavior. It can be argued that striving for an absolute system description is not practical when dealing with real-world systems for which there are essentially infinite relevant variables. But in this chapter we follow through with the assumption to see what insight can be gained.

Sect 2 describes this perspective in greater detail, and follows Ashby's argument, which he believed demystified the notion of self-organization by attributing the apparent self-organization to an opportune matching of system and environment. We move along this intellectual thread leading from Ashby's definition of self-organization to the modern theories of the BBE framework and show that a very similar approach has been converged upon, perhaps unknowingly, by an integrated BBE perspective. We will see that Ashby's insights are pervasive in this framework, and have been developed into to a rigorous research methodology. The desire to understand the relevant causal variables that come together to generate our cognitive behavior has led scientists to "extend" the mind (Clark and Chalmers 1998), and to attribute mental processes to the complex interactions that take place between many distributed components in the brain, body, and environment. The field is developing a terminology much like Ashby's, which is rooted in dynamical systems theory, and focuses on the temporal elements of cognitive behaviors that arise through distributed interaction.

In Sect 3, we dive deeper into the BBE framework by outlining Randall Beer's adaptive behavior research program (Beer 1997; Beer et al. 1996). This project extends Ashby's insights into a more rigorous methodology that focuses on minimal instances of adaptive behavior, and integrates many of the motivations behind the BBE framework. It does this by combining the insights with modern computer simulation and the mathematical toolset of dynamical systems theory. The goal of Beer's project is to simulate the entire conditions for simple adaptive systems, which includes their environment, their body, and their recurrent dynamical nervous system. He uses evolutionary algorithms to produce dynamical models of brain-body-environment systems that can engage in minimally cognitive tasks, and then analyzes their resulting dynamics to illuminate the dynamical strategies for adaptive behavior. The result of such analysis yields a similar effect as Ashby had intended, of demystifying adaptive behavior by fully reconstructing the system's conditions in a model and then studying its temporal structure.

Most models developed by this methodology have focused on the production of single actions through BBE interactions, and have uncovered the temporal patterns that allow for these particular actions to unfold. But this does not provide a complete picture of living systems, which can generate many possible actions and switch between them in a context-appropriate manner. Considering the problem of multiple actions brings up new questions about coordination between brain, body, and environment. We ask how multiple actions can arise out of a single absolute system, in which at one time a particular coordination pattern is engaged, and at a different time a completely different coordination pattern is engaged. In Sect 4, we use Beer's methodology to evolve an agent that can generate multiple different actions and smoothly switch between them. An analysis uncovers the strategy that allows it to behave in different ways that requires the coordination of different sensors, effectors, and brain regions.

For the last section we bring together many of the discussed ideas, examine their limitations and suggest improvements for future research. The brain-body-environment framework in cognitive science is a young science and still in its early stages of development. Because of this there are many assumptions left untested and many questions left unexplored. By building up a dialogue and continuing to improve our models, we may someday bring this science from its current emphasis on minimal behavior to the complexities of real living behavior.

10.2 Ashby's Self-Organization in Brain-Body-Environment Systems

In this section we review Ross Ashby's absolute system perspective, from which he believes to have demystified self-organization in his 1962 paper, "Principles of the self-organizing system" (abbreviated PSOS), (Ashby 1962), and with which he presents a scientific framework for the study of adaptive behavior in "Design for a Brain" (abbreviated DFB),(Ashby 1952). The similarity between Ashby's theoretical framework and the one suggested by an integrated BBE framework will be demonstrated by following Ashby's argument as laid out in these two publications, and comparing it to the arguments made by the various fields of research of the BBE. Where Ashby's arguments were based in purely mathematical formalisms, fields within the BBE framework have looked at the structure of particular sensori-motor interactions in the real world and have therefore extended theoretical intuitions into empirical. By following through with Ashby's arguments, we gain a better understanding of how a more integrated BBE framework might someday appear.

Ashby begins PSOS with a definition of a system as an arbitrary assignment of parts, as based on an observer's perspective and not limited to material components. This is a constructivist definition of a system, which is often described synonymously with the term "model." The parts of this system are described mathematically as variables that take a range of states and unfold through time based in the dynamical laws of system. Dynamical rules captures these laws mathematically by defining a transformation rule (evolution operator) that determines how the set of states at one point in time changes to a new set of states in the following point in time. Ashby defines organization as conditionality of variables; "As soon as the relation between two entities A and B becomes conditional on C's value or state then a necessary component of 'organization' is present"(Ashby 1962). An organized system is one with components whose states are conditional on other states, with dynamical rules that bring this conditionality into effect as the states unfold through time. Central to Ashby's perspective is that the goal of a scientist should be a description of a system as an absolute system, with a mathematical model that is free from internal contradictions. The equations that define a system's states and dynamics needs to be refined and reduced until it is described in a "machine-like way, namely, that its internal state, and the state of its surroundings, defines uniquely the next state it will go to" (Ashby 1962).

With this basis for defining systems and organization, Ashby lays the foundations for the dynamical approach in cognitive science, which is one of the cornerstones of the BBE framework. For Ashby, the ideal description of a system defines its relevant variables, their interdependences, and the dynamical rules that systematically unfold the structure through time. This is the foundational principle of "the dynamical hypothesis" (Van Gelder 1998), which received much attention many decades after Ashby. The dynamical hypothesis proposes a unifying philosophical stance in cognitive science, which insists that cognitive systems are dynamical systems, that they are best understood as dynamical systems, and that therefore scientists should thrive for dynamical explanation of such systems. This stance brings with it a certain understanding; it influences the questions asked, the analyses performed, and the interpretation of results (Beer 2007). Many separate lines of research have come upon this same line of reasoning, and have employed dynamical explanation of cognitive behavior. Dynamical systems have been used to model neuronal system (Izhikevich 2006), entire brain systems (Skarda and Freeman 1987), coordinated motor behavior (Turvey 1990), child development (Thelen and Smith 1996), language (Elman 1995), interaction between language and vision (Spivey et al. 2005), and many more.

As we will see shortly, Ashby rejects the interpretation of self-organization (in PSOS) or adaptive behavior (in DFB) as something that can come out of the internal organization of an organism on its own. He instead attributes it to an opportune matching of organism and environment. The organism alone is a non-autonomous system, whose behavior is partially dependent on its environmental situation. For Ashby, the agent and environment together make an absolute system, and so it is only on this level of description that we can truly understand adaptive behavior. "The organism affects the environment, the environment affects the organism: such a system is said to have 'feedback'" (Ashby 1952). Systems with feedback cannot be treated as if their action was controlled in a linear way; they possess properties that cannot be reproduced in systems that lack feedback. Because of this, if we are to describe an organism's behavior in a model we must bring relevant variables from the environment into the absolute system definition, or else we would miss out on the behavioral effects of feedback.

This essential pairing of organism and environment has been broken up into separate fields in cognitive science and extended in empirical study. The first related field, called situated cognition, finds its roots in the phenomenological philosophy (Heidegger 1962), in classical ethology (Tinbergen 1963), and in ecological psychology (Gibson 1986). Situated cognition concludes that cognition and behavior is a contingent on the situation in which it is enacted (Hutchins 1995; Clark and

Chalmers 1998). A cognitive system is always interacting with its environmental through sensors that perceive, and effectors that produce behavioral output. This leads to the same conclusion as Ashby; behavior is controlled via sensory-motor feedback. Perceptions trigger actions, actions produce changes in the environment, these changes are again perceived and trigger new actions that correct for or extend the effects of the previous actions. Different environmental situations will produce different perceptions, and therefore trigger different actions. This field's emphasis shifts many problems, such as memory and reasoning, from the brain to the environment. Instead of having to conceive, predict, and remember the consequences of an action, action is simply executed by reading off and reacting to information available in the environment.

Embodied cognition is a complementary field, which rather than placing emphasis on the role of the environment, places its emphasis on the role of the body. The significance of embodiment was also first described phenomenological philosophy (Merleau-Ponty 1996), and was recognized by cognitive science in the 1980's with Rodney Brooks' robotics, which emphasized the role of physical embodiment in coordinated behavior (Brooks 1991). Brooks ideas provided a radical alternative to the then-dominant computational approach. Following research in embodied cognitive science has emphasized the role of activity in perception (Noe 2004), autonomy in cognition (Pfeifer and Scheier 2001), the use of metaphor as based in sensori-motor experience (Lakoff and Johnson 1999), and in the philosophy of cognitive science (Clark 1998). These projects recognize how the physical aspects of an organism's body are crucial to its behavior and provide enormous constraints on behavior. The nervous system receives input though the embedding of sensors on its spatially extended body, and their physical properties directly affects that information available to the organism. Additionally, the particular assemblage of bones, joints, and muscles create a unique control problem for the brain. As the cognitive system develops, both physically and behaviorally, it is constrained by a body and can only learn through information provided through a body. Ashby did not place emphasis on the physical embodiment, but would certainly agree with these scientists that the body is a relevant variable, and must be accounted for when describing behavior and cognition from an absolute systems perspective.

The final definition that Ashby offers in PSOS before turning his attention to the demystification of self-organization is what he calls "good organization." This aspect of Ashby's framework has been less influential in the majority of the fields under the BBE umbrella, but has still been advanced in one of the most far-reaching fields called enactive cognition (Varela et al. 1992; Thompson 2007; Di Paolo 2005). For enactive cognition, as well as for Ashby, "good" is a relative term that is dependent on the fit between the system in question and its given environment. If the pairing is such that it acts to further the system's survival, then the system has good organization. Ashby describes what he calls "essential variables" as variables that are closely related to survival (e.g. heartbeat, core body temperature, oxygen level). A successful organism acts to maintain these variables within a narrow range, but when one of the essential variables is significantly altered, the organism dies, and the rest of the essential variables are also dramatically changed. Enactive cognition introduces a "boundary of viability," which surrounds the subset of an organism's state space within which it must remain to survive (called a viable set). It asks how adaptive agents engage with an environment in such a way that they discover possible actions, and engage in actions that bring them to increasingly robust regions of the viable set (Di Paolo 2005).

Now, back to Ashby's argument in PSOS: with the definitions of absolute system and good organization nicely laid out, Ashby proceeds with his demystification of self-organization. For Ashby, the term self-organization implies a system's ability for "changing from a bad organization to a good one," (Ashby 1962). A self-organizing system is one that at first does not have a chance of surviving as an organized system, and by dynamically unfolding through time it changes its organization and is able to persist robustly in the environment. Ashby turns the omniscient lens of the absolute system perspective onto this adaptive behavior, and attempts to describe the dynamical organization that could produce such behavior. He quickly rejects an interpretation of self-organization as something that can come out of a system's internal organization; "no machine can be self-organizing in this sense."

For Ashby, organization cannot autonomously improve itself because it is mathematical nonsense to talk of a function that is a function of the state that it defines; an evolution operator, which determines how states unfold through time, cannot be updated by the states that it transforms. There must be some additional variable that drives this apparent organizational change, but which is not contained within the organized system. "The appearance of being 'self-organizing' can be given only by a machine being coupled to another machine" (Ashby 1962). Ashby then proceeds to assert that the appearance of self-organization in systems is not only unremarkable in the sense that there are no special conditions that govern self-organization, but that it is in fact an inevitable property of large dynamic systems that have been given sufficient time to come to equilibrium. When we examine this equilibrium, we can split up the relevant components into "organism" and "environment," and will find that the organism is highly robust to perturbations from the environment, creating what Ashby calls an "adaptive fit."

This forecasts a demystification of adaptive behavior that would result if there were a complete integration of a BBE framework. By bringing focus to the dynamical approach in which a cognitive system is characterized by a set of states and dynamical operator, proponents of the BBE framework have adopted a fascination with self-organization (Kelso 1995; Maturana and Varela 1980; Thompson 2007). They have recognized that adaptive behavior is the result of feedback between brains, bodies, and environments, and that all relevant variables across this system must be integrated into our models. Just like self-organization cannot come out of a system in isolation, adaptive behavior cannot come out of an organism without an environment to couple with. If we identify the relevant components of a particular cognitive system, and bring them together in a dynamical model, cognition becomes an unfolding process that takes place between distributed components, and can be understood in a purely dynamical terminology. The coupled brain-body subsystem is called the "agent". It interacts with the environment through coupled interactions that generate feedback. Coupling that flows from the environment to the agent is called "sensory," and coupling that flows in the opposite direction is called "motor". The agent's behavior is defined by its trajectory of motor outputs (Beer 2007). Cognitive capacities such as memory, learning, attention, and recognition are predicted to fall out of this description if such an ideal model is obtained, not as intrinsic properties of a system but as patterns that emerge from the dynamical trajectories of the system.

10.3 Beer's Adaptive Behavior Program

Where Ashby had compelling terminology based in dynamics, and a complete vision for the study of adaptive behavior, he lacked on an ability to explain any cognitive behavior that we might find in the real world. Ashby's example adaptive systems were based in formal systems, which he defined and brought to life by running the equations. He was able to make his conclusions by studying the dynamical properties that arise from such simulation. But the proof that adaptive behavior cannot come out of an isolated system is not sufficient to explain the cognitive behavior we observe in real living organisms. We must ask about their particular structural properties, how these particular properties produce an adaptive fit with the properties of environment, and how the behaviors that we observe result from this opportune matching.

Meanwhile, the BBE framework has approached the cognitive process from an empirical perspective; relevant variables that signify states of real living agents are identified and brought into models that predict how they unfold through time. But the complexity of real living systems is hard to overcome. There are many components, and with their nonlinearity, accurate prediction appears futile. Ashby himself recognized this difficulty, and asserted that real systems likely have infinite variables (Ashby 1952). The difficulty in creating complete models leads to an inability to theorize about adaptive behavior in the terminology of dynamical systems.

A bridge between Ashby's theory and the BBE's empirical interest is attempted in Randall Beer's adaptive behavior project (Beer 1997). What began as the rejection of traditional artificial intelligence led Beer to set as his goal the simulation of an organism's entire capacities. His research has developed a rigorous methodology to simulate the entire conditions of a brain, body, and environment that engage in minimal instances of adaptive behavior. Beer then adopted dynamical systems theory to analyze the resulting behavior's dynamical underpinnings. These models are developed by first defining an environment and body in a computer program, and then using genetic algorithms to evolve a dynamical neural network that can control the body effectively in a way that generates the desired action. These agents are of interest to the BBE framework because their simulated behavior is easily related to behavior observed in the real world, yet they are simple enough to be analyzed and completely described with a dynamical terminology.

Beer's first examples of evolved embodied agents were designed to produce the behaviors of insect walking and chemotaxis (Beer and Gallagher 1992). The chemotaxis agent will be described in greater detail later on in this section. These examples demonstrated successful situated behavior, but drew some criticism for only studying simple sensory-motor tasks, and not addressing high-order cognitive function. Beer's next step was to extend the framework to simulated examples of minimally cognitive behavior (Beer et al. 1996), which would demonstrate higher-level cognitive behavior as the result of dynamical BBE interactions. Following from this proposal, many agents were evolved to produce behaviors such as selective attention (Slocum et al. 2000), categorical perception (Beer 2003), learning and memory (Izquierdo et al. 2008), relational categorization (Williams et al. 2008a), referential communication (Williams et al. 2008b).

Minimization allows the researcher to focus on a particular cognitive function of interest, and provide this function with a dynamical explanation. By evolving systems that specialize in specific behaviors such as learning, attention, or categorical perception, the cognitive system is reduced to a minimal organization that allows for only the behavior of interest and removes the additional functionality inherent in living systems. This is the concept Beer calls "frictionless brains"; the nervous systems are evolved to produce a well-defined function, which can then be studied without interference by other influencing factors. Real organisms don't have this specialization, and usually take part in many types of behavior who's neuronal underpinning cannot be easily teased apart (this capacity for multiple functions is further addressed in the section on action switching). Because of this, it is much more straightforward to study frictionless brains with well-defined behavioral functions because the neuronal behavior can be directly linked to the production of that particular behavior.

Simple simulated agents are also far more ideal than real living systems for full dynamical analysis. Scientists gain full access to the final, successful agents, because all of the interactions that come into producing the behavior are readily available to the scientist, and just have to be recorded and analyzed. The simulation is a full absolute system model by definition, and does not require a process of abstraction to creating a simplified model. With access to this model, behavior can be analyzed in the way dreamt about by Ashby many decades ago. Beer's framework allows him to ask questions that Ashby could not have begun to answer, such as "How do the individual components across brain, body, and environment contribute to a specific behavior?", "What classes of control mechanisms are best suited to the generation of adaptive behavior?", and "how does manipulation of the variables and parameters affect resulting behavior?"

10.3.1 CTRNNs and Genetic Algorithms

Continuous-time recurrent neural networks (CTRNNs) are adopted as the model nervous systems for these simulations. The general form of these equations is shown below. In this equation, yi is the state of the neuron, τ is the time constant, w is the weight between neurons j and i, θ is the bias term, I is external input, and $\sigma(x)=1/(1+e^{-x})$ is the standard logistic activation function.

$$\dot{y}_i = \frac{1}{\tau_i}(-y_i + \sum_{j=1}^{N} w_{ij}\sigma(y_j + \theta_j) + I_i)$$

Beer justifies this selection for a neural model with several points (Beer 1995a). First off, this model's recurrent connections allow the agent to initiate its own behaviors as a result of its internal state unlike the feed forward networks that were popular at the time. Additionally, Beer argues that they are the simplest case of a nonlinear, continuous dynamical neural network model and despite this simplicity, they are universal dynamics approximators (Funahashi and Nakamura 1993). They lend themselves to a biological interpretation, in which a state is associated with a neuron's mean membrane potential, and the output is associated with its average firing frequency. Finally, CTRNNs are computationally and analytically tractable, and they are evolvable by searching through the combinations of the CTRNN's parameter values.

Just like real-world systems produce an adaptive fit between the agent's internal control mechanism and its given environment, so too must the BBE simulation produce a fit between the dynamical nervous system, its body, and its environment. To produce this fit, Beer adopted the use of genetic algorithms (GAs) to evolve CTRNNs that optimize a fitness function by controlling a simulated body in a simulated environment (Beer and Gallagher 1992). This approach was separately developed at around the same time period by some other research groups (Cliff et al. 1993)(Nolfi et al. 2000). These GAs encode the CTRNN parameters, τ, w, and θ in genetic strings. An initial random population of such strings is created, and in each generation the fitness of each individual is evaluated by running a simulation with the individuals' CTRNNs. A new generation is created by selecting highly fit individuals and slightly mutating them to explore nearby regions of parameter space. The selection process chooses individuals with a probability proportional to their fitness, so that more highly fit individuals are represented in the next generation. A set of genetic operators modifies the selected individuals' genetic strings with mutation and crossover. Mutation randomly modifies portions of the strings with some fixed mutation probability, and crossover combines chunks of genetic strings from multiple individuals to create a whole new individual. Once a new population has been constructed, the entire process repeats, and after many such generations the population's fitness increases and converges onto some final local maximum.

The creative part of making these models lies in the experimenter's design of the agent's body and environment, and in the definition of a fitness function that can select for a behavior of interest. If the conditions are designed well, then the GA can move through the space of possible CTRNNs in a gradual way, towards regions of increased adaptive fit. With a good design, a highly successful CTRNN is results from the evolutionary search, and provides the modeler with a BBE system that can be further analyzed to uncover the dynamical strategies that generate successful behavior.

10.3.2 Dynamical Systems Theory

The evolved and well-adapted CTRNN is a system of nonlinear differential equations. Each variable makes an axis in the system's state space, which is the set of all possible states that the system can be in. Every point in this state space has an instantaneous trajectory, as determined by the equations, that leads it to a different state (or in the case of equilibrium points, there is no trajectory and the state remain constant). A particular behavior results by setting an initial condition in this space, and following the resulting trajectory through time. A phase portrait is the set of all trajectories that can result in the system. It is the BBE scientist's goal to fully describe this phase portrait, including different factors that shape this space and determine the system's behaviors.

Nonlinear systems such as CTRNNs are difficult, if not impossible, to solve analytically. This makes the characterization of their temporal structure difficult to study in traditional ways. But luckily, the mathematical field of Dynamical Systems Theory (DST) has developed various approaches for characterizing a system's given these constraints. DST has learned that it can uncover much of the systems behavior by focusing on invariant sets within the system's phase space, and on linearized behavior in these invariant sets' direct proximity. Based in this realization, a highly developed set of mathematical tools has been developed, which includes ways to identify a system's invariant sets (e.g. fixed points, limit cycles, chaotic attractors), a characterization of their local structure around these sets (e.g. stability), global structure that connects the sets (e.g. attractor basins, saddle manifolds), and the changes in qualitative structure that occurs with changing parameters (e.g. bifurcations), (Beer 1995b).

Beer adopted the DST toolset to analyze the evolved CTRNNs and uncover how particular trajectories seen in simulated behavior are guided by the system's intrinsic dynamical properties (Beer 1995a). But, as the BBE perspective emphasizes, a characterization of the brain's (or CTRNN's) dynamical landscape is not sufficient to describe the generation of behavior. The CTRNN is not autonomous, and its interactions with the environment cause perturbations of state that would not result from the CTRNN in isolation. Because of this, the CTRNN typically moves through its state space not according to a phase portrait found by a dynamical analysis of the CTRNN alone. This analysis only provides knowledge of the CTRNN's inclinations. For a full explanation of behavior the dynamics of brain, body, and environment have to be brought back together into an absolute system.

By evolving complete dynamical system models of brains, with formally defined bodies and environments, a modeler gains full access to all information about the absolute system and can investigate the underlying dynamical space to explain how a system produces adaptive behavior. Simulated experiments explore multiple scenarios' behavioral trajectories, all of which come together in the absolute phase space uncovered by DST. Each component's contribution can be directly determined with such experimentation: agents can be removed from their environments, and the effects of stimuli on motor output directly analyzed. Connections in the CTRNN can be lesioned, or states held fixed by experimenter control. By doing this, the scientists can determine how each component contributes to the overall dynamical landscape. Such analyses demonstrate that the CTRNN is attracted to equilibrium points in its state space, and that movement towards these points determines motor output. But as a result of behavioral output, the agent moves within its environment leading to changes in sensory inputs, which in turn alter the CTRNN's phase portrait and influence its next instantaneous behavior.

These ongoing interactions continue to shift the CTRNN's phase portrait either by adjusting the phase portrait slightly, or by creating bifurcations that qualitatively alter the phase space. With these ongoing changes, the system continues to chase a moving equilibrium point. We see that with DST, what were once somewhat vague descriptions of a system's dynamics are given real meaning that can be approached scientifically. Specific actions that are evolved with Beer's method can then be fully described in all their specific nuanced details. By evolving many such agents, each with unique dynamical properties, a general space of strategies is uncovered and builds up a broader picture of adaptive behavior.

10.3.3 A Simple Chemotaxis Agent

In this section we introduce Beer's chemotaxis agents, which were first described in 1992 (Beer and Gallagher 1992), and later extended with a full dynamical systems analysis (Beer 1995a). Chemotaxis is an ideal case of minimal sensori-motor goal-directed behavior, in which agents direct their movement according to chemical signals that are present in the environment. Such behavior is used often by organisms, such as bacteria and nematodes, to approach resources by moving up a chemical gradient. We introduce this basic example now to illustrate an example of the explanation made possible by Beer's framework. We also introduce the chemotaxis agent here because the next section on action switching is based on this early example, but extends it to capture a broader explanation of behavior. By first describing the simpler case, we will be prepared to extend the model later.

For the simple chemotaxis simulations, an agent is given a simple body (shown in Figure 10.1), in this example with a 6-neuron fully interconnected CTRNN. These CTRNN neurons include two spatially extended sensors that detect chemical concentrations at their location, two interneurons, and two motor neurons whose combined outputs produce a torque and thrust, which propel it through the environment. The agent is enclosed in a 2-dimensional square-shaped environment that contains a single circular resource at its center. This resource emits a chemical signal, which diffuses through the environment with intensity proportionate to the inverse square of the distance from the center. The GA's fitness function is to minimize distance between the agent and the resource, which would select for agents that can approach the resource and remain as close as possible for the duration of the simulation.

Beer found that multiple chemotactic strategies evolved under these conditions. We will discuss only the most common strategy here, which is reproduced in a novel agent shown in Figure 10.2. This agent moves forward while turning toward the side on which the chemical signal is stronger. A dynamical analysis demonstrated that

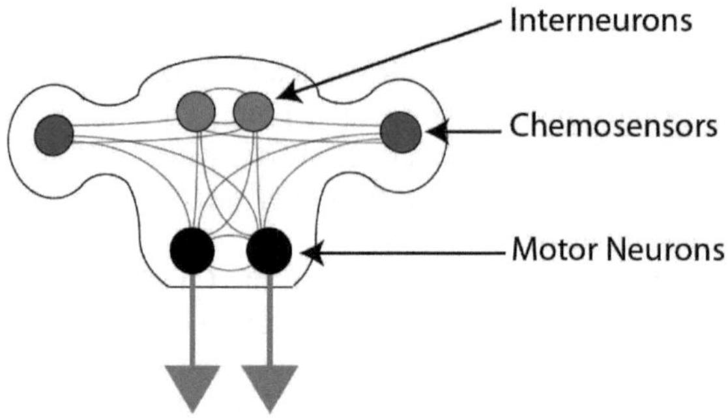

Fig. 10.1 Basic chemotaxis agent morphology, with 6 fully-interconnected CTRNN neurons

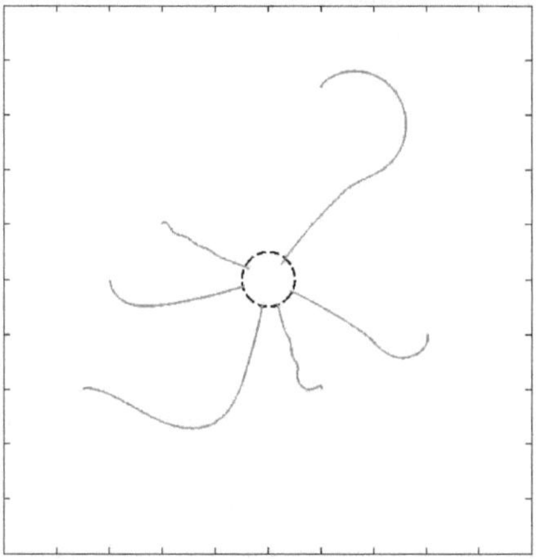

Fig. 10.2 Multiple simulated trajectories, in which the agent is initialized randomly in the environment and moves to the resource

the CTRNN has a single equilibrium point, which shifts with different levels of activation from the two chemosensors. When he examined how this equilibrium point is projected onto the left and right motor outputs, he uncovered a simple explanation for the resulting behavior. When the left chemosensor is more active than the right, the leftward orienting motor neuron becomes more active than the left. When there is higher activation of the right chemosensor, the rightward orienting effector becomes more active. This directly explains the observer turning and approach behaviors, and is rooted in dynamics

10.4 Action Switching

The minimally cognitive behavior project has described many different dynamical strategies for specific cognitive behaviors, but the emphasis on specific behavior leads to an incomplete picture of organisms' full behavioral capabilities. There are many questions that arise when we shift our focus from the dynamics of performing single actions, to the broad repertoire of actions that all organisms have accessible. Real living organisms depend on the ability flexibly switch between their possible actions. For example, a subtle movement in distant shrubs might be all the information available to a monkey to determine if a predator is on the prowl. This movement in the environment couples with the monkey's relevant sensors, and elicits a dramatic behavioral change from gathering food off of the ground to scampering up a tree for safety. Such actions require very different patterns of sensorimotor coordination; picking food might require fine finger dexterity and acute eye movements for examining food sources. Running up a tree would require full limb coordination and tactile or proprioceptive sensory input. By committing to the dynamical perspective, the scientist is obligated to describe how the many interacting components of the brain, body, and environment become engaged throughout these different actions and in the transition between them.

More traditional approaches in Artificial Intelligence assume that a higher-level mechanism must be used to determine action. This mechanism uses logical or statistical reasoning to decide upon the most beneficial action out of a repertoire of possible actions given the information it has available about the present context. After this decision-making process selects an action, the action is initiated. Some might claim that the BBE approach does contradict this depiction of higher-level mechanisms for action selection. For descriptions that attempt to bring BBE dynamics with higher-level mechanisms of action selection, self-organization and dynamics only account for feedback let loose on a one-way path towards a particular end-state or goal. The initial conditions that allow particular such actions to be instantiated are determined and initialized by a higher-level decision process.

But this misses out on the real underlying message intended by the BBE framework. This perspective aims to describe an absolute system, which has a phase space that describes the systems full range of possible behavior that result from a single, unchanging evolution operator. For this perspective, the agent's behavioral repertoire has to be completely contained within the dynamical explanation, including its movement from the state space region that defines one action to the regions that defines another. Action switching must be a product of self-organization that falls out from these dynamics, and not a higher-level mechanism that sets initial conditions. Many novel questions come up from defining the problem in this way; how is a systems phase space divided between its full repertoire of possible actions. Are there specific regions of this phase space that become responsible for each action (modularity)? How do the different types of action constrain the sensorimotor apparatus in their own unique way that produces appropriate behavior? When an action is completed, how does the system transition to a different action?

Few dynamical systems agents have explored the questions of multiple actions. It has been shown that the same CTRNN is able to implement qualitatively different behaviors when coupled to different bodies (Izquierdo and Buhrmann 2008), and that globally stable CTRNNS containing a single basin of attraction are able to sustain multiple modes of behavior (Buckley et al. 2008). In this section we introduce a new model capable of engaging in multiple different actions, which autonomously switches between these actions without a higher-level mechanism. A highly successful agent is examined in depth to reveal its dynamical organization, and how it allows for efficient action and switching.

10.4.1 Evolving an Action Switcher

To study action switching, a chemotaxis agent of the same basic design introduced in the previous section on simple chemotaxis agents was evolved, but with some additions that forced it to switch between the approaches of two different resources. Its environment was encoded as a 100x100 unit plane, with two sources of food that were held in fixed locations for each trial. The agent was given an initial position and directionality in this environment, and was allowed to move spatially by coordinating its motor neurons. Its task was to maintain two nutrient levels above zero by coming within the spatial boundaries of each food source. When it is within the resource's boundaries, the corresponding nutrient level is increased at a fixed rate to simulate the uptake of nutrients. But there is also a constant decay of nutrient levels at a fixed rate that simulates a simple metabolism. In order to sustain both nutrient levels above empty, the agent must switch between its approaches of each resource, and to spend sufficient time on each one to refill the nutrient supplies.

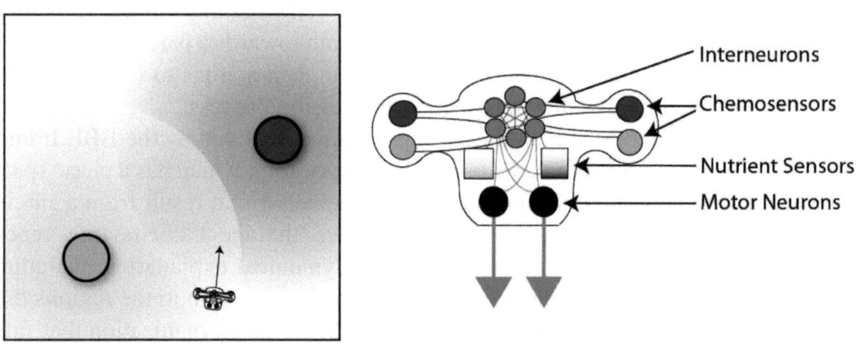

Fig. 10.3 The image on the left shows the action switcher's environment. The agent exists in a 2-D bounded environment with two different resources and their diffusing chemical signals. The image on the right shows the agent's morphology. Two types of chemosensors on each spatially-extend stalk detect the different chemicals, nutrient sensors detect internal nutrient levels, and motor neurons produce force that moves the agent.

The agent is given a body that is controlled by the CTRNN via sensors and effectors that were embedded in different locations along its body. Of the neuronal components, four are chemosensors, two are nutrient sensors, two are motor neurons, and six are interneurons. Chemosensors receive input regarding the concentration of a chemical, which just like the original chemotaxis agent, is proportionate to the inverse square of the distance between sensor and resource. The two chemosensor stalks were extended outward from the body's assigned coordinate location by 6 units, therefor spatially extending the agent's sensors within the environment. The outputs of the two effectors are used to generate movement through the environment. The output of each effector is a vector of force that pushes one side of the body. Directional change is determined by the difference between the two effectors' outputs, while magnitude of movement was determined by the sum of the effectors' outputs. Additionally, the velocity was multiplied by a friction coefficient of 0.9 at every time step, such that the agent would slow down and eventually stop if the effectors produced no force.

The two nutrient levels serve as measures of how much nutrient for each resource is stored in the agent. These values can be anywhere between 0 and 10. When the agent is within the boundary of a resource, its corresponding nutrient is replenished by 0.02 units per time step. Additionally, there is a decay of 0.004 units per time step throughout the entire simulation to simulate a constant metabolism. The simulation continues as long as both internal nutrient values are above empty, but as soon as either value dropped to 0 the agent dies and simulation ends. If the agent is successful and does not die, the simulation is halted after 5000 time steps, which is sufficient time to guarantee that the agent had to move to each resource several times. To further guarantee robustness of the agent in many different environments, it had to succeed in 11 different environments. The average length of time the agent survives in the 11 simulations defines its fitness value for the genetic algorithm.

Figure 10.4 shows the evolutionary progression of twenty different evolutionary runs that had the same morphology but different initial random seeds. The top performing agent's evolutionary run is shown in black. The evolutionary search incrementally increases their fitness as a strategy is converged upon and refined. There are many different possible strategies for action switching given the constraints provided to the GA, and the evolutionary runs can only come upon some of them. In typical experiments of this framework, researchers are recommended to investigate the full space of possible strategies by examining many of the evolved solutions, and also evolving agents with different morphologies (e.g. different amounts of sensors, interneurons, connectivity, etc.). These different morphologies produce constraints on the genetic search, and can lead to very different final strategies. By analyzing multiple strategies, the researcher attains a more general understanding of the behavior, which abstracts over the particular details of an individual instantiation. However, in this chapter only the top-performing agent is selected from the twenty evolutionary runs shown. A more comprehensive examination of different strategies will described in an article currently in preparation (Agmon and Beer 2013).

10.4.2 The Agent's Behavior and Dynamics

The top-performing agent's final fitness score is .86, which means that it survived through 86 percent of the maximum time provided in the 11 trial configurations. This indicates a high robustness to different environmental conditions, and persistence in continued action switching between the approaches and eating of the two resources. A typical behavioral trajectory is plotted in Figure 10.5, where the agent moves towards a resource, performs several loops on top of the resource (which we will call eating behavior), then after some time, it leaves the resource and approaches the other resource for a similar eating behavior. We will call the behavior in which the agent moves toward and eats resource 1 action 1, and the behavior towards resource 2 we will call action 2. Accomplishing these actions requires the engagement of different chemosensors that allows for appropriate, directed motor behavior. Somehow, between the two approaches, these different coordination patterns come into effect.

We can attempt to localize action 1 and action 2 by observing and experimenting on the agent's state during the different actions. This helps identify the relevant vari-

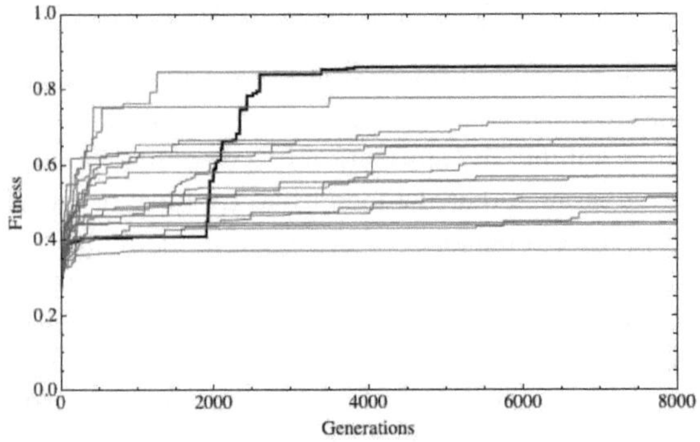

Fig. 10.4 Twenty evolutionary runs, with the top performing evolutionary run in black

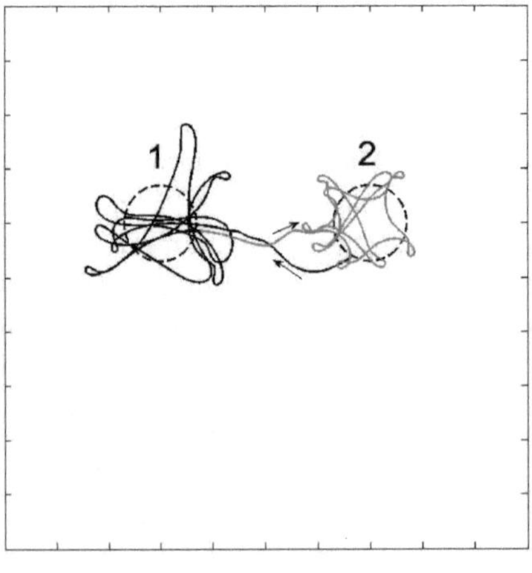

Fig. 10.5 A typical behavioral trajectory (solid line) between the two resources (outline by the dashed line)

ables for each action. Part of the trick in this agent's evolution was the introduction of nutrient sensors, which provide input to the interneurons about the two nutrient levels. It is reasonable that the evolutionary search of the GA would take advantage of this information to determine action, and this agent did just that. Figures 10.6 and 10.7 demonstrate how action is dependent on nutrient levels, by showing the behavioral effects of nutrient level manipulation. In Figure 10.6(A), nutrient 1 is held fixed at the near-empty value of 0.5 (out of 10) and nutrient 2 is held full at a value of 10. The agent is shown to approach resource 1, as if it was hungry for the resource, and remains eating the resource for the entire extent of the simulation. Figure 10.6 (B) shows the exact opposite scenario, with nutrient 1 held at 10 and nutrient 2 held at 0.5. We again clearly see that the agent approaches resource 2 and continues to engage in eating behavior for the simulation's full duration.

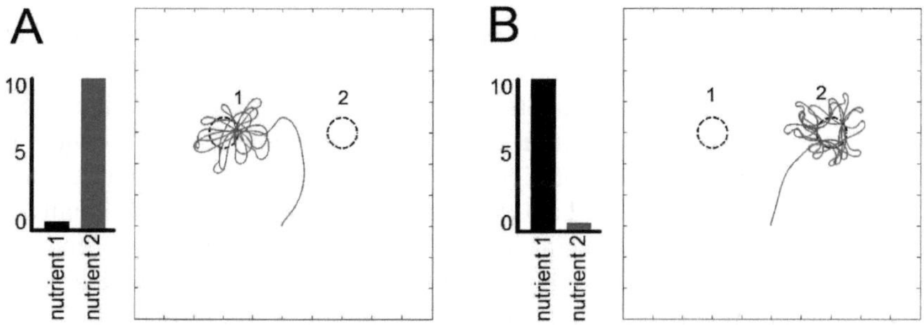

Fig. 10.6 A) Action 1, the agent moves to resource 1 and engages in eating behavior for the duration of the simulation. B) Action 2, the agent moves to resource 2 and engages in eating behavior for the duration of the simulation.

These demonstrate that the nutrient level plays a role in determining which action the agent is engaged in. But in the reality of a simulation, nutrient levels are constantly increased by eating or decreased by metabolism, such that these idealized actions shown in Figures 10.6 (A) and (B) can never happen. The agent passes through intermediate nutrient states that generate different behavioral tendencies. Figure 10.7(A) shows the behavior when both nutrient levels were held really high, which looks like the agent is not attracted to either resource but instead explores the full environment. Figure 10.7(B) shows behavior when both nutrient sensors are held at a low level, in which the agent moves around in tight circles. These figures also show behaviors that can never be achieved in simulation, because both nutrient levels can never be completely full at the same time, and can't be equally empty. Even though these behaviors are not realistic in a typical environment, they provide interesting insight into the dynamical properties of the system. We gain an understanding of the agent's behavior as ongoing transitions between more explorative modes when nutrient levels are higher, to more exploitative mode when nutrient levels are low. Additionally, off-balanced nutrient levels lead to action directed towards one of the two resources.

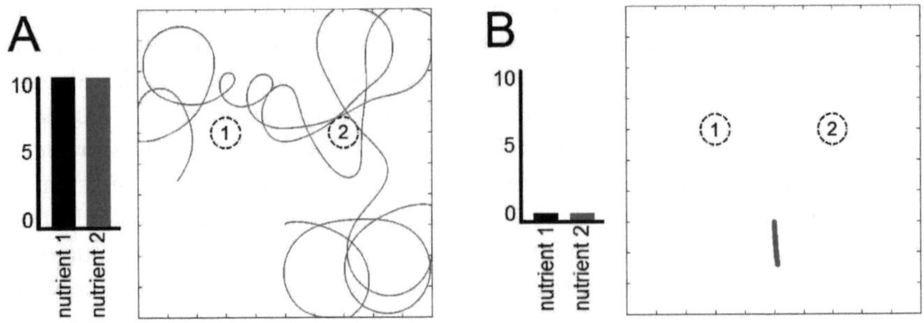

Fig. 10.7 A) The agent explores the environment without approaching either resource. B) The agent moves around in tight circles, while slowly drifting downwards.

Both actions 1 and 2 were further investigated to reveal how sensorimotor interactions are determined by the system, and how this generates the resulting observed behavior. We will focus our explanation on actions 1 by describing Figure 10.8, an identical analysis was done on action 2 and produced similar results, shown in Figure 10.9. We use the constraints of Figure 10.6(A) to approximate the nutrient levels for action 1, with nutrient 1 held at 0.5 and nutrient 2 held at 10. Additionally, it is assumed that the resource 2's chemical trace is irrelevant during action 1, and so the chemosensors for resource 2 are both held fixed at 0. We then calculated the system's equilibrium points as the left and right chemosensors for resource 1 ($C1_L$, and $C1_R$) are varied through their range of states. As it turns out, the system has only one stable equilibrium point for this range chemosensor levels. In Figure 10.8, this equilibrium point is projected onto the left and right motor neurons' outputs (M_L and M_R), which shows the outputs that the motor neurons would tend towards given the chemosensor values were held fixed.

For a reminder of how the total force on the agent is determined: the angular force is determined by the difference between the two outputs. This leads to a slightly counter-intuitive relation between effectors and behavioral output such that when there is more force coming from right motor neuron, the agent turns left. When there is more force coming from the left motor neuron, the agent turns right. When there are equal amounts of force, the agent moves forward with a thrust determined by the effector's added outputs. There is an additional friction constant that reduces the velocity at each time step. With this quick summary, we can interpret the motor surfaces of Figure 10.8. These show that when the right chemosensor is more active than the left chemosensor, the right motor becomes less active than the left motor. This turns the agent to the right. When the left chemosensor is less active than the right, there is tendency to turn left. When both chemosensors are about equally active, both motor neurons are also about equally active, which drives the agent forward. With every movement, the agent's chemosensors move, and elicit new inputs, which through ongoing feedback moves the agent successfully to the resource. These surfaces fully describe the strategy used by the agent, which successfully brings it to resource 1 as shown in Figure 10.6(A).

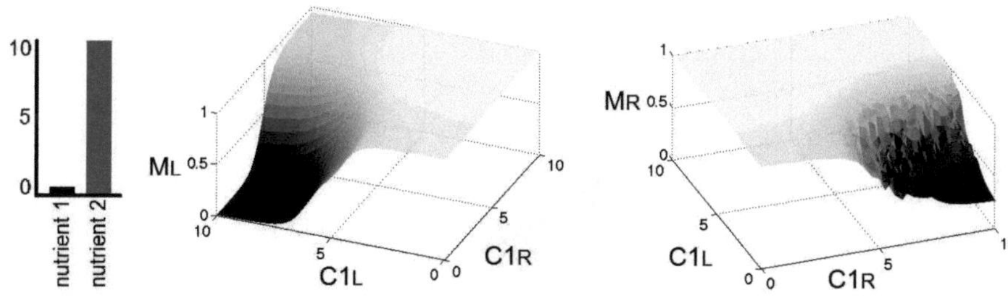

Fig. 10.8 Individual motor projections of the system's single equilibrium point for action1 (shown in Figure 10.6(A). These surfaces are functions of the two chemosensors, $C1_L$ and $C1_R$, which when held at a particular state produce the stable motor outputs in effectors M_L and M_R. Action 2, shown in Figure 10.9, but not identical motor projections.

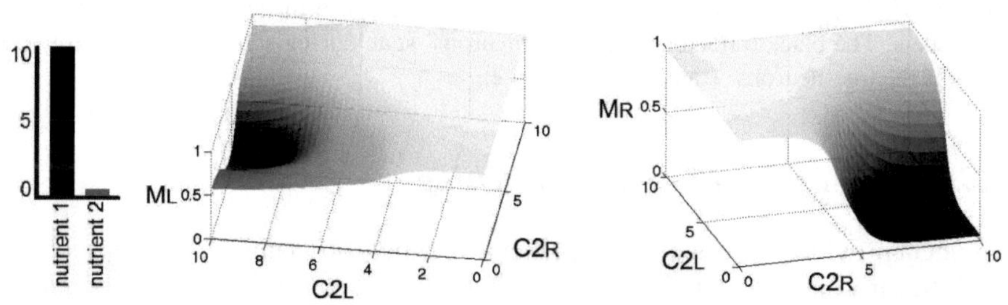

Fig. 10.9 Individual motor projections of the system's single equilibrium point for action2 (shown in Figure 10.6(B). These surfaces are functions of the two chemosensors, $C2_L$ and $C2_R$, which when held at a particular state produce the stable motor outputs in effectors M_L and M_R.

We now turn our attention to Figure 10.10, which examines the agent's internal states throughout a full simulation of both actions without any imposed constraints. This aims to provide insight into how the internal state behaves in each of the two actions and how it transitions between them. To produce this image, the states of the six interneurons are recorded throughout the same simulation of behavior previously shown in Figure 10.5. There are 6 interneurons, and so their state makes a 6-dimensional state space, which can, for obvious reasons, not be visualized in its completeness here. Instead, a principal component analysis (PCA) is performed on the data, and the top three principal components are identified to make up the axes of Figure 10.5. These components account for 92 percent of the variance of the full interneuron space. The data is transformed into the coordinates defined by these principle components, and projected into the 3-dimensional space. In this plot, the interneurons trajectories throughout two actions are colored in different colors to demonstrate the clear separation of the actions that takes place in the interneuron's activity. This separation should not be interpreted as a universal rule of action switching, but it is certainly pronounced in this particular case.

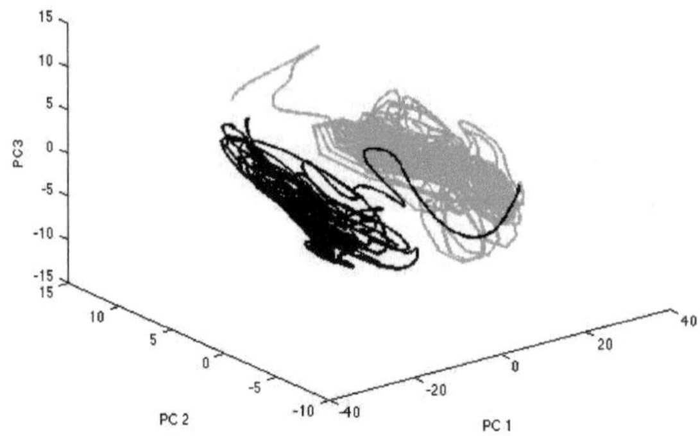

Fig. 10.10 This shows the trajectory of interneuron states throughout the two actions shown in Figure 10.5., projected onto a dimensionally reduced space made of the top 3 principle components. The black trajectory is the interneurons' state during action 1 and the gray trajectory is the interneurons' state during action 2.

10.4.3 Discussion of Action Switching

This agent provides a proof of concept that a distributed dynamical system, such as a CTRNN, can control a body in multiple different directed actions and autonomously switch between them. There is no higher-level mechanism imposed, and all the observed actions exist within the same defined absolute system. Though it is a minimal case, it can help set a basis from which dynamical systems terminology can explain more complex examples of action switching.

We gain an understanding of the system as existing in a high-dimensional state space, with specific actions resulting from the agent's state and coupled interactions with an environmental situation. The agent demonstrates that specific actions appear as reductions in the state space, in which some temporarily irrelevant variables can be removed from the analysis. The actions are approximated through projections of the high-dimensional state space onto a reduced subspace made of the temporarily relevant variables, such as $C1_L$, and $C1_R$ left during action 1. But with our analysis of this agent we recognize that these reductions are approximations, and the real system actually exists in the higher-dimensional state space that contains all variables. Action switching is movement through the higher-dimensional space, between these actions' approximated subspaces.

When analyzing how an agent traverses the state space between actions, the modeler needs to demonstrate how a trajectory moves from one projection to another, as when action 1's sensorimotor surfaces, shown in Figure 10.8, would transition to action 2's surfaces, shown in Figure 10.9, which are defined by different variables. This requires movement in more dimensions than we can illustrate directly on paper, and will require different kinds of analyses. In this agent we see the transitions in Figure 10.10, though this image does not illustrate the dynamical landscape, such as equilibrium points, which would drive the system from one action to another. A more complete description would show the dynamical properties, such as a bifurcation or moving equilibrium point. A more complete picture is provided in a later article (Agmon and Beer 2013).

10.5 The Prospect of Brain-Body-Environment Systems

There is much left to develop and discover in this sapling scientific field; what we have covered in this chapter only provides minimal depictions of what in reality are very complex phenomena. If it is true that we need to take an absolute system perspective by modeling all of the relevant variables underlying adaptive behavior, then the current dynamical system models will need to scale up many orders of magnitude if we are to describe the behavior capacity of living systems. Neuronal, biological, and ecological systems have many relevant variables that come into play in the production of an individual organism's behavior. The actions performed by simulated agents will need to diversify and complexify if they are going to describe these behaviors as seen in reality.

Real repertoires of action are typically much less symmetric than in this paper's example; whereas the agent studied here directed its behaviors towards two resources in functionally very similar ways, real organisms can engage in very qualitatively different types of actions such as reaching, peeling rind off of a fruit, chewing, or fighting an opponent. These actions are coordinated in different environmental situations, by utilizing different sets of sensors and effectors. In dynamical models of such actions, the agent will need to coordinate many degrees of freedom in morphologies that have an increased number of bones, muscles and joints. They will need to gather information about context by integrating different sources of sensory input. Environments will need to expand to include more entities with complex properties that an agent can engage with.

Not only will models of brain-body-environment systems need to scale up to more complex structures and behavior, they will also need to demonstrate adaptivity. Adaptivity is a system's ability to structurally reconfigure itself to behave in increasingly advantageous ways (Di Paolo 2005). As we know from the section on Ashby's self-organization, the change from bad organization to good organization cannot come out of the system's internal organization. Instead we must ask how an adaptive system's robustness increases through experience within the environment. Here, robustness is defined as the maintenance of essential variables (as introduced in the earlier section on Ashby's self-organization). A highly robust system maintains its essential variables far from their boundary of viability, and by doing so reduces its chance of failure or death. When at one time an adaptive system might have faired poorly and closely approached possible failure, at a later time when confronted with the same environmental situation, the system behaves in a more efficient and robust manner.

The model presented in this paper has a static behavioral repertoire, which is given to the agent at its inception by the fixing of dynamical parameters.

Additionally, its body and environment are determined a priori and are unchanged for the duration of simulation. But actions seen in living systems are often acquired through learning and development, during which the environment and body change. Organisms tune their actions and come up with new actions that engage sensors and effectors in novel ways to optimize their interactions. With the absolute systems perspective on BBE systems, learning and development become a type of self-organization that unfolds through interactions across many timescales. Future agents will need to demonstrate this capacity, and their analysis will need to uncover the underlying structures that allow for such abilities.

All of these complexifications will yield new types of behaviors and dynamics that have not yet been described with the methodology introduced here. The field is facing an explanatory gap that it must overcome if it is to provide scientific insight about the adaptive capabilities of life. By starting with minimal models, it has promised to lay a foundation that can be incrementally built up towards more realistic behavior. Minimization has allowed simple behaviors to be analyzed, and their dynamics fully unpacked. But there is no certainty that the types of analyses used in the minimal cases will transfer to more complex instances. The field will need to develop whole new approaches for grappling with these complexifications. It will need new methods for evolving models of behavior that demonstrate diversity, complexity and adaptivity, and new methodologies to analyze their structure and dynamics.

References

Agmon, E., Beer, R.D.: The dynamics of action switching in an evolved agent. Manuscript submitted for Publication (2013)

Ashby, W.R.: Design for a brain. Wiley (1952)

Ashby, W.R.: Principles of the self-organizing system. Principles of Self-organization, pp. 255–278 (1962)

Beer, R.D.: Computational and dynamical languages for autonomous agents. In: Mind as motion: Explorations in the Dynamics of Cognition, pp. 121–147 (1995a)

Beer, R.D.: A dynamical systems perspective on agent-environment interaction. Artificial Intelligence 72(1), 173–215 (1995b)

Beer, R.D.: The dynamics of adaptive behavior: A research program. Robotics and Autonomous Systems 20(2), 257–289 (1997)

Beer, R.D.: The dynamics of active categorical perception in an evolved model agent. Adaptive Behavior 11(4), 209–243 (2003)

Beer, R.D.: Dynamical systems and embedded cognition. In: The Cambridge Handbook of Artificial Intelligence (2007) (to appear)

Beer, R.D., et al.: Toward the evolution of dynamical neural networks for minimally cognitive behavior. From Animals to Animats 4, 421–429 (1996)

Beer, R.D., Gallagher, J.C.: Evolving dynamical neural networks for adaptive behavior. Adaptive Behavior 1(1), 91–122 (1992)

Brooks, R.A.: Intelligence without representation. Artificial Intelligence 47(1), 139–159 (1991)

Buckley, C., Fine, P., Bullock, S., Di Paolo, E.: Monostable controllers for adaptive behaviour. From Animals to Animats 10, 103–112 (2008)
Clancey, W.J.: Situated cognition: On human knowledge and computer representations. Cambridge University Press (1997)
Clark, A.: Being there: Putting brain, body, and world together again. MIT Press (1998)
Clark, A., Chalmers, D.: The extended mind. Analysis, 7–19 (1998)
Cliff, D., Husbands, P., Harvey, I.: Explorations in evolutionary robotics. Adaptive Behavior 2(1), 73–110 (1993)
Di Paolo, E.A.: Autopoiesis, adaptivity, teleology, agency. Phenomenology and the Cognitive Sciences 4(4), 429–452 (2005)
Dupuy, J.P.: On the origins of cognitive science. MIT Press (2009)
Elman, J.L.: Language as a dynamical system. In: Mind as Motion: Explorations in the Dynamics of Cognition, pp. 195–223 (1995)
Funahashi, K.-I., Nakamura, Y.: Approximation of dynamical systems by continuous time recurrent neural networks. Neural Networks 6(6), 801–806 (1993)
Gibson, J.J.: The ecological approach to visual perception. Lawrence Erlbaum (1986)
Heidegger, M.: Being and time. Trans. John Macquarrie and Edward Robinson. Harper, New York (1927, 1962)
Hutchins, E.: How a cockpit remembers its speeds. Cognitive Science 19(3), 265–288 (1995)
Hutchins, E.: Distributed cognition. In: Internacional Enciclopedia of the Social and Behavioral Sciences (2000)
Izhikevich, E.M.: Dynamical systems in neuroscience: the geometry of excitability and bursting. MIT Press (2006)
Izquierdo, E., Buhrmann, T.: Analysis of a dynamical recurrent neural network evolved for two qualitatively different tasks: walking and chemotaxis. In: Artificial life XI: Proceedings of the Eleventh International Conference on the Simulation and Synthesis of Living Systems, pp. 257–264. MIT Press, Cambridge (2008)
Izquierdo, E., Harvey, I., Beer, R.D.: Associative learning on a continuum in evolved dynamical neural networks. Adaptive Behavior 16(6), 361–384 (2008)
Kelso, J.S.: Dynamic patterns: The self-organization of brain and behavior. MIT Press (1995)
Lakoff, G.J., Johnson, M.: Philosophy in the Flesh. The Embodied Mind and Its Challenge to Western Thought. Basic Books, New York (1999)
Maturana, H.R., Varela, F.J.: Autopoiesis and cognition: The realization of the living, vol. 42. Springer (1980)
Merleau-Ponty, M.: Phenomenology of perception. Motilal Banarsidass Publishe (1996)
Noė, A.: Action in perception. MIT Press (2005)
Nolfi, S., Floreano, D., Floreano, D.: Evolutionary robotics: The biology, intelligence, and technology of self-organizing machines. MIT Press, Cambridge (2000)
Pfeifer, R., Scheier, C.: Understanding intelligence. MIT Press (2001)
Polani, D.: Foundations and formalizations of self-organization. Advances in Applied Self-Organizing Systems, 19–37 (2008)
Port, R.F., van Gelder, T.: Mind as motion: Explorations in the dynamics of cognition. MIT Press (1995)
Skarda, C.A., Freeman, W.J.: How brains make chaos in order to make sense of the world. Behavioral and Brain Sciences 10(2), 161–195 (1987)
Slocum, A.C., Downey, D.C., Beer, R.D.: Further experiments in the evolution of minimally cognitive behavior: From perceiving affordances to selective attention. In: From Animals to Animats 6: Proceedings of the Sixth International Conference on Simulation of Adaptive Behavior, pp. 430–439 (2000)

Spivey, M.J., Grosjean, M., Knoblich, G.: Continuous attraction toward phonological competitors. Proceedings of the National Academy of Sciences of the United States of America 102(29), 10393–10398 (2005)

Thelen, E., Smith, L.B.: A dynamic systems approach to the development of cognition and action. MIT Press (1996)

Thompson, E.: Mind in life: Biology, phenomenology, and the sciences of mind. Belknap Press (2007)

Tinbergen, N.: On aims and methods of ethology. Zeitschrift für Tierpsychologie 20(4), 410–433 (1963)

Turvey, M.T.: Coordination. American Psychologist 45(8), 938 (1990)

Van Gelder, T.: The dynamical hypothesis in cognitive science. Behavioral and Brain Sciences 21(05), 615–628 (1998)

Varela, F.J., Thompson, E.T., Rosch, E.: The embodied mind: Cognitive science and human experience. MIT Press (1992)

Williams, P.L., Beer, R.D., Gasser, M.: An embodied dynamical approach to relational categorization. In: Proceedings of the 30th Annual Conference of the Cognitive Science Society, pp. 223–228 (2008a)

Williams, P.L., Beer, R.D., Gasser, M.: Evolving referential communication in embodied dynamical agents. In: ALife XI: Proceedings of the Eleventh International Conference on Artificial Life, pp. 702–709 (2008b)

Chapter 11
Guided Self-Organization of Input-Driven Recurrent Neural Networks*

Oliver Obst and Joschka Boedecker

11.1 Introduction

To understand the world around us, our brains solve a variety of tasks. One of the crucial functions of a brain is to make predictions of what will happen next, or in the near future. This ability helps us to anticipate upcoming events and plan our reactions to them in advance. To make these predictions, past information needs to be stored, transformed or used otherwise. How exactly the brain achieves this information processing is far from clear and under heavy investigation. To guide this extraordinary research effort, neuroscientists increasingly look for theoretical frameworks that could help explain the data recorded from the brain, and to make the enormous task more manageable. This is evident, for instance, through the funding of the billion-dollar "Human Brain Project", of the European Union, amongst others. Mathematical techniques from graph and information theory, control theory, dynamical and complex systems (Sporns 2011), statistical mechanics (Rolls and Deco 2010), as well as machine learning and computer vision (Seung 2012; Hawkins and Blakeslee 2004), have provided new insights into brain structure and possible function, and continue to generate new hypotheses for future research.

A marked feature of brain networks is the massive amount of recurrent connections between cortical areas, especially on a local scale (Douglas et al. 2004). Since information in these recurrent connections, or loops, can circulate between many neurons in a given circuit, they are ideally suited to provide a time-context for computations leading to predictions about future events. One particular mathematical

Oliver Obst
CSIRO Computational Informatics PO Box 76, Epping, NSW 1710, Australia
e-mail: oliver.obst@csiro.au

Joschka Boedecker
Machine Learning Lab, University of Freiburg, Germany

* Both authors contributed equally to this work.

M. Prokopenko (ed.), *Guided Self-Organization: Inception*,
Emergence, Complexity and Computation 9,
DOI: 10.1007/978-3-642-53734-9_11, © Springer-Verlag Berlin Heidelberg 2014

model that is used to investigate the consequences of loops for computation and optimization in neuronal circuits are *recurrent neural networks* (RNNs).

In RNNs, many detailed properties of real neurons are abstracted for the sake of tractability, but important general concepts are kept. Elements of these networks are simple nodes that combine inputs from other nodes in the network, usually in a non-linear fashion, to form their outputs. They are connected in a directed graph, which may contain cycles. In *input-driven* RNNs, a constant stream of input data drives the dynamics of the network. Dedicated output units can then use this dynamics to compute desired functions, for instance for prediction or classification tasks. Since they can make use of the temporal context provided by the recurrent connections, RNNs are very well suited for time-series processing, and are in principle able to approximate any dynamical system (Maass et al. 2007).

While the recurrent connections of RNNs enable them to deal with time-dependencies in the input data, they also complicate training procedures compared to algorithms for networks without loops (e.g., Backpropagation (Rumelhart et al. 1986) or R-Prop (Riedmiller and Braun 1993)). Notably, training RNNs with traditional training methods suffer from problems like slow convergence and vanishing gradients. This slow convergence is due to the computational complexity of the algorithms training all of the connections in a network (such as BPTT (Werbos 1990) or RTRL (Williams and Zipser 1989)), as well as to bifurcations of network dynamics during training, which can render gradient information unusable (Doya 1992). Also, derivatives of loss functions need to be propagated over many time steps, which leads to a vanishing error signal (Bengio et al. 1994; Hochreiter 1998).

The realization of these fundamental issues led to alternative ways of using and training RNNs, some of which can be summarized in the field of *Reservoir Computing* methods (see, e.g., a recent overview by Lukovsevivcius and Jaeger 2009), specialized architectures like the Long Short Term Memory (LSTM) networks (Hochreiter and Schmidhuber 1997) or training by evolutionary algorithms as in the *Evolino* approach (Schmidhuber et al. 2007). The Reservoir Computing field has been an active part of RNN research over the last decade, while there was less activity in gradient-descent-like methods which appear to generate renewed interest only recently (Bengio et al. 2012), partially due to the development of more efficient training techniques as in (Martens and Sutskever 2011).

Reservoir methods implement a fixed high-dimensional reservoir of neurons, using random connection weights between the hidden units, chosen small enough to guarantee dynamic stability. Input weights into this reservoir are also selected randomly, and reservoir learning procedures train only the output weights of the network to generate target outputs. A particular appeal of reservoir methods is their simplicity, and that the computation required for training is relatively low.

Taking the echo state network approach as a specific example of a typical reservoir network (see Fig. 11.1), it will consist of the following components: A random input-matrix $\mathbf{W}_{in}$ combines input values $\mathbf{u}$ linearly and sends them to the units in the high-dimensional hidden layer, referred to as the *reservoir*. The units in the reservoir also have recurrent connections amongst each other, collected in the matrix $\mathbf{W}_{\text{res}}$. These loops implement a fading memory, so information can remain in the

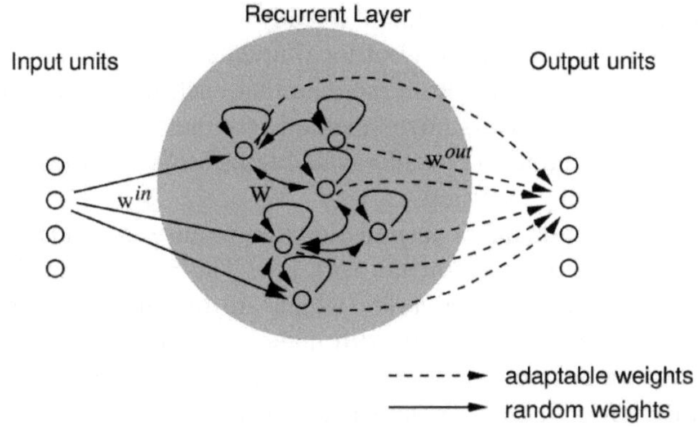

Fig. 11.1 The architecture of a typical Echo State Network (ESN), which belongs to the class of reservoir computing networks. In ESNs, the input and recurrent hidden layer (reservoir) connections are fixed randomly, and only output weights are trained. The reservoir projects the input stream nonlinearly into a high-dimensional representation, which can then be used by a linear readout layer. An important precondition for the approach to work is that the reservoir implements a fading memory, i.e. that reservoir states do not amplify, but fade out over time if no input is presented to the network.

system for some time. In this context, the metaphor of a reservoir is often used since the hidden layer can be seen as a water reservoir that gets disturbed by a drop, but slowly returns to its initial state after the ripples from the input have decayed. This reservoir state $\mathbf{x}$ is mapped at time step $t+1$ by an activation function $f()$ such as a hyperbolic tangent, in the following way:

$$\mathbf{x}_{t+1} = f(\mathbf{W}_{\text{res}} * \mathbf{x}_t + \mathbf{W}_{\text{in}} * \mathbf{u}_{t+1}) \tag{11.1}$$

The input and hidden layer connections, $\mathbf{W}_{\text{in}}$ and $\mathbf{W}_{\text{res}}$, are not trained in reservoir computing approaches. It is also possible to introduce feedback connections from outputs back into the reservoir (Lukovsevivcius and Jaeger 2009). To approximate a specific target function, only the output weights $\mathbf{w}_{\text{out}}$ are trained with a simple linear regression. This drastically simplifies the training procedure compared to previous approaches, while leading to excellent performance on time-series processing tasks (Jaeger and Haas 2004). It also avoids the problems of vanishing gradient information and disruptive bifurcations in the dynamics since no error gradients have to be propagated into the fixed, random parts of the network.

This approach works very well in practice. However, results will depend on the particular random set of weights that is drawn. In fact, there is considerable variability in performance between runs of networks with equal parameter settings, but different reservoir weights drawn each time (Ozturk et al. 2007). Striking a balance between the two extremes of fully trained RNNs and reservoir methods, it is interesting to retain some of the simplicity and efficiency of reservoir methods, but at the same time avoid some of the variability that comes with randomly created reservoirs. Self-organized methods are of interest here, because the initial random configuration of the reservoir is in general already useful to perform the task. Each unit or connection then could, by way of local updates, contribute to an improved version of the reservoir, dependent on the data that each unit or weight processes over time. Advantages of self-organized methods are their potential for scalability, since they usually rely mainly on locally available information, making them good candidates for distributed processing.

Driving self-organization into a desired direction requires an understanding what properties a good RNN or reservoir network has. The mathematical tools to understand computation in these networks (which are instances of the larger class of input-driven dynamical systems) are still under active development (Manjunath et al. 2012). However, different perspectives, e.g., from functional analysis, dynamical systems, information theory, or statistical learning theory already offer insights towards this goal. They can provide answers to questions such as: how well can a given RNN approximate a given class of functions? How does it implement a certain function within the collective of its distributed nodes? How much memory does an RNN provide about past inputs and where is this information stored? How does information flow through the system at different time points in time and space? How well can it separate different inputs into different network states, and how well will it generalize on data that has not been seen during training? All of these aspects contribute to the successful performance of a network on a given task (or class of tasks). Understanding how to improve them will provide possible target signals to enrich and guide the self-organized optimization process of an RNN.

In this chapter, we review attempts that have been made towards understanding the computational properties and mechanisms of input-driven dynamical systems like RNNs, and reservoir computing networks in particular. We provide details on methods that have been developed to give quantitative answers to the questions above. Following this, we show how self-organization may be used to improve reservoirs for better performance, in some cases guided by the measures presented before. We also present a possible way to quantify task performance using an information-theoretic approach, and finally discuss promising future directions aimed at a better understanding of how these systems perform their computations and how to best guide self-organized processes for their optimization.

11.2 Assessing the Computational Power and Mechanisms of Information Processing of Reservoirs

In many cases, artificial neural networks are created with a specific goal in mind, for example to approximate a particular function or system. Training success and computational capability of the network with respect to this task are usually assessed on data that have not been used for training. Similarly to the training data, these are expected to match properties of the (yet unknown) application data well enough. A loss functional like the mean square error (MSE) or the cross-entropy is used to assess the quality of the trained system. For specific applications of the network, this is a standard approach that usually delivers meaningful results. When a neural network is trained for a single purpose, it is not necessary to determine its general computational power, and the loss on the validation data or during its application may be the only relevant property.

The loss on a specific class of problems does not express the general computational power of the network, though. This property becomes more interesting when a part of the system is used for more than one task: relevant cases would be dynamical reservoirs that are used for multiple applications, networks that are trained "online" when the task changes, or to set up or to compare generic microcircuits. One of our motivations to evaluate mechanisms of information processing is to compare self-organized approaches within reservoirs. Ideally, self-organization leads to measurable effects in the reservoir which positively affect the performance of the system. In this section, we present a number of measures for different qualities of dynamical systems that are useful in this evaluation. These measures can be roughly divided into approaches that are based on or related to information theory, approaches that relate to learning theory, and dynamical systems theory.

11.2.1 Information-Theory Related Measures

Information theory and (Shannon) entropy have been used in a number of ways in neural network and complex systems research. One particular heuristic to measure (and eventually improve) RNN is to estimate and influence the entropy distribution of firing rates of a neuron. In individual biological neurons, for example, an approximate exponential distribution of the firing rate has been observed in visual cortical neurons (Baddeley et al. 1997). Under the constraint of a fixed energy budget, i.e., a fixed mean firing rate, an exponential distribution maximizes the potentially available information: it is the maximum entropy distribution for positive random values with a given mean. Triesch (2005) uses this idea to adapt the intrinsic excitability of a neuron with an online adaption. In this approach, the Kullback-Leibler divergence is used to measure the difference between the sample distribution of an individual neuron's output, and the exponential distribution. Target distributions different from the exponential distribution are plausible dependent on specific circumstances. For example, in reservoir networks with real-valued units, normal distributions have been used (Schrauwen et al. 2008) to reflect the use of both negative and positive values. In both cases, the mechanism attempts to maximize the information per available energy unit locally at each neuron. Since energy constraints in reservoirs of artificial neural networks are typically not an issue, the maximum entropy distribution for these would in fact be the uniform. Without an energy constraint, the approach resembles the Infomax principle (Linsker 1987), where the average mutual information between input and output is maximized. As Bell and Sejnowski point out in their approach to maximize the mutual information for non-linear units (Bell and Sejnowski 1995), for invertible continuous deterministic mappings this mutual information is maximized by maximizing the output entropy alone. Due to the limited degrees of freedom of the approach, the desired target distribution cannot be approximated for every kind of input (Boedecker et al. 2009). Intrinsic plasticity as well as its particular limitation can be related to Ashby's law of requisite variety (Ashby 1956) in that by increasing variety available in the reservoir a larger variety of outputs can be successfully approximated on one hand. On the other hand, the lack of variety in the mechanism adapting the individual neurons is also responsible for the difficulty in increasing the entropy for a variety of inputs.

The field of information dynamics (Lizier et al. 2007, 2012) provides information-theoretic measures that explicitly deal with processes or time-series. *Information storage*, as one of the tools, quantifies how much of the stored information is actually in use at the next time step when the next process value is computed. $A(X)$ is expressed as the mutual information between the semi-infinite past of the process X and its next state X', with $X^{(k)}$ denoting the last k states of that process:

$$A(X) = \lim_{k\to\infty} A^{(k)}(X) \tag{11.2}$$

$$A(X,k) = I(X^{(k)};X') \tag{11.3}$$

Information transfer, expressed as transfer entropy (Schreiber 2000), quantifies the influence of another process on the next state (for a formal definition, see Sect. 11.3.3 below). Boedecker et al. (2011) use these measures, to gain a better understanding of computational mechanisms inside the reservoir, and how they increase task performance for certain tasks at the phase transition between ordered and unstable dynamics. For a rote-memory task, a sharp peak can be observed in both information storage and information transfer near this phase transition, and suggests that a maximized capacity for information storage and information transfer correlates well with task performance in this dynamics regime, and this particular task. Prokopenko et al. (2011) suggest the Fisher information matrix as a way to detect phase transitions, but this has, to our knowledge, not been applied to RNN yet.

Fisher information also plays a role in quantifying the memory stored in a dynamical reservoir: Information about the recent input in reservoir networks is stored in the transient dynamics, rather than in attractor states. To quantify this information storage, Ganguli et al. (2008) use Fisher information as basis for a measure of memory traces in networks. The measure is applicable for systems with linear activations $\mathbf{f}(\mathbf{x}) = \mathbf{x}$, subject to Gaussian noise $\mathbf{z}$, and input $v(t)$:

$$\mathbf{x}(t) = \mathbf{f}(\mathbf{W}^{\text{in}} v(t) + \mathbf{W}\mathbf{x}(t-1) + \mathbf{z}(t)). \tag{11.4}$$

The Fisher Memory Matrix (FMM) between the present state of the system $\mathbf{x}$ and the past signal is defined as

$$J_{k,l}(\mathbf{v}) = \left\langle -\frac{\delta^2}{\delta_{\mathbf{v_k}} \delta_{\mathbf{s_l}}} \log P(\mathbf{x}(t)|\mathbf{v}) \right\rangle_{P(\mathbf{x}(t)|\mathbf{v})}. \tag{11.5}$$

Diagonal elements $J(k) \equiv \mathbf{J}_{k,k}$ are the Fisher information that the system keeps in $\mathbf{x}(t)$ about a pulse at k steps back in the past, i.e., the decay of the memory trace of a past input. $J(k)$ is called the Fisher memory curve (FMC). Tino and Rodan (2013) investigate the relation between $J(k)$ and the short term memory capacity

MC (Jaeger 2001) (details on MC in the following subsection), and show that the two are closely related in linear systems. For these, $J(k)$ is independent of the actual input used. In the general, nonlinear case that is interesting for us, however, the FMC depends on the input signal, as the memory capacity MC does, and is hard to analyze.

A measure for Active Information Storage in input-driven systems has been proposed to quantify storage capabilities of a nonlinear system independent of particular inputs (Obst et al. 2013). The measure is an Active Information Storage (Lizier et al. 2012) where the current input u_{n+1} is conditioned out:

$$A_X^U(k) = \langle a_X^U(n+1,k) \rangle_n \text{, with} \tag{11.6}$$

$$a_X^U(n+1,k) = \log \frac{p(x_n^{(k)}, x_{n+1} | u_{n+1})}{p(x_n^{(k)})\, p(x_{n+1} | u_{n+1})} \tag{11.7}$$

$$= \log \frac{p(x_{n+1} | x_n^{(k)}, u_{n+1})}{p(x_{n+1} | u_{n+1})}. \tag{11.8}$$

The idea for this measure is to remove influences of structure in input data, and to only characterize the system itself, rather than a combination of system and input data. In theory, this influence would be removed by having the history sizes in computing the information storage converge to infinity. Large history sizes, however, require large amounts of data to estimate the involved joint probabilities, and this data, and the time required for the estimation is often not available. Active information Storage for input-driven systems assesses one aspect of the computational capabilities of a dynamical system, others, like the information transfer, would need to be defined for input-driven systems in a similar way.

11.2.2 Measures Related to Learning Theory

Legenstein and Maass (2007) propose two measures to quantify the computational capabilities of reservoirs in the context of liquid state machines, one of the two main flavors of reservoir computing networks: the linear separation property and the generalization capability. The linear separation property quantifies the ability of a computational system to map different input streams to significantly different internal states. This is useful because only then will the system be able to (linearly) map internal states to different outputs. The measure is based on the rank of an $n \times m$ matrix M whose columns are state vectors $\mathbf{x}_{u_i}(t_0)$ of circuit C after having been driven by input stream u_i up to a fixed time t_0. These state vectors are collected for m different input streams, i.e., $u_1, \ldots, u_m$. If the rank of M is m, then C, together with a linear readout, is able to implement any assignment of output units $y_i \in \mathbb{R}$ at time t_0 for inputs u_i.

For the generalization ability, they propose to approximate the VC-dimension of class $\mathcal{H}_C$ of the reservoir, which includes all maps from a set S_{univ} of inputs u into $\{0,1\}$ which can be implemented by a reservoir C. They present a theorem (and corresponding proof sketch) stating that under the assumption that S_{univ} is finite and contains s inputs, the rank r of a $n \times s$ matrix whose columns are the state vectors $\mathbf{x}_u(t_0)$ for all inputs u in S_{univ} approximates the VC-dimension($\mathcal{H}_C$), specifically $r \leq \text{VC-dimension}(\mathcal{H}_C) \leq r+1$.

According to (Legenstein and Maass 2007), a simple difference of both (normalized) measures leads to good predictions about which reservoirs perform well on a range of tasks.

The loss or the success on a set of test functions is another possibility to characterize the systems from a learning point of view. One such measure is the short term memory capacity MC (Jaeger 2001) that we briefly mentioned above. To compute the MC, a network is trained to generate delayed versions $v(t-k)$ of a single channel input $v(k)$. The measure then is the sum of the precisions for all possible delays, expressed as a correlation coefficient:

$$\text{MC} = \sum_{k=1}^{\infty} \text{MC}_k \tag{11.9}$$

$$\text{MC}_k = \max_{\mathbf{w}_k^{\text{out}}} \frac{\text{cov}^2(v(t-k), y_k(t))}{\sigma^2(v(t))\,\sigma^2(y_k(t))}, \text{ with} \tag{11.10}$$

$$y_k(t) = \mathbf{w}_k^{\text{out}} \begin{pmatrix} v(t) \\ \mathbf{x}(t) \end{pmatrix}, \text{ and } \mathbf{x}(t) = \mathbf{f}(\mathbf{W}^{\text{in}} v(t) + \mathbf{W}\mathbf{x}(t-1)).$$

The symbols cov and σ^2 denote covariance and variance, respectively. Each coefficient takes values between 0 and 1, and expresses how much of the variance in one signal is explainable by the other. As shown in (Jaeger 2001), for i.i.d. input and linear output units, the MC of N-unit RNN is bounded by N. The measure is related to the Fisher memory matrix approach above.

Another approach in this area is the *information processing capacity* of a dynamical system (Dambre et al. 2012). It is a measure based on the mean square error MSE in reconstructing a set of functions $z(t)$. The idea is to distinguish from approaches that view dynamical systems merely providing some form of memory for a – possibly nonlinear – readout. In (Dambre et al. 2012), systems are regarded as both providing memory and performing nonlinear computation. The readouts then only combine states of the system linearly, attempting to minimize the MSE for a function $z(t)$, so that all essential aspects of computation have to be covered by the dynamical system. The capacity of the system for approximating the desired output is computed using the (normalized) MSE of the optimal linear readout,

$$C_T[X,z] = 1 - \frac{\min_W \text{MSE}_T[\hat{z}]}{\langle z^2 \rangle_T} \tag{11.11}$$

This computed capacity is dependent on the input. In order to avoid an influence of structure in the input on the results, i.i.d. input is required for the purpose of measuring the capacity. To measure information processing capacity, several functions z have to be evaluated. The idea is that if z and z' are orthogonal, $C_T[X,z]$ and $C_T[X,z']$ measure independent properties of the system. The total capacity, on the other hand, is limited by the number of variables x_i, so that a finite number of output functions is sufficient. A possible choice of output functions are Legrende polynomials, which are orthogonal over $(-1,1)$.

The proposed approach has been used to compare different implementations of dynamical systems, like reaction-diffusion systems and reservoirs. An interesting idea that is also mentioned in (Dambre et al. 2012) would be to extend the approach so that the underlying system adapts to provide specific mappings. One possibility might be to adjust the number of internal units in an online-learning setting, e.g., when the task changes. The requirement for i.i.d. input is a limitation of the current approach, though it appears that even in the non-i.i.d. input case useful information about the system can be gathered. It might also be interesting to compare how the approach relates to the information-dynamics framework (Lizier et al. 2007, 2012) to quantify computation in non-linear systems.

11.2.3 Measures Related to Dynamical Systems Theory

To gain understanding of the internal operations that enable high-dimensional RNNs solving a given task, a recent effort by Sussillo and Barak (2013) draws on tools from dynamical systems theory. Using numerical optimization techniques, the authors identify fixed points and points of only gradual change (also called *slow points*) in the dynamics of the networks. Linearization around these points then reveals computational mechanisms like fixed points that implement memories, saddle points (fixed points with one unstable dimension) that enable switching between memory states, and approximate plane attractors that organize the storage of two consecutive input values to be memorized. For the tasks that were looked at in this work, the computational mechanisms could be inferred from the linearized dynamics around the set of fixed and slow points, and task performance of the trained networks was well explained.

In (Williams and Beer 2010), the authors argue for a complementary role of dynamical analysis, which involves, e.g., looking at attractors and switching between attractor states, and also an information-theoretic analysis when trying to understand computation in dynamical systems (including input-driven ones – even though the input might simply be considered as part of the environment and is assumed to be distributed uniformly). They evolve agents that are controlled by small continuous-time recurrent neural networks (CTRNNs) and evaluate their behavior in a relational categorization task. This involves keeping a memory about different objects the agent can sense, and reacting with avoid or catch behaviors based on the relation of both objects. Dynamical analysis shows that the state of a specific neuron in the CTRNN is correlated with the size of the first object, and switching on or off a different neuron determines whether the agent catches or avoids the second object. Both features are found to be connected through a saddle-node bifurcation in the CTRNN dynamics whose timing and location depends on properties of the second object. The desire to understand the flow of information through the brain-body-environment system between these events leads the authors to information-theoretic measures unrolled over time (similar to the motivation and approach in (Lizier et al. 2007)). By considering the temporal evolution of measures like conditional mutual information, they are able to measure *information gain* or *information loss* of a state variable at specific time points. Similarly, they can quantify the *specific information* that a state variable carries about a particular stimulus at each time step. The behavior of the agent can then be explained by a sudden gain and then loss of information about object sizes in the first neuron, and then a rapid gain of information about relative size of the objects. In summary, the authors state that the two different ways to look at the computational mechanisms of the RNN differ, but provide coherent and even complementary information on how the agent solves the task that would be difficult to get with either approach alone.

Another approach from dynamical systems theory to understand and predict computational capabilities in RNNs builds on the concept of Lyapunov exponents. Although these concepts are only defined for autonomous dynamical systems, an analogous idea is to introduce a small perturbation into the state of one of two identical networks but not the other, and observe the time evolution of the state difference while the networks are driven with identical input. In case the perturbation fades out, the network is assumed to be in the stable phase of the dynamics. If it amplifies, the network is in the unstable, and possibly, the chaotic dynamics regime. If it approximately persists, the network is arguably at the phase-transition between stable and unstable regimes. Example applications of this approach can be found in (Bertschinger and Natschläger 2004; Boedecker et al. 2011). In (Bertschinger and Natschläger 2004), it was observed that the performance of binary threshold unit RNNs is maximized at this phase-transition for a task that requires memory and nonlinear processing to be solved successfully. This result was later replicated for analogue Echo State Networks in (Boedecker et al. 2011) for a rote-memory task; however, it was also found that some tasks do not benefit from reservoirs at the phase-transition, as observed before in the complex systems literature (e.g., Mitchell et al. 1993).

11.3 Improving Reservoir Information Processing Capabilities through Self-Organized Adaptation

A pragmatic way to evaluate the quality of a reservoir is to train the output, and evaluate it on a training or validation set (Lukovsevivcius 2012a). In most circumstances, training is fast so that a number of hyper-parameter settings can be tested. Lukovsevivcius (2012a) proposes a number of invaluable recipes to reservoir production. The recipes are very helpful for creating a good enough reservoir before output weights are trained. They show up promising directions for exploration, but are intended to be used as a guide rather than hard and fast rules, as some of them are mutually exclusive. The approaches selected for this section are intended to improve the reservoir itself in a self-organized way after it was created or selected. Possibly, this might happen simultaneously in combination with online learning of output weights, or, alternatively, as a self-organized pre-training approach followed by the standard offline output weight training.

11.3.1 SORN: Self-Organized Optimization Based on 3 Local Plasticity Mechanisms

One approach that has demonstrated how self-organization can be leveraged to optimize a reservoir network can be found in (Lazar et al. 2009). SORN is a self-organizing recurrent network architecture using discrete-time binary units. The three plasticity mechanisms are: a variant of spike-time dependent plasticity (STDP), adjusting certain weights in the reservoir, a synaptic normalization rule (SN) responsible to keep the sum of afferent weights of a neuron constant, and intrinsic plasticity (IP) learning to adapt the unit firing threshold. The network state evolves using the following update functions:

$$R_i(t+1) = \sum_{j=1}^{N^E} W_{ij}^{EE}(t)x_j(t) - \sum_{k=1}^{N^I} W_{ik}^{EI} y_k(t) - T_i^E(t) \tag{11.12}$$

$$x_i(t+1) = \Theta(R_i(t+1) + v_i^U(t)) \tag{11.13}$$

$$y_i(t+1) = \Theta(\sum_{j=1}^{N^E} W_{ij}^{IE}(t)x_j(t) - T_j^I) \tag{11.14}$$

T^E and T^I are threshold values, drawn randomly from positive intervals for excitatory units and inhibitory units, respectively. Θ is the Heaviside step function, and $v_i^U(t)$ the network input drive. Matrices W^{IE} and W^{EI} are fully connected, and represent connections between inhibitory and excitatory units, and vice versa. W^{EE} holds connections between excitatory units. These are random, sparse, and without self-recurrence. Inhibitory units are not directly connected to each other. All weights are drawn from the interval $[0,1]$, and the three matrices W^{IE}, W^{EI}, and W^{EE} are normalized, i.e., $\sum_j W_{ij} = 1$. The network state at time t is given by the two binary vectors $x(t) \in \{0,1\}^{N^E}$, and $y(t) \in \{0,1\}^{N^I}$, representing activity of the N^E excitatory and the N^I inhibitory units, respectively.

STDP and synaptic scaling update connections of excitatory units of the reservoir, while IP changes their thresholds. Inhibitory neurons and their connections remain unchanged. In the SORN the STDP for some small learning constant η_{stdp} is formalized as:

$$\Delta W_{ij}^{EE}(t) = \eta_{stdp}(x_i(t)x_j(t-1) - x_i(t-1)x_j(t)). \tag{11.15}$$

Synaptic scaling normalizes the values to sum up to one:

$$\Delta W_{ij}^{EE}(t) = W_{ij}^{EE}(t) / \sum_j W_{ij}^{EE}(t). \tag{11.16}$$

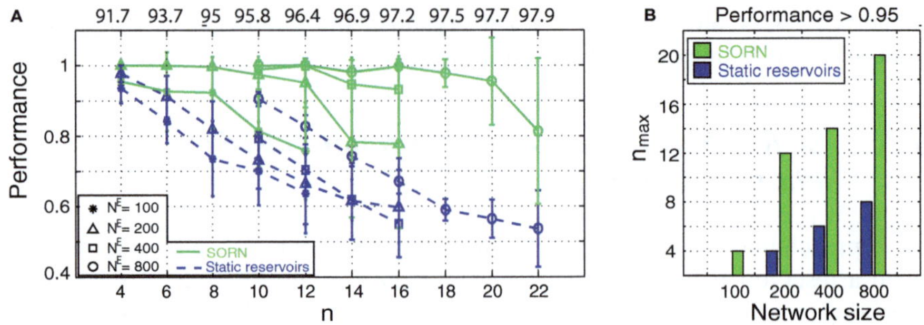

Fig. 11.2 (left) Normalized performance versus task difficulty as indicated by n, the number of repeated characters of a word which the network should predict. Different network sizes were tested. The numbers on top indicate the maximum possible performance – which is limited by the inherent randomness of the first character of a word within the sequence. Standard deviation among trials is indicated by the error bars. (right) Highest value of n for which a network achieved more than 95% of maximum performance as function of network size. The plastic SORN networks are able to deal with significantly harder tasks than the static reservoirs at this performance level. Graphs reproduced from (Lazar et al. 2009).

IP learning is responsible for spreading activations more evenly, using a learning rate η_{ip}, and a target firing rate of H_{IP}:

$$T_i^E(t+1) = T_i^E(t) + \eta_{ip}(x_i(t) - H_{IP} \tag{11.17}$$

Lazar et al. (2009) show that the SORN outperforms static reservoir networks using a letter prediction task. The network has to predict the next letter in a sequence of two different artificial words of length $n+2$. These words are made up of three different characters, with the second character repeated n times. The first character of a word is random and the network cannot do better than randomly guessing which one will come up. If the reservoir is able to efficiently separate the repeated character in the middle part of the word, though, the network can learn to count these characters and predict the rest of the sequence correctly. Figure 11.2 compares the normalized performance of SORNs and static reservoir networks of different sizes on instances of the task with increasing n (increased difficulty). SORNs are able to outperform static reservoirs clearly on this task. A PCA analysis in (Lazar et al. 2009) reveals that the SORN indeed shows a much better separation property and maps repeated inputs to distinct network states, while the states of static reservoirs are much more clustered together and thus harder to distinguish by the linear readout.

The combination of the three mechanisms appears to be a key to successful self-organization in an RNN. Figure 11.3 illustrates that the dynamics of SORN reservoir become sub-optimal if only two of the three plasticity mechanisms are active. Without synaptic normalization, the network units become highly synchronized. This severely restricts the representational power of the reservoirs. If IP learning is switched off, the activity of neurons in the network becomes unbalanced. Some

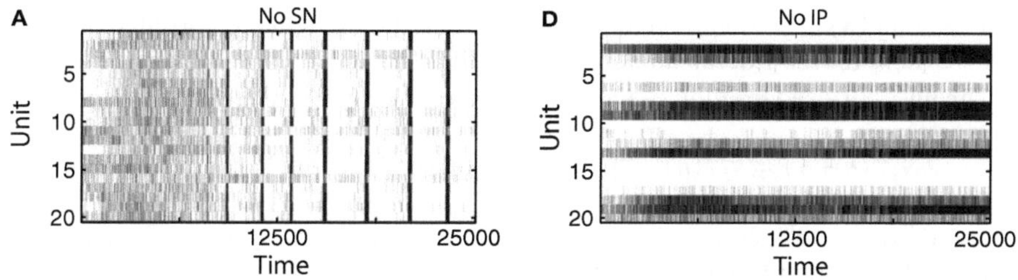

Fig. 11.3 Activity snapshots for 50 randomly selected reservoir neurons. (left) A reservoir without synaptic scaling develops highly synchronized, seizure-like firing patterns. (right) Without the IP mechanism, neuron activity is unevenly distributed with some neurons firing almost constantly while others are nearly silent. Graphs reproduced from (Lazar et al. 2009).

neurons fall nearly silent while others are active almost all the time. This is in contrast to the case with IP where activity is more evenly distributed, enabling a richer representation of information in the reservoir.

Though some of the self-organizing mechanisms like STDP are biologically plausible, there are not too many examples of successful applications for training RNNs, or, as Lazar et al. (2009) states, "Understanding and controlling the ensuing self-organization of network structure and dynamics as a function of the network's inputs is a formidable challenge". For time-series prediction and system identification tasks, an extension of the approach to analog units would be required. Also, an investigation of the information dynamics during and after adaptation may provide insights, for example into the relation between reservoir configuation and information transfer.

11.3.2 Hierarchical Self-Organizing Reservoirs

A different approach based on self-organized optimization of reservoirs is presented in (Lukovsevivcius 2012b). The author compares classical ESNs and recurrent RBF-unit based reservoir networks (called Self-Organizing Reservoirs, SORs) which resemble Recurrent Self-Organizing Maps (RSOMs) (Voegtlin 2002). The input and reservoir weights of the SOR are adapted by learning rules traditionally used for Self-Organizing Maps (SOMs) (Kohonen 2001) and NeuralGas networks (Martinetz and Schulten 1991).

The update equations for the SOR are:

$$\tilde{x}_i(n) = \exp(-\alpha\|\mathbf{v}_i^{in} - \mathbf{u}(n)\|^2 - \beta\|\mathbf{v}_i - \mathbf{x}(n-1)\|^2), \quad i = 1,\ldots,N_x, \qquad (11.18)$$

$$\mathbf{x}(n) = (1-\gamma)\mathbf{x}(n-1) + \gamma\tilde{\mathbf{x}}(n). \qquad (11.19)$$

Here, the internal reservoir neuron states at time n are collected in vector $\mathbf{x} \in \mathbb{R}^{N_x}$ and their update in vector $\tilde{\mathbf{x}} \in \mathbb{R}^{N_x}$. The factor $\gamma \in (0,1]$ is the leak-rate. The vector $\mathbf{u} \in \mathbb{R}^{N_u}$ contains the input-signal, while matrices $\mathbf{V}^{in}$ and $\mathbf{V}$ are the input and recurrent weight matrix, respectively, whose ith column vectors are denoted by

$\mathbf{v}_i^{in} \in \mathbb{R}^{N_x}$ and $\mathbf{v}_i \in \mathbb{R}^{N_x}$. Parameters α and β scale the input and recurrent distances, and $\|\cdot\|$ denotes the Euclidean norm.

The unsupervised training of the SOR updates the input and recurrent weights as:

$$\mathbf{v}_i^{all}(n+1) = \mathbf{v}_i^{all}(n) + \eta(n)h(i,n)([\mathbf{u}(n);\mathbf{x}(n)] - \mathbf{v}_i^{all}(n)), \tag{11.20}$$

where $\mathbf{v}_i^{all}(n) \equiv [\mathbf{v}^{in};\mathbf{v}]$ and $\eta(n)$ is a time-dependent learning rate. The learning-gradient distribution function h is defined either as:

$$h(i,n) = \exp(-d_h(i,\mathrm{bmu}(n))^2/b_h(n)^2), \tag{11.21}$$

where $d_h(i,j)$ is the distance between reservoir units i and j on a specific topology, $\mathrm{bmu}(n) = \arg\max_i(x_i(n))$ is a function returning the index of a *best matching unit* (BMU), and $b_h(n)$ is the time-dependent of the learning gradient distribution. With this definition of h, the learning proceeds according to the SOM algorithm. To implement NeuralGas-like learning, it suffices to change this definition to:

$$h_{ng}(i,n) = \exp(-d_{ng}(i,n)/b_h(n)), \tag{11.22}$$

where $d_{ng}(i,n)$ denotes the index of node i in the descending ordering of activities $x_i(n)$ (see (Lukovsevivcius 2012b) for additional details). Both algorithms were found to be similarly effective to improve the pattern separation capability of reservoirs compared to standard ESNs when tested on detection of certain signal components on a synthetic temporal pattern benchmark, and on classification of handwritten digits from a stream of these characters. Further improvements are reported if these SORs are stacked on top of each other in a hierarchy, trained in a layer-by-layer fashion. However, results only improve if enough time is given for the self-organization process to find suitable representations. If layers are stacked with very little training time for each of them, performance actually worsens.

11.3.3 Guided Self-Organization of Reservoir Information Transfer

In (Obst et al. 2010), the information transfer between input data and desired outputs is used to guide the adaptation of the self-recurrence in the hidden layer of a reservoir computing network. The idea behind this step is to change the memory within the system with respect to the inherent memory in input and output data (see Section 11.4 below for a a discussion which develops these ideas further).

The network dynamics is updated as:

$$\mathbf{x}(k+1) = diag(\mathbf{a})\mathbf{W}\mathbf{y}(k) + (\mathbf{I} - diag(\mathbf{a}))\mathbf{y}(k) + \mathbf{w}^{in}\mathbf{u}(k) \tag{11.23}$$
$$\mathbf{y}(k+1) = \mathbf{f}(\mathbf{x}(k+1)), \tag{11.24}$$

where $x_i, i = 1,\ldots,N$ are the unit activations, $\mathbf{W}$ is the $N \times N$ reservoir weight matrix, $\mathbf{w}^{in}$ the input weight vector, $\mathbf{a} = [a_1,\ldots,a_N]^T$ a vector of local decay factors,

$\mathbf{I}$ is the identity matrix, and k denotes the discrete time step. As a nonlinearity, $f(x) = \tanh(x)$ is used. The a_i represent the coupling of a unit's previous state with the current state, and are computed as:

$$a_i = \frac{2}{1+m_i},$$

where m_i represents the memory length of unit i ($m_i \in \{1,2,3,\ldots\}$), initialized to $m_i = 1$. Increasing individual m_i through adaptation increases the influence of a unit's past states on its current state. The information transfer is quantified as a conditional mutual information or transfer entropy (Schreiber 2000):

$$T_{X\to Y} = \lim_{k,l\to\infty} T^{(k,l)}_{X\to Y}, \text{with} \tag{11.25}$$

$$T^{(k,l)}_{X\to Y} = I(X^{(l)}; Y' | Y^{(k)}). \tag{11.26}$$

Parameters k and l are history sizes, which lead to finite-sized approximations of the transfer entropy for finite values.

In a first step, the required history size l is determined which maximizes the information transfer $T_{u\to v}$ from input u to output v. This value will increase for successively larger history sizes, but the increases are likely to level off for large values of l. Therefore, l is determined as the smallest value which is still able to increase $T_{u\to v}$ by more than a threshold ε:

$$T_{\mathbf{u}\to\mathbf{v}}(1,\hat{l}+1) \leq T_{\mathbf{u}\to\mathbf{v}}(1,\hat{l}) + \varepsilon \quad \text{and} \tag{11.27}$$

$$T_{\mathbf{u}\to\mathbf{v}}(1,l) > T_{\mathbf{u}\to\mathbf{v}}(1,l-1) + \varepsilon \quad \text{for all } l < \hat{l}. \tag{11.28}$$

In a second step, the local couplings of the reservoir units are adapted so that the transfer entropy from the input of each unit to its respective output is optimized *for the particular input history length* $\hat{l}$, as determined in step one. Over each epoch θ of length ℓ, we compute the transfer entropy from activations $x_i^{(\ell)}$ to output $y_i^{(\ell)}$ for each unit i:

$$te_i^\theta = T_{x_i^{(\ell)}\to y_i^{(\ell)}}(1,\hat{l}).$$

If the information transfer during the current epoch θ exceeds the information transfer during the past epoch by a threshold (i.e., $te_i^\theta > te_i^{\theta-1} + \varepsilon$), the local memory length m_i is increased by one. In case $te_i^\theta < te_i^{\theta-1} - \varepsilon$, the local memory length is decreased by one, down to a minimum of 1. The decay factors a_i are fixed once they stabilize, which ends the pre-training phase.

In (Obst et al. 2010), the method is tested on a one-step ahead prediction of unidirectionally coupled maps and of the Mackey-Glass time series benchmark. Showing results for the former task as an example, Figure 11.4 (left) displays the mean square errors of the prediction over the test data for different coupling strengths e and fixed order parameter ω for both echo state learning with and without adaptation of information transfer in the reservoir (averages over 50 trials). For each individual trial the same reservoir and time series have been used once with and without adaptation.

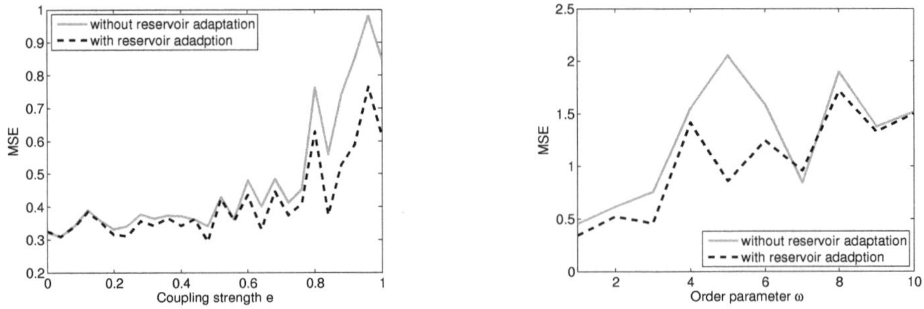

Fig. 11.4 (left) Mean squared errors of the prediction over the test data for different coupling strengths and fixed $\omega = 0$. (right) Mean squared error for different ω using a fixed coupling of $e = 0.75$.

The prediction using the reservoir adaptation is better over almost the entire range of e, with the improvement becoming more significant as the influence of the input time series becomes larger. Figure 11.4 (right) is a plot of the mean square error for different ω using a fixed coupling e. In all but one cases the reservoir adaptation improves results.

11.4 Quantifying Task Complexity

Most currently existing measures capture some of the generic computational properties of recurrent neural networks (as an important class of input-driven system), such as memory capacity or entropy at the neuron-level, but do not take task complexity into account. Optimization of the network properties based on these generic measures therefore will only do a "blind" adjustment of parameters while no optimality guarantee for the task at hand can be given. More positively put, the philosophy behind these measures is that a maximization of some of them leads to reservoirs that are capable to solve a variety of tasks. The self-organizing mechanisms in Sect. 11.3 are one way to achieve this maximization. In situations where no teaching signal is given, e.g., in clustering tasks, one can do no better than that; however, if the desired output signal is available, it can be used to quantify the task complexity as a relationship between inherent difficulty of predicting the output based on its own history, and to what extent the input data can contribute to improve these predictions. This would inform us how achievable a task is, and may also be used to trade off complexity of the system against the expected quality of the solution.

It is possible to use some of the tools that we introduced above, and take an information-theoretic approach to tackle this problem. Essentially we are interested in quantifying how difficult it is for a system to produce its next output. The systems we are interested in take a time series X as an input, and have the goal to generate output Y, another time series. To produce the next state y_{t+1}, both the output's past $(y_1 \ldots y_t)$ as well as the input up to the current step $(y_1, \ldots, y_{t+1})$ can be considered.

The Active Information Storage A_Y (Lizier et al. 2012) can be used to capture the influence of previous outputs in producing the next output: how much information is contained in the past of Y that can be used to compute its next state? This is expressed as the average mutual information between past $Y^{(k)}$ of the output and the next state Y':

$$A_Y = \lim_{k\to\infty} A_Y(k), \text{with} \tag{11.29}$$

$$A_Y(k) = I(Y^{(k)};Y'). \tag{11.30}$$

We use $A_Y(k)$ to represent finite-k estimates. Now, A_Y and $A_Y(k)$ allow for two kind of measurements: (a) higher values for A_Y indicate better predictability of Y from its own past, i.e., A_Y is one component of the overall task difficulty. (b) With increasing values of $k_i = 1,2,...,n$, estimates $A_Y(k_i)$ indicate the amount of memory that is in use. As the information that can be used to predict the next state increases with larger values of k_i, $A_Y(k_i)$ will monotonically increase with k_i, and asymptotically converge to some maximum. Finding k^* so that $A_Y(k^*) \geq A_Y(k) + \varepsilon$, for some small $\varepsilon > 0$ and $k \to \infty$, thus gives us a useful quantity for the amount of memory required, and at the same time $A_Y(k^*) \approx A_Y$ quantifies the difficulty in predicting Y' from its own past.

The other component that plays a role in the task is the input X. Its contribution to producing the next output Y' of the system, too, can be quantified, using the transfer entropy introduced above. The transfer entropy indicates how much information the input X contributes to the next state Y', given that the past of Y is known. Increasing the input history size l increases the information available in computing Y', for fixed output history size k^*. Large $T_{X\to Y}$ suggest that the input X helps in computing the next output, i.e., the task for the system is less difficult than for smaller transfer entropies. Finding an input history size l^* so that $T_{X\to Y}^{(k^*,l^*)} \approx \lim_{l\to\infty} T_{X\to Y}^{(k^*,l)}$ gives us another useful quantity for the amount of memory required.

Unfortunately, using these quantities to compare tasks or to design systems is not entirely straightforward, for a number of reasons. For continuous-valued time series X and Y, estimating mutual informations is cumbersome, and requires larger amounts of data in particular for larger history sizes l and k. To compare task difficulties, it would also be helpful to normalize both quantities, e.g., for the Active Information Storage by the joint entropy $H(Y;Y')$, to values between 0 and 1. The true output history Y may also be simply not available to the system, dependent on how it operates. For example, in batch mode, the only information that is available is the input X and the *estimated* output $\hat{Y}$. The true history of Y is usually only accessible if the system operates online. Finally, the two components A_Y and $T_{X\to Y}$ cannot be simply added to specify the overall task difficulty since input and output may redundantly share some information.

We will reserve a detailed investigation of applying both measures to a later publication. As a concept to explain contributions of input and output history, they can be an indicator for how complex the information processing system needs to be. It will also be interesting to see how other measures relate to them, and to show which aspect of the computation they measure. As an example, the memory capacity MC, as a sum of correlation coefficients can be seen as a linear measure of the potential information transfer between input X and the desired output Y. Mutual information expresses a nonlinear relationship between two variables, and so does the transfer entropy, a conditional mutual information between X and Y. In contrast to MC, the $T_{X \to Y}$ measures actual information transfer between input and desired output, i.e., $T_{X \to Y}$ is a purely a property of the task. MC is a property of the RNN, but as it is using task specific input, it combines the properties of the RNN with properties of the input. The two quantities could be used to adjust the architecture of a neural network for better performance on a specific task.

As another example, measuring the individual distributions of unit activations in the reservoir and their divergence from a maximum entropy distribution capture properties of the input combined with properties of the network. On the other hand, Active Information Storage for input-driven systems, applied to reservoirs or individual reservoir units, expresses the amount of information in the system that is in use to predict the next state, and is meant to measure capabilities of the system only.

Related work on complexity measures includes Grassberger's forecast complexity (Grassberger 1986, 2012), which considers the difficulty of making an optimal prediction of a sequence created by a stochastic process. A sequence can be compressed up to its entropy rate, and the forecast complexity is the computational complexity of the algorithm responsible for the decompression. The probability of the next symbol is needed for this decompression. Related ideas can be found in Minimum Description Length (MDL) approaches (Rissanen 1978) and Kolmogorov complexity (Kolmogorov 1965) as measures for complex objects. Also introduced by Grassberger (1986, 2012) is the Effective Measure Complexity (EMC), the relative memory required to calculate the probability distribution of the next symbol of a sequence. The EMC is a lower bound on the forecast complexity. Both forecast complexity and EMC look at sequences, e.g., the output of an autonomous system without regard to its input, whereas we are interested in systems that produce an output based on some input. More complexity measures can be found in a special issue on "Measures of Complexity from Theory to Applications", with (Crutchfield and Machta 2011) as an introductory article.

11.5 Conclusion

We presented methods to assess different computational properties of input-driven RNNs, and reservoir computing networks in particular, in the first part of the paper. These methods were drawn from information-theory, statistical learning theory, and dynamical systems theory, and provided different perspectives on important aspects of information processing in these systems. They help to quantify properties like the memory capacity a certain network provides, the flow of information through the system and its modification over time, the ability to separate similar inputs and generalize to new, unseen data, and others. In addition to their usefulness in their own right when trying to understand how RNNs implement the functions they are trained for, they also have the potential to be used as target signals to guide self-organized optimization procedures aimed at improving the quality of reservoirs for a specific task over random initialization.

In the second part of the paper, we presented some recent efforts at implementing self-organized optimization for reservoir computing networks. One approach combined different plasticity mechanisms to improve coding quality and separation ability of the network, while a different approach was using methods similar to recursive self-organizing maps with SOM and NeuralGas-like learning rules. The final approach we presented proceeds in two phases: determining a learning goal in terms of information transfer between input and desired output, and using this quantity to guide local adjustments to the self-recurrence of each reservoir unit. All of these methods showed the potential of self-organized methods to improve network performance over standard, random reservoirs while avoiding problems associated with back propagation of error-gradients throughout the whole networks.

As a next step towards methods that are able to automatically generate or optimize recurrent neural networks for a specific task (or class of tasks), it seems worthwhile to combine measures for network properties and task complexity, and devise algorithms that adjust the former based on the latter. The approach taken in (Dambre et al. 2012) of using orthogonal functions to measure information processing capacity could be extended to construct suitable dynamical systems for a task when the requirements for a specific task can be measured in a similar way.

A comparison of how the measures of the information dynamics framework (Lizier et al. 2007, 2012), the information processing capacity for dynamical systems (Dambre et al. 2012), measures of criticality (Bertschinger and Natschläger 2004; Prokopenko et al. 2011) or of memory capacity (Jaeger 2001; Ganguli et al. 2008) relate to each other should reveal some interesting insights (see, e.g., Tino and Rodan 2013), since they all cover some aspects of dynamical systems. Establishing the relation between the information dynamics framework, with recent extension for input-driven systems, and information processing capacity, for example, could help to overcome requirements for i.i.d. input in the latter, to better understand dynamical systems with arbitrary input.

References

Ashby, W.R.: An Introduction to Cybernetics. Chapman & Hall, London (1956)

Baddeley, R., Abbott, L.F., Booth, M.C.A., Sengpiel, F., Freeman, T., Wakeman, E.A., Roll, E.T.: Responses of neurons in primary and inferior temporal visual cortices to natural scenes. Proc. R. Soc. Lond. B 264, 1775–1783 (1997)

Bell, A.J., Sejnowski, T.J.: An information-maximization approach to blind separation and blind deconvolution. Neural Computation 7(6), 1129–1159 (1995)

Bengio, Y., Boulanger-Lewandowski, N., Pascanu, R.: Advances in Optimizing Recurrent Networks. arXiv preprint 1212.0901, arXiv.org (2012)

Bengio, Y., Simard, P., Frasconi, P.: Learning long-term dependencies with gradient descent is difficult. IEEE Transaction on Neural Networks 5(2), 157–166 (1994)

Bertschinger, N., Natschläger, T.: Real-time computation at the edge of chaos in recurrent neural networks. Neural Computation 16(7), 1413–1436 (2004)

Boedecker, J., Obst, O., Lizier, J.T., Mayer, N.M., Asada, M.: Information processing in echo state networks at the edge of chaos. Theory in biosciences Theorie in den Biowissenschaften 131(3), 1–9 (2011)

Boedecker, J., Obst, O., Mayer, N.M., Asada, M.: Initialization and self-organized optimization of recurrent neural network connectivity. HFSP Journal 3(5), 340–349 (2009)

Crutchfield, J.P., Machta, J.: Introduction to focus issue on "Randomness, Structure, and Causality: Measures of complexity from theory to applications". Chaos 21(3), 037101 (2011)

Dambre, J., Verstraeten, D., Schrauwen, B., Massar, S.: Information processing capacity of dynamical systems. Scientific Reports 2, 514 (2012)

Douglas, R., Markram, H., Martin, K.: Neocortex. In: Shepherd, G. (ed.) Synaptic Organization In the Brain, pp. 499–558. Oxford University Press (2004)

Doya, K.: Bifurcations in the learning of recurrent neural networks. In: IEEE International Symposium on Circuits and Systems, pp. 2777–2780. IEEE (1992)

Ganguli, S., Huh, D., Sompolinsky, H.: Memory traces in dynamical systems. Proceedings of the National Academy of Sciences 105(48), 18970–18975 (2008)

Grassberger, P.: Toward a quantitative theory of self-generated complexity. International Journal of Theoretical Physics 25(9), 907–938 (1986)

Grassberger, P.: Randomness, information, and complexity. Technical Report 1208.3459, arXiv.org (2012)

Hawkins, J., Blakeslee, S.: On Intelligence. Times Books (2004)

Hochreiter, S.: The vanishing gradient problem during learning recurrent neural nets and problem solutions. International Journal of Uncertainty, Fuzziness and Knowledge-Based Systems 6(2), 107–116 (1998)

Hochreiter, S., Schmidhuber, J.: Long short-term memory. Neural Computation 9(8), 1735–1780 (1997)

Jaeger, H.: Short term memory in echo state networks. Technical Report 152, GMD – German National Research Institute for Computer Science (2001)

Jaeger, H., Haas, H.: Harnessing Nonlinearity: Predicting Chaotic Systems and Saving Energy in Wireless Communication. Science 304(5667), 78–80 (2004)

Kohonen, T.: Self-Organizing Maps, 3rd, extended edn. Springer (2001)

Kolmogorov, A.N.: Three approaches to the quantitative definition of information. Problemy Peredachi Informatsii 1(1), 3–11 (1965)

Lazar, A., Pipa, G., Triesch, J.: SORN: a self-organizing recurrent neural network. Frontiers in Computational Neuroscience 3, 23 (2009)

Legenstein, R., Maass, W.: What makes a dynamical system computationally powerful. In: Haykin, S., Principe, J.C., Sejnowski, T., McWhirter, J. (eds.) New Directions in Statistical Signal Processing: From Systems to Brains, pp. 127–154. MIT Press (2007)

Linsker, R.: Towards an organizing principle for a layered perceptual network. In: Anderson, D.Z. (ed.) NIPS, pp. 485–494. American Institute of Physics (1987)

Lizier, J.T., Prokopenko, M., Zomaya, A.Y.: Detecting non-trivial computation in complex dynamics. In: Almeida e Costa, F., Rocha, L.M., Costa, E., Harvey, I., Coutinho, A. (eds.) ECAL 2007. LNCS (LNAI), vol. 4648, pp. 895–904. Springer, Heidelberg (2007)

Lizier, J.T., Prokopenko, M., Zomaya, A.Y.: Local measures of information storage in complex distributed computation. Information Sciences 208, 39–54 (2012)

Lukovsevivcius, M.: A practical guide to applying echo state networks. In: Montavon, G., Orr, G.B., Müller, K.-R. (eds.) Neural Networks: Tricks of the Trade, 2nd edn. LNCS, vol. 7700, pp. 659–686. Springer, Heidelberg (2012a)

Lukoʋseviʋcius, M.: Self-organized reservoirs and their hierarchies. In: Villa, A.E.P., Duch, W., Érdi, P., Masulli, F., Palm, G. (eds.) ICANN 2012, Part I. LNCS, vol. 7552, pp. 587–595. Springer, Heidelberg (2012b)

Lukoʋseviʋcius, M., Jaeger, H.: Reservoir computing approaches to recurrent neural network training. Computer Science Review 3(3), 127–149 (2009)

Maass, W., Joshi, P., Sontag, E.D.: Computational aspects of feedback in neural circuits. PLOS Computational Biology 3(1), e165 (2007)

Manjunath, G., Tino, P., Jaeger, H.: Theory of Input Driven Dynamical Systems. In: dice.ucl.ac.be, pp. 25–27 (April 2012)

Martens, J., Sutskever, I.: Learning recurrent neural networks with hessian-free optimization. In: Proceedings of the 28th International Conference on Machine Learning, vol. 46, p. 68. Omnipress Madison, WI (2011)

Martinetz, T., Schulten, K.: A "neural-gas" network learns topologies. Artificial Neural Networks 1, 397–402 (1991)

Mitchell, M., Hraber, P.T., Crutchfield, J.P.: Revisiting the edge of chaos: Evolving cellular automata to perform computations. Complex Systems 7, 89–130 (1993)

Obst, O., Boedecker, J., Asada, M.: Improving Recurrent Neural Network Performance Using Transfer Entropy. Neural Information Processing Models and Applications 6444, 193–200 (2010)

Obst, O., Boedecker, J., Schmidt, B., Asada, M.: On active information storage in input-driven systems. preprint 1303.5526v1, arXiv.org (2013)

Ozturk, M.C., Xu, D., Príncipe, J.C.: Analysis and design of echo state networks. Neural Computation 19(1), 111–138 (2007)

Prokopenko, M., Lizier, J.T., Obst, O., Wang, X.R.: Relating Fisher information to order parameters. Physical Review E 84(4), 041116 (2011)

Riedmiller, M., Braun, H.: A direct adaptive method for faster backpropagation learning: the rprop algorithm. In: IEEE International Conference on Neural Networks, vol. 1, pp. 586–591 (1993)

Rissanen, J.: Modeling by shortest data description. Automatica 14(5), 465–471 (1978)

Rolls, E.T., Deco, G.: The Noisy Brain - Stochastic Dynamics as a Principle of Brain Function. Oxford University Press (2010)

Rumelhart, D., Hinton, G., Williams, R.: Learning representations by back-propagating errors. Nature 323(6088), 533–536 (1986)

Schmidhuber, J., Wierstra, D., Gagliolo, M., Gomez, F.: Training recurrent networks by evolino. Neural Computation 19(3), 757–779 (2007)

Schrauwen, B., Wardermann, M., Verstraeten, D., Steil, J.J., Stroobandt, D.: Improving reservoirs using intrinsic plasticity. Neurocomputing 71(7-9), 1159–1171 (2008)

Schreiber, T.: Measuring information transfer. Physical Review Letters 85(2), 461–464 (2000)

Seung, H.S.: Connectome: How the Brain's Wiring Makes Us Who We Are. Houghton Mifflin Harcout, New York (2012)

Sporns, O.: Networks Of the Brain. The MIT Press (2011)

Sussillo, D., Barak, O.: Opening the black box: low-dimensional dynamics in high-dimensional recurrent neural networks. Neural Computation 25(3), 626–649 (2013)

Tino, P., Rodan, A.: Short term memory in input-driven linear dynamical systems. Neurocomputing (2013)

Triesch, J.: A gradient rule for the plasticity of a neuron's intrinsic excitability. In: Duch, W., Kacprzyk, J., Oja, E., Zadrożny, S. (eds.) ICANN 2005. LNCS, vol. 3696, pp. 65–70. Springer, Heidelberg (2005)

Voegtlin, T.: Recursive self-organizing maps. Neural Networks 15(8-9), 979–991 (2002)
Werbos, P.J.: Backpropagation through time: what it does and how to do it. Proceedings of the IEEE 78(10), 1550–1560 (1990)
Williams, P.L., Beer, R.D.: Information dynamics of evolved agents. From Animals to Animats 11, 38–49 (2010)
Williams, R.J., Zipser, D.: A learning algorithm for continually running fully recurrent neural networks. Neural Computation 1(2), 270–280 (1989)

Part IV
Swarms and Networks of Agents

Chapter 12
Measuring Information Dynamics in Swarms

Jennifer M. Miller, X. Rosalind Wang, Joseph T. Lizier,
Mikhail Prokopenko, and Louis F. Rossi

12.1 Introduction

We propose a novel, information theoretic characterization of dynamics within swarms, through explicitly measuring the extent of collective communications and tracing collective memory. These elements of distributed computation provide complementary views into the capacity for swarm coherence and reorganization. The approach deals with both global and local information dynamics ultimately discovering diverse ways in which an individual's location within the group is related to its information processing role.

Many animals display examples of spatial aggregations such as schools of fish, swarms of locusts, herds of wildebeest, and flocks of birds (Lissaman and Shollenberger 1970; Parrish and Edelstein-Keshet 1999; Sinclair and Norton-Griffiths 1979; Uvarov 1928). A group is better equipped to assist individuals with protection, mate choices, foraging, habitat assessment, migratory route information,

Jennifer M. Miller · Louis F. Rossi
Department of Mathematical Sciences,
University of Delaware, 501 Ewing Hall, Newark DE 19716, USA
e-mail: {millerje,rossi}@math.udel.edu

X. Rosalind Wang
CSIRO Computational Informatics
PO Box 76, Epping, NSW 1710, Australia
e-mail: rosalind.wang@csiro.au

Mikhail Prokopenko
CSIRO Computational Informatics PO Box 76, Epping, NSW 1710, Australia
& School of Physics The University of Sydney NSW 2006, Australia
& Department of Computing Macquarie University NSW 2109, Australia
e-mail: mikhail.prokopenko@csiro.au

Joseph T. Lizier
CSIRO Computational Informatics, PO Box 76, Epping, NSW 1710, Australia
& School of Information Technologies, The University of Sydney NSW 2006, Australia
& Max Planck Institute for Mathematics in the Sciences,
Inselstraße 22, D-04103, Leipzig, Germany

M. Prokopenko (ed.), *Guided Self-Organization: Inception*,
Emergence, Complexity and Computation 9,
DOI: 10.1007/978-3-642-53734-9_12, © Springer-Verlag Berlin Heidelberg 2014

etc. (Camazine et al. 2003; Giraldeau et al. 2002; Partridge 1982). These groups are self-organized through local interactions, and so a disturbance, such as detection of a predator by some individuals, can quickly spread through the entire group (Treherne and Foster 1981). Complex large-scale patterns and structures emerge within a swarm through individual decisions based on perception of local conditions. For example, in response to a predator, many schools of fish display complex collective patterns, including compression, 'hourglass', 'vacuole', 'flash expansion', or form highly parallel translating groups (Parrish et al. 2002).

Even a few individuals may strongly bias the motion of an entire group. For instance, if a certain number of fishes in close proximity turn together, this may result in a wave of turning across the whole group (Radakov 1973). Studies have shown that formation of waves is a wide spread phenomenon observed in insect swarms, bird flocks, fish schools and mammalian herds (Couzin et al. 2006; Treherne and Foster 1981; Kastberger et al. 2008; Potts 1984; Procaccini et al. 2011). Crucially, these waves serve to transfer information rapidly within a group. For example, Treherne and Foster (1981) showed that marine isopods, *Halobates robustus*, increased velocity in response to a model predator and this change of motion spreads across the group faster than the speed of approaching predator, and labeled this phenomenon the 'Trafalgar effect'. Procaccini et al. (2011) quantified "wave collective behavior in flocks", observing that the velocity of propagation of the perturbation inside the group is greater than the velocity of the predator.

12.1.1 Background on Information Cascades

Information that is propagated within swarms can be understood both as Shannon information ("reduction in uncertainty") and semantic information ("meaningful data"). In this paper, we adopt characterization of swarm dynamics in Shannon information-theoretic terms. Specifically, we utilize measures for average and local information storage as well as directed information transfer, used by Wang et al. (2012). This allows us to directly measure information properties of swarms, and directly compare complex aspects of swarm dynamics such as collective memory and collective communications across different models. Before describing details of this framework, we briefly review some previous related works that attempted to deal with information cascades in groups.

Couzin et al. (2006) pointed out that information cascades in collective systems often result in a rapid autocatalytic adaptive response to changing conditions. They argued that this heightened response allows the group to be extremely sensitive to weak or ambiguous external stimuli, while keeping it fairly susceptible to noise and false alarms. They called for new mathematical approaches that link the behavior of individuals to the resultant higher-order group/population properties, specifically hoping for new insights into how information is acquired, processed and storage. In a later review, Couzin (2009) posed a specific challenge: "How does spatial position within groups reflect informational status, or information-processing capability, of individuals?"

Sumpter et al. (2008) used self-propeled particle (SPP) simulations to examine the transfer of information from a leader or a threat. They measured the information transfer by the alignment of individuals. This measure indicated that highest transfer occurred when the group was close to a phase transition between ordered and random motion. They also used pairs of homing pigeons to collect data on how an individual might combine its own information with transferred information. Dall et al. (2005) mentioned that public information favors group cohesion, argued that information implies utility as well as uncertainty reduction, and proposed an explicit statistical decision theory framework.

Information cascades may induce long series of incorrect decisions (e.g. false alarms). For example, Galef and Giraldeau (2001) explored the applicability of informational cascades to social-foraging systems trying to estimate the extent to which the short-term profitability of social learning can lead to potential costly errors. Giraldeau et al. (2002) highlighted that cascades are extremely sensitive to the initial sequence of events and may occasionally lead to errors.

As pointed out by Katz et al. (2011), important questions include how animals integrate information from widely disparate sources in real time (Couzin 2007) and how this nonlinear integration translates into higher-order computational capabilities that emerge at the level of the collective.

12.1.2 Motivation and Objectives

In previous work, we attempted to answer these questions from a generic information theoretic viewpoint on distributed computation (Wang et al. 2012). Specifically, we proposed to measure swarm coherence as Active Information Storage (AIS), and information cascades as changes in Transfer Entropy (TE). This approach allowed us to formalize the study of information transmitted within a swarm in 'waves', as well as relate swarm coherence/synchrony to dampening down incorrect decisions carried out by some cascades.

In this work, we investigate differences in information dynamics brought about by constraints imposed on the system, e.g. by limiting the speed of individuals within the swarm. This investigation contrasts dynamics created by two different swarm models: variable-speed model and constant-speed model. The variable-speed model allows individuals to respond to the local influences by changing speed and/or direction, and the constant-speed model allows only changes in the direction. This comparative analysis demonstrates how one may guide self-organisation in a swarm.

Guided Self-Organization (GSO) typically has the following features: (i) an increase in organization (structure and/or functionality) over some time; (ii) the local interactions are not *explicitly* guided by any external agent; (iii) task-independent objectives are combined with task-dependent constraints (Ay et al. 2011). The constant-speed model exemplifies the second feature, by constraining local interactions without any explicit interference with the individuals' decision-making mechanism. The main focus of our study is to observe how such changes affect generic information-theoretic dynamics. For instance, we explore whether a (possibly task-dependent) constraint for constant speed affects (task-independent) distributed computation: collective memory and collective long-range communications within the swarm.

To investigate changes brought about by constraints on speed, we explore two scenarios in each of the two swarm models. Our first scenario reveals how different local initial perturbations affect a single swarm. The second scenario introduces a different type of perturbation, introduced by three separate but merging swarms. The experiments trace global and local AIS and TE over time. These measures identify multiple information cascades, changes to global coherence, local processing of incorrect decisions, and eventual synchrony. Furthermore, the observed differences in information dynamics between these two models reveal that under constraints, the global coherence is harder to achieve and sustain, and the information cascades are less well-formed. At the same time, some generic principles are observed for both models: e.g. maximal information transfer tends to follow the stage with maximal collective memory.

12.2 Three Zones Model for Swarms

In this study, we use an individual-based model for modeling and simulating aggregations of discrete individuals interacting with three groups of neighbors. Each individual responds to its neighbors in three concentric zones with either repulsion, orientation, or attraction, respectively (Aoki 1982; Couzin et al. 2002; Huth and Wissel 1992; Lukeman et al. 2010; Vicsek et al. 1995). The contribution from the repulsion response results in an individual moving away from a weighted average of the relative positions of neighbors with velocity $\mathbf{v}_r$. Similarly, attraction causes an individual to move toward a weighted average of the relative positions of neighbors, $\mathbf{v}_a$. In addition, an individual aligns its velocity with a weighted average of its own velocity and neighbors' velocities, $\mathbf{v}_o$.

We explore a three-zone swarming model that uses continuous, overlapping zones with smooth transitions (see (Miller et al. 2011)). This model has a rigorous continuum limit in both space and time so that it captures the behavior of arbitrarily large swarms as a system of PDE's. The individual based model that we use in this paper is a discretization of the continuum system. The response of individual i to nearby individuals is determined by a weighted sum of contributions from the three zones:

$$\mathbf{v}_{r,i} = \sum_{j=1}^{N} -\frac{1}{8\pi\sigma_1^4}\mathbf{s}_{ij}\exp(-|\mathbf{s}_{ij}|^2/4\sigma_1^2), \tag{12.1}$$

$$\mathbf{v}_{a,i} = \sum_{j=1}^{N} \frac{1}{64\pi\sigma_3^6}\mathbf{s}_{ij}|\mathbf{s}_{ij}|^2\exp(-|\mathbf{s}_{ij}|^2/4\sigma_3^2), \tag{12.2}$$

$$\mathbf{v}_{o,i} = \frac{\sum_{j=1}^{N}\frac{1}{4\pi\sigma_2^2}\exp(-|\mathbf{s_{ij}}|^2/4\sigma_2^2)\mathbf{v}_j}{\sum_{j=1}^{N}\frac{1}{4\pi\sigma_2^2}\exp(-|\mathbf{s_{ij}}|^2/4\sigma_2^2)}, \tag{12.3}$$

where $\mathbf{s}_{ij} := \mathbf{s}_j - \mathbf{s}_i$, the relative position of individuals i and j. The lengths $\sigma_1 < \sigma_2 < \sigma_3$ represent the sizes of the repulsion, orientation and attraction zones, respectively. The three responses are combined to determine the desired velocity, $\mathbf{v}_d$:

$$\mathbf{v}_{d,i} = \mathbf{v}_{r,i} + \mathbf{v}_{o,i} + c_a \mathbf{v}_{a,i},$$

for the ith individual, and c_a specifies the relative importance of attraction to orientation and repulsion. The desired velocity vector is used to control the change in velocity.

We examine the information transfer and storage in two similar models for swarms. The first is a *variable-speed model*, which allows individuals to respond to the local influences by changing speed and/or direction. The velocity is updated by

$$\mathbf{v}_i^{n+1} = \mathbf{v}_i^n + \delta\tau \cdot \kappa \left(\mathbf{v}_{d,i}^n - \mathbf{v}_i^n \right),$$

where $\delta\tau$ is the time step length, n is the discrete time step index and κ is the turning rate.

The second swarm model is a *constant-speed model*, in which only the direction can change. Without loss of generality, each individual moves with unit speed. We update the direction θ_i by

$$\theta_i^{n+1} = \theta_i^n + \delta\tau \cdot \kappa (\mathbf{v}_i^n)^\perp \cdot \mathbf{v}_{d,i}^n$$

where $(\mathbf{v}_i^n)^\perp = [-\sin\theta_i^n, \cos\theta_i^n]^T$. The velocity becomes

$$\mathbf{v}_i^{n+1} = [\cos\theta_i^{n+1}, \sin\theta_i^{n+1}]^T.$$

The models differ in that the first allows individuals' speeds to vary over time while the second requires that all individuals have unit velocity vectors.

12.3 Information Dynamics in Swarms

We utilized a framework for local information dynamics developed by Lizier et al. (2007, 2008b, 2010, 2012) (and see Lizier (2013)). The framework precisely quantifies information storage, transfer and modification at each spatiotemporal point in a complex system. We measure the information dynamics of a swarm by computing the local and average values of *active information storage* (Lizier et al. 2012, 2007) and *conditional transfer entropy* (Lizier et al. 2008b, 2010).

The local active information storage of an agent in the system is the amount of information in its past that is used in predicting its next state. AIS for agent X is defined as the local mutual information between its semi-infinite past $x_n^{(k)} = \{x_n, x_{n-1}, \ldots, x_{n-k+1}\}$ (as $k \to \infty$) and its next state x_{n+1} at time step $n+1$:

$$a_X(n+1) = \lim_{k\to\infty} \log_2 \frac{p(x_n^{(k)}, x_{n+1})}{p(x_n^{(k)})p(x_{n+1})}, \tag{12.4}$$

with $a_X(n,k)$ representing an approximation with finite history length k. Note that AIS captures both the information stored internally by an agent, and also that stored in a distributed fashion in its neighbors and retrieved at a later time point. We characterize the collective memory within the swarm that is used for computation by the swarm's average AIS over all agents at each time step: $A(n,k) = \langle a_X(n,k) \rangle_X$.

The information transfer between a source and a destination agent is defined as the information provided by the source about the destination's next state, in the context of the past of the destination. This concept is quantified by the transfer entropy (Schreiber 2000), which properly measures a directed, dynamic flow of information. The local transfer entropy (Lizier et al. 2008b) from a source agent Y to a destination agent X is the local mutual information between the previous value of the source y_n and the next value of the destination x_{n+1}, *conditioned* on the semi-infinite past of the destination $x_n^{(k)}$ (as $k \to \infty$):

$$t_{Y\to X}(n+1) = \lim_{k\to\infty} \log_2 \frac{p(x_{n+1}|x_n^{(k)}, y_n)}{p(x_{n+1}|x_n^{(k)})}. \tag{12.5}$$

Again, $t_{Y\to X}(n,k)$ represents finite-k approximation, and we characterize overall transfer within the swarm as the average over all causally connected pairs $Y \to X$ at each time step: $T(n,k) = \langle t_{Y\to X}(n,k) \rangle_{Y\to X}$. We also make a single characterization of the local transfer into an agent at a given time step by averaging over the transfer from each of its causal sources: $t_X(n,k) = \langle t_{Y\to X}(n,k) \rangle_Y$.

Note that one can also condition the TE on another information contributor W to form the *conditional transfer entropy* (Lizier et al. 2010):

$$t_{Y\to X|W}(n+1) = \lim_{k\to\infty} \log_2 \frac{p(x_{n+1}|x_n^{(k)}, w_n, y_n)}{p(x_{n+1}|x_n^{(k)}, w_n)}. \tag{12.6}$$

It is this form of transfer entropy that we used for measuring local information transfer and overall/collective communications (see below).

To apply information dynamics to swarms, we accumulated the observations across agents and measured the state transitions with *relative variables* as described by Wang et al. (2011). For local information storage, the variables in the equation are composed of the change in velocity and the speed of the swarm particle. That is, for a particle p and $k = 1$:

$$x_n = \{\mathbf{v}_p^n - \mathbf{v}_p^{n-1}, |v|^n\},$$
$$x_{n+1} = \{\mathbf{v}_p^{n+1} - \mathbf{v}_p^n, |v|^{n+1}\}.$$

For transfer entropy, we condition on the current speed of the destination variable (since this could modulate the effect of the source on the destination) as $w_n = |v|^n$. We do not take into account the speed in the destination variable's next state; thus the relevant variables for the conditional transfer entropy in Equation 12.6 are:

$$\begin{aligned} y_n =& \{\mathbf{s}_p^n - \mathbf{s}_{p'}^n, \mathbf{v}_p^n - \mathbf{v}_{p'}^n\}, \\ w_n =& |v|^n, \\ x_n =& \mathbf{v}_p^n - \mathbf{v}_p^{n-1}, \\ x_{n+1} =& \mathbf{v}_p^{n+1} - \mathbf{v}_p^n \end{aligned}$$

where p is the destination particle and p' is the source particle.

12.4 Results and Discussion

12.4.1 Variable-Speed Swarm Model

Results for the variable-speed swarm model were described in detail by Wang et al. (2012), here we reproduce the results for completeness.

12.4.1.1 Experiment 1: One Group, Variable Speed

Initially, the individuals are positioned in a square configuration 49×49 in size. This experiment models different local initial perturbations which are absorbed by the swarm. The simulation is run until the swarm reaches a steady state.

Figure 12.1 shows the local active information storage of individual particles at some key steps during the simulation. These figures show that local AIS can be positive and negative. Positive local information storage indicates that the next state can be accurately predicted from past states, while negative values indicate that the past misinforms about the next state (Lizier et al. 2008b, 2010). For swarm particles, negative local storage means the particle's movement is unusually strongly influenced by other neighboring particles (via high transfer) at this point in time, given the past history of that particle.

As argued previously (Wang et al. 2012), negative local storage represents processing of incorrect decisions propagated from the periphery. For example, at $T = 19.0$ (Figure 12.1 (e)) some of individuals in the center are trying to compute their next state while being influenced by competing signals from their neighbors. In such situations, their past is misinformative about the next state.

Similar to local AIS, local TE can also be positive or negative (see plots of $t_X(n,k)$ in Figure 12.2). Positive local transfer entropy means that the source agent is informative about the next state, while negative local TE indicates that the source misleads an observer about the next state of the destination given the destination's history (Lizier et al. 2008b, 2010). Here, positive local TE shows the swarm individual's movement is strongly coherently affected by individual neighbors, while negative local TE shows the particle is either exhibiting strong independent motion or is under the collective influence of several neighbors (rather than the coherent influence of individual neighbors) (e.g. $T = 19.0$). This understanding of information storage and transfer for swarm dynamics suggests low information transfer may often be found with high storage and vice versa. Indeed, Figures 12.1 and 12.2 show

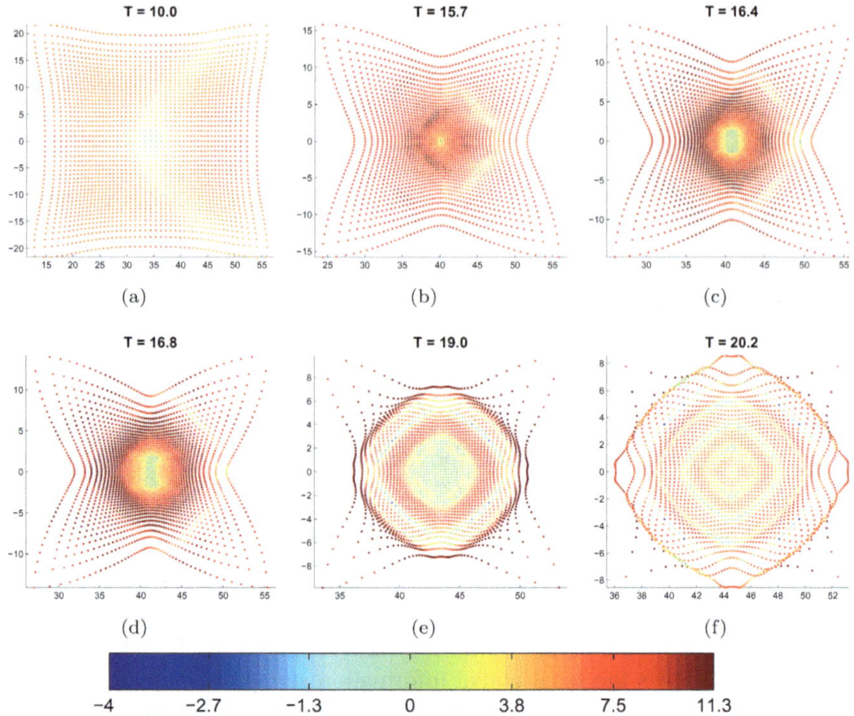

Fig. 12.1 Local information storage through the variable-speed swarm at key time steps. The swarm has an initial configuration of one large square. Note, to visualize the range of values, we map 0 to the center of the colorbar and the maximum and minimum values to the extremes of the colorbar. Values vary linearly on the left half and the right half, but the scales on the two halves are different.

that in most cases, areas of high local storage often have low or negative local transfer and vice versa.

The next interesting stage can be observed at $T = 20.2$ when a front of a cascade is formed by the individuals that begin to move coherently, i.e. have comparably high local TE.

We turn our attention now to collective aspects of swarm dynamics.

The collective memory within the swarm that is used for computation is modeled by average AIS. At early stages, the particles in the center of the swarm are not affected by changes at the swarm's periphery, i.e. they do not participate in collective computation. As the changes propagate deeper, more and more particles get engaged and the collective memory used in computation grows. This computation is needed to create coordinated motion. As the swarm reaches a state where the majority of particles are dynamically coordinated, AIS reaches its maximum (blue solid curve in Fig. 12.3 at $T = 16.8$). This results in a 'bell' shaped curve of average AIS shown in Fig. 12.3. Such a curve characterising a complex collective behavior is not dissimilar

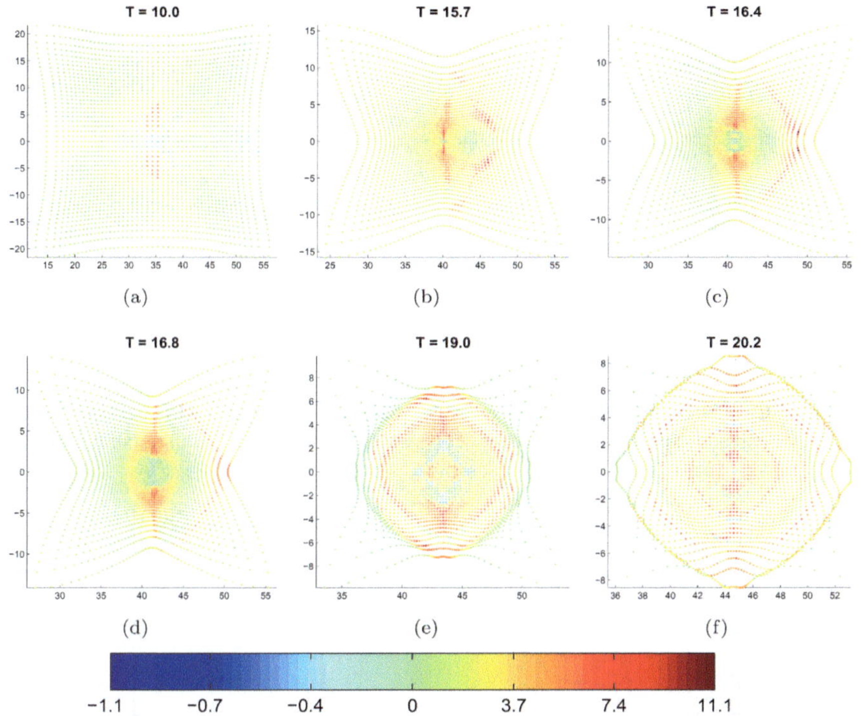

Fig. 12.2 Local information transfer through the variable-speed swarm at key time steps. The swarm has an initial configuration of one large square. Note, as with the previous figure, the scales on the positive and negative halves of the colorbar are different.

to many complexity curves (Mathews et al. 2005; Lizier et al. 2008a; Prokopenko et al. 2009).

The communication aspect of computation, that is, information cascades is modeled by average TE (cf. blue solid curve in Fig. 12.4). We can observe several waves. The first wave occurs when the initial changes originated at the periphery are 'absorbed' by the swarm. This slightly increases the overall TE ($T < 10.0$). When the wave reaches the center we can observe some individuals with high local TE. This is followed by an outward wave spreading in the opposite direction away from the center. The overall TE achieves a local maximum at $T = 15.7$ followed by a local minimum when it dissipates at $T = 16.8$. As has been observed by Wang et al. (2012), at this time AIS attained its global maximum. In other words, at this point the computation is non-trivial involving both memory and communication.

At $T = 20.2$, average TE reaches its own global maximum because at this time the formed cascades dominate incoherent individuals. Interestingly, this stage has *followed* the time when the collective memory (AIS) was highest. The cascades help to stablize and coordinate the swarm, supported by a steady decrease of TE during this process.

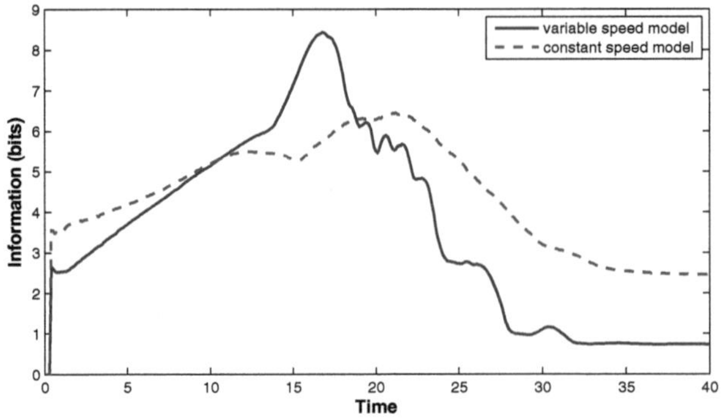

Fig. 12.3 Comparison of global average active information storage (AIS) between variable-speed and constant-speed model swarms. The swarm is initially in a square configuration.

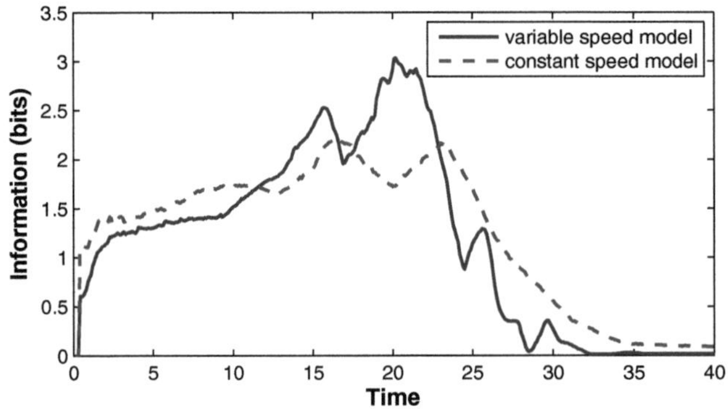

Fig. 12.4 Comparison of global average transfer entropy (TE) between variable-speed and constant-speed model swarms. The swarm is initially in a square configuration.

12.4.1.2 Experiment 2: Three Groups, Variable Speed

The second experiment for the variable-speed model (Wang et al. 2012) had the swarm individuals in three squares of checker configuration initially, each square is 28×28 in size, with each individual having an initial velocity of 1 (i.e., being set in motion towards the right). Initially the three groups simply form three individual swarms much like that in the previous case, while generally moving towards the right. Eventually, these groups start interacting with each other more noticeably. Thus, this experiment allows us to model different boundary perturbations caused by the interactions between the swarms. It is also interesting to observe how the initial asymmetry in the swarms' position and motion is propagated during the process.

The detailed observations are described by Wang et al. (2012). Here we point out only some of the local dynamics (see Fig. 12.5 for local AIS, and Fig. 12.6 for local TE), and note two general stages, each characterized by a 'bell' shaped curve for both average AIS and TE. The first AIS maximum (cf. blue solid curve in Fig. 12.7)

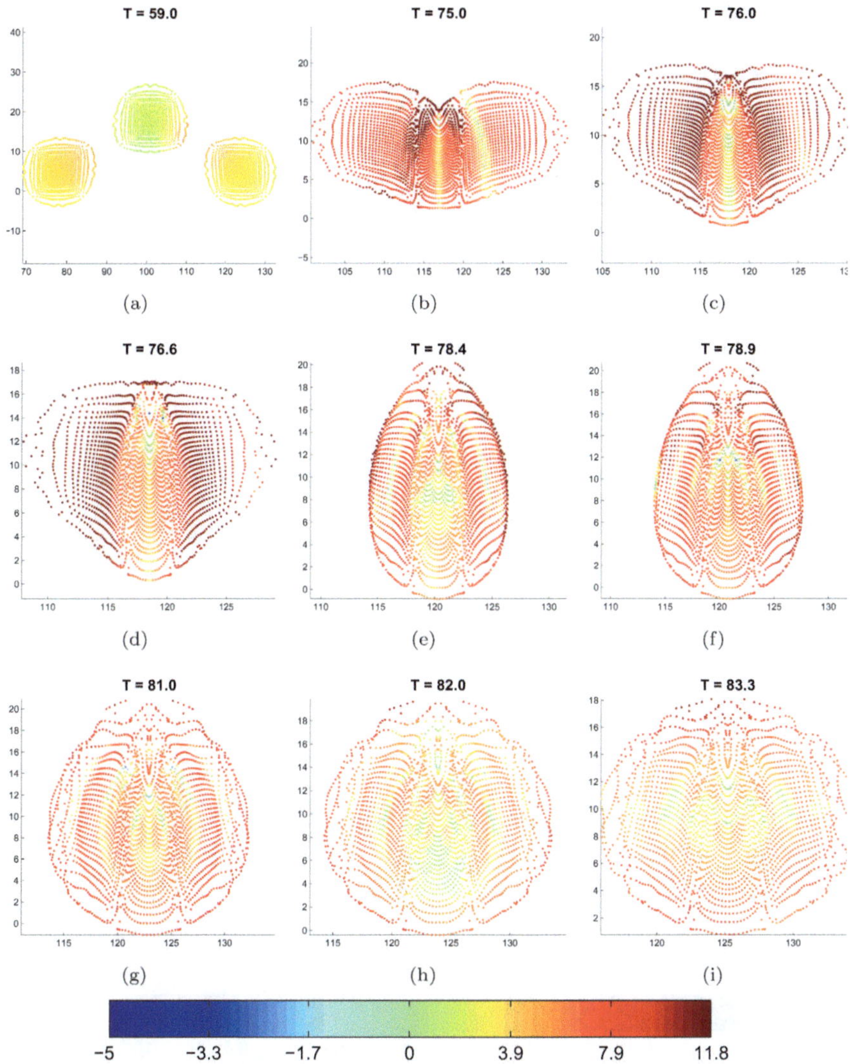

Fig. 12.5 Local information storage in a variable-speed swarm at key time steps. The swarm initially consists of three square groups in a checker configuration. Note, to visualize the range of values, we map 0 to the center of the colorbar and the maximum and minimum values to the extremes of the colorbar. Values vary linearly on the left half and the right half, but the scales on the two halves are different.

corresponds to dynamic coordination within each swarm, while the second, global, AIS maximum at $T = 76.0$ indicates the moment when all three swarms merge into a single coordinated entity (Fig. 12.5 (c)), confined within a spatial extent that will not change significantly in size past this point. The TE maxima follow shortly after the corresponding AIS maxima (cf. blue solid curve in Fig. 12.8).

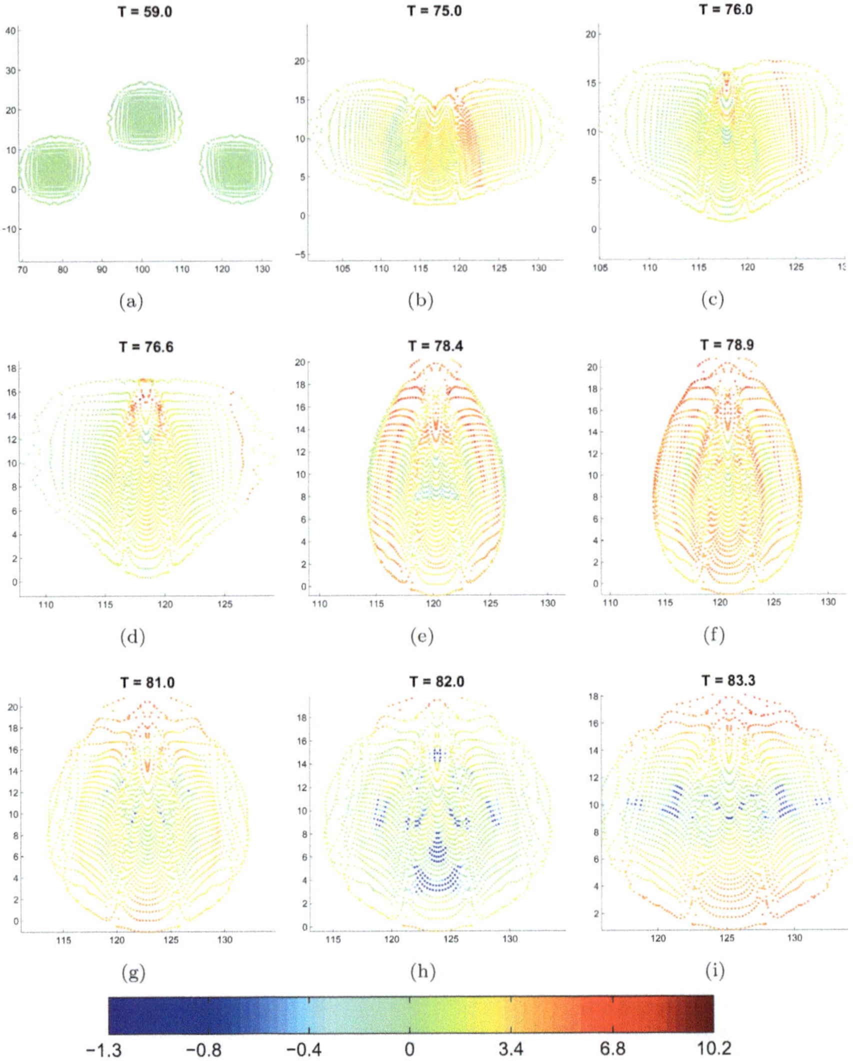

Fig. 12.6 Local information transfer in a variable-speed swarm at key time steps. The swarm initially consists of three square groups in a checker configuration. Note, as with the previous figure, the scales on the positive and negative halves of the colorbar are different.

The study of Wang et al. (2012) pointed out the asymmetry in local values (e.g., Fig. 12.6 (b)-(c)). This is due to an initial asymmetry in the swarm propagated during the process. The eventual decline in information storage and transfer is similar to the one observed by Ceguerra et al. (2011) when the system of coupled oscillators approached a synchronized (steady) state. Both of these phenomena indicate that the distributed computation is complete (if and) when the dynamics within the system become synchronized.

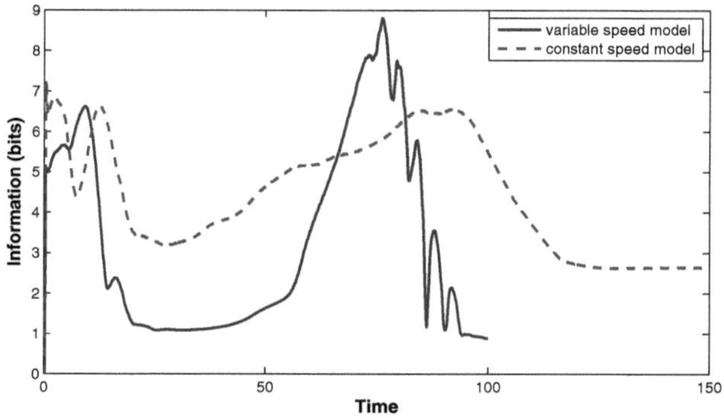

Fig. 12.7 Comparison of global average active information storage (AIS) between variable-speed and constant-speed model swarms. The swarm initially consists of three square groups in a checker configuration.

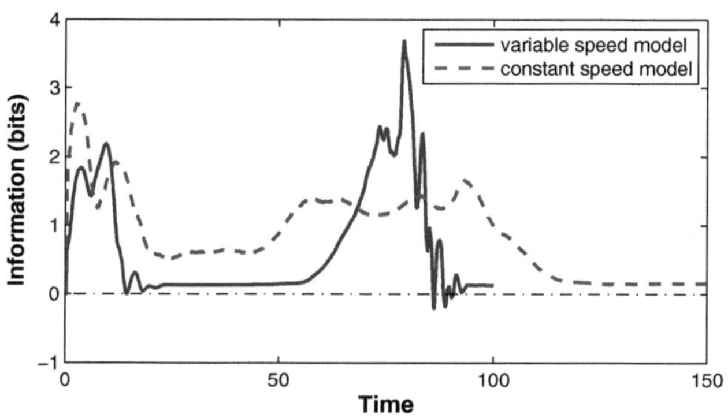

Fig. 12.8 Comparison of global average transfer entropy (TE) between variable-speed and constant-speed model swarms. The swarm initially consists of three square groups in a checker configuration.

12.4.2 Constant-Speed Swarm Model

12.4.2.1 Experiment 3: One Group, Constant Speed

We use the same initial conditions in our investigation of the constant-speed model. We begin with a square configuration 49×49 in size with individuals placed on a 1×1 grid. The red dashed curves in Figures 12.3 and 12.4 show the global AIS and TE values for this system over time, notable time steps and their individual local dynamics are shown in Fig. 12.9 and Fig. 12.10.

We use the AIS to measure memory in use for computation. Again, the individuals in the middle of the swarm do not feel the boundary effects at first and do not participate in computation. However, those on the top and bottom edges are actively

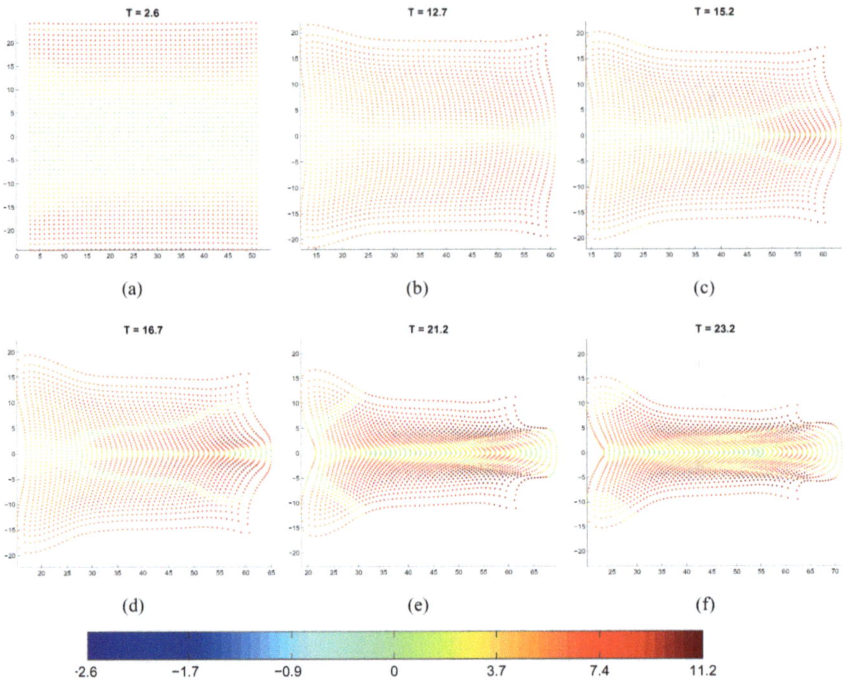

Fig. 12.9 Local information storage in a constant-speed swarm at key time steps: (b) first local maximum average AIS; (c) local minimum average AIS; (d) first local maximum TE; (e) second local maximum average AIS; (f) second local maximum average TE. The swarm is initially in a square configuration. Note, to visualize the range of values, we map 0 to the center of the colorbar and the maximum and minimum values to the extremes of the colorbar. Values vary linearly on the left half and the right half, but the scales on the two halves are different.

participating. The TE has a similar spatial pattern early on, with higher values at the edge. The center is negative because information from neighbors is misleading.

Note that the individuals in the front and back of the swarm (the left and right sides) have low AIS even though they are near the edge. This is because the direction of the desired velocity is opposite the current velocity (for those in the front) or coincides with the current velocity (for those in the back), and so the net effect is zero. No change in velocity (i.e. a static state) means AIS and TE values are close to zero.

At first, waves of TE and AIS in the constant-speed model tend to travel from front to back and from the top and bottom edges. This is different from the variable-speed model where the waves traveled from every edge inward or from the center outward in a circle.

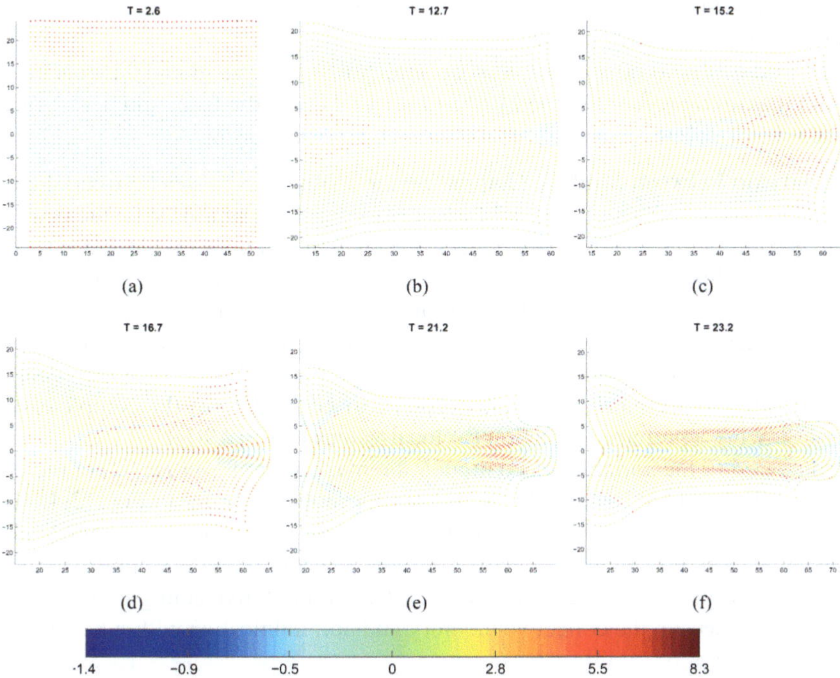

Fig. 12.10 Local information transfer in a constant-speed swarm at key time steps: (b) first local maximum average AIS; (c) local minimum average AIS; (d) first local maximum average TE; (e) second local maximum average AIS; (f) second local maximum average TE. The swarm is initially in a square configuration. Note, as with the previous figure, the scales on the positive and negative halves of the colorbar are different.

The group has a wave of low local TE in a similar place to the AIS wave (cf. Figures 12.10(c) and 12.9(c)). However, either side of the wave of negative TE has some of the highest values of TE at these time steps ($T = 16.7$ and $T = 23.2$). The individuals with negative TE are receiving conflicting information from their neighbors. On either side of this wave with low TE, individuals with higher TE have more coordination with their neighbors.

The global AIS increases steadily before dropping briefly and then attaining its absolute maximum. When the average AIS drops a little, we observe a wave of low or negative local AIS starting from the front center of the swarm and moving out and back. On either side of this wave, the individuals are coordinated, and the wave of low AIS marks the place where the individuals also have the smallest change in direction.

Each local maximum of average AIS is followed by a local maximum of average TE. The first local maximum of average AIS at $T = 12.7$ is before the waves of low AIS begin. As the wave of low AIS moves through the group, the overall AIS drops. However, the AIS rises again once the wave nears the back of the group until it achieves its absolute maximum at $T = 21.2$. The average AIS decreases for $T > 21.2$ because first the middle of the swarm and then the edges reach a static state (cf. red dashed line in Fig. 12.3).

We would like to point out that soon after this stage the global TE reaches its second maximum (cf. red dashed line in Fig. 12.4), and the front and center of the group is settling into synchrony, as indicated by the decreasing AIS.

Comparison between Models. Comparing the global AIS values between the two models in Fig. 12.3, there are two differences in the maximum AIS: (i) the peak average value for constant-speed model is lower than that in the variable-speed model, which shows that a system under constraints will have lower maximum memory; (ii) it occurs at a considerably later time, in other words, it is harder for this system to achieve coherence. In addition, we point out that the spatiotemporal representation of local AIS for the variable-speed model was symmetrical (Fig. 12.1(f)), while the corresponding representation for the constant-speed model lost the complete symmetry showing a preferred direction for local values (Fig. 12.9(f)).

Compared to the variable-speed model, the maximum of average TE in the constant speed model has lower value and is less pronounced (refer to Fig. 12.4). This indicates that communications within the system are less informative, thus leading to less well-formed cascades. Finally, we note that the final configurations of the models are very different (see Miller et al. (2011) for details), configurations driven by the differences in the collective memory and communication within the swarms.

12.4.2.2 Experiment 4: Three Groups, Constant Speed

We also consider the constant-speed model with initial placement in three squares in a checker configuration and examine the TE and AIS as the three groups interact. The red dashed curves in Figures 12.7 and 12.8 show the overall information dynamics in the swam over time, with key time steps shown in Fig. 12.11 and Fig. 12.12.

In general, the AIS is higher in the center and leftmost groups, since they have the most directional change. As the center group turns downward with fixed speed, the left group begins to 'catch up' to the center group. This means that the center and left groups feel a stronger influence from one another than from the right group. The right group has little change in its shape or direction until after $T = 49.6$ (Fig. 12.11(c)). As the left and center groups combine, more and more individuals join in the collective computation. The maximum AIS at $T = 91.5$ (Fig. 12.11(g)) occurs when all the individuals in this new, larger group are actively participating in the collective computation. As this larger group begins to settle into its steady state, the AIS decreases. The AIS in the right group is close to zero the whole time. By $T = 105.0$, the two groups are moving away from each other since the distance between them is great enough that they have only a small influence on each other.

The local TE values in the groups increased very slowly prior to $T = 42.7$ (Fig. 12.12). Notice that at $T = 56.9$, the rightmost group has an attraction to the other two groups, but can not slow down to merge with them. We can also observe a wave of negative TE on the left side of the rightmost group, since the neighbors

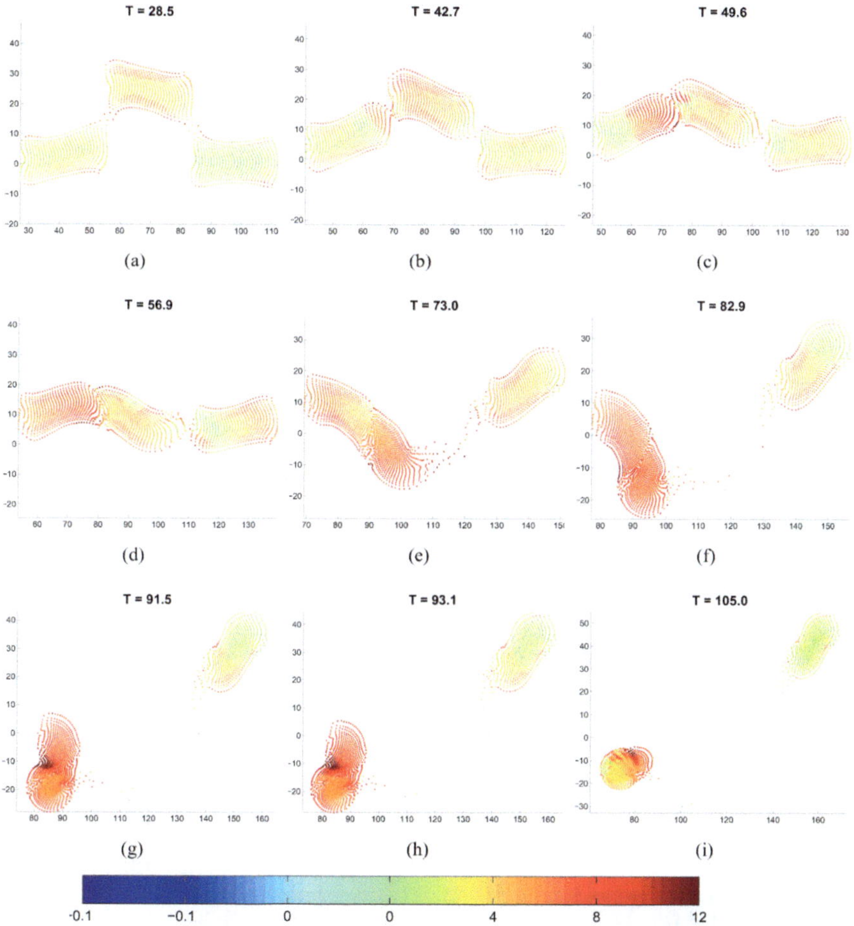

Fig. 12.11 Local information storage (AIS) in a constant-speed swarm at key time steps: (b) local minimum average TE; (d) local maximum average TE; (e) local minimum average TE; (f) local maximum average TE; (g) maximum average AIS; (h) maximum average TE. The swarm initially consists of three square groups in a checker configuration. Note, to visualize the range of values, we map 0 to the center of the colorbar and the maximum and minimum values to the extremes of the colorbar. Values vary linearly on the left half and the right half, but the scales on the two halves are different.

(in the leftmost and center groups) misinform about the next state of the nearest individuals in the rightmost group.

At $T = 73.0$, there is a high spot of AIS (Fig. 12.11 (e)) and a low spot of TE (Fig. 12.12 (e)) where the leftmost and center groups are merging. The groups are

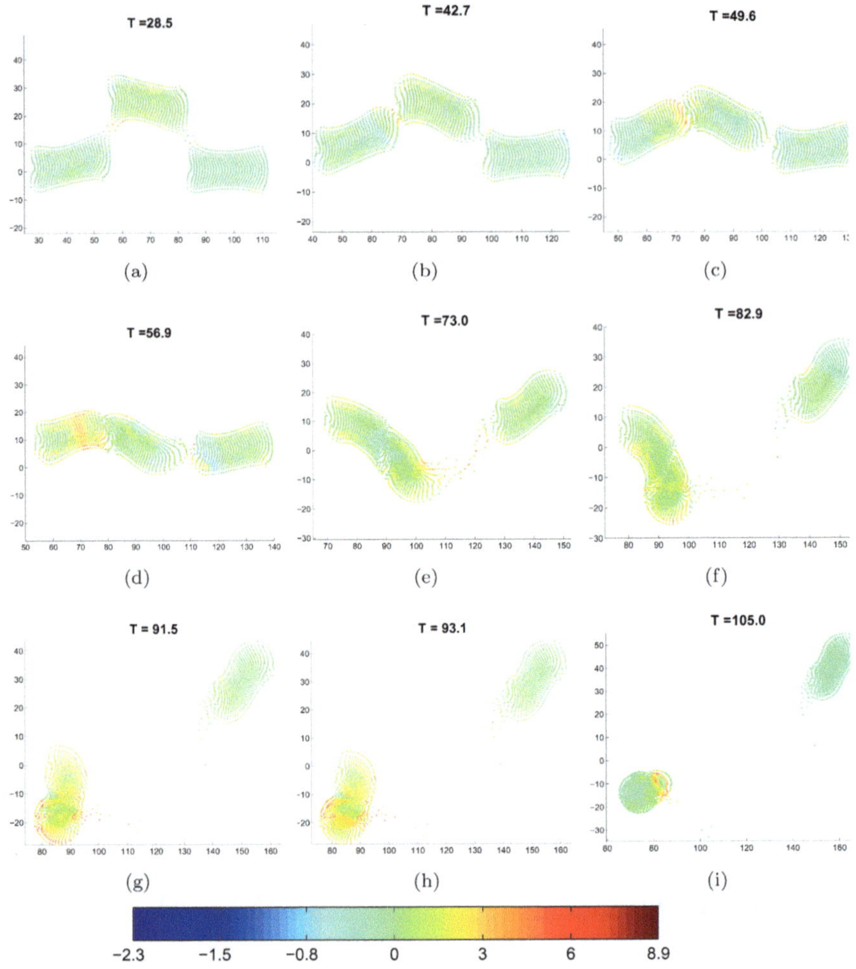

Fig. 12.12 Local information transfer (TE) in a constant-speed swarm at key time steps: (b) local minimum average TE; (d) local maximum average TE; (e) local minimum average TE; (f) local maximum average TE; (g) maximum average AIS; (h) maximum average TE. The swarm initially consists of three square groups in a checker configuration. Note, as with the previous figure, the scales on the positive and negative halves of the colorbar are different.

still moving toward each other. This means that the neighbors are misinforming each other since they are not aligned.

Comparison between Models. Comparing to the corresponding experiment in the variable-speed model (refer to Fig. 12.7 for AIS, and Fig. 12.8 for TE), we note that the constraints imposed by the constant speed model prevented formation of a single fully merged swarm. Information-theoretically, the differences in global AIS and TE amounted to the following: (i) both information dynamics have lower maximum values here; (ii) the maximum AIS and TE occurs at a later time; (iii) the TE peaks are less pronounced.

12.5 Conclusions

This paper applies a novel information theoretic characterization of information dynamics within swarms in the context of GSO. We argue that collective communication and memory are two necessary elements of distributed computation, and study these elements in two different models (one being more constrained/guided than the other).

The proposed framework measures cascades in precise information theoretic terms using well defined notions of conditional transfer entropy. The results confirm that the information cascades occur in waves propagating through the swarm and can be observed via coherent changes in local transfer entropies of individual particles. Local transfer entropy was recently given a thermodynamic interpretation as external entropy production (Prokopenko et al. 2013). This interpretation relates higher local TE (brought about by information cascades) to negentropy, the entropy that the system exports (dissipates) to keep its own entropy low (Schrödinger 1944; Prokopenko 2013).

Our characterization deals with weak and ambiguous external stimuli by incorporating both positive and negative local information transfer values. Information cascades are not just observed as changes in behaviors and activities, but rather are rigorously determined and computed as coherent collective communication. On the other hand, collective memory is identified with active information storage — another novel but well defined concept. Higher values of AIS are associated with higher levels of dynamic coordination.

The comparative analysis shows that there are significant differences between the information dynamics that emerge from our two models. Firstly, the maxima of the global AIS and TE are lower for the constrained constant-speed model. This indicates that a system under constraints (which therefore has lower uncertainty or entropy of its state) tends to have a lower collective memory, and that communications within such a system are less informative. Secondly, the overall maximum TE in the constrained system is less pronounced, leading to less well-formed cascades. Thirdly, the global maxima of AIS and TE are attained at a later stage for the swarm constrained by constant speed. In other words, it is harder for a constrained system to self-organize into a coherent, swarming state. However, the constrained system was observed to produce a different pattern in the spatiotemporal representation of local AIS, rather than being isotropic. In general, the principle of maximal information transfer following the stage with maximal collective memory is observed under both models. This analysis exemplifies the use of information-theoretic tools for GSO studies.

The overall study, following the work of Wang et al. (2012), demonstrates the strength of an information-theoretic approach that allows us to directly compare different systems.

Acknowledgements. We thank the CSIRO Advanced Scientific Computing (`http://www.hpsc.csiro.au`) for the use of their supercomputer clusters in performing the experiments for this paper. LFR and JMM acknowledge the support of US National Science Foundation grants CCF-0726556 and CCF-0829748.

References

Aoki, I.: A simulation study on the schooling mechanism in fish. Bulletin of the Japanese Society of Scientific Fisheries 48, 1081–1088 (1982)

Ay, N., Bernigau, H., Der, R., Prokopenko, M.: Information driven self-organization: The dynamical system approach to autonomous robot behavior. Theory in Biosciences 131(3), 161–179 (2011)

Camazine, S., Deneubourg, J.-L., Franks, N.R., Sneyd, J., Theraulaz, G., Bonabeau, E.: Self-Organization in Biological Systems. Princeton University Press (2003)

Ceguerra, R.V., Lizier, J.T., Zomaya, A.Y.: Information storage and transfer in the synchronization process in locally-connected networks. In: 2011 IEEE Symposium on Artificial Life (ALIFE), pp. 54–61. IEEE (2011)

Couzin, I.: Collective minds. Nature 445(7129), 715–715 (2007)

Couzin, I.D.: Collective cognition in animal groups. Trends in Cognitive Sciences 13(11), 36–43 (2009)

Couzin, I.D., James, R., Croft, D.P., Krause, J.: Social organization and information transfer in schooling fishes. In: Brown, C., Laland, K., Krause, J. (eds.) Fish and Aquatic Resources, pp. 166–185. Blackwell Publishing (2006)

Couzin, I.D., Krause, J., James, R., Ruxton, G.D., Franks, N.: Collective memory and spatial sorting in animal groups. Journal of Theoretical Biology 218, 1–11 (2002)

Dall, S.R.X., Giraldeau, L.-A., Olsson, O., McNamara, J.M., Stephens, D.W.: Information and its use by animals in evolutionary ecology. Trends in Ecology & Evolution 20(4), 187–193 (2005)

Galef, J. B.G., Giraldeau, L.-A.: Social influences on foraging in vertebrates: causal mechanisms and adaptive functions. Animal Behaviour 61(1), 3–15 (2001)

Giraldeau, L.-A., Valone, T.J., Templeton, J.J.: Potential disadvantages of using socially acquired information. Philosophical Transactions of the Royal Society London B Biological Sciences 357(1427), 1559–1566 (2002)

Huth, A., Wissel, C.: The simulation of the movement of fish schools. Journal of Theoretical Biology 156, 365–385 (1992)

Kastberger, G., Schmelzer, E., Kranner, I.: Social waves in giant honeybees repel hornets. PLoS One 3(9), e3141 (2008)

Katz, Y., Tunstrøm, K., Ioannou, C.C., Huepeb, C., Couzin, I.D.: Inferring the structure and dynamics of interactions in schooling fish. Proceedings of the National Academy of Sciences 108(46), 18720–18725 (2011)

Lissaman, P.B.S., Shollenberger, C.A.: Formation flight of birds. Science 168(3934), 1003–1005 (1970)

Lizier, J.T.: The Local Information Dynamics of Distributed Computation in Complex Systems. Springer Theses. Springer, Heidelberg (2013)

Lizier, J.T., Prokopenko, M., Zomaya, A.Y.: Detecting non-trivial computation in complex dynamics. In: Almeida e Costa, F., Rocha, L.M., Costa, E., Harvey, I., Coutinho, A. (eds.) ECAL 2007. LNCS (LNAI), vol. 4648, pp. 895–904. Springer, Heidelberg (2007)

Lizier, J.T., Prokopenko, M., Zomaya, A.Y.: The information dynamics of phase transitions in random boolean networks. In: Bullock, S., Noble, J., Watson, R., Bedau, M.A. (eds.) Proceedings of the Eleventh International Conference on the Simulation and Synthesis of Living Systems (ALife XI), Winchester, UK, pp. 374–381. MIT Press, Cambridge (2008a)

Lizier, J.T., Prokopenko, M., Zomaya, A.Y.: Local information transfer as a spatiotemporal filter for complex systems. Physical Review E 77(2), 026110 (2008b)

Lizier, J.T., Prokopenko, M., Zomaya, A.Y.: Information modification and particle collision in distributed computation. Chaos 20(3), 037109 (2010)

Lizier, J.T., Prokopenko, M., Zomaya, A.Y.: Local measures of information storage in complex distributed computation. Information Sciences 208, 39–54 (2012)

Lukeman, R., Li, Y.-X., Edelstein-Keshet, L.: Inferring individual rules from collective behavior. Proceedings of the National Academy of Sciences of the United States of America 107(28), 12576–12580 (2010)

Mathews, G., Durrant-Whyte, H.F., Prokopenko, M.: Measuring global behaviour of multi-agent systems from pair-wise mutual information. In: Khosla, R., Howlett, R.J., Jain, L.C. (eds.) KES 2005. LNCS (LNAI), vol. 3684, pp. 587–594. Springer, Heidelberg (2005)

Miller, J., Kolpas, A., Juchem Neto, J., Rossi, L.: A continuum three-zone model for swarms. Bulletin of Mathematical Biology 74(3), 1–26 (2011)

Parrish, J.K., Edelstein-Keshet, L.: Complexity, pattern, and evolutionary trade-offs in animal aggregation. Science 284(5411), 99–101 (1999)

Parrish, J.K., Viscido, S.V., Grünbaum, D.: Self-organized fish schools: An examination of emergent properties. The Biological Bulletin 202, 296–305 (2002)

Partridge, B.L.: The structure and function of fish schools. Scientific American 246(6), 114–123 (1982)

Potts, W.K.: The chorus-line hypothesis of manoeuvre coordination in avian flocks. Nature 309, 344–345 (1984)

Procaccini, A., Orlandi, A., Cavagna, A., Giardina, I., Zoratto, F., Santucci, D., Chiarotti, F., Hemelrijk, C.K., Alleva, E., Parisi, G., Carere, C.: Propagating waves in starling, sturnus vulgaris, flocks under predation. Animal Behaviour 82(4), 759–765 (2011)

Prokopenko, M.: Information dynamics at the edge of chaos: Measures, examples, and principles. In: 2013 IEEE Symposium on Artificial Life (ALIFE), pp. 148–152. IEEE (2013)

Prokopenko, M., Boschetti, F., Ryan, A.J.: An information-theoretic primer on complexity, self-organization, and emergence. Complexity 15(1), 11–28 (2009)

Prokopenko, M., Lizier, J.T., Price, D.C.: On thermodynamic interpretation of transfer entropy. Entropy 15(2), 524–543 (2013)

Radakov, D.V.: Schooling in the ecology of fish. John Wiley & Sons Inc, New Year (1973), Translated from Russian by Mills, H.

Schreiber, T.: Measuring information transfer. Physical Review Letters 85(2), 461–464 (2000)

Schrödinger, E.: What is life? The Physical Aspect of the Living Cell. Cambridge University Press (1944)

Sinclair, A.R.E., Norton-Griffiths, M. (eds.): Serengeti: Dynamics of an Ecosystem. University of Chicago Press (1979)

Sumpter, D., Buhl, J., Biro, D., Couzin, I.: Information transfer in moving animal groups. Theory in Biosciences 127(2), 177–186 (2008)

Treherne, J.E., Foster, W.A.: Group transmission of predator avoidance behaviour in a marine insect: The Trafalgar effect. Animal Behaviour 29(3), 911–917 (1981)

Uvarov, B.P.: Grasshoppers and Locusts. Imperial Bureau of Entomology (1928)

Vicsek, T., Czirok, A., Ben-Jacob, E., Cohen, I., Shochet, O.: Novel Type of Phase Transition in a System of Self-Driven Particles. Physical Review Letters 75, 1226–1229 (1995)

Wang, X.R., Miller, J.M., Lizier, J.T., Prokopenko, M., Rossi, L.F., 2011, E.C.A.L.: Measuring information storage and transfer in swarms. In: Proceedings of the Eleventh European Conference on the Synthesis and Simulation of Living Systems (ECAL 2011), pp. 838–845. MIT Press, Paris (2011)

Wang, X.R., Miller, J.M., Lizier, J.T., Prokopenko, M., Rossi, L.F.: Quantifying and tracing information cascades in swarms. PLoS One 7(7), e40084 (2012)

Chapter 13
Guiding Designs of Self-Organizing Swarms: Interactive and Automated Approaches

Hiroki Sayama

13.1 Introduction

Engineering design has traditionally been a top-down process in which a designer shapes, arranges and combines various components in a specific, precise, hierarchical manner, to create an artifact that will behave deterministically in an intended way (Minai et al. 2006; Pahl et al. 2007). However, this process does not apply to complex systems that show self-organization, adaptation and emergence. Complex systems consist of a massive amount of simpler components that are coupled locally and loosely, whose behaviors at macroscopic scales emerge partially stochastically in a bottom-up way. Such emergent properties of complex systems are often very robust and dynamically adaptive to the surrounding environment, indicating that complex systems bear great potential for engineering applications (Ottino 2004).

In an attempt to design engineered complex systems, one of the most challenging problems has been how to bridge the gap between macro and micro scales. Some mathematical techniques make it possible to analytically show such macro-micro relationships in complex systems (e.g., those developed in statistical mechanics and condensed matter physics (Bar-Yam 2003; Boccara 2010)). However, those techniques are only applicable to "simple" complex systems, in which: system components are reasonably uniform and homogeneous, their interactions can be approximated without losing important dynamical properties, and/or the resulting emergent patterns are relatively regular so that they can be characterized by a small number of macroscopic order parameters (Bar-Yam 2003; Doursat et al. 2012). Unfortunately, such cases are exceptions in a vast, diverse, and rather messy compendium of

Hiroki Sayama
Collective Dynamics of Complex Systems Research Group,
Binghamton University, State University of New York
Binghamton, NY 13902-6000, USA
e-mail: sayama@binghamton.edu

M. Prokopenko (ed.), *Guided Self-Organization: Inception*,
Emergence, Complexity and Computation 9,
DOI: 10.1007/978-3-642-53734-9_13, © Springer-Verlag Berlin Heidelberg 2014

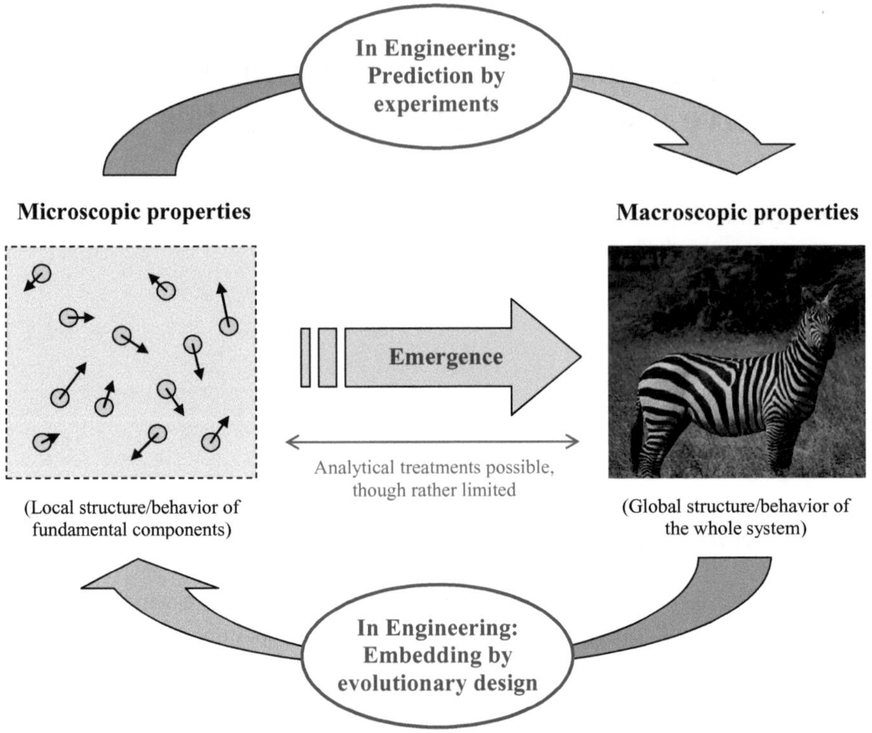

Fig. 13.1 Relationships of macroscopic and microscopic properties in complex systems and how complex systems engineering has been handling the gap between them

complex systems dynamics (Camazine 2003; Sole and Goodwin 2008). To date, the only generalizable methodology available for predicting macroscopic properties of a complex system from microscopic rules governing its fundamental components is to conduct experiments—either computational or physical—to let the system show its emergent properties by itself (Fig. 13.1, top).

More importantly, the other way of connecting the two scales—embedding macroscopic requirements the designer wants into microscopic rules that will collectively achieve those requirements—is by far more difficult. This is because the mapping between micro and macro scales is highly nonlinear, and also the space of possible microscopic rules is huge and thus hard to explore. So far, the only generalizable methodology available for macro-to-micro embedding in this context is to acquire microscopic rules by evolutionary means (Bentley 1999) (Fig. 13.1, bottom). Instead of trying to derive local rules analytically from global requirements, evolutionary methods let better rules spontaneously arise and adapt to meet the requirements, even though they do not produce any understanding of the macro-micro relationships. The effectiveness of such "blind" evolutionary search (Dawkins 1996) for complex systems design is empirically supported by the fact that it has been the primary mechanism that has produced astonishingly complex, sophisticated, highly emergent machinery in the history of real biological systems.

The combination of these two methodologies—experiment and evolution—that connect macro and micro scales in two opposite directions (the whole cycle in Fig. 13.1) is now a widely adopted approach for guiding systematic design of self-organizing complex systems (Minai et al. 2006; Anderson 2006). Typical design steps are to (a) create local rules randomly or using some heuristics, (b) conduct experiments using those local rules, (c) observe what kind of macroscopic patterns emerge out of them, (d) select and modify successful rules according to the observations, and (e) repeat these steps iteratively to achieve evolutionary improvement of the microscopic rules until the whole system meets the macroscopic requirements.

Such experiment-and-evolution-based design of complex systems is not free from limitations, however. In typical evolutionary design methods, the designer needs to explicitly define a performance metric, or "fitness", of design candidates, i.e., how good a particular design is. Such performance metrics are usually based on relatively simple observables easily extractable from experimental results (e.g., the distance a robot traveled, etc.). However, simple quantitative performance metrics may not be suitable or useful in evolutionary design of more complex structures or behaviors, such as those seen in real-world biological systems, where the key properties a system should acquire could be very diverse and complex, more qualitative than quantitative, and/or even unknown to the designer herself beforehand.

In this chapter, we present our efforts to address this problem, by (1) utilizing and enhancing interactive evolutionary design methods and (2) realizing spontaneous evolution of self-organizing swarms within an artificial ecosystem.

13.2 Model: Swarm Chemistry

We use Swarm Chemistry (Sayama 2007, 2009) as an example of self-organizing complex systems with which we demonstrate our design approaches. Swarm Chemistry is an artificial chemistry (Dittrich et al. 2001) model for designing spatio-temporal patterns of kinetically interacting heterogeneous particle swarms using evolutionary methods. A swarm population in Swarm Chemistry consists of a number of simple particles that are assumed to be able to move to any direction at any time in a two- or three-dimensional continuous space, perceive positions and velocities of other particles within its local perception range, and change its velocity in discrete time steps according to the following kinetic rules (adopted and modified from the rules in Reynolds' Boids (Reynolds 1987); see Fig. 13.2):

- If there are no other particles within its local perception range, steer randomly (*Straying*).
- Otherwise:
  - Steer to move toward the average position of nearby particles (*Cohesion*, Fig. 13.2(a)).

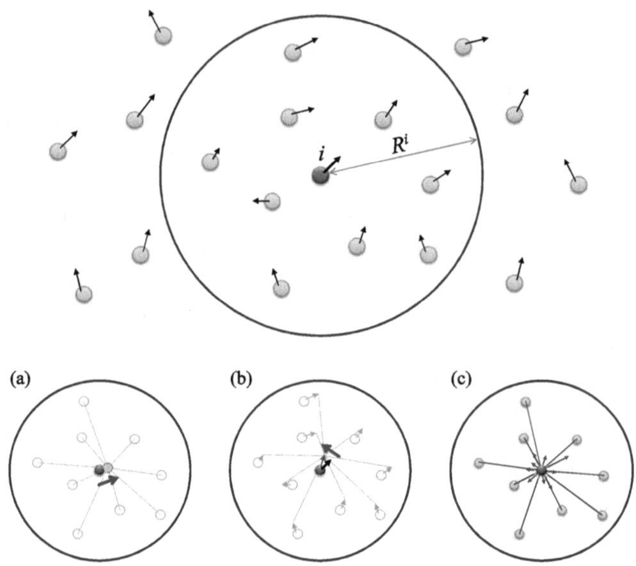

Fig. 13.2 Kinetic interactions between particles. Top: Particle i senses only positions and velocities of nearby particles within distance R^i. Bottom: (a) Cohesion. Particle i accelerates toward the center of mass of nearby particles. (b) Alignment. Particle i steers to align its orientation to the average orientation of nearby particles. (c) Separation. Particle i receives repulsion forces from each of the nearby particles whose strength is inversely related to distance.

  - Steer toward the average velocity of nearby particles (*Alignment*, Fig. 13.2(b)).
  - Steer to avoid collision with nearby particles (*Separation*, Fig. 13.2(c)).
  - Steer randomly with a given probability (*Randomness*).

- Approximate its speed to its own normal speed (*Self-propulsion*).

These rules are implemented in a simulation algorithm that uses kinetic parameters listed and explained in Table 13.1 (see (Sayama 2009, 2010) for details of the algorithm). The kinetic interactions in our model uses only one omni-directional perception range (R^i), which is much simpler than other typical swarm models that use multiple and/or directional perception ranges (Reynolds 1987; Couzin et al. 2002; Kunz and Hemelrijk 2003; Hemelrijk and Kunz 2005; Cheng et al. 2005; Newman and Sayama 2008). Moreover, the information being shared by nearby particles is nothing more than kinetic one (i.e., relative position and velocity), which is externally observable and therefore can be shared without any specialized communication channels[1]. These features make this system uniquely simple compared to other self-organizing swarm models.

[1] An exception is local information transmission during particle recruitment processes, which will be discussed later.

Table 13.1 Kinetic parameters involved in the simulation of particle behavior. Unique values are assigned to these parameters for each particle i as its own kinetic properties.

Name	Min	Max	Meaning	Unit
R^i	0	300	Radius of local perception range	pixel
V_n^i	0	20	Normal speed	pixel step^{-1}
V_m^i	0	40	Maximum speed	pixel step^{-1}
c_1^i	0	1	Strength of cohesive force	step^{-2}
c_2^i	0	1	Strength of aligning force	step^{-1}
c_3^i	0	100	Strength of separating force	pixel^2 step^{-2}
c_4^i	0	0.5	Probability of random steering	—
c_5^i	0	1	Tendency of self-propulsion	—

```
97 * (226.76, 3.11, 9.61, 0.15, 0.88, 43.35, 0.44, 1.0)
38 * (57.47, 9.99, 35.18, 0.15, 0.37, 30.96, 0.05, 0.31)
56 * (15.25, 13.58, 3.82, 0.3, 0.8, 39.51, 0.43, 0.65)
31 * (113.21, 18.25, 38.21, 0.62, 0.46, 15.78, 0.49, 0.61)
```

Fig. 13.3 Example of a recipe, formatted as a list of kinetic parameter sets of different types within a swarm. Each row represents one type, which has a number of particles of that type at the beginning, followed by its parameter settings in the format of $(R^i, V_n^i, V_m^i, c_1^i, c_2^i, c_3^i, c_4^i, c_5^i)$.

Each particle is assigned with its own kinetic parameter settings that specify preferred speed, local perception range, and strength of each kinetic rule. Particles that share the same set of kinetic parameter settings are considered of the same type. Particles do not have a capability to distinguish one type from another; all particles look exactly the same to themselves.

For a given swarm, specifications for its macroscopic properties are indirectly and implicitly woven into a list of different kinetic parameter settings for each swarm component, called a *recipe* (Fig. 13.3) (Sayama 2009). It is quite difficult to manually design a specific recipe that produces a desired structure and/or behavior using conventional top-down design methods, because the self-organization of a swarm is driven by complex interactions among a number of kinetic parameters that are intertwined with each other in highly non-trivial, implicit ways.

In the following sections, we address this difficult design problem using evolutionary methods. Unlike in other typical evolutionary search or optimization tasks, however, in our swarm design problem, there is no explicit function or algorithm readily available for assessing the quality (or fitness) of each individual design. To meet with this unique challenge, we used two complementary approaches: The interactive approach, where human users are actively involved in the evolutionary design process, and the automated approach, where spontaneous evolutionary dynamics of artificial ecosystems are utilized as the engine to produce creative self-organizing patterns.

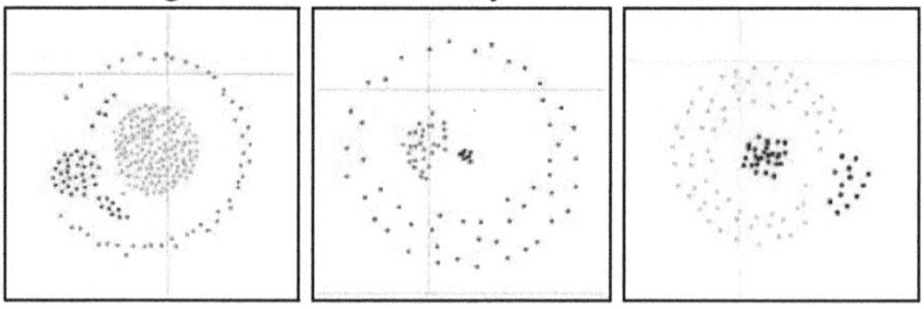

Fig. 13.4 Examples of swarms designed using IEC methods. Their recipes are available on the Swarm Chemistry website (`http://bingweb.binghamton.edu/~sayama/SwarmChemistry/`).

13.3 Interactive Approach

The first approach is based on interactive evolutionary computation (IEC) (Banzhaf 2000; Takagi 2001), a derivative class of evolutionary computation which incorporates interaction with human users. Most IEC applications fall into a category known as "narrowly defined IEC" (NIEC) (Takagi 2001), which simply outsources the task of fitness evaluation to human users. For example, a user may be presented with a visual representation of the current generation of solutions and then prompted to provide fitness information about some or all of the solutions. The computer in turn uses this fitness information to produce the next generation of solutions through the application of a predefined sequence evolutionary operators.

Our initial work, Swarm Chemistry 1.1 (Sayama 2007, 2009), also used a variation of NIEC, called Simulated Breeding (Unemi 2003). This NIEC-based application used discrete, non-overlapping generation changes. The user selects one or two favorable swarms out of a fixed number of swarms displayed, and the next generation is generated out of them, discarding all other unused swarms. Selecting one swarm creates the next generation using perturbation and mutation. Selecting two swarms creates the next generation by mixing them together (similar to crossover, but this mixing is not genetic but physical). Figure 13.4 shows some examples of self-organizing swarms designed using Swarm Chemistry 1.1.

As a design tool, NIEC has some disadvantages. One set of disadvantage stems from the confinement of the user to the role of selection operator (Fig. 13.5, left). Creative users who are accustomed to a more highly involved design process may find the experience to be tedious, artificial, and frustrating. Earlier literature suggests that it is important to instill in the user a strong sense of control over the entire evolutionary process (Bentley and O'Reilly 2001) and that the users should be the initiators of actions rather than simply responding to prompts from the system (Shneiderman et al. 2009).

These lines of research suggest that enhancing the level of interaction and control of IEC may help the user better guide the design process of self-organizing swarms. Therefore, we developed the concept of hyper-interactive evolutionary computation (HIEC) (Bush and Sayama 2011), a novel form of IEC in which a human user actively chooses when and how to apply each of the available evolutionary

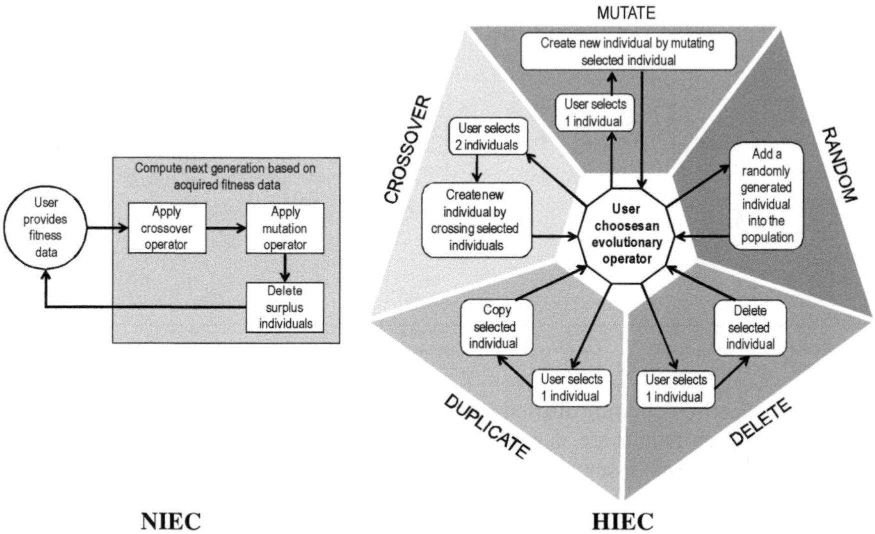

Fig. 13.5 Comparison of control flows between two interactive evolutionary computation (IEC) frameworks (redrawn based on figures in (Bush and Sayama 2011)). Left: Narrowly defined IEC (NIEC). Right: Hyper-interactive IEC (HIEC).

operators, playing the central role in the control flow of evolutionary search processes (Fig. 13.5, right). In HIEC, the user directs the overall search process and initiates actions by choosing when and how each evolutionary operator is applied. The user may add a new solution to the population through the crossover, mutate, duplicate, or random operators. The user can also remove solutions with the delete operator. This naturally results in dynamic variability of population size and continuous generation change (like steady-state strategies for genetic algorithms).

We developed Swarm Chemistry 1.2 (Sayama et al. 2009; Bush and Sayama 2011), a redesigned HIEC-based application for designing swarms. This version uses continuous generation changes, i.e., each evolutionary operator is applied only to part of the population of swarms on a screen without causing discrete generation changes. A mutated copy of an existing swarm can be generated by either selecting the "Mutate" option or double-clicking on a particular swarm. Mixing two existing swarms can be done by single-clicking on two swarms, one after the other. The "Replicate" option creates an exact copy of the selected swarm next to it. One can also remove a swarm from the population by selecting the "Kill" option or simply closing the frame. More details of HIEC and Swarm Chemistry 1.2 can be found elsewhere (Sayama et al. 2009; Bush and Sayama 2011).

We conducted the following two human-subject experiments to see if HIEC would produce a more controllable and positive user experience, and thereby better swarm design outcomes, than those with NIEC.

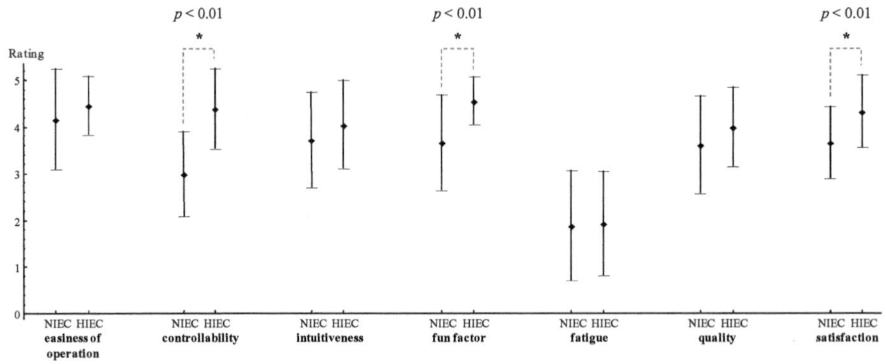

Fig. 13.6 Comparison of rating distribution between the NIEC and HIEC applications across seven factors. Mean ratings are shown by diamonds, with error bars around them showing standard deviations. Significant differences are indicated with an asterisk and corresponding t-test p-values.

13.3.1 User Experience

In the first experiment, individual subjects used the NIEC and HIEC applications mentioned above to evolve aesthetically pleasing self-organizing swarms. We quantified user experience outcomes using questionnaire, in order to quantify potential differences in user experience between the two applications.

Twenty-one subjects were recruited from students and faculty/staff members at Binghamton University. Each subject was recruited and participated individually. The subject was told to spend five minutes using each of two applications to design an "interesting and lifelike" swarm. Each of these two applications ran on their own dedicated computer station. After completing two sessions, each of which used either NIEC or HIEC application, the subject filled out a survey, rating each of the two platforms on the following factors: easiness of operation, controllability, intuitiveness, fun factor, fatigue level, final design quality, and overall satisfaction. Each factor was rated on a 5-point scale.

The results are shown in Fig. 13.6. Of the 7 factors measured, 3 showed statistically significant difference between two platforms: controllability, fun factor, and overall satisfaction. The higher controllability ratings for HIEC suggest that our original intention to re-design an IEC framework to grant greater control to the user was successful. Our results also suggest that this increased control may be associated with a more positive user experience, as is indicated by the higher overall satisfaction and fun ratings for HIEC. In the meantime, there was no significant difference detected in terms of perceived final design quality. This issue is investigated in more detail in the following second experiment.

13.3.2 Design Quality

The goal of the second experiment was to quantify the difference between HIEC and NIEC in terms of final design quality. In addition, the effects of mixing and mutation operators on the final design quality were also studied. The key feature of this experiment was that design quality was rated not individually by the subjects who designed them, but by an entire group of individual subjects. The increased amount of rating information yielded by this procedure allowed us to more effectively detect differences in quality between designs created using NIEC and designs created using HIEC.

Twenty-one students were recruited for this experiment. Those subjects did not have any overlap with the subjects of experiment 1. The subjects were randomly divided into groups of three and instructed to work together as a team to design an "interesting" swarm design in ten minutes using either the NIEC or HIEC application, the latter of which was further conditioned to have the mixing operator, the mutation operator, or both, or none. The sessions were repeated so that five to seven swarm designs were created under each condition. Once the sessions were over, all the designs created by the subjects were displayed on a large screen in the experiment room, and each subject was told to evaluate how "cool" each design was on a 0-to-10 numerical scale. Details of the experimental procedure and data analysis can be found elsewhere (Sayama et al. 2009; Bush and Sayama 2011).

The result is shown in Fig. 13.7. There was a difference in the average rating scores between designs created using NIEC and HIEC (conditions 0 and 4), and the rating scores were higher when more evolutionary operators were made available. Several final designs produced through the experiment are shown in Fig. 13.8 (three with the highest scores and three with the lowest scores), which indicate that highly evaluated swarms tended to maintain coherent, clear structures and motions without dispersal, while those that received lower ratings tended to disperse so that their behaviors are not appealing to students.

To detect statistical differences between experimental conditions, a one-way ANOVA was conducted. The result of the ANOVA is summarized in Table 13.2. Statistically significant variation was found between the conditions ($p < 0.005$). Tukey's and Bonferroni's post-hoc tests detected a significant difference between conditions 0 (NIEC) and 4 (HIEC), which supports our hypothesis that the HIEC is more effective at producing final designs of higher quality than NIEC. The post-hoc tests also detected a significant difference between conditions 1 (HIEC without mixing or mutation operators) and 4 (HIEC). These results indicate that the more active role a designer plays in the interactive design process, and the more diverse evolutionary operators she has at her disposal, the more effectively she can guide the evolutionary design of self-organizing swarms.

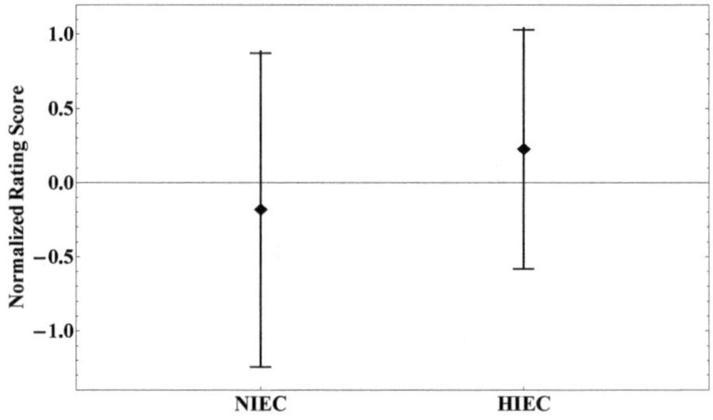

Fig. 13.7 Comparison of normalized rating score distributions between swarms produced using NIEC and HIEC. Average rating scores are shown by diamonds, with error bars around them showing standard deviations.

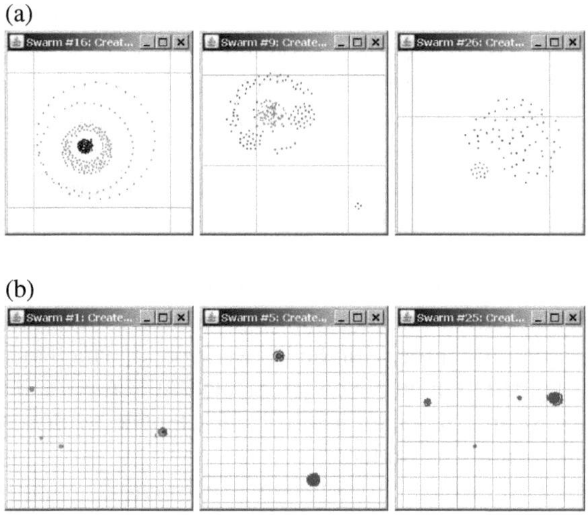

Fig. 13.8 Samples of the final swarm designs created by subjects. (a) Best three that received the highest rating scores. (b) Worst three that received the lowest rating scores.

Table 13.2 Results of one-way ANOVA on the rating scores for five conditions obtained in experiment 2 (from (Bush and Sayama 2011)). Significant difference is shown with an asterisk.

Source of variation	Degrees of freedom	Sum of squares	Mean square	F	F-test p-value
Between groups	4	14.799	3.700	4.11	0.003*
Within groups	583	525.201	0.901		
Total	587	540			

13.4 Automated Approach

The second approach we took was motivated by the following question: *Do we really need human users in order to guide designs of self-organizing swarms?* This question might sound almost paradoxical, because designing an artifact implies the existence of a designer by definition. However, this argument is quite similar to the "watchmaker" argument claimed by the English theologist William Paley (as well as by many other leading scientists in the past) (Dawkins 1996). Now that we know that the blind evolutionary process did "design" quite complex, intricate structures and functions of biological systems, it is reasonable to assume that it should be possible to create automatic processes that can spontaneously produce various creative self-organizing swarms without any human intervention.

In order to make the swarms capable of spontaneous evolution within a simulated world, we implemented several major modifications to Swarm Chemistry (Sayama 2010, 2011; Sayama and Wong 2011), as follows:

1. There are now two categories of particles, active (moving and interacting kinetically) and passive (remaining still and inactive). An active particle holds a recipe of the swarm (a list of kinetic parameter sets) (Fig. 13.9(a)).
2. A recipe is transmitted from an active particle to a passive particle when they collide, making the latter active (Fig. 13.9(b)).
3. The activated particle differentiates randomly into one of the multiple types specified in the recipe, with probabilities proportional to their ratio in it (Fig. 13.9(c)).
4. Active particles randomly and independently re-differentiate with small probability, r, at every time step ($r = 0.005$ for all simulations presented in this chapter).
5. A recipe is transmitted even between two active particles of different types when they collide. The direction of recipe transmission is determined by a competition function that picks one of the two colliding particles as a source (and the other as a target) of transmission based on their properties (Fig. 13.9(d)).
6. The recipe can mutate when transmitted, as well as spontaneously at every time step, with small probabilities, p_t and p_s, respectively (Fig. 13.9(e)). In a single recipe mutation event, several mutation operators are applied, including duplication of a kinetic parameter set (5% per set), deletion of a kinetic parameter set (5% per set), addition of a random kinetic parameter set (10% per event; increased to 50% per event in later experiments), and a point mutation of kinetic parameter values (10% per parameter).

These extensions made the model capable of showing morphogenesis and self-repair (Sayama 2010) and autonomous ecological/evolutionary behaviors of self-organized "super-organisms" made of a number of swarming particles (Sayama 2011; Sayama and Wong 2011). We note here that there was a technical problem in the original implementation of collision detection in an earlier version of evolutionary Swarm Chemistry (Sayama 2011), which was fixed in the later implementation (Sayama and Wong 2011).

In addition, in order to make evolution occur, we needed to confine the particles in a finite environment in which different recipes compete against each other.

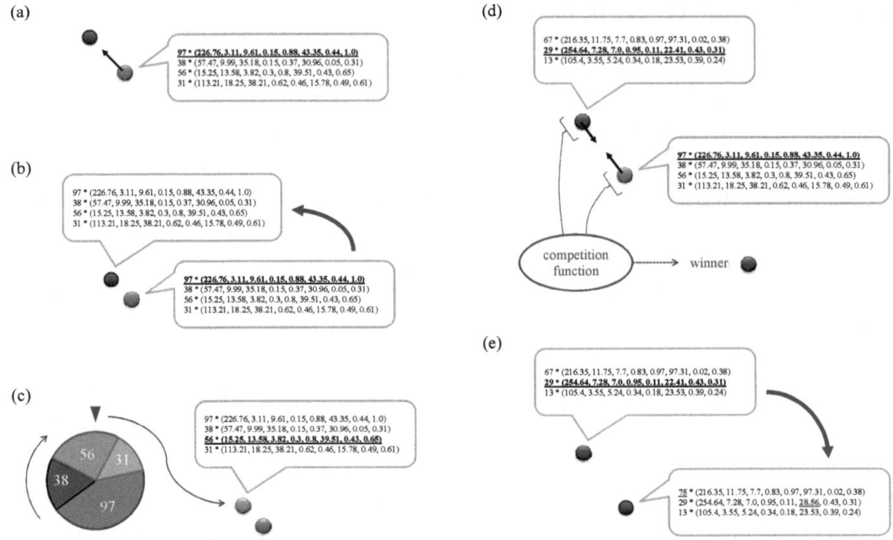

Fig. 13.9 How particle interactions work in the revised Swarm Chemistry (from (Sayama 2010b)). (a) There are two categories of particles, active (blue) and passive (gray). An active particle holds a recipe of the swarm in it (shown in the call-out). Each row in the recipe represents one kinetic parameter set. The underline shows which kinetic parameter set the particle is currently using (i.e., which kinetic type it is differentiated into). (b) A recipe is transmitted from an active particle to a passive particle when they collide, making the latter active. (c) The activated particle differentiates randomly into a type specified by one of the kinetic parameter sets in the recipe given to it. (d) A recipe is transmitted between active particles of different types when they collide. The direction of recipe transmission is determined by a competition function that picks one of the two colliding particles as a source (and the other as a target) of transmission based on their properties. (e) The recipe can mutate when transmitted with small probability.

We thus conducted all the simulations with 10,000 particles contained in a finite, $5{,}000 \times 5{,}000$ square space (in arbitrary units; for reference, the maximal perception radius of a particle was 300). A "pseudo"-periodic boundary condition was applied to the boundaries of the space. Namely, particles that cross a boundary reappear from the other side of the space just like in conventional periodic boundary conditions, but they do not interact across boundaries with other particles sitting near the other side of the space. In other words, the periodic boundary condition applies only to particle positions, but not to their interaction forces. This specific choice of boundary treatment was initially made because of its simplicity of implementation, but it proved to be a useful boundary condition that introduces a moderate amount of perturbations to swarms while maintaining their structural coherence and confining them in a finite area.

In the simulations, two different initial conditions were used: a *random* initial condition made of 9,900 inactive particles and 100 active particles with randomly generated one-type recipes distributed over the space, and a *designed* initial condition consisted of 9,999 inactive particles distributed over the space, with just one active particle that holds a pre-designed recipe positioned in the center of the space. Specifically, recipes of "swinger", "rotary" and "walker-follower" (shown in Fig. 13.4) patterns were used.

13.4.1 Exploring Experimental Conditions

Using the evolutionary Swarm Chemistry model described above, we studied what kind of experimental conditions (competition functions and mutation rates) would be most successful in creating self-organizing complex patterns (Sayama 2011).

The first experiment was to observe the basic evolutionary dynamics of the model under low mutation rates ($p_t = 10^{-3}$, $p_s = 10^{-5}$). Random and designed ("swinger") initial conditions were used. The following four basic competition functions were implemented and tested:

- *faster*: The faster particle wins.
- *slower*: The slower particle wins.
- *behind*: The particle that hit the other one from behind wins. Specifically, if a particle exists within a 90-degree angle opposite to the other particle's velocity, the former particle is considered a winner.
- *majority*: The particle surrounded by more of the same type wins. The local neighborhood radius used to count the number of particles of the same type was 30. The absolute counts were used for comparison.

Results are shown in Fig. 13.10. The results with the "behind" competition function were very similar to those with the "faster" competition function, and therefore omitted from the figure. In general, growth and replication of macroscopic structures were observed at early stages of the simulations. The growth was accomplished by recruitment of inactive particles through collisions. Once a cluster of active particles outgrew maximal size beyond which they could not maintain a single coherent structure (typically determined by their perception range), the cluster spontaneously split into multiple smaller clusters, naturally resulting in the replication of those structures. These growth and replication dynamics were particularly visible in simulations with designed initial conditions. Once formed, the macroscopic structures began to show ecological interactions by themselves, such as chasing, predation and competition over finite resources (i.e., particles), and eventually the whole system tended to settle down in a static or dynamic state where only a small number of species were dominant. There were some evolutionary adaptations also observed (e.g., in faster & designed ("swinger"); second row in Fig. 13.10) even with the low mutation rates used.

It was also observed that the choice of competition functions had significant impacts on the system's evolutionary dynamics. Both the "faster" and "behind" competition functions always resulted in an evolutionary convergence to a homogeneous cloud of fast-moving, nearly independent particles. In contrast, the "slower" competition function tended to show very slow evolution, often leading to the emergence

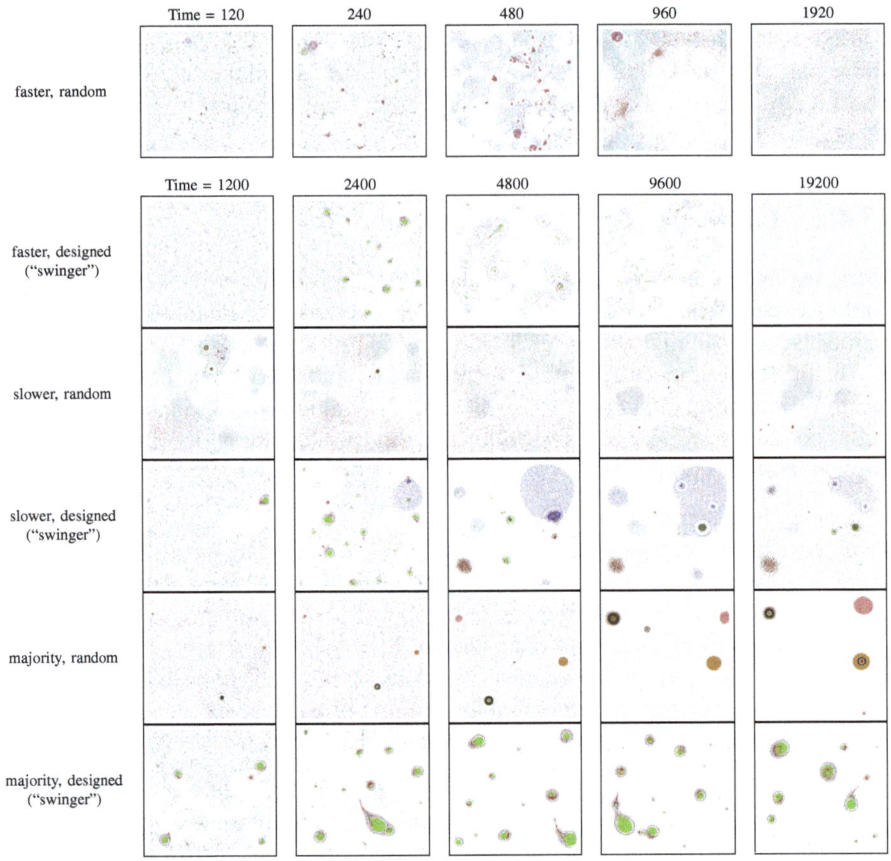

Fig. 13.10 Evolutionary processes observed in the evolutionary Swarm Chemistry model. Each image shows a snapshot of the space in a simulation, where dots with different colors represent particles of different types. Labels on the left indicates the competition function and the initial condition used in each case. Snapshots were taken at logarithmic time intervals.

of crystallized patterns. The "majority" competition function turned out to be most successful in creating and maintaining dynamic behaviors of macroscopic coherent structures over a long period of time, yet it was quite limited regarding the capability of producing evolutionary innovations. This was because any potentially innovative mutation appearing in a single particle would be lost in the presence of local majority already established around it.

Based on the results of the previous experiment, the following five more competition functions were implemented and tested. The last three functions that took recipe length into account were implemented in the hope that they might promote evolution of increasingly more complex recipes and therefore more complex patterns:

- *majority (probabilistic)*: The particle surrounded by more of the same type wins. This is essentially the same function as the original "majority", except that the winner is determined probabilistically using the particle counts as relative probabilities of winning.
- *majority (relative)*: The particle that perceives the higher density of the same type within its own perception range wins. The density was calculated by dividing the number of particles of the same type by the total number of particles of any kind, both counted within the perception range. The range may be different and asymmetric between the two colliding particles.
- *recipe length*: The particle with a recipe that has more kinetic parameter sets wins.
- *recipe length then majority*: The particle with a recipe that has more kinetic parameter sets wins. If the recipe length is equal between the two colliding particles, the winner is selected based on the "majority" competition function.
- *recipe length* $\times$ *majority*: A numerical score is calculated for each particle by multiplying its recipe length by the number of particles of the same type within its local neighborhood (radius = 30). Then the particle with a greater score wins.

Results are summarized in Fig. 13.11. As clearly seen in the figure, the majority-based rules are generally good at maintaining macroscopic coherent structures, regardless of minor variations in their implementations. This indicates that interaction between particles, or "cooperation" among particles of the same type to support one another, is the key to creating and maintaining macroscopic structures. Experimental observation of a number of simulation runs gave an impression that the "majority (relative)" competition function would be the best in this regard, therefore this function was used in all of the following experiments.

In the meantime, the "recipe length" and "recipe length then majority" competition functions did not show any evolution toward more complex forms, despite the fact that they would strongly promote evolution of longer recipes. What was occurring in these conditions was an evolutionary accumulation of "garbage" kinetic parameter sets in a recipe, which did not show any interesting macroscopic structure. This is qualitatively similar to the well-known observation made in Tierra (Ray 1992).

The results described above suggested the potential of evolutionary Swarm Chemistry for producing more creative, continuous evolutionary processes, but none of the competition functions showed notable long-term evolutionary changes yet. We therefore increased the mutation rates to a 100 times greater level than those in the experiments above, and also introduced a few different types of exogenous perturbations to create a dynamically changing environment (for more details, see (Sayama 2011)). This was informed by our earlier work on evolutionary cellular automata (Salzberg et al. 2004; Salzberg and Sayama 2004), which demonstrated that such dynamic environments may make evolutionary dynamics of a system more variation-driven and thus promote long-term evolutionary changes.

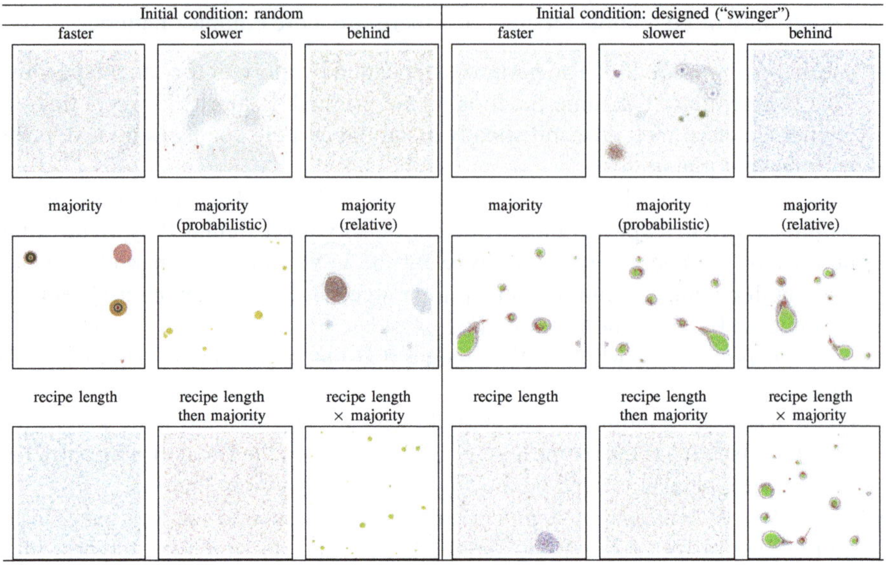

Fig. 13.11 Comparison between several different competition functions. The nine cases on the left hand side started with random initial conditions, while the other nine on the right hand side started with designed initial conditions with the "swinger" recipe. Snapshots were taken at time = 22,000 for all cases.

With these additional changes, some simulation runs finally demonstrated continuous changes of dominant macroscopic structures over a long period of time (Fig. 13.12). A fundamental difference between this and earlier experiments was that the perturbation introduced to the environment would often break the "status quo" established in the swarm population, making room for further evolutionary innovations to take place. A number of unexpected, creative swarm designs spontaneously emerged out of these simulation runs, fulfilling our intension to create automated evolutionary design processes. Videos of sample simulation runs can be found on our YouTube channel (`http://youtube.com/ComplexSystem`).

13.4.2 Quantifying Observed Evolutionary Dynamics

The experimental results described above were quite promising, but they were evaluated only by visual inspection with no objective measurements involved. To address the lack of quantitative measurements, we developed and tested two simple measurements to quantify the degrees of evolutionary exploration and macroscopic structuredness of swarm populations (Sayama and Wong 2011), assuming that the evolutionary process of swarms would look interesting and creative to human eyes if it displayed patterns that are clearly visible and continuously changing. These

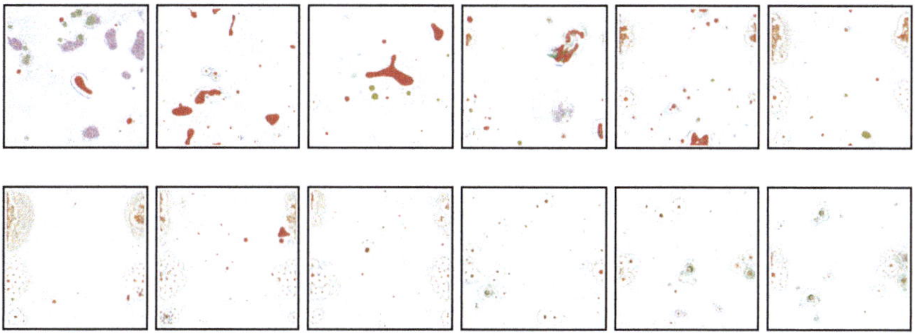

Fig. 13.12 An example of long-term evolutionary behavior seen under dynamic environmental conditions with high mutation rates. Snapshots were taken at constant time intervals (2,500 steps) to show continuous evolutionary changes.

Table 13.3 Four conditions used for the final experiment to quantify evolutionary dynamics

Name	Mutation rate	Environmental perturbation	Collision detection algorithm
original-low	low	off	original
original-high	high	on	original
revised-low	low	off	revised
revised-high	high	on	revised

measurements were developed so that they can be easily calculated a posteriori from a sequence of snapshots (bitmap images) taken in past simulation runs, without requiring genotypic or genealogical information that was typically assumed available in other proposed metrics (Bedau and Packard 1992; Bedau and Brown 1999; Nehaniv 2000).

Evolutionary exploration was quantified by counting the number of new RGB colors that appeared in a bitmap image of the simulation snapshot at a specific time point for the first time during each simulation run (Fig. 13.13, right). Since different particle types are visualized with different colors in Swarm Chemistry, this measurement roughly represents how many new particle types emerged during the last time segment. Macroscopic structuredness was quantified by measuring a Kullback-Leibler divergence (Kullback and Leibler 1951) of a pairwise particle distance distribution from that of a theoretical case where particles are randomly and homogeneously spread over the entire space (Fig. 13.13, left). Specifically, each snapshot bitmap image was first analyzed and converted into a list of coordinates (each representing the position of a particle, or a colored pixel), then a pair of coordinates were randomly sampled from the list 100,000 times to generate an approximate pairwise particle distance distribution in the bitmap image. The Kullback-Leibler divergence of the approximate distance distribution from the homogeneous case is larger when the swarm is distributed in a less homogeneous manner, forming macroscopic structures.

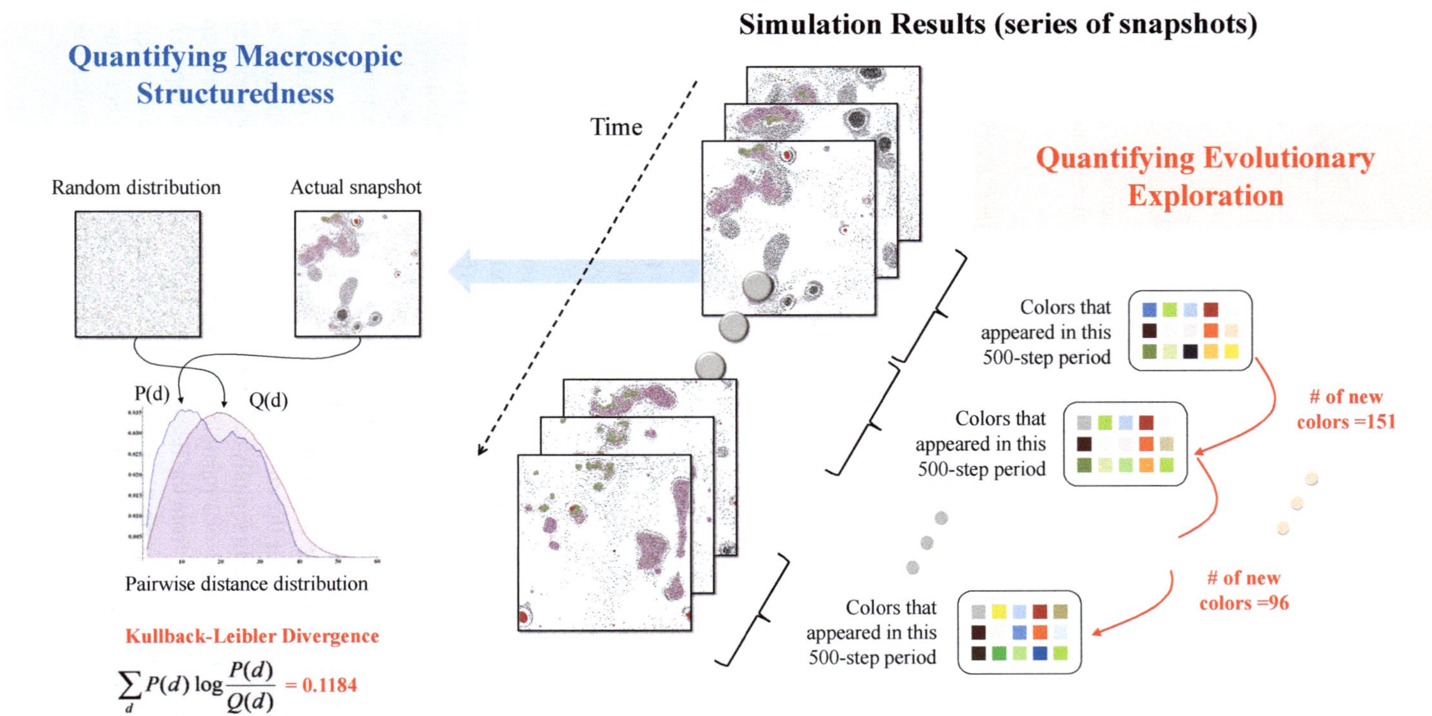

Fig. 13.13 Methods to quantify evolutionary exploration (right) and macroscopic structuredness (left) directly from a sequence of snapshots (bitmap images, center).

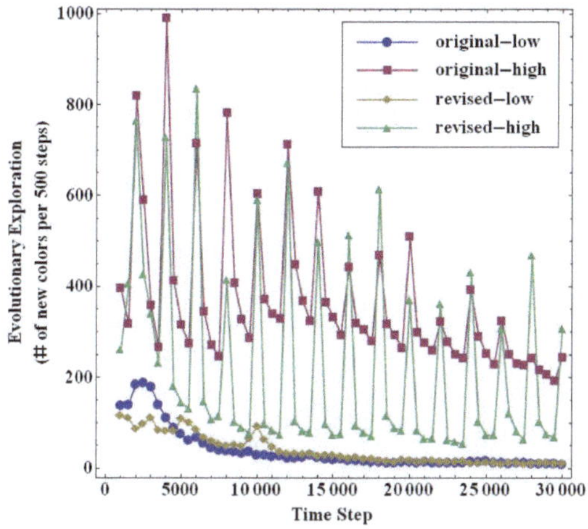

Fig. 13.14 Temporal changes of the evolutionary exploration measurement (i.e., number of new colors per 500 time steps) for four different experimental conditions, calculated from snapshots of simulation runs taken at 500 time step intervals (from (Sayama and Wong 2011)). Each curve shows the average result over 12 simulation runs (3 independent runs × 4 different initial conditions given in (Sayama 2011)). Sharp spikes seen in "high" conditions were due to dynamic exogenous perturbations.

We applied these measurements to simulation runs obtained under each of the four conditions shown in Table 13.3. Results are summarized in Figs. 13.14 and 13.15. Figure 13.14 clearly shows the high evolutionary exploration occurring under the conditions with high mutation rates and environmental perturbations. In the meantime, Figure 13.15 shows that the "original-high" condition had a tendency to destroy macroscopic structures by allowing swarms to evolve toward simpler, homogeneous forms. Such degradation of structuredness over time was, as mentioned earlier, due to a technical problem in the previous implementation of collision detection (Sayama 2011; Sayama and Wong 2011) that mistakenly depended on perception ranges of particles. The "revised" conditions used a fixed collision detection algorithm. This modification was found to have an effect to maintain macroscopic structures for a prolonged period of time (Fig. 13.15). Combining these results together (Fig. 13.16), we were able to detect automatically that the "revised-high" condition was most successful in producing interesting designs, maintaining macroscopic structures without losing evolutionary exploration. This conclusion also matched subjective observations made by human users.

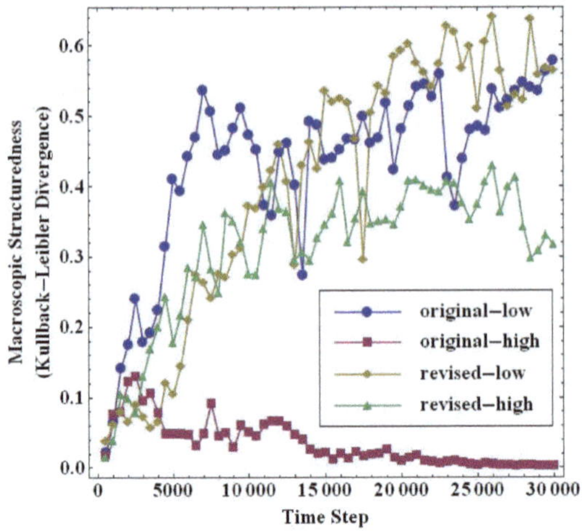

Fig. 13.15 Temporal changes of the macroscopic structuredness measurement (i.e., Kullback-Leibler divergence of the pairwise particle distance distribution from that of a purely random case) for four different experimental conditions, calculated from snapshots of simulation runs taken at 500 time step intervals (from (Sayama and Wong 2011)). Each curve shows the average result over 12 simulation runs (3 independent runs with 4 different initial conditions). The "original-high" condition loses macroscopic structures while other conditions successfully maintain them.

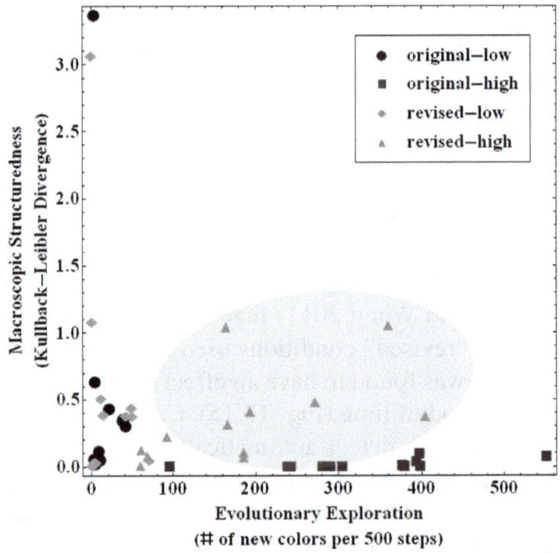

Fig. 13.16 Evolutionary exploration and macroscopic structuredness averaged over $t = 10,000 - 30,000$ for each independent simulation run (from (Sayama and Wong 2011), with slight modifications). Each marker represents a data point taken from a single simulation run. It is clearly observed that the "revised-high" condition (shaded in light blue) most successfully achieved high evolutionary exploration without losing macroscopic structuredness.

13.5 Conclusions

In this chapter, we have reviewed our recent work on two complementary approaches for guiding designs of self-organizing heterogeneous swarms. The common design challenge addressed in both approaches was the lack of explicit criteria for what constitutes a "good" design to produce. In the first approach, this challenge was solved by having a human user as an active initiator of evolutionary design processes. In the second approach, the criteria were replaced by low-level competition functions (similar to laws of physics) that drive spontaneous evolution of swarms in a virtual ecosystem.

The core message arising from both approaches is the unique power of evolutionary processes for designing self-organizing complex systems. It is uniquely powerful because evolution does not require any macroscopic plan, strategy or global direction for the design to proceed. As long as the designer—this could be either an intelligent entity or a simple unintelligent machinery—can make local decisions at microscopic levels, the process drives itself to various novel designs through unprescribed evolutionary pathways. Designs made through such open-ended evolutionary processes may have a potential to be more creative and innovative than those produced through optimization for explicit selection criteria.

We conclude this chapter with a famous quote by Richard Feynman. At the time of his death, Feynman wrote on a blackboard, *"What I cannot create, I do not understand."* This is a concise yet profound sentence that beautifully summarizes the role and importance of constructive understanding (i.e., model building) in scientific endeavors, which hits home particularly well for complex systems researchers. But research on evolutionary design of complex systems, including ours discussed here, has illustrated that the logical converse of the above quote is not necessarily true. That is, evolutionary approaches make this also possible—*"What I do not understand, I can still create."*

Acknowledgments. We thank the following collaborators and students for their contributions to the research presented in this chapter: Shelley Dionne, Craig Laramee, David Sloan Wilson, J. David Schaffer, Francis Yammarino, Benjamin James Bush, Hadassah Head, Tom Raway, and Chun Wong. This material is based upon work supported by the US National Science Foundation under Grants No. 0737313 and 0826711, and also by the Binghamton University Evolutionary Studies (EvoS) Small Grant (FY 2011).

References

Anderson, C.: Creation of desirable complexity: strategies for designing selforganized systems. In: Complex Engineered Systems, pp. 101–121. Springer (2006)
Banzhaf, W.: Interactive evolution. Evolutionary Computation 1, 228–236 (2000)
Bar-Yam, Y.: Dynamics of complex systems. Westview Press (2003)
Bedau, M.A., Brown, C.T.: Visualizing evolutionary activity of genotypes. Artificial Life 5(1), 17–35 (1999)

Bedau, M.A., Packard, N.H.: Measurement of evolutionary activity, teleology, and life. In: Artificial Life II, pp. 431–461. Addison-Wesley (1992)
Bentley, P.: Evolutionary design by computers. Morgan Kaufmann (1999)
Bentley, P.J., O'Reilly, U.-M.: Ten steps to make a perfect creative evolutionary design system. In: GECCO 2001 Workshop on Non-Routine Design with Evolutionary Systems (2001)
Boccara, N.: Modeling complex systems. Springer (2010)
Bush, B.J., Sayama, H.: Hyperinteractive evolutionary computation. IEEE Transactions on Evolutionary Computation 15(3), 424–433 (2011)
Camazine, S.: Self-organization in biological systems. Princeton University Press (2003)
Cheng, J., Cheng, W., Nagpal, R.: Robust and self-repairing formation control for swarms of mobile agents. In: AAAI, vol. 5, pp. 59–64 (2005)
Couzin, I.D., Krause, J., James, R., Ruxton, G.D., Franks, N.R.: Collective memory and spatial sorting in animal groups. Journal of Theoretical Biology 218(1), 1–11 (2002)
Dawkins, R.: The blind watchmaker: Why the evidence of evolution reveals a universe without design. WW Norton & Company (1996)
Dittrich, P., Ziegler, J., Banzhaf, W.: Artificial chemistries – a review. Artificial Life 7(3), 225–275 (2001)
Doursat, R., Sayama, H., Michel, O.: Morphogenetic engineering: Reconciling self-organization and architecture. In: Morphogenetic Engineering, pp. 1–24. Springer (2012)
Hemelrijk, C.K., Kunz, H.: Density distribution and size sorting in fish schools: an individual-based model. Behavioral Ecology 16(1), 178–187 (2005)
Kullback, S., Leibler, R.A.: On information and sufficiency. The Annals of Mathematical Statistics 22(1), 79–86 (1951)
Kunz, H., Hemelrijk, C.K.: Artificial fish schools: collective effects of school size, body size, and body form. Artificial Life 9(3), 237–253 (2003)
Minai, A.A., Braha, D., Bar-Yam, Y.: Complex engineered systems: A new paradigm. Springer (2006)
Nehaniv, C.L.: Measuring evolvability as the rate of complexity increase. In: Artificial Life VII Workshop Proceedings, pp. 55–57 (2000)
Newman, J.P., Sayama, H.: Effect of sensory blind zones on milling behavior in a dynamic self-propelled particle model. Physical Review E 78(1), 011913 (2008)
Ottino, J.M.: Engineering complex systems. Nature 427(6973), 399–399 (2004)
Pahl, G., Wallace, K., Blessing, L.: Engineering design: a systematic approach, vol. 157. Springer (2007)
Ray, T.S.: An approach to the synthesis of life. In: Artificial Life II, pp. 371–408. Addison-Wesley (1992)
Reynolds, C.W.: Flocks, herds and schools: A distributed behavioral model. ACM SIGGRAPH Computer Graphics 21(4), 25–34 (1987)
Salzberg, C., Antony, A., Sayama, H.: Evolutionary dynamics of cellular automata-based self-replicators in hostile environments. BioSystems 78(1), 119–134 (2004)
Salzberg, C., Sayama, H.: Complex genetic evolution of artificial self-replicators in cellular automata. Complexity 10(2), 33–39 (2004)
Sayama, H.: Decentralized control and interactive design methods for large-scale heterogeneous self-organizing swarms. In: Almeida e Costa, F., Rocha, L.M., Costa, E., Harvey, I., Coutinho, A. (eds.) ECAL 2007. LNCS (LNAI), vol. 4648, pp. 675–684. Springer, Heidelberg (2007)
Sayama, H.: Swarm chemistry. Artificial Life 15(1), 105–114 (2009)

Sayama, H.: Robust morphogenesis of robotic swarms. IEEE Computational Intelligence Magazine 5(3), 43–49 (2010)
Sayama, H.: Swarm chemistry evolving. In: Artificial Life XII: Proceedings of the Twelfth International Conference on the Synthesis and Simulation of Living Systems, pp. 32–33. MIT Press (2010b)
Sayama, H.: Seeking open-ended evolution in swarm chemistry. In: 2011 IEEE Symposium on Artificial Life (ALIFE), pp. 186–193. IEEE (2011)
Sayama, H.: Swarm-based morphogenetic artificial life. In: Morphogenetic Engineering, pp. 191–208. Springer (2012)
Sayama, H., Dionne, S., Laramee, C., Wilson, D.S.: Enhancing the architecture of interactive evolutionary design for exploring heterogeneous particle swarm dynamics: An in-class experiment. In: IEEE Symposium on Artificial Life (ALife 2009), pp. 85–91. IEEE (2009)
Sayama, H., Wong, C.: Quantifying evolutionary dynamics of swarm chemistry. In: Advances in Artificial Life, ECAL 2011: Proceedings of the Eleventh European Conference on Artificial Life, pp. 729–730 (2011)
Shneiderman, B., Plaisant, C., Cohen, M., Jacobs, S.: Designing the User Interface: Strategies for Effective Human-Computer Interaction, 5th edn. Prentice Hall (2009)
Sole, R., Goodwin, B.: Signs of life: How complexity pervades biology. Basic books (2008)
Takagi, H.: Interactive evolutionary computation: Fusion of the capabilities of ec optimization and human evaluation. Proceedings of the IEEE 89(9), 1275–1296 (2001)
Unemi, T.: Simulated breeding–a framework of breeding artifacts on the computer. Kybernetes 32(1/2), 203–220 (2003)

Chapter 14
Mutual Information As a Task-Independent Utility Function for Evolutionary Robotics

Valerio Sperati, Vito Trianni, and Stefano Nolfi

14.1 Introduction

The design of the control system for a swarm of robots is not a trivial enterprise. Above all, it is difficult to define which are the individual rules that produce a desired swarm behaviour without an *a priori* knowledge of the system features. For this reason, evolutionary or learning processes have been widely used to automatically synthesise group behaviours (see, for instance, Matarić 1997; Quinn et al. 2003; Baldassarre et al. 2007). In this paper, we investigate the use of information-theoretic concepts such as *entropy* and *mutual information* as task-independent utility functions for mobile robots, which adapt on the basis of an evolutionary or learning process. We believe that the use of implicit and general purpose utility functions—fitness functions or reward/error measures—can allow evolution or learning to explore the search space more freely, without being constrained by an explicit description of the desired solution. In this way, it is possible to discover behavioural and cognitive skills that play useful functionalities, and that might be hard to identify beforehand by the experimenter without an *a priori* knowledge of the system under study. Such task-independent utility functions can be conceived as universal intrinsic drives toward the development of useful behaviours in adaptive embodied agents.

In this paper, we investigate whether information-theoretic measures can be used to drive the evolution of coordinated behaviours in groups of evolving robots. In particular, we demonstrate how the use of a utility function that maximises the mutual information between the motor states of wheeled robots leads to the evolution of a variety of effective coordinated behaviours.

In the present study, three robots driven by identical neural controllers prove capable of displaying behaviours that are both structured and coordinated. Looking at

Valerio Sperati · Vito Trianni · Stefano Nolfi
Laboratory of Autonomous Robotics and Artificial Life,
Institute of Cognitive Sciences and Technologies, CNR,
via S. Martino della Battaglia, 44 - 00185 Rome, Italy

M. Prokopenko (ed.), *Guided Self-Organization: Inception, Emergence, Complexity and Computation* 9,
DOI: 10.1007/978-3-642-53734-9_14, © Springer-Verlag Berlin Heidelberg 2014

the individual level, we define a "structured" behaviour as a temporal sequence of several elementary behaviours, where the latter are sequences of atomic actions that produce a well-defined outcome (e.g., "move-straight", "move-to-light", "avoid-obstacle", etc.). For instance, an oscillatory behaviour in which a single robot moves back and forth from a light bulb is structured as it can be described as a periodic sequence of "move-to-light" and "move-away-from-light" behaviours. In contrast, sequences of random atomic actions would not be considered structured. Looking at the collective level, we define a "coordinated" behaviour as a situation in which the behaviours produced by the individuals are correlated as for example, in the case of the production of similar and synchronized oscillatory movements or as in the case of alternated turn-taking behaviours.

We present two sets of experiments, which differ by the environmental cues available to the robots. In the first experiment, referred to as E_l, robots evolve within an arena presenting a clearly distinguishable cue, that is, a light bulb perceivable from every location. In the second experiment, referred to as E_d, there is no light bulb to provide exploitable environmental cues, and the robots have to autonomously create the conditions required to perform structured and coordinated behaviours. We show how the proposed measure leads to the evolution of a rich—non trivial—repertoire of coordinated behaviours. Moreover, the paper assesses the effectiveness of the proposed methodology through the use of realistic simulations and through the test of the solutions evolved in simulation on the physical robots.

The rest of the paper is organised as follows. In the next section, we briefly review the relevant aspects of information theory. In Sec. 14.3 we briefly review related literature. In Sec. 14.4 and 14.5, we describe the experimental setup and the results obtained. Finally, in Sec. 14.6 we discuss the main contributions of the paper and we draw our conclusions.

14.2 Short Introduction to Information Theory

In this section, we briefly discuss the information theory concepts and measures first introduced by Shannon (1948), used in the definition of the task-independent utility function described in Sec. 14.4.3. Regarding notation, we follow Feldman's style: we use capital letters to indicate a random variable, and lowercase letters to indicate a particular value of that variable (Feldman 2002). For example, let X be a discrete random variable. The variable X may take on the values $x \in \mathcal{X}$. Here, $\mathcal{X}$ is the finite set of M possible values (or states) for X, referred to as the *alphabet* of X. The probability that X takes on the particular value x is written $p(X = x)$, or just $p(x)$ (first order probability density function). We may also form joint and conditional probabilities. Let Y be another random variable with $Y = y \in \mathcal{Y}$. The probability that $X = x$ and $Y = y$ is written $p(X = x, Y = y)$, or simply $p(xy)$ (second order probability density function), and is referred to as a joint probability. The conditional probability that $X = x$ given $Y = y$, is written $p(X = x|Y = y)$ or simply $p(x|y)$. Now, we can introduce the Shannon entropy equation, which is formally defined as:

$$H[X] = -\sum_{x \in \mathcal{X}} p(x) \cdot \log_2 p(x). \tag{14.1}$$

The entropy $H[X]$ - or *marginal entropy* - is equal to zero if the variable X always takes on the same value. The maximum value is equal to $log_2 M$, and it is obtained when X takes on all M possible values in *alphabet* $\mathcal{X}$ with the same probability ($\frac{1}{M}$). There are many interpretations about the meaning of Shannon entropy. In our case, we consider entropy as "a precise measure of the amount of freedom of choice in an object; an object with many possible states has high entropy" (see Prokopenko and Wang 2003). The same formula and interpretation is applicable to a joint distribution:

$$H[XY] = -\sum_{x \in \mathcal{X}} \sum_{y \in \mathcal{Y}} p(xy) \cdot \log_2 p(xy). \tag{14.2}$$

Note that, by definition, $H[XY] \leq H[X] + H[Y]$. The equality is obtained if and only if X and Y are statistically independent. Given a conditional distribution we can define the conditional entropy:

$$H[X|Y] = -\sum_{x \in \mathcal{X}} \sum_{y \in \mathcal{Y}} p(xy) \cdot \log_2 p(x|y). \tag{14.3}$$

The conditional entropy quantifies the remaining entropy about X, given that the value of Y is known. Note that $H[X|Y] = 0$ if and only if the value of X is completely determined by the value of Y. Conversely, $H[X|Y] = H[X]$ if and only if X and Y are statistically independent. It is quite useful to see that the equation of joint entropy can be re-expressed in terms of marginal entropy and conditional entropy:

$$H[XY] = H[X] + H[Y|X] = H[Y] + H[X|Y]. \tag{14.4}$$

Finally, we present the Mutual Information (*MI*), which is formally defined as:

$$MI[X;Y] = -\sum_{x \in \mathcal{X}} \sum_{y \in \mathcal{Y}} p(xy) \cdot \log_2 \frac{p(x) \cdot p(y)}{p(xy)}. \tag{14.5}$$

The properties of *MI* are more evident if we re-express the above formula in terms of marginal entropy and joint entropy and in terms of marginal entropy and conditional entropy:

$$MI[X;Y] = H[X] + H[Y] - H[XY]. \tag{14.6}$$

$$MI[X;Y] = H[X] - H[X|Y] = H[Y] - H[Y|X]. \tag{14.7}$$

Looking at eq. (14.5), it is possible to notice that $MI[X;Y] = 0$ if the two variables are statistically independent. On the other hand, eq. (14.6) shows that $MI[X;Y] = 0$ if the two variables have zero entropy. The measure is symmetric, namely $MI[X;Y] = MI[Y;X]$.

The interpretation of *MI* is quite clear looking at eq. (14.7). Feldman describes *MI* as "the reduction in uncertainty of one variable due to knowledge of another.

If knowledge of Y reduces our uncertainty of X, then we say that Y carries information about X" (Feldman 2002). In other words, if X and Y are independent variables, the mutual information that one variable brings about the other is null. On the other extreme, mutual information is maximised if the knowledge of one variable is sufficient to completely describe the other variable. When this happens, we can imagine that a bidirectional communication channel through whom the information flows, establishes between the two variables. In practice, MI can be used as a powerful index of correlation: the greater the value of MI, the more correlate are two variables. The great advantage of MI is that it takes into account both linear and nonlinear dependencies (Lungarella and Pfeifer 2001).

14.3 Related Work

The above measures and related derivations have been successfully used as analytic tools in different fields. In ethology for example, information theory was used to describe the interplay between pheromone molecules and ants' movements. By observing ants' foraging behaviour, Van Dyke Parunak and Brueckner (2001) showed that the increase in entropy at the micro-level of the chemical particles is compatible with the reduction of disorder at the macro-level of the ants' movements. Brenner et al. (2000) used information entropy to describe the behaviour of the visual system of the fly. The authors showed how the fly's response to the environmental features is dynamically adapted in order to maximise the information inflow. In neurosciences, the dynamics observable in the human brain have been studied under the light of information theory (Tononi et al. 1994, 1996, 1998; Sporns et al. 2000). A measure called *neural complexity* (C_N) captures some aspects of the interplay between the functional segregation of different cortical areas and their global integration during perception and behaviour. C_N is shown to be high when functional segregation coexists with global integration, and to be low when the components of a system are either completely independent (segregated) or completely dependent (integrated).

In robotics, Olsson et al. (2005) proved that the perceptions of a robot can be treated in an efficient and computationally economic way if sensors can adapt to the statistical properties of the incoming signals. Tarapore et al. (2004, 2006) applied entropy and mutual information to the sensory channels of a two wheeled simulated robot: These measures are used to classify different behaviours, such as exploring the environment, searching for red objects and tracking them. The authors argued that information theory can provide useful methods to discover the "fingerprints" of particular agent-environment interactions. Similarly, Lungarella and Pfeifer (2001) used entropy and mutual information to analyse the input data obtained by a robotic arm holding a colour camera. The authors compared coordinated movements (e.g., foveation on a red object), with uncoordinated ones (e.g., random movements), detecting clear informational structures in the first case. Comparable results were obtained by Lungarella and Sporns (2005) and Lungarella et al. (2005), using a robotic setup very similar to the previous work. The authors argued that coordinated sensory-motor activity induces information structures in the sensory experience.

This idea has been further elaborated by Klyubin et al. (2005a) with the notion of *Empowerment*, an information based quantity that allows to characterise the efficiency of the perception-action loop of an organism model. This quantity measures the potential of the organism to imprint information on the environment via its actuators in a way that can be recaptured by its sensors. A generalization of this measure for continuous domains has been later proposed by Jung et al. (2011).

Finally, other recent works investigated the use of information-based measures to characterise collective behaviour. Wang et al. (2011) investigate how information propagate in groups of coordinated individuals. In particular the authors showed how crucial phases in collective behaviours (corresponding to clustering, merging, and separation) are characterised by well-marked peaks of the active information storage and transfer entropy measures. Harder et al. (2010) instead demonstrated how the mutual information among the actions displayed by a couple of coordinated individuals can be used to characterise the autonomy level of the individuals and whether the couple can be characterised as a unique coherent entity. Harder et al. (2011) demonstrated how information theoretic measures can be used to quantify the ability of individual agents to extract information locally about global features. Finally, information theoretic measures have been used to characterise the emergence of self-organizing collective properties, e.g. the abrupt formation of a dynamic chain pattern within a swarm of robots evolved for the ability to navigate between two target areas (Sperati et al. 2011).

More importantly, from the point of view of the objectives of this paper, several recent works demonstrated how information theoretic measures can be used to synthesize sensory-motor coordination capabilities and/or co-ordinated behaviours. Sporns and Lungarella (2006) demonstrated how the maximisation of the information structure of the sensory states experienced by embodied and situated agents might lead to the development of useful behavioural skills. The agent is a simulated arm provided with visual and tactile sensors, placed in an environment including an object that moves in a random direction at constant speed. The object is characterised by a uniform colour which can be distinguished from the randomly coloured pixels of the background. By selecting evolving agents on the basis of the information structure of their experienced sensory states, the authors observed the development of useful behavioural skills consisting in the ability to foveate and to touch the moving object. Prokopenko et al. (2006) demonstrated how the maximisation of *Excess Entropy* (a measure of apparent memory or structure in a system), can lead to useful coordinated behaviour. In particular, the authors showed how a simulated snake-like modular robot, evolved on the basis of this measure, displays an effective locomotion behaviour: the linked actuators composing the robot get coordinated and produce a forward motion which interestingly adapts to the environment features, and makes the robot capable to face challenging terrains characterised by obstacles, narrow passages and ragged textures. In a subsequent work involving the same robotic setup, the authors used as fitness function the *Transfer Entropy*, i.e. a measure of information transfer (Lizier et al. 2008). In the experiment reported in this paper the authors observed a propagation of information between the head and the tail of the robot. The observed information transfer structures are analogous to gliders in cellular automata, which have been demonstrated to represent the coherent transfer of information across space and time, and play an important role in facilitating distributed computation.

Zahedi et al. (2010) proposed a learning method based on the maximisation of the predictive information, i.e. the mutual information between past and future sensors states, in the sensory-motor loop. The method was evaluated on a series of experiments involving robot constituted by chains of individually controlled elements of varying length. The *Empowerment* measure mentioned above has been used as a selection criteria for evolving the properties of the agent sensors and actuators (Klyubin et al. 2008, 2005b). As observed by the authors this triggers a process in which the morphology of the sensors and of the actuators adapts to the characteristics of the environment in which the agents are situated. More generally the authors claim that empowerment, being a measure of what the agents could do rather than a measure of what they actually do, can constitute a general adaptive drive that could enable the development of survival-relevant behaviour even in the absence of behaviour specific drives (Klyubin et al. 2008). Empowerment and *Infotaxis*, a measure that encodes the expected reduction of entropy achieved by selecting actions, have also been successfully used to synthesize coordinated collective behaviours (Capdepuy et al. 2007; Salge and Polani 2011).

Most of these works—as much as the study presented in this chapter—belong to a novel methodology in evolutionary robotics called *information-driven evolution*, in which information based measures that are task-independent are used as utility functions.

14.4 Experimental Setup

As mentioned in Sec. 14.2 , $MI[X;Y]$ can be seen as a powerful measure to grasp the correlation between two stochastic processes X and Y. Moreover, maximising *MI* also corresponds to maximising the entropy of the single processes $H[X]$ and $H[Y]$,[1] which is related to an higher probability of observing X or Y in multiple states. In this paper, we study whether *MI* can be used to evolve coordinated behaviours in a group of robots (see also Sperati et al. (2008)). The application of such a measure as utility function for an evolutionary robotics experiment is not straightforward. Given the experimental setup, it is necessary to define which are the stochastic processes under observation, discretise them in a suitable way and compute the desired utility functions. We chose to maximise the mutual information of the *motor states* observed in a group of autonomous robots (see Section 14.4.3). In particular, we focus on the wheels' speed, which characterise the robot movements in the environment. In this way, we aim at evaluating the quality of the individual and group behaviour, without any reference to the sensory pattern perceived by the robots.

The experimental setup involves three wheeled robots provided with a neural controller and different types of actuators and sensors (see Section 14.4.1). Robots are placed in a square arena of 1x1 m in side surrounded by walls. In the

[1] This is true if the joint entropy is kept constant, see eq. (14.6).

Fig. 14.1 Left: The e-puck robot developed at the EPFL in Lausanne, Switzerland (Mondada and Bonani 2007). Right: A close up view of the environment with the light bulb in the centre and three robots.

experiment E_l, a light bulb is placed in the centre of the arena. The intensity of the light decreases quadratically with the distance from the light bulb, but it is anyway perceivable by the robots from every location in the arena. Therefore, the light bulb provides a clearly distinguishable environmental cue to be exploited by the robots for coordination. In the experiment E_d, such environmental cue is not present, making the coordination between the robots more difficult to achieve.

We performed 20 *evolutionary runs* per experiment, in order to establish the viability of the approach varying the initial population of genotypes. Each evolutionary run lasts 200 generations. In each generation, the population is evaluated and genotypes are selected for reproduction on the basis of an estimate of their fitness (see Section 14.4.2). This estimate is obtained by testing each genotype 10 times—i.e., we perform 10 independent *trials* randomly varying the initial conditions (see Section 14.4.3). The best evolved genotypes resulting from each evolutionary run is then selected for a qualitative and quantitative analysis, presented in Section 14.5.

14.4.1 The Robot and the Neural Controller

The experiments presented in this paper are performed using the *e-puck* robots (see Fig. 14.1 left), which are wheeled robots with a cylindrical body having a diameter of 70 mm (Mondada and Bonani 2007; Cianci et al. 2007). A rich subset of the sensory-motor features of the *e-puck* has been exploited, as detailed in the following sections. In fact, by using an implicit and task-independent fitness function, we do not define a particular goal to be pursued by the robots. As a consequence, we do not know in advance which are the sensory-motor features that can be exploited to maximise the fitness function. We therefore decided to provide the robots with a rich set of sensors and actuators in order to leave the evolutionary process free to explore a wide set of possible solutions.

Each robot is provided with various sensory systems to perceive the environment, including the other robots. Among these, we make use of infrared proximity sensors, ambient light sensors and a VGA camera pointing in the direction of forward motion. Moreover, the robots can communicate with their neighbours in two different ways. They can light up the 8 red LEDs distributed around their body, in order to be detected by the camera of the other robots. Additionally, robots can exploit their wireless bluetooth interface to send and receive short messages (see Fig. 14.2).

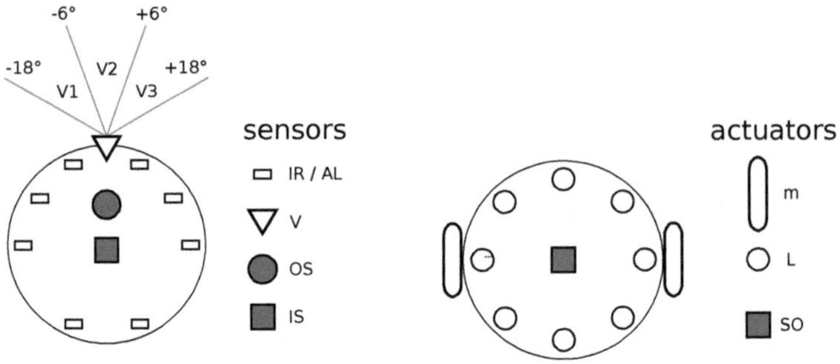

Fig. 14.2 Robot sensors and actuators. Left: sensors "IR" (8 infrared sensors), "AL" (8 light sensors), "V" (3 fields of view from pre-processed camera data), "IS" (average group signal), "OS" (own signal). Right: actuators "m" (wheel velocity and direction), "L" (leds ring on/off), "SO" (signal). Note that gray symbols refer to virtual actuators/sensors not present in the physical robot, but implemented through the wireless bluethoot interface.

The robots are controlled by artificial neural networks, whose parameters are set by an evolutionary algorithm. A single genotype is used to create a group of robots with an identical control structure—a homogeneous group. Each robot is controlled by a fully connected two layer neural network with fixed topology, characterised by an input layer with leaky integrator neurons and by an output layer of motor neurons (see Fig. 14.3). The activation of the output neurons is computed as the weighted sum of all input units and the bias, filtered through a sigmoid function:

$$O_j(t) = \sigma\left(\sum_i w_{ij} I_i(t) + \beta_j\right), \qquad \sigma(z) = \frac{1}{1+e^{-z}}, \tag{14.8}$$

where $I_i(t)$ corresponds to the activation of the ith sensory neuron at time t, w_{ij} is the weight of the synaptic connection between the sensory neuron i and the motor neuron j, and β_j is a bias term. Sensory neurons are leaky integrators, that is, they hold a certain amount of their activation from the previous time step, and the effect of the previous state on their current state is determined by their time constant:

$$I_i(t) = \tau_i \cdot I_i(t-1) + (1-\tau_i) \cdot \mathcal{I}_i(t), \tag{14.9}$$

where τ_i is the time constant of the ith neuron, and $\mathcal{I}(t)$ is the sensory input at time t.

The activations of the output neurons are real valued numbers in the range $[0.0, 1.0]$, and are used to control the actuators of the robot (see Fig. 14.3).

Two motor neurons (m_1 and m_2) encode the desired speed of the two motors which control the two corresponding wheels. The activation of each neuron is linearly scaled in the range $[-2\pi, 2\pi]$ rad/sec, and used to set the desired angular speed of the corresponding motor. One motor neuron (L) controls the red LEDs: all eight LEDs are switched on or off depending on whether the activation of the motor neuron is above or below an arbitrary threshold of 0.9. Finally, one motor neuron (SO) encodes the value of the communication signal produced by the robot at each cycle, which varies in the range [0.0, 1.0]. This signal is transmitted to the other robots through the wireless bluetooth interface.

Concerning the sensory inputs, they are set by the robot sensors after normalising their value onto the range $[0.0, 1.0]$. Eight sensory inputs are dedicated to the infrared proximity sensors ($IF_i, i = 1, \ldots, 8$), which can detect an obstacle up to a distance of approximately 25 mm (see Fig. 14.3). Three sensory inputs ($V_i, i = 1, \ldots, 3$) encode the presence of nearby robots—provided that they have their red LEDs switched on—as detected by the camera: the image that is grabbed at each cycle is pre-processed, in order to extract the percentage of pixels that have a predominant red colour within the following three vertical visual sectors: $[-18°, -6°]$, $[-6°, +6°]$, and $[+6°, +18°]$. The two remaining sensory inputs are dedicated to the communication signal: one input (IS) encodes the average signal produced by all the robots placed in the arena, the other input (OS) encodes the signal produced by the robot itself during the previous cycle. Additionally, in the experimental setup that includes the light bulb, the robots are provided with eight further sensory inputs, which are dedicated to the ambient light sensors ($AL_i, i = 1, \ldots, 8$), shown in grey in Fig. 14.3.

In the experiments performed in simulation, the state of the infrared and ambient light sensors has been simulated through a sampling technique (Miglino et al. 1995). The visual sensors have been simulated through a ray tracing technique, by using 36 rays uniformly distributed over the camera range. All sensors have been subjected to noise implemented as a random value with a uniform distribution in the range

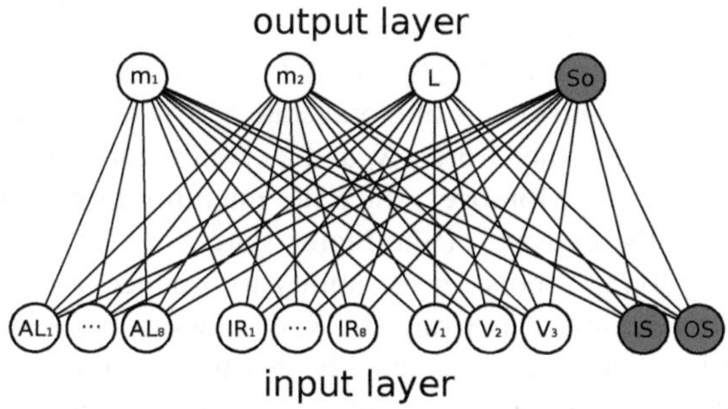

Fig. 14.3 The architecture of the neural controller. Note that gray neurons refers to virtual actuator/sensors not present in the physical robot, but implemented through the wireless bluethoot interface. Neurons "AL" are used in the experiment E_l only, while the other neurons are common to both setups.

$[-0.05, 0.05]$, added to the state of each simulated sensor. The use of simulated noise should favour the portability of the controllers evolved in simulation to the physical robots (see Jakobi 1997, for a detailed discussion about this topic).

14.4.2 The Evolutionary Process

The free parameters of the robot's neural controller are adapted through an evolutionary process (Nolfi and Floreano 2000). The initial population consists of 100 randomly generated binary genotypes, that encode the connection weights, the bias terms and the time constants of 100 corresponding neural controllers. Each parameter is encoded by 8 bits, and its value is linearly scaled from the range $[0, 255]$ to the range $[-5.0, 5.0]$ in the case of connection weights and bias terms, and in the range $[0.0, 0.95]$ in the case of time constants. The 20 best genotypes of each generation were allowed to reproduce by generating five copies each, with 4% of their bits replaced with a new randomly selected value, excluding one copy (elitism). The evolutionary process lasted 200 generations.

Each genotype is translated into three identical neural controllers which are downloaded onto three identical robots (i.e., the robots are homogeneous). Each team was tested for 10 trials, lasting 200 seconds (i.e., 2000 simulation cycles of 100 ms each). The performance of the genotype is the average fitness, as computed by eq. (14.12), over 10 trials. At the beginning of each trial, the three robots are placed in the arena with a random position and orientation. In case of collision the team is repositioned randomly again. The evolutionary process has been conducted in simulation.[2] The best evolved neural controllers have been tested with physical robots.

14.4.3 The Fitness Function

Evolving individuals are selected on the basis of a fitness function which measures the Mutual Information *MI* between the motor states X_i of all possible robot pairs. The maximisation of *MI* should drive evolution towards the development of coordinated behaviours. In fact, high values of *MI* correspond to motor states that are positively correlated: the knowledge of motor state X_i gives information about motor state X_j and vice versa. In other words, X_i and X_j result from processes that we can describe as "coordinated". Moreover, since the maximisation of *MI* also requires the maximisation of the entropy of the motor state X_i of each robot, this fitness function rewards evolving robots for the ability to produce structured behaviours. In fact, entropy is maximised not only by very random behaviours, but also by very structured behaviours that systematically vary through time. In particular, periodic sequences of equally frequent elementary behaviours such as "move-forward",

[2] Using a similar setup, a single evolutionary run—i.e., 200 generations, 100 individuals, 10 trials per individual, 200 seconds per trial—performed with physical robots would last longer than one year.

"move-backward", "turn-left" and "turn-right" allow the robot to uniformly cover many possible motor states, therefore maximising entropy.

For the purpose of computing the fitness function as the *MI* between the motor states of a robot pair, we need to define a discrete variable X that accounts for the current motor state—the wheels' speed. To avoid that motor state variations are caused by the random noise injected in the simulation, we filter the motor state through a slow moving average. In this way, for robots not having internal dynamics, systematic variations of X_i can solely be produced by exploiting the dynamics of the robot/environment interaction (i.e., by exploiting sensory-motor coordination). The activation values $m_j, j = 1,2$ of the two motor neurons controlling the wheels has been averaged through time into the variables M_j:

$$M_j(t) = \gamma \cdot M_j(t-1) + (1-\gamma) \cdot m_j(t), \qquad j = 1,2 \tag{14.10}$$

where $m_j(t) \in [0.0, 1.0]$ indicates the current activation of the motor neuron j and $\gamma = 0.9$ is a fixed time constant that represents the rate at which $M_j(t) \in [0.0, 1.0]$ changes over time. This moving average also channels the evolutionary process towards the synthesis of behaviours that extend for sensible time durations.[3] The overall motor state X of a robot is a discrete variable computed according to the following equation:

$$X = \lfloor M_1 \cdot 5 \rfloor + \lfloor M_2 \cdot 5 \rfloor \cdot 5, \tag{14.11}$$

where $\lfloor M_j \cdot 5 \rfloor$ means that the value M_j has been discretised into the integer range $[0,4]$, encoding all possible activation values of the motor neuron into five discrete states.[4] As a consequence, X takes on integer values in the range $[0,24]$.

In order to compute the *MI* of a robot pair, the value X_i of each robot $i = 1,2,3$ is recorded in every cycle, obtaining statistics about the states encountered during a trial. On the basis of these statistics, it is possible to estimate the probability distribution $p(X_i = x)$ and the joint distribution $P(X_i = x, X_j = y)$ needed to compute $MI[X_i;X_j]$, according to equation (14.5). Having estimated the probability distribution, the fitness function F of the group of robots in a trial is calculated on the basis of the following equation:

$$F = \frac{\sum_{i=1}^{N} \sum_{j=i+1}^{N} MI[X_i;X_j] \cdot \frac{20-c}{20}}{a} \qquad a = \binom{N}{2} \tag{14.12}$$

[3] Preliminary experiments conducted without the moving average produced behaviours that were coordinated but neither periodic nor structured (result not shown). In these experiments, we observed that the motor state of each robot varied in a quasi-random way (e.g., alternating at each time-step very different actions such us move-forward, move-backward, turn-right, turn-left), therefore maximising the individual entropy without actually being structured or periodic. Such variations were produced by achieving and maintaining a given relative position with respect to an obstacle or to the other robots, so that each movement resulted in a large variation of the sensory pattern.

[4] The activation value equal to $M_j = 1.0$ is considered as state 4.

where N is the number of robots, c is the number of times in which one of the robots collided against a wall or against another robot, truncated to 20, a is the binomial coefficient for couples of robots. In other worlds, this equation computes the average Mutual Information calculated between all possible pairs of robots. The second term of the fitness function has been introduced in order to reward robots for the ability to avoid collisions. All robots are randomly repositioned whenever a collision is detected: in this way, we bypass the problem of accurately simulating the physical interactions during a collision, offering the robots further possibilities to coordinate. Moreover, a maximum of 20 collisions per trial is allowed before the trial is stopped. These choices channel the evolution of good collision avoidance behaviours.

The maximum value of F is obtained when no collisions are detected and all robot pairs have maximum MI. Since X can assume 25 different states, the fitness takes values in the range $[0.0, \log_2 25]$. It is worth noting, however, that the maximum value cannot be achieved in practice. The main reason for this is that the individual entropy cannot be maximised because robots are embodied and their dynamical interaction with the environment—as it is defined by the neural controller—constrains the number of motor states visited during the robot's lifetime, and their relative frequency. Moreover, the motor state X is the result of a moving average with a fixed time constant, which influences X's variability. Finally, the computation of the MI includes the initial transitory phase during which the robots try to achieve a coordinated behaviour.

14.5 Results

In this section, we describe the results obtained in the two experiments E_l and E_d. As detailed in the following sections, in both experiments the evolved robots display behaviours that are structured (i.e., they consist of a sequence of atomic movements with varying time durations), periodic (i.e., the sequence of atomic movements is repeated through time), and coordinated (i.e., the different robots tend to produce the same structured behaviour in a synchronised manner). From a qualitative point of view, the evolved behaviours vary considerably between the two experiments, and also across the different evolutionary runs of the same experiment.

14.5.1 Experiment E_l

In the experiment E_l, the robots are situated in in square arena of 1x1 m in side presenting a light bulb, which can be perceived by means of the robot's ambient light sensors. As mentioned above, we performed 20 evolutionary runs, each time starting with a different randomly generated population. After the evolutionary process, we selected the best individual of each run for post-evaluation. In this case, the fitness of each individual was further evaluated for 500 trials, using eq. (14.12). The results obtained are summarised in Table 14.1, in which we show mean and standard deviation over the 500 trials of the fitness F, of the average mutual information $\widehat{MI}$ over all possible robot pairs, and of the average entropy $\widehat{H}$ computed over all robots.

The results of the post-evaluation show that the average fitness varies between 1.70 and 3.24, respectively obtained in run 1 and 16. Given that F has been explicitly constructed as a task independent and implicit utility function, the absolute value of F is not very informative about the quality of the evolved behaviour. Recall that the absolute value of F is mainly given by the $\widehat{MI}$. The latter is constrained by the robots' embodiment which limits the number of possible motor states actually visited during the robot's lifetime. A qualitative analysis revealed that 18 out of 20 evolutionary runs resulted in controllers that produce structured and coordinated behaviours (see the runs indicated by a black dot in Table 14.1). This is a first result proving that the proposed methodology is viable: mutual information can be exploited as a generic utility function to obtain task-less adaptation in a group of robots.

Table 14.1 Experiment E_l: fitness F, average mutual information $\widehat{MI}$ and average entropy $\widehat{H}$ computed by testing in simulation the best evolved controller of each evolutionary run for 500 trials of 2000 cycles. Mean value and standard deviation are shown. The symbol • indicates a run in which the best evolved individuals clearly show behaviours that an external observer can judge as structured and coordinated.

run	F	$\widehat{MI}$	$\widehat{H}$	run	F	$\widehat{MI}$	$\widehat{H}$
1	1.70±0.27	1.72±0.28	2.55±0.39	11 •	2.73±0.17	2.75±0.13	3.51±0.09
2 •	2.81±0.14	2.84±0.11	3.55±0.05	12 •	2.27±0.15	2.29±0.13	3.52±0.19
3 •	1.91±0.20	1.93±0.17	2.97±0.08	13 •	2.38±0.22	2.39±0.21	3.19±0.25
4 •	2.99±0.21	3.02±0.18	3.96±0.04	14 •	2.72±0.13	2.75±0.09	3.55±0.06
5 •	2.97±0.13	2.99±0.11	3.84±0.08	15 •	2.47±0.18	2.51±0.12	3.23±0.04
6 •	2.50±0.07	2.50±0.07	3.24±0.13	16 •	3.24±0.14	3.25±0.12	4.01±0.06
7 •	2.41±0.14	2.42±0.12	3.26±0.15	17 •	2.49±0.16	2.50±0.13	3.55±0.02
8 •	2.19±0.19	2.24±0.16	3.43±0.08	18	1.72±0.17	1.75±0.16	3.05±0.21
9 •	2.40±0.18	2.43±0.14	3.32±0.05	19 •	2.99±0.17	3.01±0.14	3.96±0.04
10 •	2.17±0.20	2.18±0.19	3.07±0.13	20 •	3.12±0.17	3.14±0.14	4.08±0.06

14.5.1.1 Behavioural Analysis

The qualitative inspection of the results obtained indicates that the robots always display structured and coordinated behaviours. Generally, the environmental cue offered by the light bulb is exploited by the robots to achieve the same relative position and to display a periodic, structured behaviour. Moreover, robots perform a coordinated behaviour through the synchronisation of their movements. Synchronisation is generally achieved through the exploitation of the communication signal only. Infrared sensors are generally exploited to avoid collisions with walls and with other robots, while visual information is often ignored.

A particularly interesting example of structured and coordinated behaviour is produced by the controller evolved in run 16, characterised by the highest mean performance (see Table 14.1). In this case, robots circle anticlockwise around the light bulb maintaining a distance of about 20 cm (see the trajectories of the robots shown in Fig. 14.4 and the video "replication 16 E_l" in the online supplementary

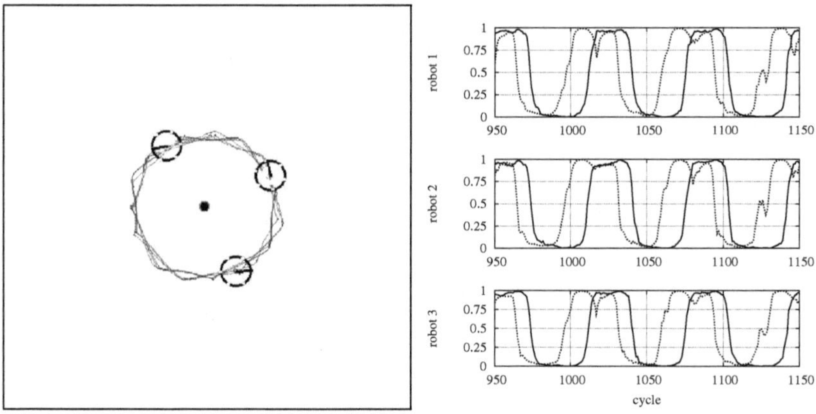

Fig. 14.4 Analysis of the behaviour produced by the best evolved controller in run 16 of experiment E_l. Left: trajectories of the robots. Right: activation of the motor neurons of each robot, plotted from cycle 950 to cycle 1150 to highlight the periodic motion of the robots. The solid and dotted lines indicate respectively the left and right motor neurons.

material[5]). While circling around the light bulb, robots display a structured behaviour composed of four atomic movements: (i) forward motion on the circle, (ii) clockwise turn on the spot, (iii) backward motion on the circle, and (iv) anticlockwise turn on the spot. These atomic movements can be clearly identified looking at the plots in Fig. 14.4 right, in which we show the activation of the motor neurons that control the two wheels. Recall that maximum forward rotation corresponds to 1, while maximum backward rotation corresponds to 0. Starting at cycle 950, both wheels present forward rotation, resulting in forward movement on the circle. Afterwards, the activation of the right motor neuron sharply decreases to 0, leading to a clockwise rotation on the spot. Then, the left motor activations also drops to 0, resulting in backward motion. Finally, the right motor activation increases to 1, producing an anticlockwise rotation on the spot. After this, the robot starts again with forward motion.

The above description accounts for the structure of the evolved behaviour. The coordination between the robots can be appreciated by observing how the motor activations of the three robots coincide in time (see Fig. 14.4 right). In short, robots are synchronised as they perform the same movements at the same time. The mechanism that the robots exploit to achieve and maintain synchronisation is based on communication, and on the fact that robots are homogeneous. An individual robot mainly signals during forward motion, and stops signalling as soon as the clockwise movement starts. All robots perform the same individual movements, which synchronise on the basis of the mutual interactions through communication. If an external signal is perceived, the robot keeps moving forward until signalling stops. As a consequence, the clockwise movement cannot start until all robots are

[5] See `http://laral.istc.cnr.it/esm/sperati-etal-GSO_2012.html` for videos and other supplementary material.

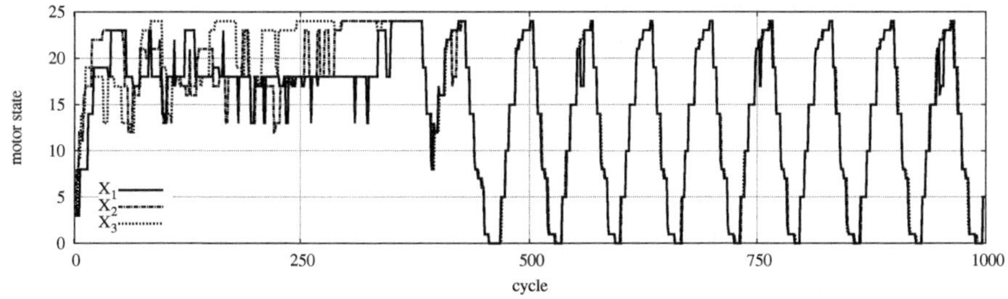

Fig. 14.5 The motor states of the three robots—computed using eq. (14.11)—are plotted against the number of cycles. Notice the initial coordination phase, followed by synchronised movements.

performing forward motion. When this happens, synchronisation is achieved. This simple mechanism—already observed by Trianni and Nolfi (2009) —is based on a simple reaction to the perception of a signal, that allows a robot to achieve and maintain a certain sensory-motor condition—referred to as *reset configuration* by Trianni and Nolfi (2009) —waiting for the other robots. Synchronised movements start when all robots achieve the reset configuration.

Having described qualitatively the evolved behaviour, the questions remain: how did this behaviour evolve? In what way is *MI* maximised? To answer these questions, it is necessary to observe the motor states X_i and to analyse their statistics. Figure 14.5 shows how the motor states vary through time. First of all, it is possible to notice how the initial coordination phase is followed by a phase in which the group behaviour is perfectly synchronised. Moreover, it is possible to observe how, during the coordinated phase, the motor states take on many different values. In other words, the motor states of the robots vary considerably through time, which corresponds to a high individual entropy. Besides, once robots are synchronised, the motor states are highly correlated. This means that the joint entropy is minimised and the mutual information maximised.

Similar conclusions can be drawn looking at Fig. 14.6. In the left part, the histograms represent the probability $p(X_i = x), x \in [0:24]$ estimated on a single trial. It is possible to notice how X_i takes on many different values with relatively high probability. As a consequence, the individual entropy $H[X_i]$ is rather high (see the individual values shown above the plot). Similarly, in the right part of Fig. 14.6, the 3D histogram represents the probability $p(X_1 = x_1, X_2 = x_2)$ estimated on the same trial.[6] Here, it is worth noting that the joint distribution takes values mainly on the diagonal $X_1 = X_2$, meaning that the probability of having $X_1 \neq X_2$ is rather low. As a consequence, we observe a small value for the joint entropy $H[X_1X_2]$, and a high value for the mutual information $MI[X_1;X_2]$.

Owing to the above analysis, it is possible to claim that (i) structured behaviours maximise the individual entropy, because they are characterised by motor states that

[6] The histograms for the other pairs $\langle X_1, X_3 \rangle$ and $\langle X_2, X_3 \rangle$ are extremely similar and have been omitted for space reasons.

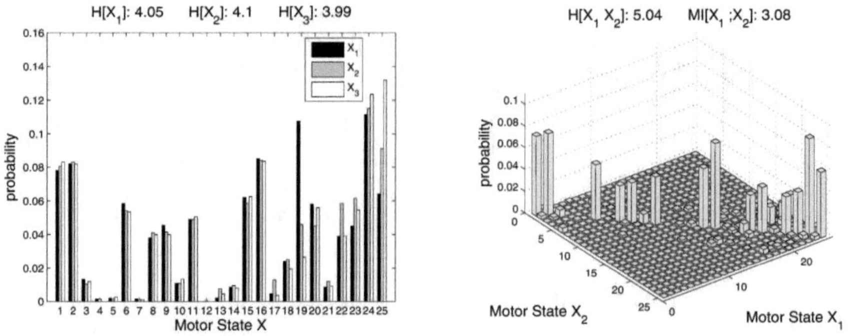

Fig. 14.6 Left: Probability distribution for the motor states X_i of each robot $i = 1,2,3$. Right: Probability distribution of the joint state $\langle X_1, X_2 \rangle$.

Table 14.2 Experiment E_I: average mutual information ($\widehat{MI}$) and average entropy ($\widehat{H}$) computed by testing the evolved controllers on physical robots for 5 trials of 2000 cycles each. We show here only the evolutionary runs that successfully transfer to reality from a qualitative standpoint. The column labelled 'ratio' indicates the ratio between the performance observed in hardware with respect to the performance observed in simulation.

run	$\widehat{MI}$	$\widehat{H}$	ratio	run	$\widehat{MI}$	$\widehat{H}$	ratio
2	2.29 ± 0.07	3.55 ± 0.03	0.81	12	1.83 ± 0.12	3.65 ± 0.22	0.81
3	1.34 ± 0.24	3.23 ± 0.16	0.70	13	1.78 ± 0.51	2.89 ± 0.28	0.75
5	2.62 ± 0.17	3.87 ± 0.05	0.88	16	2.82 ± 0.05	3.96 ± 0.01	0.87
6	2.24 ± 0.04	3.14 ± 0.09	0.90	17	1.89 ± 0.13	3.45 ± 0.06	0.76
9	2.22 ± 0.12	3.44 ± 0.05	0.92	19	2.42 ± 0.17	3.54 ± 0.10	0.81
11	1.93 ± 0.05	3.31 ± 0.07	0.71	20	2.55 ± 0.12	4.17 ± 0.05	0.82

have sensible time duration and vary systematically across the range of possible values; (ii) coordinated behaviours maximise the mutual information, because they ensure that a certain motor state of one robot is correlated with the motor state of other robots; (iii) the homogeneity of the robots results in synchronisation behaviours that ensure the one-to-one correspondence of the motor states between robots.

14.5.1.2 Porting to Reality

By testing with physical robots all controllers that proved successful in simulation, we observed qualitatively similar behaviours with respect to simulation in the majority of the evolutionary runs (12 out of 18 runs).[7] In all other cases, we observed a fairly good correspondence with simulation for individual behaviours, but not for coordination among robots. In fact, coordination was difficult to achieve and to maintain throughout a whole trial.

[7] See videos in the online supplementary material.

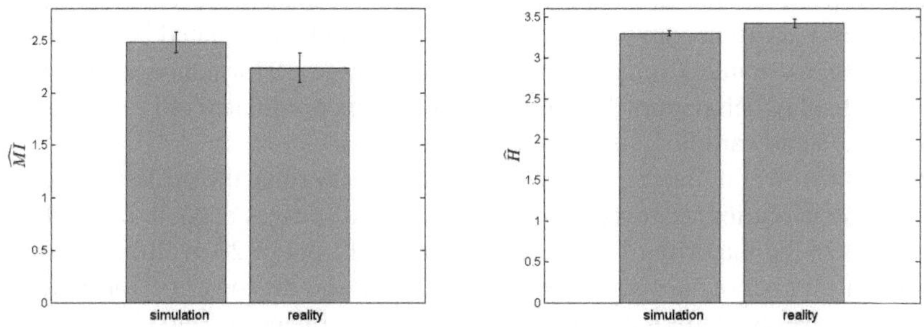

Fig. 14.7 Average mutual information ($\widehat{MI}$) and entropy ($\widehat{H}$) computed by testing the best evolved controller of run 9 of experiment E_l in simulation and in reality for 20 trials of 2000 cycles. During the tests in hardware, the robots were situated in the same randomly generated positions and orientations that were used for the tests in simulation.

In order to quantitatively determine the correspondence between tests with simulated and physical robots, we tested the evolved controllers by placing three real robots in locations randomly chosen from a set of 32 possible initial positions and 8 possible rotations. We performed 5 trials for each evolutionary run, and we measured the average mutual information computed among all possible robot pairs. The results obtained are shown in Table 14.2, along with the ratio with the average mutual information resulting from simulation. It is worth noting that the ratio between the mutual information observed in simulation and in the real environment is generally quite high, indicating that the behaviours tested in reality correspond fairly well to those observed in simulation.

After this preliminary test was performed on all evolutionary runs, we analysed in detail the best individual of run 9 (i.e., the individual with the highest ratio between the performance observed in simulation and in reality[8]). We performed 20 further evaluations keeping exactly the same initial conditions in both simulated and real tests. We observed a good correspondence between the mean mutual information observed in simulation and in reality, as shown in Fig. 14.7 left. Similarly, the mean entropy over 20 trials computed on the tests with physical robots corresponds to the value obtained in simulation (see Fig. 14.7 right).

14.5.2 Experiment E_d

In the second set of experiments, the robots are situated in an arena without a light bulb. Moreover, robots are not provided with ambient light sensors. Also in this case, we performed 20 evolutionary runs, we selected the best individual of each run and we post-evaluated it in 500 different trials. As shown in Table 14.3, the evolved controllers present lower fitness values compared to the results obtained in experiment E_l. In this case, in fact, the fitness varies between 1.24 and 2.93,

[8] See video "replication 9 E_l" in the online supplementary material.

obtained respectively in run 14 and 10. The qualitative analysis revealed that 11 out of 20 evolutionary runs converge toward structured and coordinated behaviours. In other two cases—namely runs 17 and 19—the average performance is rather high but robots display behaviours that are structured and coordinated only initially, and later degenerate toward non-structured behaviours.

Despite the lower number of successful evolutionary runs, the proposed methodology for the evolution of coordinated behaviour still proves capable of producing good results in the majority of the tests performed (11 out of 20 evolutionary runs). The smaller number of successful evolutionary runs and the lower performance obtained in the average within experiment E_d is a consequence of the absence of the environmental cue that characterises experiment E_l. Indeed, all evolutionary runs of experiment E_l exploit such environmental cue, which gives a reference point that can be perceived from far away and that can be used by the robots to initiate and maintain a structured and coordinated behaviour. In contrast, the absence of the environmental cue forces the robots to search for other regularities that can be exploited for coordination. Given that the environment does not offer such obvious regularities, they must be extracted from the sensory-motor experience of the robots interacting with the *social environment*. Clearly, solutions of this kind are more difficult to evolve, because they are based on dynamical interactions among robots. However, as we show in the next section, a number of possible strategies exist to solve this problem.

Table 14.3 Experiment E_d: fitness F, average mutual information $\widehat{MI}$ and average entropy $\widehat{H}$ computed by testing in simulation the best evolved controller of each evolutionary run for 500 trials of 2000 cycles. Mean value and standard deviation are shown. The symbol • indicates the runs in which the best evolved individuals display structured and coordinated behaviours. The symbol ○ indicates the runs characterised by behaviours that degenerate with time.

run	F	$\widehat{MI}$	$\widehat{H}$	run	F	$\widehat{MI}$	$\widehat{H}$
1 •	2.56±0.15	2.57±0.13	3.42±0.03	11	1.45±0.27	1.48±0.25	2.88±0.26
2 •	2.66±0.12	2.71±0.06	3.35±0.08	12	1.46±0.33	1.49±0.33	2.22±0.51
3 •	1.75±0.15	1.77±0.14	2.53±0.13	13 •	1.85±0.08	1.88±0.07	2.99±0.09
4	1.82±0.13	1.84±0.12	3.44±0.25	14	1.24±0.22	1.31±0.21	2.72±0.39
5 •	1.98±0.11	1.99±0.10	3.16±0.06	15 •	2.59±0.12	2.61±0.09	3.31±0.04
6 •	2.69±0.15	2.72±0.11	3.55±0.04	16	1.42±0.12	1.42±0.12	2.35±0.23
7	1.54±0.07	1.54±0.07	1.76±0.06	17 ○	2.22±0.15	2.22±0.14	2.55±0.12
8 •	1.92±0.14	1.94±0.12	2.62±0.12	18	1.27±0.28	1.27±0.28	2.07±0.39
9 •	2.17±0.12	2.19±0.10	3.18±0.19	19 ○	1.93±0.28	1.94±0.28	2.31±0.21
10 •	2.93±0.07	2.95±0.04	3.54±0.04	20 •	2.02±0.08	2.03±0.08	2.72±0.07

14.5.2.1 Behavioural Analysis

As mentioned before, the qualitative inspection of the evolved controllers allowed us to identify 11 evolutionary runs that produce structured and coordinated behaviours. Also in this case, after an initial transitory phase, robots perform synchronised movements. Communication is exploited to achieve and maintain synchrony.

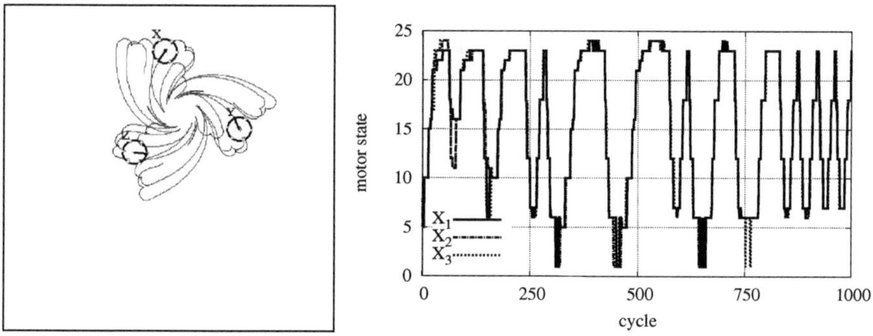

Fig. 14.8 Left: Trajectories of the robots produced by the best evolved controller in run 6 of experiment E_d. Right: The motor states of the three robots are plotted against the number of cycles.

The behaviours produced by the evolved controllers can be grouped into three strategies, described as follows.

The first strategy—the most common one—encompasses the controllers evolved in runs 1, 3, 5, 6, 9, 15 and 20. An interesting example of this strategy is given by run 6, which presents the highest average fitness within its group. This strategy is characterised by robots that periodically aggregate and disband, performing oscillatory movements around the centre of mass of the group and faraway from the walls (see the trajectories in Fig. 14.8 left and video "replication 6 E_d" in the online supplementary material). To do so, robots exploit vision, infrared proximity sensors and communication. Vision is mainly exploited in the aggregation phase, during which robots get close to one other assuming a triangular formation. When robots are close enough to perceive each other through the infrared proximity sensors, they disband moving backward. Due to relative differences in robots positions and orientations with respect to the centre of mass of the group, the behaviour of the three robots is not well coordinated during the first oscillatory movements. However, the robots quickly converge toward a well coordinated behaviour, as is apparent looking at the motor states plotted in Fig.14.8 right. Notice also how the motor states vary through time, taking on many different values: this corresponds to a very structured behaviour, which is also well coordinated as the robots perform the same actions at the same time. Moreover, the oscillations have different amplitude and duration during a trial, as can be noticed in Figure 14.8 right. This fact indicates that robots are able to perform a variety of atomic movements, which can be triggered depending on the particular contingency the robots experience. Nevertheless, they prove capable of maintaining coordination even when switching between different oscillation modalities.

The second strategy encompasses the controllers evolved in runs 2, 10 and 13. The highest average fitness within this group is obtained by run 10 (see Fig. 14.9 and video "replication 10 E_d" in the online supplementary material). In this case, robots do not interact visually or through their proximity sensors. They mainly produce a

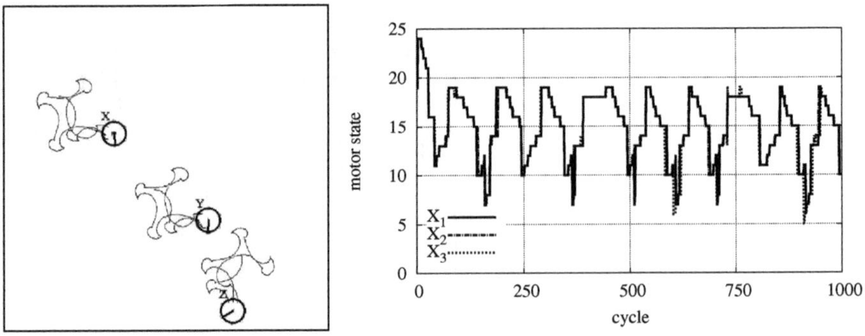

Fig. 14.9 Left: Trajectories of the robots produced by the best evolved controller in run 10 of experiment E_d. Right: The motor states of the three robots are plotted against the number of cycles.

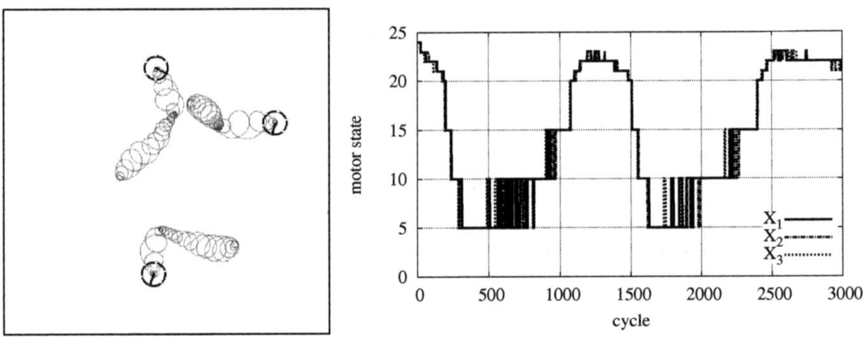

Fig. 14.10 Left: Trajectories of the robots produced by the best evolved controller in run 8 of experiment E_d. Right: The motor states of the three robots are plotted against the number of cycles.

behaviour structured in a sequence of atomic movements, such as backward motion on a large circle followed by forward motion on a small circle. These movements are performed without any reference to the position and orientation of the other robots or to the position and orientation of the robot in the arena, provided that robots are located far enough from walls. Robots exploit only the communication signal to coordinate, and the robots display synchronised movements without keeping any relation between their relative positions in the arena. As a consequence, coordinated movements are performed from the very beginning of the trial, because there is no need to achieve a particular spatial formation (see the motor states plotted in Fig.14.9 right).

Finally, the last strategy includes only the controller evolved in run 8 (see Fig. 14.10 and video "replication 8 E_d" in the online supplementary material). This controller produces a peculiar behaviour characterised by four atomic movements that last from 10 to 40 seconds—i.e., a time span considerably longer than those observed in other evolutionary runs, which can be appreciated by looking at the motor states in Fig. 14.10 right—which are periodically repeated: (i) rotating several times to produce a nearly circular trajectory with a diameter of about 8 cm, (ii) rotating several times to produce a spiral trajectory with a diameter decreasing to 0 cm, (iii) rotating several times on the spot at full speed, (iv) rotating several times to produce a spiral trajectory with a diameter increasing from about 0 to about 8 cm. Also in this case, the movements of the robot are performed without any reference to the position and orientation of the other robots. However, we observed that visual information is exploited to switch between different rotating modes. Synchronisation of movements also characterises this behaviour (see the coordinated motor states in Fig. 14.10 right), and it is achieved and maintained exploiting communication only.

14.5.2.2 Porting to Reality

By testing with physical robots all controllers that proved successful in simulation, we observed good generalisation only in 5 out of 11 cases, namely runs 2, 8, 9, 10, 13[9] The main reason to explain the limited generalisation ability of these controllers is likely to be found in the fine grained interactions between robots that take place by means of the infrared proximity sensors. We found that proximity sensors differ significantly in sensitivity and perceptual range among different physical robots. Similar inter-robot differences were not systematically simulated, reducing the portability in hardware of the results obtained in simulation. Indeed, the evolutionary runs that produce qualitatively similar behaviour in simulation and in reality are characterised by limited interactions through infrared sensors.

For all evolutionary runs that properly generalise to the physical setup, the comparison of the mean mutual information $\widehat{MI}$ and mean entropy $\widehat{H}$ measured in simulation and in reality reveals a very good correspondence, as indicated by the high values of the ratio between the measures in the two conditions (see Table 14.4).

Table 14.4 Experiment E_d: average mutual information ($\widehat{MI}$) and average entropy ($\widehat{H}$) computed by testing the evolved controllers on physical robots for 5 trials of 2000 cycles each. We show here only the evolutionary runs that successfully transfer to reality from a qualitative standpoint. The column labelled 'ratio' indicates the ratio between the performance observed in hardware and in simulation.

run	$\widehat{MI}$	$\widehat{H}$	ratio
2	$2,69 \pm 0,06$	$3,36 \pm 0,07$	0.99
8	$1,91 \pm 0,03$	$2,65 \pm 0,06$	0.94
9	$2,06 \pm 0,10$	$3,21 \pm 0,26$	0.95
10	$2,88 \pm 0,05$	$3,51 \pm 0,06$	0.98
13	$1,66 \pm 0,06$	$2,90 \pm 0,20$	0.92

[9] See videos in the online supplementary material.

14.6 Conclusion

In this paper, we investigated the use of information theoretic measures for the evolution of coordinated behaviours in groups of homogeneous robots. In particular, we defined a fitness function based on the average mutual information between the motor states of all possible robot pairs within a group of three robots. The results obtained show that evolution is able to find solutions that maximise the mutual information. This corresponds, in qualitative terms, to controllers that produce *structured* and *coordinated* behaviours. This is mainly the result of two different evolutionary drives. On the one hand, the maximisation of the mutual information corresponds to the maximisation of the individual entropy (see eq. (14.6)). This favours the evolution of individual behaviours that allow the robot to produce different actions during its lifetime. The embodiment of the robot, and the particular way we defined the computation of the motor state—as defined by eq. (14.10) and (14.11)—favour the evolution of behaviours in which the motor state varies smoothly with time, producing sequences of atomic movements with varying time duration. These sequences are also periodic, due to the necessity to visit as many motor states as possible for multiple times. On the other hand, the maximisation of the mutual information corresponds to the minimisation of the joint entropy between the motor states of two robots, which also corresponds to the observation of motor states that are positively correlated. The homogeneity of the robotic group ensures that this positive correlation leads to coordinated synchronous behaviours.

We presented the results of two experiments that differ mainly in the characteristics of the environment, which may or may not offer obvious regularities to be exploited for coordination among the robots. We observed that, when these regularities are present, artificial evolution finds a way to exploit them to produce structured behaviours and to support the achievement of coordination among the robots. The situation is more complicated when the environment does not provide such regularities. In this case, the robots exploit the possibility to generate the required regularities through social behaviours (i.e. by aggregating and/or by communicating). Moreover we observed how the obtained results can be validated in hardware. More specifically, we demonstrated how several of the controllers evolved in simulation work also with physical robots (12 out of 18 in the E_l setup, 5 out of 11 in the E_d setup). Overall this demonstrates how the proposed measure is able to synthesize robust solution that can overcome the problems caused the simulation-reality gap.

We believe that the proposed methodology is particularly relevant for swarm robotics research, as it can efficiently synthesise self-organising, coordinated behaviours for a robotic swarm. In fact, there is a fundamental problem—referred to as the *design problem*—that arises in the development of self-organising behaviours for a group of robots (see also Funes et al. 2003; Trianni et al. 2008, for a detailed discussion of this topic). This problem consists in defining the appropriate individual rules that will lead to a certain global pattern, and it is particularly challenging due to the indirect relationship between control rules and individual behaviour, and between interacting individuals and the desired global pattern. In this respect, evolutionary robotics is particularly suited to synthesise self-organising behaviours

(Trianni and Nolfi 2012). In fact, it bypasses the design problem as it relies on the automatic generation, test and selection of control solutions for the robotic system as a whole, without the need of an arbitrary decomposition of the given control problem into sub-problems (e.g., the desired global behaviour into individual behaviours and inter-individual interactions, as well as the individual behaviour in a set of control rules). The methodology we propose in this paper goes a step further in this direction: it promotes the evolution of coordinated behaviours without any constraint imposed by an explicit description of the desired solution. As a consequence, the proposed approach does not require a thorough knowledge of the system under study to devise the individual control rules, neither does it need a description of the desired solution to promote cooperative behaviours, as it can benefit of a task-independent, implicit utility function.

The proposed methodology represents a first step towards the evolution of self-organising behaviours for robotic swarms. In future work, we plan to exploit information theoretic measures in support of the evolution of task-oriented group behaviours. So far, we obtained synchrony without any constraint on the characteristics of the individual behaviour. We believe that a task-independent function can be successfully used in combination with a task-oriented one (on this issue, see also Prokopenko et al. 2006). The former should provide the drives to synthesise structured and coordinated behaviour. The latter should simply channel the evolutionary process towards individual and group behaviours that serve specific functionalities. Another possible extension over the work presented in this paper concerns the use of heterogeneous robots. Using different controllers and/or different sensory-motor apparatus, it should be possible to observe coordination among the robots that does not forcedly limit to synchronisation of the movements. Turn taking, entrainment and other forms of coordination become possible whenever the robots may have access to different sensory-motor experiences. Finally, we intend to investigate also the possibility of exploiting different information theoretic measures and different neural controllers, such as recurrent neural networks.

Acknowledgements. This research has been supported by CNR under the European Science Foundation project Hierarchical Heterogeneous SWARM (H2Swarm).

References

Baldassarre, G., Trianni, V., Bonani, M., Mondada, F., Dorigo, M., Nolfi, S.: Self-organised coordinated motion in groups of physically connected robots. IEEE Transactions on Systems, Man and Cybernetics - Part B: Cybernetics 37(1), 224–239 (2007)

Brenner, N., Bialek, W., de Ruyter van Steveninck, R.: Adaptive rescaling maximizes information transmission. Neuron 26, 695–702 (2000)

Capdepuy, P., Polani, D., Nehaniv, C.: Maximization of potential information flow as a universal utility for collective behaviour. In: Proceedings of the 2007 IEEE Symposium on Artificial Life (CI-ALife 2007), pp. 207–213. IEEE Press, Piscataway (2007)

Cianci, C.M., Raemy, X., Pugh, J., Martinoli, A.: Communication in a swarm of miniature robots: The e-puck as an educational tool for swarm robotics. In: Şahin, E., Spears, W.M., Winfield, A.F.T. (eds.) SAB 2006 Ws 2007. LNCS, vol. 4433, pp. 103–115. Springer, Heidelberg (2007)

Feldman, D.: A brief introduction to: Information theory, excess entropy and computational mechanics. Technical report, College of the Atlantic, Bar Harbor, ME (2002)

Funes, P., Orme, B., Bonabeau, E.: Evolving emergent group behaviors for simple humans agents. In: Banzhaf, W., Christaller, T., Dittrich, P., Kim, J.T., Ziegler, J. (eds.) ECAL 2003. LNCS (LNAI), vol. 2801, pp. 76–89. Springer, Heidelberg (2003)

Harder, M., Polani, D., Nehaniv, C.: Two agents acting as one. In: Proc. of the Alife XII Conference (2010)

Harder, M., Polani, D., Nehaniv, C.: Think globally, sense locally: From local information to global features. In: IEEE Symposium on Artificial Life (2011)

Jakobi, N.: Evolutionary robotics and the radical envelope of noise hypothesis. Adaptive Behavior 6, 325–368 (1997)

Jung, T., Polani, D., Stone, P.: Empowerment for continuous agent-environment systems. Adaptive Behavior 19, 16–39 (2011)

Klyubin, A.S., Polani, D., Nehaniv, C.L.: All else being equal be empowered. In: Capcarrère, M.S., Freitas, A.A., Bentley, P.J., Johnson, C.G., Timmis, J. (eds.) ECAL 2005. LNCS (LNAI), vol. 3630, pp. 744–753. Springer, Heidelberg (2005a)

Klyubin, A., Polani, D., Nehaniv, C.: Empowerment: A universal agent-centric measure of control. In: Proceedings of the 2005 IEEE Congress on Evolutionary Computation, pp. 128–135. IEEE Press, Piscataway (2005b)

Klyubin, A., Polani, D., Nehaniv, C.: Keep your options open: An information-based driving principle for sensorimotor systems. PLoS ONE 3(12) (2008)

Lizier, J., Prokopenko, M., Tanev, I., Zomaya, A.: Emergence of glider-like structures in a modular robotic system. In: Proc. Eleventh International Conference on the Simulation and Synthesis of Living Systems (ALife XI), pp. 366–373 (2008)

Lungarella, M., Pegors, T., Bulwinkle, D., Sporns, O.: Methods for quantifying the information structure of sensory and motor data. Neuroinformatics 3(3), 243–262 (2005)

Lungarella, M., Pfeifer, R.: Robots as cognitive tools: Information theoretic analysis of sensory-motor data. In: Proceedings of the 2nd International IEEE/RSJ Conference on Humanoid Robotics, pp. 245–252. IEEE Press, Piscataway (2001)

Lungarella, M., Sporns, O.: Information self-structuring: Key principle for learning and development. In: Proceedings of the 4th International Conference on Development and Learning, pp. 25–30. IEEE Press, Piscataway (2005)

Matarić, M.: Learning social behavior. Robotics and Autonomous Systems 20, 191–204 (1997)

Miglino, O., Lund, H., Nolfi, S.: Evolving mobile robots in simulated and real environments. Artificial Life 2(4), 417–434 (1995)

Mondada, F., Bonani, M.: The e-puck education robot (2007), `http://www.e-puck.org/`

Nolfi, S., Floreano, D.: Evolutionary Robotics: The Biology, Intelligence, and Technology of Self-Organizing Machines. MIT Press/Bradford Books, Cambridge, MA (2000)

Olsson, L., Nehaniv, C., Polani, D.: Sensor adaptation and development in robots by entropy maximization of sensory data. In: Proceedings of the 6th IEEE International Symposium on Computational Intelligence in Robotics and Automation (CIRA 2005), pp. 587–592. IEEE Computer Society Press, Piscataway (2005)

Prokopenko, M., Gerasimov, V., Tanev, I.: Evolving spatiotemporal coordination in a modular robotic system. In: Nolfi, S., Baldassarre, G., Calabretta, R., Hallam, J.C.T., Marocco, D., Meyer, J.-A., Miglino, O., Parisi, D. (eds.) SAB 2006. LNCS (LNAI), vol. 4095, pp. 558–569. Springer, Heidelberg (2006)

Prokopenko, M., Wang, P.: Evaluating team performance at the edge of chaos. In: Polani, D., Browning, B., Bonarini, A., Yoshida, K. (eds.) RoboCup 2003. LNCS (LNAI), vol. 3020, pp. 89–101. Springer, Heidelberg (2004)

Quinn, M., Smith, L., Mayley, G., Husbands, P.: Evolving controllers for a homogeneous system of physical robots: Structured cooperation with minimal sensors. Philosophical Transactions of the Royal Society of London, Series A: Mathematical, Physical and Engineering Sciences 361, 2321–2344 (2003)

Salge, C., Polani, D.: Local information maximisation creates emergent flocking behaviour. In: Advances in Artificial Life, ECAL 2011: Proceedings of the Eleventh European Conference on Artificial Life (2011)

Shannon, C.: A mathematical theory of communication. The Bell System Technical Journal 27, 379–423, 623–656 (1948)

Sperati, V., Trianni, V., Nolfi, S.: Evolving coordinated group behaviours through maximisation of mean mutual information. Swarm Intelligence 2(2-4), 73–95 (2008)

Sperati, V., Trianni, V., Nolfi, S.: Self-organised path formation in a swarm of robots. Swarm Intelligence 5(2), 97–119 (2011)

Sporns, O., Lungarella, M.: Evolving coordinated behavior by maximizing information structure. In: Rocha, L., Yaeger, L., Bedau, M., Floreano, D., Goldstone, R., Vespignani, A. (eds.) Artificial Life X: Proceedings of the Tenth International Conference on the Simulation and Synthesis of Living Systems, pp. 323–329. MIT Press, Cambridge (2006)

Sporns, O., Tononi, G., Edelman, G.: Connectivity and complexity: the relationship between neuroanatomy and brain dynamics. Neural Networks 13, 909–922 (2000)

Tarapore, D., Lungarella, M., Gomez, G.: Fingerprinting agent-environment interaction via information theory. In: Groen, F., Amato, N., Bonarini, A., Yoshida, E., Kröse, B. (eds.) Intelligent Autonomous Systems, vol. 8, pp. 512–520. IOS Press, Amsterdam (2004)

Tarapore, D., Lungarella, M., Gomez, G.: Quantifying patterns of agent-environment interaction. Robotics and Autonomous Systems 54(2), 150–158 (2006)

Tononi, G., Edelman, G., Sporns, O.: Complexity and coherency: integrating information in the brain. Trends in Cognitive Sciences 2(12), 474–484 (1998)

Tononi, G., Sporns, O., Edelman, G.: A measure for brain complexity: Relating functional segregation and integration in the nervous system. Proceedings of the National Academy of Sciences 91, 5033–5037 (1994)

Tononi, G., Sporns, O.: G. Edelman, G. A complexity measure for selective matching of signals by the brain. Proceedings of the National Academy of Sciences 93, 3422–3427 (1996)

Trianni, V., Nolfi, S.: Self-organising sync in a robotic swarm. a dynamical system view. IEEE Transactions on Evolutionary Computation 13(4), 722–741 (2009)

Trianni, V., Nolfi, S.: chapter Evolving collective control, cooperation and distributed cognition. In: The Handbook of Collective Robotics - Fundamentals and Challenges, pp. 168–189. Pan Stanford Publishing, Singapore (2012)

Trianni, V., Nolfi, S., Dorigo, M.: Evolution, self-organisation and swarm robotics. In: Blum, C., Merkle, D. (eds.) Swarm Intelligence. Introduction and Applications. Natural Computing Series. Springer, Berlin (2008)

Van Dyke Parunak, H., Brueckner, S.: Entropy and self-organization in multi-agent systems. In: Proceedings of the Fifth International Conference on Autonomous Agents, pp. 124–130. ACM Press, New York (2001)

Wang, X., Miller, J., Lizier, J., Prokopenko, M., Rossi, L.: Measuring information storage and transfer in swarms. In: Proc. Eleventh European Conference on the Synthesis and Simulation of Living Systems (ECAL 2011), pp. 838–845 (2011)

Zahedi, K., Ay, N., Der, R.: Higher coordination with less control - a result of information maximization in the sensorimotor loop. Adaptive Behavior 18, 338–355 (2010)

Chapter 15
Evolution of Complexity and Neural Topologies

Larry S. Yaeger

15.1 Introduction

One of the grandest and most intriguing self-organizing systems is nature itself. Whether couched in terms of evolutionary theory (Darwin 1859), information theory (Avery 2003), or thermodynamics and maximum physical entropy (Jaynes 1957a,b; Swenson 1989) natural processes have yielded a remarkable diversity of behavioral and organizational levels of complexity ranging from microbes to man.

Though one could reasonably argue that single-celled organisms are as suited to their ecological niches as human beings are to theirs, no one would argue that microorganisms are as complex as humans. And looking at the fossil record, it is clear that complexity, by any metric, has increased over geological time scales (Carroll 2001), from algae to plants, from ediacarans to arthropods to insects to mammals. Understanding the nature and sources of that complexity will yield insights into all of biology, including ourselves.

The fundamental, causal forces that give rise to the observed trends in organizational and behavioral complexity are much debated. Darwin (1871) argued for a gradualist interpretation of the emergence and evolution of intelligence that suggests an increasing arrow of complexity, and many since have offered theoretical support for this idea and even some limited physical evidence from the paleobiological record (Rensch 1960a,b; Waddington 1969; Saunders and Ho 1976; Kimura 1983; Katz 1987; Bonner 1988; Arthur 1994; Huynen 1996; Newman and Engelhardt 1998). However others have argued that complexity growth has been nothing more than the result of a random walk away from a brick wall of minimum complexity (Raup et al. 1973; Gould 1989, 1994, 1996; McShea and Brandon 2010).

In order to understand the mechanisms that give rise to complexity, one must first define exactly what is meant by complexity. However, finding a way to assess complexity in a formal, quantitative fashion has proved to be problematic. Humans have,

Larry S. Yaeger
School of Informatics & Computing, Indiana University,
919 E. 10th St., Bloomington, Indiana 47408, United States
and Google Inc., 1600 Amphitheatre Parkway, Mountain View, CA 94043, United States
e-mail: larry.yaeger@gmail.com

M. Prokopenko (ed.), *Guided Self-Organization: Inception,*
Emergence, Complexity and Computation 9,
DOI: 10.1007/978-3-642-53734-9_15, © Springer-Verlag Berlin Heidelberg 2014

at times, demonstrated an anthropocentric chauvinism in defining degree of complexity, very loosely, as little more than similarity to humans below and a human-like deity above, as in the "Great Chain of Being". More scientifically, physiological characteristics, such as size, number of distinct cell types, and morphological measurements, have been used as proxies for complexity in the field of evolutionary biology.

Recently, information theory has been used to define a more rigorous measure of complexity specifically for neural systems (Tononi et al. 1994; Lungarella et al. 2005), though the mathematics apply equally to an arbitrary collection of processes. This "TSE Complexity" metric, to be discussed in some detail later, provides a rigorous measure of the complexity of the dynamics of a neural network. However, there currently is no way to apply such a metric to biological organisms, simply because it is impossible to make the detailed measurements of the states of many neurons that would be required.

By contrast, it is relatively straightforward to record the activations of the units in an artificial neural network (ANN). And artificial life simulations offer the possibility of evolving such ANNs while they serve as artificial brains controlling the behaviors of agents in a computational ecosystem. We are thus able to carry out repeated evolutionary experiments, varying only the random elements of a system or carrying out careful parameter sweeps, to begin to dissect the structures and processes that give rise to complexity and guide its evolution.

Once we are able to make these kinds of assessments of neural complexity, it is natural to ask what kinds of network topologies give rise to dynamical complexity—what network structures foster complex network functionality. It turns out that normal evolutionary pressures on behavioral adaptation to an environment induce pressures on complexity and on network topologies that are consistent with observations of biological systems and that work in concert with purely physical constraints, such as brain volume and wiring length, to shape the design of our brains.

In this chapter we review work carried out primarily between 2004 and 2013, though the artificial life simulator used to perform the work, Polyworld, was originally designed and built between 1990 and 1992, was first presented at the Artificial Life III conference in Santa Fe, NM in 1992, and first saw publication in the proceedings of that conference in 1994 (Yaeger 1994). More background and detail on the work reported here may be found in a series of articles published in Artificial Life, ECAL (European Conference on Artificial Life), and GECCO (Genetic and Evolutionary Computation Conference) proceedings, and the HFSP and Advances in Complex Systems journals (Yaeger and Sporns 2006; Griffith and Yaeger 2006; Yaeger et al. 2008; Lizier et al. 2009; Yaeger 2009; Yaeger et al. 2010; Murdock and Yaeger 2011b,a; Yaeger 2013).

15.2 Complexity

To be useful scientifically, complexity must be quantifiable. Complexity for the purposes of evolutionary biology has taken many definitions and forms over the years, from organism size (Cope 1871) to distinct cell types (Bonner 1988; Valentine et al. 1994) to morphology (Thomas and Reif 1993; McShea 1993) to ecological webs of interaction (Knoll and Bambach 2000). Though somewhat subjective, McShea (1996) has defined four types of complexity that may be fairly broadly applicable: (1) the number of different parts (such as cell types), (2) the number of different interactions among those parts, (3) the number of levels to be found in a causal, hierarchical description of a system or organism, and (4) the number of parts or interactions between those parts at different spatial and temporal scales.

Information theory (Shannon 1948) has produced various methods for defining and quantifying complexity that might be suitable for the study of biological and evolutionary processes, but most such measures share an undesirable trait with Shannon entropy, which is that they are maximized by random processes in which all states are equiprobable. By contrast, biological complexity corresponds to neither minimal nor maximal entropy, but something in between.

The fourth of McShea's complexity types mentioned previously shares an insight with Chaitin (1979): The way information flows across multiple scales relates directly to the internal structure of a system and thence to its structural and functional complexity. Chaitin also defined a "measure of mutual information for n-tuples" that presages a key component of the TSE complexity measure discussed below (integration, aka multi-information), however he stopped short of applying these insights to his algorithmic entropy measure of complexity which, like Shannon entropy, is maximized by random processes.

Tononi et al. (1994) proposed an information-theoretic measure for quantifying the complexity of neural dynamics which avoids maximization by randomness, instead capturing quantitatively the manner in which information is processed at multiple scales. Acknowledged as an important conceptual milestone in the measurement and understanding of complexity, the Tononi-Sporns-Edelman measure is often referred to as "TSE complexity" (for the authors' initials). A simpler, more computationally efficient measure was subsequently proposed in (Tononi et al. 1998), explored computationally in (Sporns et al. 2000), and refined in (Lungarella et al. 2005). Implementations of both the full TSE "neural complexity" and the simplified TSE "description complexity" are publicly available in a MATLAB Complexity Toolbox at `http://www.indiana.edu/~cortex/complexity.htm`. In order to be able to process the neural dynamics for hundreds of time steps from hundreds of neurons in thousands of agents, from each of many simulations, the simplified TSE complexity measure was re-implemented in C++ and is what we use and refer to throughout the bulk of this chapter as "complexity" or C. This software is available with the simulation software; see below.

Though non-trivial to derive and implement, the intuition behind TSE complexity is straightforward: Cooperation amongst various elements of a network, called *integration* and measured by a multivariate extension to mutual information, increases network complexity, to a point. But specialization of network subunits, called *segregation*, also increases network complexity. Complex network dynamics, then, require both cooperation and specialization—integration and segregation—amongst the nodes of the network.

Specialization at multiple scales (subsets of units) means that the integration one measures at these different scales varies non-linearly, and differs from what would be measured in a system where integration was distributed uniformly at all scales. It is the area between the actual integration curve as a function of scale and a line that corresponds to the uniform distribution of integration over all scales that defines TSE complexity. Thus, maximal complexity is achieved in networks that simultaneously maximize the opposing tensions of integration and segregation to the extent possible.

The original TSE complexity is given by:

$$C_N(X) = \sum_{k=1}^{n} [\ (k/n)I(X) - \langle I(X_k) \rangle\ ] \tag{15.1}$$

where X is a system of n variables, $I(X)$ is defined as the *integration* over the full system of variables, $I(X_k)$ is the integration over a subset of variables of size k, and $\langle \cdot \rangle$ denotes an ensemble average over all subsets of size k.

Integration is a multivariate form of mutual information that captures the degree to which nodes in the system share information and may be computed as:

$$I(X) = \sum_{i=1}^{n} H(X_i) - H(X) \tag{15.2}$$

where $H(X_i)$ is the Shannon entropy (Shannon 1948) of the ith variable and $H(X)$ is the entropy of the entire system of variables.

Cast in terms of entropy, TSE complexity may also be written as:

$$C_N(X) = \sum_{k=1}^{n} [\ \langle H(X_n^k) \rangle - \frac{k}{n} H(X)\ ] \tag{15.3}$$

where $H(X)$ is again the Shannon entropy of the entire system of n variables, k is again the size of a subset of variables, and the ensemble average $\langle H(X_n^k) \rangle$ is to be taken over all $n!/(k!(n-k)!)$ subsets of size k.

The simplified approximation looks only at a single subset term, involving mutual information between individual variables and the rest of the system, or cast in terms of entropies:

$$C(X) = H(X) - \sum_{i=1}^{n} H(x_i | X - x_i) \tag{15.4}$$

where $H(X)$ is again the entropy of the entire system and the $H(x_i|X - x_i)$ terms are the conditional entropy of each of the variables x_i given the entropy of the rest of the system (excluding x_i).

As discussed in (Ay et al. 2006), modulo a constant the simplified version of TSE complexity is equivalent to *excess entropy* (Crutchfield and Feldman 2003) (aka *effective measure complexity* (Grassberger 1986) and *dual total correlation* (Han 1978)).

15.3 Simulation Software

The simulation software used for these studies is Polyworld (Yaeger 1994), an evolutionary model of a computational ecosystem, populated by haploid agents with a suite of primitive, neurally controlled behaviors (move, turn, eat, mate, attack, light, focus). The Artificial Neural Networks (ANNs) that control these agents use "summing and squashing" (aka "firing rate") neurons and perform Hebbian learning at the synapses.

The topologies of the ANNs are derived from the agents' genomes, expressed as a number of neural groups of arbitrary excitatory and inhibitory neuron counts, along with connection density, "topological distortion" (degree of disorderdness between sequentially indexed neurons of two groups), and learning rate between each of the neural groups and types. Acting as a generative model, a given genetic encoding produces a family of related topologies sharing the same statistics. Thus evolution is forced to select for statistics of connectivity, rather than specific topologies.

It may also be worth noting that the network's weights (synaptic efficacies) are *not* encoded in the genome. Rather, they are initialized randomly within a genetically determined range at the time of an agent's birth. Since these synaptic strengths are updated using a Hebbian learning rule, they vary over the course of an agent's life. So while the underlying, binary topology of these brain graphs does not change during an agent's life, the weighted graphs do. All graph theoretical analyses presented here are carried out on the final state of an agent's neural network, as it existed at the time of the agent's death. All information theoretic complexity analyses here are calculated from the temporal histories of an agent's neurons over its full lifespan.

Input to the ANN consists of a row of pixels from a rendering of the scene from each agent's point of view, like light falling on a retina. Though agent morphologies are simple and static, agents interact with the world and each other in fairly complex ways. Their body color varies continuously and is determined by mapping the activation level of the *fight* neuron to the red color component, the activation of the *mate* neuron to the blue color component, and a specific *ID* gene to the green color component. Since they can "see" each other in the same color spectrum that we observe, they are thus, in principle, able to observe and respond to both a couple of key, behavioral neural states and a limited measure of the genetic identity of other agents. They reproduce through the simultaneous expression of a mating behavior by two collocated agents.

Agents' energy stores are normally depleted by all actions, including neural activity, with stronger actions (higher behavioral neuron activations) and larger neuron and synapse counts depleting more energy. To simplify the analysis and to remove all pseudo-physical constraints on the evolution of network topologies, per-neuron and per-synapse neural costs have been eliminated for this series of experiments; behavior-based costs have been retained. Energy is also depleted when one agent is attacked by another, and agents die when their energy level reaches zero. Normally dying agents leave a carcass that becomes a food object containing a suitable amount of energy (for details see (Yaeger 1994)), however for these simulations dying agents are simply removed from the simulation. Energy must therefore be replenished by seeking out and consuming food in the environment. (Normally, killing and eating other agents would also be an option.) Population size is controlled by proportionately decreasing energy costs as the population dwindles and increasing energy costs as the population swells.

In its normal mode of operation all evolutionary pressures emerge from a process of natural selection—the survival and reproduction of the agents—and there is no influence of any kind from a fitness function (though an *ad hoc* heuristic fitness function is typically computed for purely informational purposes). Polyworld can be run, however, in a steady-state Genetic Algorithm mode, as has been done for a subset of the simulations discussed here using neural complexity as the fitness function. There is also a "passive" mode in which a new simulation can be forced to run in lockstep with a previous normal simulation but with reproduction, births, and deaths controlled randomly (rather than behaviorally by the agents) to produce a kind of *null model*. More on these alternative simulation modes later.

Polyworld was originally designed to explore the possibility of open-ended evolution of artificially intelligent agents. Figure 15.1 shows the Polyworld evolutionary ecosystem and various attendant graphical elements. More detailed information may be found in (Yaeger 1994). The software is open source and the latest version may be downloaded at `http://sourceforge.net/projects/polyworld/`. The information theoretic tools used to carry out the complexity analyses are included in the Polyworld source. The graph theoretical tools used to perform the network topology data analyses are open source and may be downloaded at `http://code.google.com/p/bct-cpp/`.

15.4 Natural Selection vs. Random Drift

Many good arguments have been put forward in support of the popular view that evolution is biased towards increasing complexity. Rensch (1960a,b) and Bonner (1988) argued that more parts will allow a greater division of labor among parts, and since this greater division of labor will confer an evolutionary advantage, evolution naturally favors a continuing increase in complexity as measured by a count of distinct components. Waddington (1969) and Arthur (1994) have suggested that due to increasing diversity, niches become more complex, and are thus filled with more complex organisms. Knoll and Bambach (2000) echo and expand on this argument, arguing that evolution into an ever-increasing "ecospace" will confer a continual growth in complexity. Saunders and Ho (1976) and Katz (1987) suggest that component additions are more likely than deletions, because additions are less likely to disrupt normal function, again resulting in an evolutionary increase in complexity as measured by parts. Kimura (1983), Huynen (1996), and Newman and Engelhardt (1998) have demonstrated the value of neutral mutations in bridging gulfs in fitness landscapes, allowing evolution to select for novel functions in previously neutral changes. These neutral mutations frequently take the form of gene duplications—a common and straightforward mechanism for increasing genetic complexity.

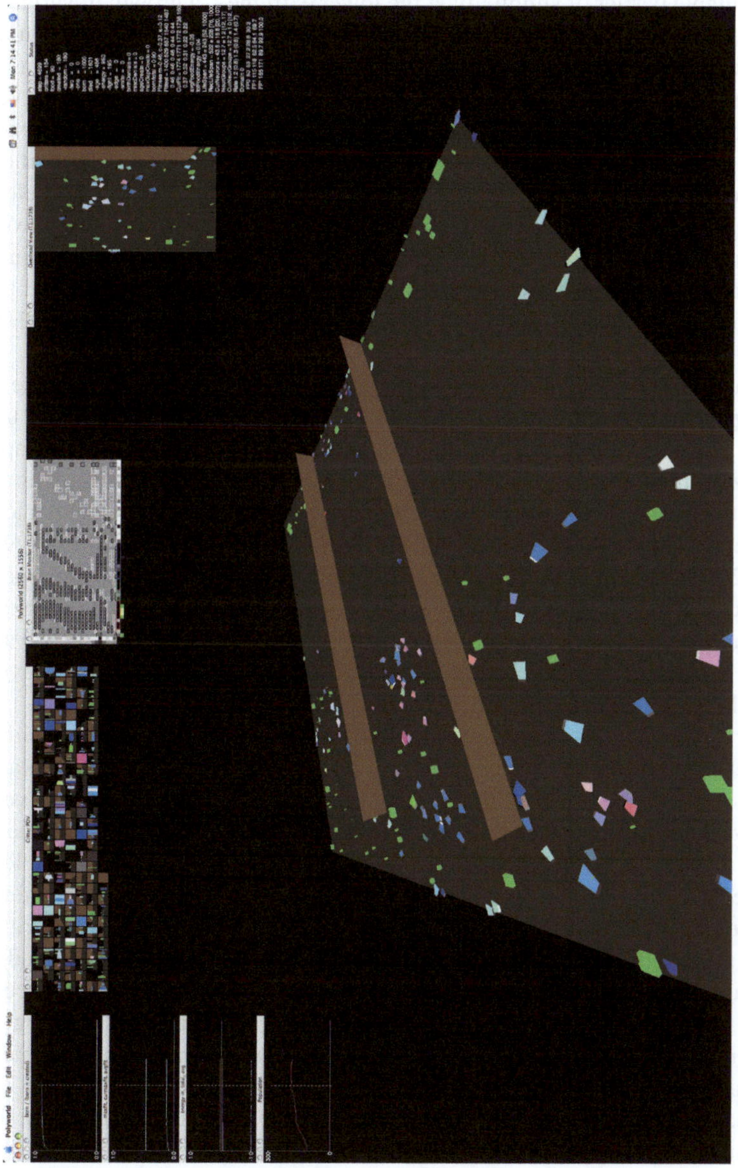

Fig. 15.1 The Polyworld evolutionary ecosystem. The multi-colored trapezoids are agents (with color denoting certain behaviors and genetics); bright green blocks are food; brown walls are barriers the agents cannot pass through; all action takes place on the dark green ground plane. Also shown along the top are certain graphical statistics (left), the point of view of all agents in the world (left-center), a brain connection matrix (center), an overhead detail view (right-center), and a few numerical statistics (right).

However, Maynard Smith (1970), Raup et al. (1973), Gould (1989, 1994, 1996), McShea (1994) and others have questioned whether the observed growth in complexity has been the outcome of natural selection or simply, in Maynard Smith's words, the "obvious and uninteresting explanation" of a sort of random walk away from an immutable barrier of simplicity at the lower extreme—a growth in variance relative to the necessarily low complexity at the origin of life. Gould, in particular, has argued extensively that random chance not only plays a greater role in evolution than previously understood, but also is entirely sufficient to explain the observed increases in biological complexity over geological time scales. McShea and Brandon (2010) have proposed a "Zero Force Evolutionary Law (ZFEL)" that predicts a strong drive toward complexity, but based solely on random genetic drift (selection may play a role, but only for the specialization of parts, not for complexity per se). As part of an extended debate with Gould, Dawkins (1997) counter-argued that evolution must be progressive and biased, even if that bias is not always in the same direction at the micro-evolutionary scale (which turns out to be very much in keeping with our experimental results).

Bedau et al. (1997) and Rechsteiner and Bedau (1999) have provided evidence of an increasing and accelerating "evolutionary activity" in biological systems that until Channon (2001) had not been observed in an artificial system. However, attempts to characterize complexity trends in the paleontological record have produced mixed results at best (McShea 1996; Heylighen 2000; Carroll 2001), leaving us with a less than clear picture of the influence of natural selection on complexity. McShea (1994, 1996, 2001, 2005) has, for almost two decades, attempted to clarify and, where possible, empirically address the debate, by identifying distinct classes of complexity and, importantly, by distinguishing between "driven" trends, in which evolution actively selects for complexity, and "passive" trends, in which increases in complexity are due simply to Gould's asymmetric random walk.

A number of researchers have used simple computational models of branching clade lineages to study driven vs. passive trends in the evolution of complexity. Raup et al. (1973) looked exclusively at branching patterns. Raup and Gould (1974) extended that work using a 10-parameter vector to represent morphological characters. In McShea's 1994 model, a single numerical parameter takes the place of complexity, and biased or unbiased anagenetic (intra-species) and cladogenetic (branching) random variation produces different populations. In all of these efforts, the distribution and statistics of the resulting populations are then at least roughly compared to the distribution and statistics of biological species over time in order to try to discern the presence or absence of any inherent bias in biological evolution. They largely conclude there is reason to doubt any systematic bias towards increasing complexity. Raup and Gould go so far as to suggest that "undirected selection may be the rule rather than the exception in nature." However, they base their arguments on the ability of random systems such as theirs to produce clade histories and temporal lengths of continuously changing characters that are only vaguely similar to biological systems. They acknowledge that directional selection exists, but emphasize that since the temporal geometry of evolutionary phenomena can also be produced within purely random systems, such geometric patterns are insufficient to unequivocally imply directed (or undirected) causes.

McShea uses his model to conclude that a biased system will necessarily exhibit an increase in the minimum complexity present in the system, and this observation has become common wisdom (Wagner 1996; McShea 2001; Carroll 2001). The presence of a multitude of minimum-complexity, single-celled organisms today is then taken as evidence against a biased evolutionary trend in complexity. However, even though an increase in the minimum is evidenced in the most commonly reported version of his model, McShea notes that it is not always possible to use this observation to distinguish a passive system from a weakly driven one. Also, in a more realistic evolutionary system, in which fitness at smaller, less complex scales is relatively independent of fitness at larger, more complex scales, the system could possess a substantial positive bias at larger scales without eliminating or even disadvantaging organisms at the lower end of the spectrum. Indeed, in real biology, the largest, most complex organisms are themselves extremely hospitable niches for single-celled organisms, and, as opposed to competing with microorganisms in any way, significantly increase their available resources.

There are some specific details and assumptions in McShea's model that make conclusions drawn from it suspect, yet which could mostly be easily addressed. One is the presence of anagenetic (within lineage) change with the same magnitude and likelihood as the model's cladogenetic (between lineages) change. While anagenetic change—fine-tuning some morphological or behavioral character in a species—is certainly observable in nature, substantive complexity changes are more likely to be associated with speciation events—cladogenetic change. Thus anagenetic complexity changes should be of a smaller magnitude than cladogenetic changes, but are not. Perhaps of greater concern, this kind of anagenetic change is formally equivalent to a cladogenetic branch followed by an immediate extinction event of the ancestor lineage. It is as if half of all speciation events resulted in fierce competition for the same resources and the parent lineage always lost. But speciation is usually associated with adaption to a different niche and need not imply the extinction of the ancestral species, so this equal magnitude, equal probability anagenetic change component of the model is guaranteed to produce unrealistic trends in population statistics. Significantly, McShea (1994) acknowledges a much lower, nearly unmeasurable rate of growth in the observed minimum complexity in a purely cladogenetic model as compared to the widely reported, mixed anagenetic plus cladogenetic model. Thus it is clear that reducing or eliminating this anagenetic component would significantly reduce the observed growth in the minimum for a positively biased system.

Another key aspect of McShea's model that is likely to skew his results is his choice of extinction rates. He makes what seems an unbiased choice—a small, constant extinction rate that applies equally to all lineages. But, intuitively, one might expect complex systems to break down more easily than simple ones. Farmer and Griffith (2007) studied the robustness of self-reproducing machines and derived an intriguing functional relationship between viability and complexity that is consistent with that intuition. Separately, population sizes are likely to have an even greater effect on extinction rates. A small population is obviously more easily extinguished than a large one. McShea does not model lineage populations, but integrating any reasonable growth rate (linear?, exponential?, varying with complexity?) over time since its inception would provide such an estimate which could then be used to modulate the extinction rate of species. We strongly suspect that modification of extinction rates according to these viability and population principles would substantially reduce any growth in the minimum for driven systems in McShea's model. Taken in combination with the problematic anagenetic change discussed previously, it raises considerable doubt about the common wisdom derived from such models.

Given the shortcomings of these simple models and the difficulties and ambiguities one finds when studying the paleontological record, it makes sense to turn to computer models that actually employ evolution to investigate the question of evolutionary trends in complexity. Turney (1999, 2000) has used a simple evolutionary model to suggest that increasing evolvability is central to progress in evolution and predicts an accelerating increase in biological systems that may correlate with complexity growth. Adami (Adami et al. 2000; Adami 2002) has defined complexity as the information that an organism's genome encodes about its environment, and calculates a complexity metric that is inversely proportional to the entropy of its genetic elements (bits in simulation, base pairs in biological genomes). He has used the Avida software to show that asexual agents in a fixed, single niche always evolve toward greater complexity of this uniquely defined type. Though we believe there is little doubt that Adami's observation regarding genes encoding information about the environment is both true and important, we suspect that his "complexity" metric is better viewed as a measure of a species' adaptation to its environment and is ultimately just a measure of genetic consistency in the population. Its inapplicability to multiple species, multiple and variable niches, or niche creation, limits its value as a measure of complexity. Adami's measure also suffers the inverse malady of Shannon entropy: Whereas entropy is maximized by randomness, Adami's measure is maximized by perfect uniformity. This assignment of zero value to biodiversity probably rules it out as a useful measure of complexity.

In (Yaeger and Sporns 2006) we used our computational ecosystem software, Polyworld, together with our information-theoretic measure of neural complexity to demonstrate that complexity of this particular type increases over evolutionary time scales. That early work, however, did not distinguish contributions from natural selection versus contributions from Gould's random walk. Basically, there was no "null model" to compare with, and since the seed population used to initialize those simulations consisted of very low-complexity agents, we could not rule out Maynard Smith's obvious and uninteresting explanation.

Though it would be difficult, if not impossible, to eliminate natural selection from a biological ecosystem, artificial life software is more manipulable. In order to distinguish between driven trends in complexity due to natural selection vs. passive trends due to random genetic drift, we implemented a "lockstep" mode of operation for Polyworld, in which a passive run, where natural selection is disabled, is tied to a driven run, where natural selection operates like normal. First the normal, driven simulation is carried out, during which all agent births and deaths are recorded (along with all relevant neural data). Then the lockstep, passive simulation is carried out, during which agent births and deaths are forced to mirror those in the normal, driven run. In the passive run, agents cannot reproduce or die as a result of their own behaviors. Instead, at each moment in time that an agent died in the driven run, a randomly selected agent in the passive run is killed. Similarly, any time two agents reproduced in the driven run, two randomly selected agents in the passive run are bred and their offspring placed at a random location. We thus allow genetic changes (and resulting neural topologies) to follow their own separate courses in the two simulations while synchronizing key reproductive and population aspects of the simulations at every time step.

Using this method, population statistics between the driven and passive runs are held identical. As a result, the statistics of the genetic operations—crossover and mutation—are comparable between the paired runs. The number of crossover points and mutation rates are themselves embedded in the genome (Yaeger 1994), so the genetic operations are the same statistically, but not in detail. Similarly, the "life experiences" of a given agent—its trajectory through the world and the inputs to its visual system—are comparable statistically between paired runs, though not identical. Due to the resulting consistency in visual environment, contributing statistics, such as the entropy and mutual information in the visual inputs, should have comparable influences on neural complexity in the two simulation modes. Because complexity is affected by agent behaviors and their resulting sensory inputs, agents in lock-step runs must be allowed to control their own actions in order to obtain sensible measures of their neural complexity. Even though we expect observed complexity differences to be primarily the result of differences in neural topology, subtle differences due to agent behaviors and resulting life experiences will then naturally be included in the calculations.

The effective result is that natural selection is "turned on" in the driven runs, and "turned off" in the passive runs. That is, gene states are subject to natural selection, based on the evolutionary viability—the fitness—of the agents' behaviors, in the driven runs, while gene states are subject only to the same degree of variation, with no evolutionary fitness consequences or effects, in the passive runs. The passive run serves as our null model, thus allowing us to distinguish between complexity growth due to natural selection vs. that due to a random walk in gene (and neural topology) space.

Note that while genetic changes in both the driven and passive cases produce different neural topologies, neural dynamics, and agent behaviors, in the passive case (only) these changes have absolutely no impact on the reproductive success of agents. Thus this random walk in gene space produces a random walk in neural topology space and in complexity space as measured by neural dynamics, without regard for the evolutionary value of that complexity or the corresponding agent behaviors. By contrast, in the driven case the variations in neural structure and function and resulting behavioral adaptations produced by these genetic changes directly affect the agents' reproductive success, hence any sensible measure of their fitness.

15.5 Data Generation and Acquisition

A simple world was set up in Polyworld that accommodated between 90 and 300 agents, grew food uniformly and randomly in two patches at opposite ends of the world (80% at one end, 20% at the other), and positioned two barriers that ran 90% of the depth of the world and divided the world into thirds across its width. (See Figure 15.1.) The agents were allowed to evolve between zero and five internal neural groups, with up to 16 excitatory and 16 inhibitory neurons per group. Input neural groups, internal neural groups, and output/behavior neural groups were connected according to evolved connection densities specified in the agents' genomes. This allows up to 217 neurons and 45,584 synapses. Each simulation was allowed to run for 30,000 time steps, which corresponds to about 400 generations. For each run, the neural topologies at birth and death plus the neural activations of every neuron, at every time step, for every agent are recorded to disk, yielding approximately 10 GB of data. It is this data to which we subsequently apply our complexity and graph theoretical calculations, thus allowing us to determine the neural complexity of each individual agent and the characteristics of the network graph that produced that complexity. Note that an agent's complexity is only fully determinable upon its death—when its neural activation time series are complete. We therefore examine mean trends in the population's complexity for all agents that have died during a particular time interval—typically every 1,000 time steps. And, as mentioned previously, due to Hebbian learning in the network during an agent's life and the resulting ongoing synaptic weight changes, we carry out our graph theoretical analyses on the state of the network graph at the time of the agent's death.

Ten driven runs were carried out, in which natural selection operated normally, varying only the initial seed to the random number generator. A corresponding passive "lockstep" run, in which natural selection was disabled, was carried out for each of the driven runs, yielding 20 simulations in all.

The runs were all seeded with a population of agents based on a uniform genome. This seed genome was designed to produce modest dispositions towards potentially useful behaviors, such as running towards green (food) and away from red (attacking agents). However, these seed agents are not a viable species. That is, test runs have demonstrated that, without the ability to evolve, the seed agents will fail to survive and reproduce in numbers sufficient to sustain their population, as a result of which their population will dwindle and become extinct. There must therefore be a significant evolutionary value to some kind of rewiring of the genome and resulting neural network architectures as a result of natural selection, at least until such point as the agents' behaviors are sufficiently adapted to the environment to sustain their numbers through their foraging and mating behaviors. The genome used in this seed population produces neural architectures that are nearly minimal in size and complexity, so one should expect the random variations in gene space associated with the passive, lockstep runs to also yield increases in complexity, as indeed they do. It is primarily the differences in the rate of change of complexity between driven and passive runs that are used to assess the role of natural selection in complexity increase.

Complexity may be calculated for various subsets of neurons—all neurons, just the input neurons, just the behavior neurons, or the "processing" neurons (all neurons except inputs). As the state of the input neurons are determined entirely by the environment and agent, rather than the dynamics of the network, all complexities presented here are based on the processing neurons only. Although, in general, there are few differences in complexity trends between all neurons and processing neurons. As evolution progresses, changes to the parts of the genome that specify neural topology produce different network graphs, which produce different internal neural dynamics and agent behaviors, which yield different complexities.

Both complexity and graph theoretical metrics were calculated for each agent and averaged to produce a population mean in each temporal bin, for each driven and passive run. In addition, for each agent's actual neural network, 10 graphs with an identical node count, edge count, and distribution of weights were generated randomly, and the means of the graph theoretical measures for these networks were used to characterize the structure of a random graph corresponding to each actual graph.

As mentioned, as configured for these runs, a maximum of 217 neurons and 45,584 edges were possible. Actual, evolved neuron counts ranged from 12 to 187, with a mean of 56. Evolved edge counts ranged from 33 to 13,081, with a mean of 1,077. In all, over half a million evolved graphs were analyzed using 42 different metrics (counting metrics for different neuron subsets and graph types as distinct), and in excess of five million random graphs were analyzed using 24 of those metrics.

The data shown and discussed here are from a set of runs first presented in (Yaeger 2013), and are based on the world structure and general methodology appearing in (Yaeger et al. 2008) and (Yaeger et al. 2010).

15.6 Complexity As a Fitness Function

Given a system such as Polyworld that is capable of evolving a wide range of neural topologies and using them to control a fairly broad range of agent behaviors, and given the ability to then calculate an information-theoretic measure of complexity for these agents' neural networks, an intriguing possibility presents itself: Why not dispense with natural selection altogether and, instead, directly employ neural complexity as a fitness function, in the traditional genetic algorithm (GA) sense? To investigate the fruitfulness of such an approach, in addition to the paired driven vs. passive runs just described, we carried out a suite of simulations using neural complexity as a fitness function. This required us to run Polyworld in a special steady-state GA mode. In this mode the simulator retains a list of the N (typically 30) best agents that have ever lived, according to a specified fitness function. For these runs we computed neural complexity of the processing units for each agent upon its death and used this complexity directly as a measure of the agent's fitness. Then, only in this steady-state GA mode, upon the death of each agent the software uses one of a couple of different selection and reproduction strategies. One strategy selects the parents by taking an incremental step through the list of N best agents, using a pair of indexes, one of which only increments when the other index finishes its run and wraps around. The other strategy uses tournament selection from amongst the N fittest with a small tournament size, typically 3. Offspring are then produced from the selected parents employing the usual mechanisms of crossover and mutation. A simple experiment using a fitness function of 1 / (velocity + 1), in which agents were successfully evolved to slow to a near stand-still population-wide, was used to validate these techniques. Results of these simulations will be discussed below.

15.7 Evolutionary Trends of Complexity

The resulting population-mean complexities for all 20 simulations are shown with light lines (solid for driven, dashed for passive) in Figure 15.2, along with additional, bold lines corresponding to the means of the two classes of runs (driven vs. passive). In addition to the simulation results, a measure of statistical significance is shown at the bottom of the graph. The fine dotted line shows 1−p-value for the dependent Student's T-test. The horizontal dashed line at 0.95 thus corresponds to a $p>0.05$ critical threshold. Where the dotted line rises above the dashed line, the difference between the driven and passive runs may be considered statistically significant; where it falls below, the difference is not significant. Where the driven and passive mean complexity values cross (t=7000) significance drops almost to chance (0.5), as one would expect, but values below 0.8 have been cropped for clarity.

The first thing to note is that during the early stages of the simulation evolution is actively selecting for increased complexity. This is consistent with our observation that the seed population is too simple to sustain its numbers, and must evolve or become extinct. As long as increasing complexity offers an evolutionary advantage, natural selection will behave in this biased, driven fashion with regard to complexity.

However, note that any statistical significance has disappeared by about t=5000, and, indeed, the mean passive complexity reaches the same level as the mean driven complexity by about t=7000. At roughly the point that statistically significant differences disappear, the driven complexities produced by natural selection begin to plateau. By this stage, a "good enough" solution has emerged and begun to spread throughout the population, as evidenced by the peakedness of the temporal histogram of complexity in Figure 15.3a at later times. From this time forward, in the majority of the simulations, natural selection tends to act as a weakly stabilizing force, maintaining a nearly constant mean level of fitness in the population. Since most genetic variation at this point is likely to be of a deleterious nature, evolution acts to weakly maintain this good-enough solution. Meanwhile, the random walk continues in the passive runs, thus increasing variance in the genes and in complexity, as evidenced by the flat and broad complexity distributions of Figure 15.3b. The growing variance produces networks of greater and greater complexity—now significantly exceeding that of the driven runs, on average—but this complexity was not produced by natural selection and confers no behaviorally adaptive advantage on the agents.

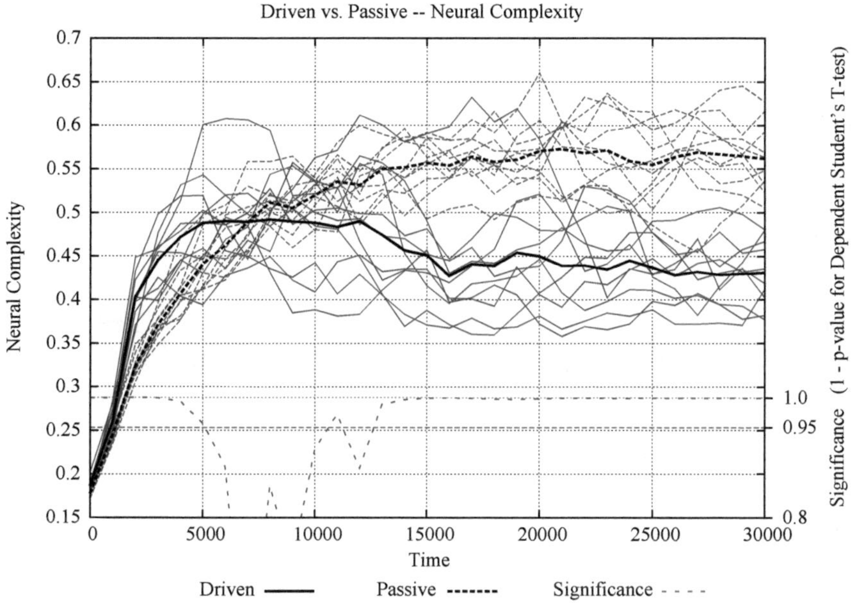

Fig. 15.2 Driven and passive complexity vs. time. Light solid lines show population mean complexity for individual driven, natural selection runs. Light dashed lines show population mean complexity for individual passive, lockstep runs. Heavy lines show means of all ten runs for corresponding line style. Light dotted line at bottom shows 1−p-value for a dependent Student's T-test, with a horizontal $p>0.05$ T-critical dashed line at 0.95.

The stabilization effect consistently observed in driven runs is, however, only a weak effect. In similar experiments not shown here, including those reported in (Yaeger et al. 2008), attention to the individual driven runs reveals a kind of punctuated equilibrium in some of them, where genetic variation eventually produces a network that is sufficiently better adapted to the environment that its genes spread throughout the population causing the mean level of complexity in the population to rise again. These new driven solutions reach approximately the same level of complexity as the passive runs and then plateau again at this new, higher level. With more evolutionary time it is possible that additional driven runs would make a secondary leap in complexity, bringing the overall mean driven level of complexity up to the level of complexity seen in the passive runs or higher.

In runs that do not exhibit these secondary leaps in complexity, such as the ones seen here, there often appears to be a slow and modest reduction in complexity after the initial growth spurt and plateauing. We speculate that this may result from a selection for robustness to mutation, causing networks to give up any complexity not absolutely required to sustain the agents' evolutionarily useful behaviors. However, we are not sure how robust this effect is and have not yet devised a method to confirm the source of this gradual reduction in complexity.

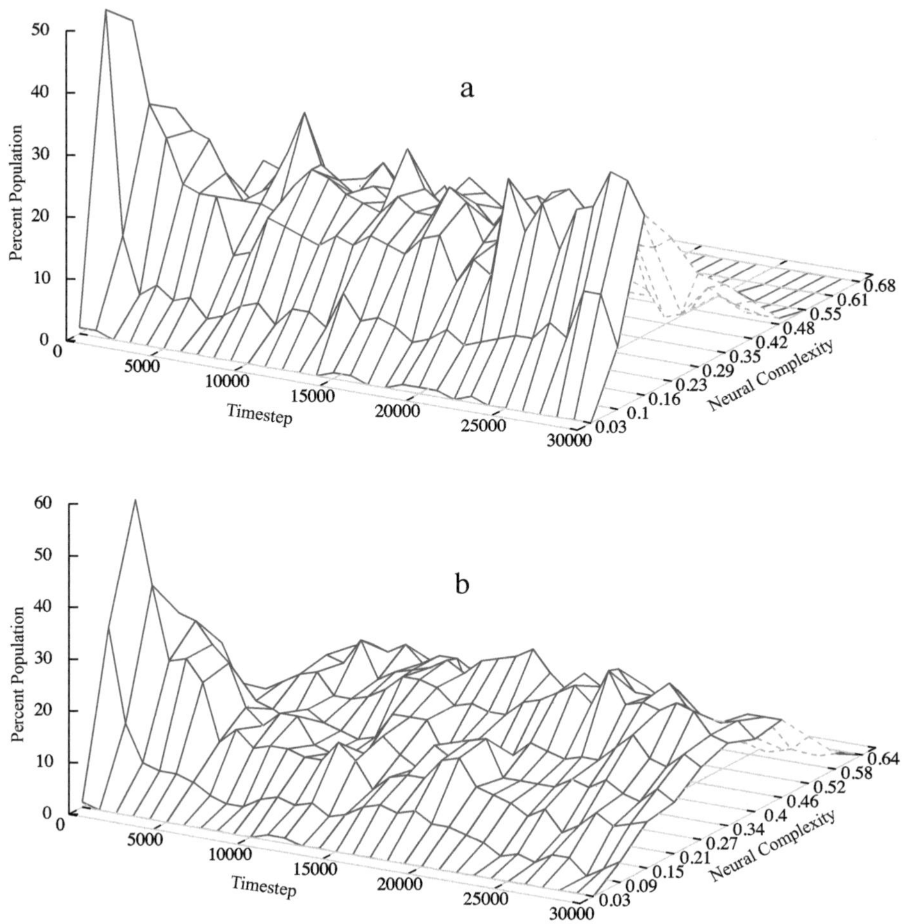

Fig. 15.3 Population histograms of complexity over time, for a driven run (a) and a passive run (b)

In other work not presented here, we used a clustering algorithm developed for studying biological genes to analyze genetic clustering of Polyworld agents in runs similar to these and both confirmed and extended this "good enough" interpretation of events (Murdock and Yaeger 2011b). Those results made it clear that alternative genetic "solutions to the world" were emerging, spreading briefly, and either failing or, rarely, succeeding in displacing the dominant population. Such branching and extinction of sub-populations, that might reasonably be considered species, correlated well with subtle bumps and dips in the whole-population means as presented here.

The passive complexity also plateaus in these runs, but for a much less interesting reason. By about t=10000 the bits of the agent genomes have been fully randomized to roughly 50% on and 50% off. At this point, genetic variance is at a maximum in the model, and further randomization merely shuffles bits around without any change in the mean values of the genes that control the topology of the neural networks. It is due to the fixed size genome and fundamentally linear interpretation of genes in Polyworld that such an upper bound on variance exists, and once maximum variance is reached the passive runs cease to characterize a truly random walk. Complexity would most likely continue to increase in an unbounded random walk, at least until a maximum complexity for the given parameter regime was achieved, which would only increase any statistical significance of differences between passive and driven trends at later times.

Note that final complexities range from about 0.44 to 0.56. Compare this to the approximately 1.0 level of complexity achieved in a suite of 10 steady-state GA simulations using complexity as a fitness function, as shown in Figure 15.4. The factor of two multiplier shows that the random walk was indeed limited by bit randomization rather than any limit imposed by the simulation parameters. It also demonstrates that this method for evolving larger values of complexity is quite effective. However, the behaviors of the agents in this complexity-as-fitness run are not well suited to the normal evolutionary constraints of life in Polyworld. The agents do not forage for food. There is no indication that they seek out other agents for reproduction. This run was seeded with agents based on the same seed genome as the driven and passive runs, so initially they are not particularly well suited to their environment. After about 6,000 or so time steps, at a moderate level of complexity (around 0.75),

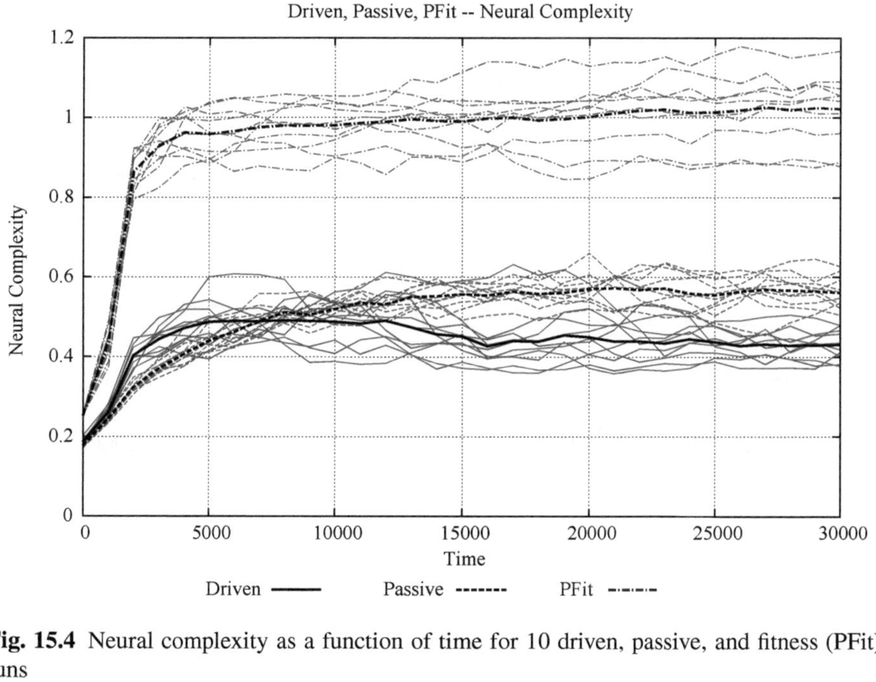

Fig. 15.4 Neural complexity as a function of time for 10 driven, passive, and fitness (PFit) runs

most of the agents have adopted a stereotypical tight turning behavior, all spinning in small loops, and ignoring food and other agents. We note that this behavior produces fairly high entropy while maintaining a significant degree of mutual information in the sensory units (due to adjacent pixels being correlated and some degree of frame-to-frame coherency), thus producing a fairly high degree of complexity throughout the network. Later, near peak complexity the agents' behaviors have become more interesting and more varied, but they still do not explicitly seek food, and there is still no evidence of any substantial degree of agent-agent interactions, such as would be required for reproduction.

The driven vs. passive simulation results clearly demonstrate that during the time the agents are behaviorally adapting to their environment natural selection is actively driving complexity increases, in what McShea would refer to as a "biased" fashion (McShea 1994). This makes sense intuitively, because this period of behavioral adaptation is precisely the period of time when increases in complexity are likely to confer an evolutionary advantage. While the driven results show that TSE complexity is well correlated with the evolution of adaptive behaviors in an ecosystem subject to natural selection, the fact that complexity in the passive case ultimately surpasses that in the driven case demonstrates that such a correlation does not exist when natural selection is absent. This is consistent with the complexity-as-fitness-function results, where, again, complexity growth in the absence of natural selection is shown to be uncorrelated with evolutionarily useful behaviors.

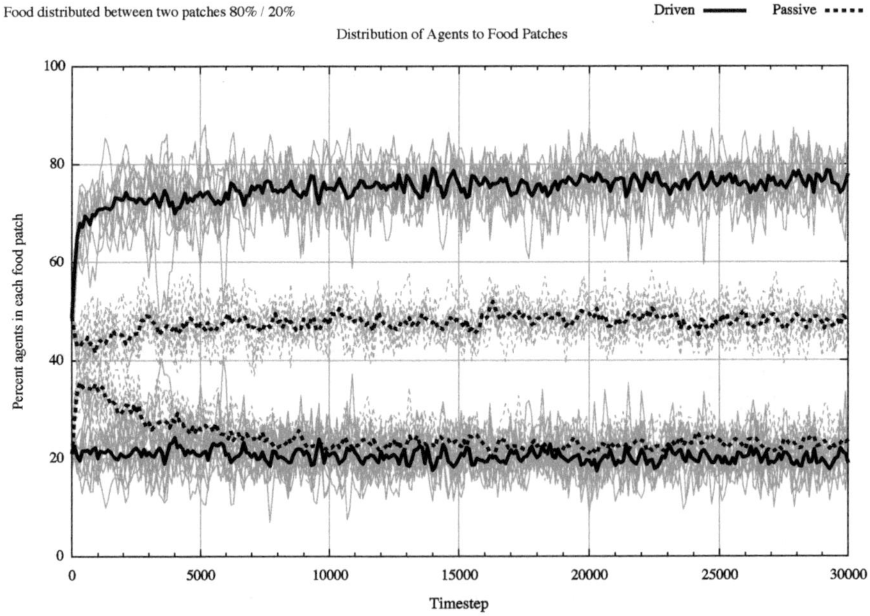

Fig. 15.5 Distribution of agents to heterogeneous resources. In the driven case (solid lines) agents approach an Ideal Free Distribution (IFD), over the same time period that evolution actively selects for complexity growth. In the passive case (dashed lines) they do not.

Prior work (Griffith and Yaeger 2006) demonstrated that Polyworld agents evolve a near Ideal Free Distribution (IFD) (Fretwell and Lucas 1970; Fretwell 1972) of agents to resources, for a variety of different distributions of food in patches. In these simulations agents have evolved to mostly inhabit the food patches, in IFD proportions, by approximately t=5000 and exhibit little change in food patch occupation beyond about t=7000. (See Figure 15.5.) These times correspond to the period when driven complexity reaches its peak and plateaus, and after which passive complexity surpasses driven complexity. The period of behavioral adaptation is made evident by this approach to IFD conditions and corresponds directly to the period of increasing neural complexity when natural selection drives genetic change. When genetic and neural changes are random, however, as in the passive runs, agents never adopt an IFD, further illuminating the disconnect between neural complexity and behavioral adaptation in the absence of natural selection.

15.8 Evolutionary Trends of Network Topology

The dynamics of a neural network depend upon its neural activation functions, the inputs to the network, and the topological graph of connections between neurons. The previously discussed neural complexity analysis was based on those dynamics—on the time series of the neural activations of each agent's neural network over the course of its life. In (Lizier et al. 2009; Yaeger et al. 2010; Yaeger 2013) we turned to the topology of the graphs underlying and carrying out the calculations that produced those dynamics. Our intention was to illuminate the relationship between neural structure and function, asking what network designs give rise to complexity and which do not.

The inputs to these neural networks are the agents' visual inputs, their current health (level of their internal energy store), and a randomly firing neuron. Since all of these inputs are determined by the state of the world and the agent, independent of the other neurons and network dynamics, we left them out of our complexity analysis. Topologically they also have an unusual, zero in-degree constraint. In recognition of that and for consistency with our complexity analysis we also leave them out of our graph theoretical analysis, using graphs defined by the non-sensory or "processing" neurons only—the internal and output/behavior neurons, but not the input neurons.

In general, graphs may be treated as undirected or directed, and binary or weighted. As neural networks are normally directed and synaptic strengths act as a kind of edge weight, we carry out these network analyses using weighted, directed graphs. Since edge weights are treated as distances when calculating some of these metrics, synaptic strength is inverted to provide our graphs' edge weights, so that a large synaptic weight corresponds to a short distance, consistent with its strong influence. Neural network synaptic strengths are also typically signed—positive for excitatory connections, negative for inhibitory connections. Since few graph theoretical metrics extend well to signed graphs, we have made the less than desirable, but simple and common approximation of using the absolute values of the network weights on the graph edges.

We start by examining the evolutionary histories of several key graph theoretical metrics, calculated for each agent's neural network topology and averaged over the full population. As discussed previously, we use the state of the network graph at the end of each agent's life for our analyses. We examine these population means in multiple pairs of driven versus passive simulations, as we did for complexity, to see which, if any, of the graph metrics appear to be subject to natural selection during the period of behavioral adaptation and which, if any, correspond to high and low complexity neural dynamics. Specifically, given prior evidence for a connection between small-world networks and complexity (Sporns et al. 2000), we will look at four metrics known to be associated with small-world networks in general (Watts and Strogatz 1998) and biological brains in particular (Bullmore and Sporns 2009)—clustering coefficient, characteristic path length, global efficiency, and small-world index.

15.8.1 Clustering Coefficient

One of the most common metrics for characterizing graph structure is *clustering coefficient* (*CC*). Watts and Strogatz (1998) cited large values of *CC* as one of the two defining characteristics of a small-world network (short path length being the other). *CC* measures cliquishness in a graph—the degree to which one's friends are also friends of each other. It may be calculated by taking the ratio of the number of edges actually present between neighbors of a given node to the maximum possible number of edges between those neighbors, and averaging over all nodes.[1] Conceptually it corresponds to identifying clusters of neurons that function as a group, likely engaging in coordinated, specialized activities. As such we expect it to be correlated with the *segregation* component of neural complexity. Thus we expect high *CC* to correspond to high complexity, at least to a point, and hypothesize that natural selection will favor larger values of *CC* during the early period of behavioral adaptation.

Figure 15.6 shows the evolutionary trajectory of *CC* for 10 natural selection (driven) and 10 null model (passive) runs. The relationship between driven and passive *CC* trajectories is similar to that seen for neural complexity (Figure 15.2), with the values in the driven case rising significantly more rapidly than in the passive case initially, but with the passive values eventually overtaking and surpassing the driven values. Thus, as hypothesized, higher values of *CC* appear to confer an evolutionary advantage during periods of behavioral adaptation to the environment and neural complexity growth. The time to overtake is substantially longer for *CC* than it is for complexity, with significance not dropping below 0.95 until about t=16000,

[1] Actual and maximum edge counts vary depending upon whether edges are treated as directed or undirected. All analyses here are based on directed edges. For weighted graphs, such as these, the adjacency matrix is replaced with the weights matrix^1/3.

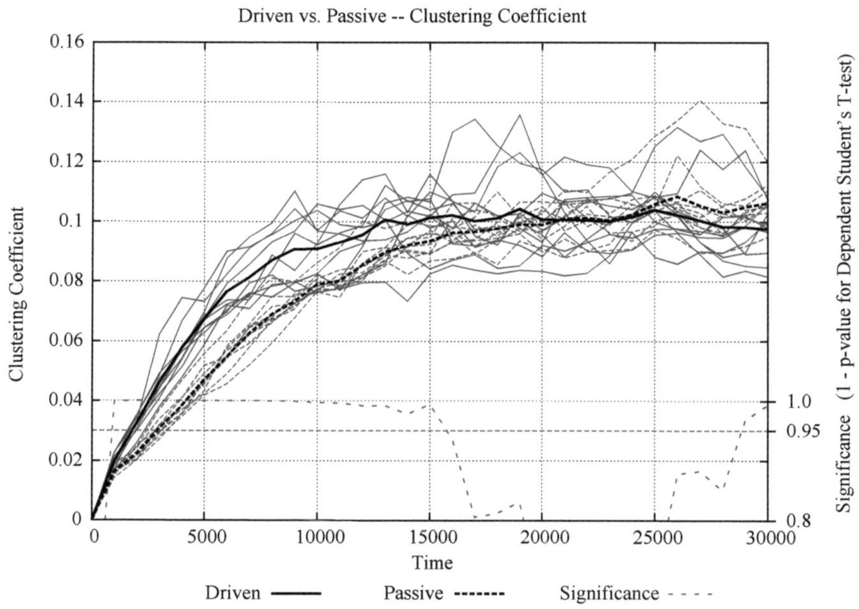

Fig. 15.6 Clustering coefficient as a function of time for 10 paired driven and passive simulations

compared to t=5000 for complexity. We suspect this is due, at least in part, to complexity's dependence on both clustering and our next metric.

15.8.2 Characteristic Path Length

Another widely used metric for characterizing the structure of a graph is its *characteristic path length* (*CPL*). A low value of *CPL* is the other defining characteristic of a small-world network (Watts and Strogatz 1998). *CPL* quantifies the distance between nodes in a graph. It can be calculated by determining the shortest path from a given node to all other nodes, averaging those lengths to produce a mean path length for that node, and then averaging these mean path lengths over all nodes.[2] Conceptually it corresponds to the ease or difficulty with which a signal may propagate from one part of a graph to another, and, therefore, the likelihood of cooperation amongst nodes. As such we expect it to be correlated with the *integration* component of neural complexity, only inversely, since smaller values of *CPL* should correspond to higher integration. Thus we expect low *CPL* to correspond to high complexity, to a

[2] The presence or absence of paths is significantly affected by whether edges are treated as directed or undirected; again, we use directed edges in our analyses. Edge length is obtained by inverting synaptic weight, so high efficacy corresponds to a short distance, consistent with such a synapse's strong influence.

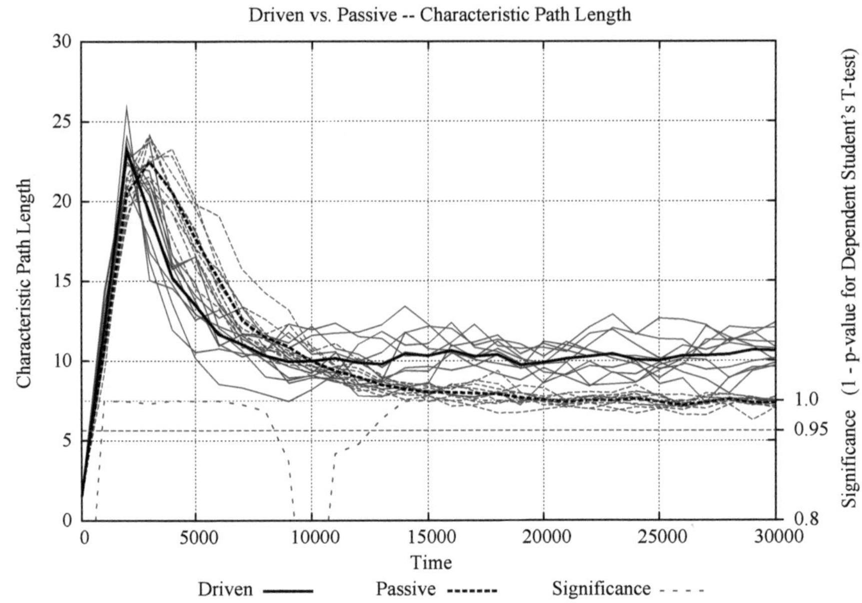

Fig. 15.7 Characteristic Path Length as a function of time for 10 paired driven and passive simulations

point, and hypothesize that lower values of *CPL* will be evolutionarily selected for during the early period of behavioral adaptation.

Figure 15.7 shows the evolutionary trajectory of *CPL* for 10 natural selection (driven) and 10 null model (passive) runs. The relationship between driven and passive *CPL* trajectories mirrors that seen in neural complexity and *CC*, only inverted—with shorter path lengths proving to be of an evolutionary advantage during the period of behavioral adaptation and complexity growth, as hypothesized. Passive runs overtake driven runs (1 – p-value drops below 0.95) at about t=8000, slightly later than neural complexity, but substantially earlier than *CC*. The brief excursion of driven over passive around t=2000 is due to problems intrinsic to calculating *CPL* for graphs that are allowed to contain disjoint subgraphs and even disconnected nodes. In fact, *CPL* is typically only calculated for the largest connected subgraph. In earlier work (Yaeger et al. 2010) we examine alternative measures of path length that are more well-behaved in this regard, none of which exhibit this behavior. For this investigation, however, we chose to stick to a "community standard" metric.

15.8.3 Global Efficiency

A commonly used measure of graph structure in biological neural networks is *global efficiency* (*E*) (Latora and Marchiori 2001; Bullmore and Sporns 2009). Just as *CPL* is the mean minimum path length, *E* is the mean *inverse* minimum path length.

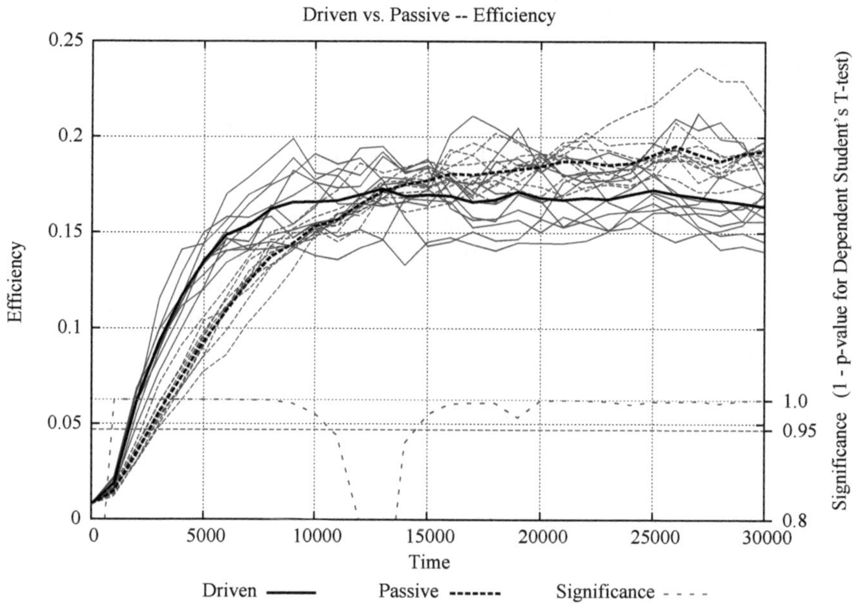

Fig. 15.8 Efficiency as a function of time for 10 paired driven and passive simulations

A benefit of using efficiency is that it applies naturally to sparse and disjoint graphs, since nodes that are disconnected have zero inverse distances between them, rather than the infinite distances that prevent their graceful accommodation in *CPL*. When applied to the whole graph it conceptually captures the efficiency with which communication between nodes in the graph is achieved globally, like an inverse of *CPL*.[3] As such, we expect global efficiency to be correlated with the *integration* component of neural complexity.[4] Thus we expect high E to correspond to high complexity, at least to a point, and hypothesize that evolution will select for larger values of E during the early period of behavioral adaptation.

Figure 15.8 shows the evolutionary trajectory of E for 10 natural selection (driven) and 10 null model (passive) runs. The relationship between driven and passive E trajectories is again similar to that seen for neural complexity, as greater efficiency proves to be of an evolutionary advantage and is correlated with complexity growth. Passive runs overtake driven runs at about t=10500. Note that, unlike *CPL*, there is no temporary inverted relationship between driven and passive data around t=2000, due to E's better handling of nodes lacking connective paths.

[3] When calculated for neighboring nodes only, it captures the local efficiency with which neighbors communicate, and quantifies the fault tolerance of neighbors of a node to the removal of that node (Latora and Marchiori 2003), but we will focus only on global efficiency.

[4] Local efficiency would be correlated with *segregation*.

15.8.4 Small-World Index

Humphries et al. (2006) proposed what has come to be called a graph's *small-world index* (*SWI*), that simultaneously captures the two characteristics of small-world networks identified by Watts and Strogatz (1998). It is a ratio of ratios, with a numerator that is the ratio of the *CC* of the graph in question to the *CC* of a comparable random graph (typically with the same number of nodes, number of edges, and degree distribution) and a denominator that is the ratio of the *CPL* of the actual graph to a comparable random graph. Humphries' original definition involved dividing by the statistic of a single random graph. Recognizing the variability inherent in graph randomization, we used the mean statistic of 10 comparable random graphs, resulting in the following definition:

$$\gamma = CC/\langle CC_r\rangle \tag{15.5}$$

$$\lambda = CPL/\langle CPL_r\rangle \tag{15.6}$$

$$s = \gamma/\lambda \tag{15.7}$$

where $\langle CC_r\rangle$ and $\langle CPL_r\rangle$ are the ensemble averages of *CC* and *CPL* over some number of comparable random graphs, and s is the desired *SWI*.

By definition, then, a small-world network will normally have a numerator greater than one and a denominator less than or near to one.[5] Hence, small-world networks may be identified by values of *SWI* near to or greater than one. Since *SWI* captures both increased *segregation* (in the numerator) and increased *integration* (in the denominator), we hypothesized a strong, possibly even monotonic, correlation between *SWI* and TSE neural complexity. As we will discuss later, a strong correlation between the two measures proved to be correct, but monotonicity was very much not the case.

Figure 15.9 shows the evolutionary trajectory of *SWI* for 10 natural selection (driven) and 10 null model (passive) runs. Despite a period of statistically significant difference between driven and passive data comparable to that seen in complexity and the other graph metrics, we find that the expected relationship is reversed. *SWI* grows more slowly and maintains a lower value in the driven runs than in the passive runs. Thus over this important period of behavioral adaptation and neural complexity growth, *SWI* appears to be inversely related to complexity, contrary to expectations. We will discuss this further and get at some of the underlying reasons

[5] The ratio of path lengths in the denominator can be greater than one, even for small-world networks, because the randomized version of a graph can have a smaller path length than even a small-world graph. Random graphs generally have very short path lengths, approaching that of a fully connected graph—1.0. Thus even for graphs with relatively short path lengths, the randomized versions of those graphs may have even shorter path lengths, just not very much shorter. Hence the denominator may be greater than one, just not very much greater. As a result, *SWI* for a small-world network may be less than one, just not very much less.

for this discrepancy when we analyze the relationship between complexity and these graph metrics in more detail later. Though the expected relationship is inverted, of the metrics analyzed, *SWI*'s period of statistical significance is most similar to that of complexity, likely because the two measures share a dependency on both *CC* and *CPL*.

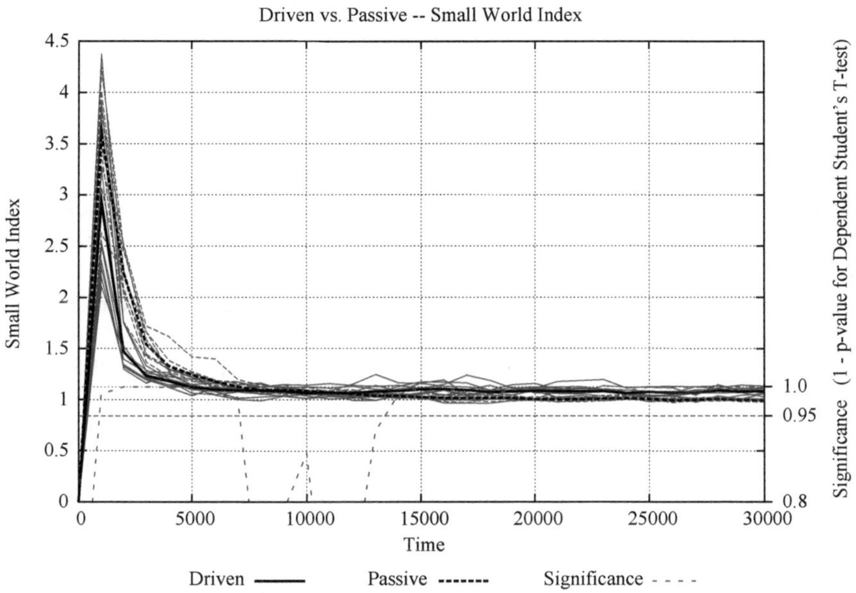

Fig. 15.9 Small-world index as a function of time for 10 paired driven and passive simulations

15.9 Relating Neural Complexity to Network Topology

The evolutionary trends observed in the aforementioned graph theoretical metrics mostly align with our expectations about their relationship to neural complexity. However, the behavior of *SWI* runs counter to those expectations, and variations in the period of statistical significance of driven-versus-passive differences suggests these relationships may be somewhat complicated. So we will now turn to detailed, individual (per-agent) distributions of complexity as a function of the various graph metrics (rather than population means as in the previous results) to try to tease apart these relationships more carefully.

Studying the kinds of results presented and discussed below for individual simulations (not shown here) reveals that there is variation from run to run, as might reasonably be expected, in terms of the range of values explored and details of the relationships between the metrics being examined. So in order to identify more universal regularities, a sampling—20% of the data points chosen at random—from the results of all 10 driven runs, all 10 passive runs, and all 10 fitness runs are combined in each of the following figures. This yielded 182,292 data points in each graph (51,903 driven, 51,869 passive, and 78,520 fitness) for these results. The data from each run are given a subtly different, but unique color within a given part of the spectrum—driven data are various shades of green, passive data are blue, and fitness data are red. If viewed in grayscale, the driven data are lighter shades, generally towards the bottom; the passive data are medium gray, typically ringing the driven data and sometimes difficult to distinguish; the fitness data are darker shades that extend to much higher complexity values.

15.9.1 Clustering Coefficient

Figure 15.10 shows neural complexity as a function of clustering coefficient. For all sources of data—driven, passive, and fitness—higher values of complexity are mostly (though not entirely, see below) found within a limited band of values of *CC*, and exhibit a peaked distribution. All 10 driven simulations explored roughly the same part of complexity space, exhibiting a peak around $CC = 0.065$. The passive runs explore much the same part of the space as each other, and though they extend to slightly broader ranges of both complexity and *CC* than the driven runs (easier to see if viewed in color), peak complexity occurs at about the same value of *CC*. The fitness runs extend to a much higher range of complexity values, but occupy a generally narrower range of *CC* values. Most of them exhibit distributions that are similar to the driven and passive results, including peaking at about the same value of *CC*, however, a single fitness run got stuck in a low-*CC* regime (to the left, at about 0.025), and another single fitness run showed a huge variability over larger values of *CC* (the sparse data points to the right, generally between 0.1 and 0.25).

Though high complexity is not entirely restricted to the vicinity of the obvious peak at about 0.065, the vast majority of networks exhibiting high complexity are found in a modest band around it. Despite that fact, possession of a value of *CC* near the optimum is not sufficient to guarantee high complexity, as evidenced by the full range of complexity values being found in that band.

As evidenced by the lack of very high-*CC* points in the PFit data, these fitness runs are so quickly pushed towards high complexity by their steady-state GA that they never explore many high-*CC* parts of the space. (Refer back to Fig. 15.4 to see how rapidly complexity increases in these fitness runs.) Though difficult to see because of obscuring driven and passive data, the PFit runs also do not explore much of the lower complexity parts of the space. However, they clearly do explore higher complexity parts of the space not visited by the other simulations.

15.9.2 Characteristic Path Length

Figure 15.11 shows neural complexity as a function of characteristic path length. Except for a single fitness simulation, there is a clear optimal range of *CPL* values, centered around 6.5 or 7. Interestingly, that unique fitness run, peaking around a

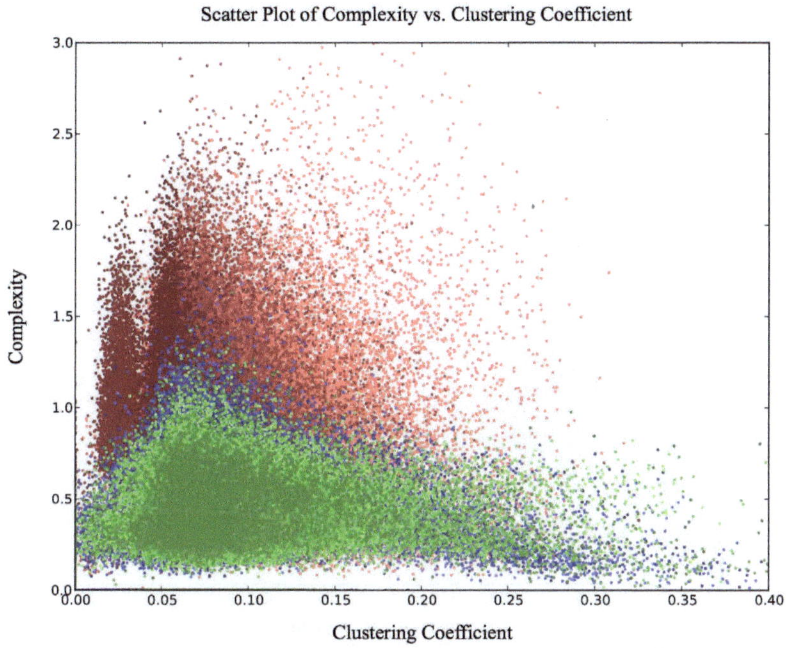

Fig. 15.10 Neural complexity as a function of clustering coefficient. (For a discussion of color-coding of this data, see Section 15.9 of the text.)

very low *CPL* of about 3 to 3.5, is the same run that exhibits the extremely variable set of unusually large values of *CC*. So that one fitness run evolved graphs exhibiting very tight clusters, possibly including disjoint subgraphs, which would explain both the high *CC* and the low *CPL*, since *CPL* will only be calculated for the largest connected subgraph. By contrast, the other unusual fitness run, that peaked at a lower value of *CC*, exhibits the largest values of *CPL*, explained at least in part by the fact that this simulation exhibited normal node counts, but the lowest number of edges seen in any of the fitness runs (data not shown).

Here again, though high complexity is largely confined to a modest range of *CPL* values, an optimal value of *CPL* is insufficient to guarantee high complexity. Also as before, the complexity-as-fitness data never extends to very large values of *CPL* (and correspondingly low values of complexity) because the GA accelerates so rapidly to high values of complexity and stays there.

15.9.3 Global Efficiency

Figure 15.12 shows neural complexity as a function of global efficiency. Efficiency exhibits a relatively gentle peak in the main body of data, and displays one of the

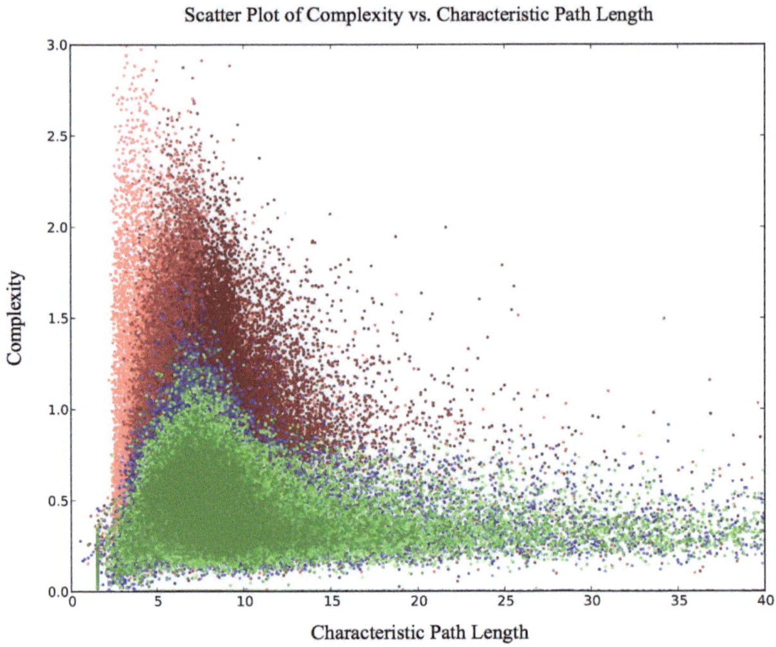

Fig. 15.11 Neural complexity as a function of characteristic path length. (For a discussion of color-coding of this data, see Section 15.9 of the text.)

outlier fitness run's high-*CC*, low-*CPL* cloud of points at larger values of E. Given what we have observed in the other graph metrics, there are no surprises here.

15.9.4 Small-World Index

Figure 15.13 shows neural complexity as a function of small-world index. Prior to these experiments we hypothesized that complexity might be highly correlated with SWI, possibly even monotonically so, due to the well documented relationship between complexity and small-world-ness. Interestingly, however, SWI exhibits by far the most peaked distribution of any of the graph theoretical metrics we analyzed. Driven, passive, and fitness runs all exhibit a sharp peak in complexity around an SWI of 1.0. Even the outlier fitness runs, that showed up as exploring particularly high or low values of *CC* and *CPL*, when normalized by equivalent random graphs and taken as a ratio, as here, show up as just another part of the main distribution.

We believe the extreme sharpness of this peak explains the unexpected relationship between driven and passive evolutionary trends in SWI seen in Fig. 15.9. Even though the full spectrum of complexity appears to be possible near an SWI value of 1.0, high values of complexity are almost never found for much larger (or smaller) values of SWI. Thus during the period of behavioral adaptation and strong pressure

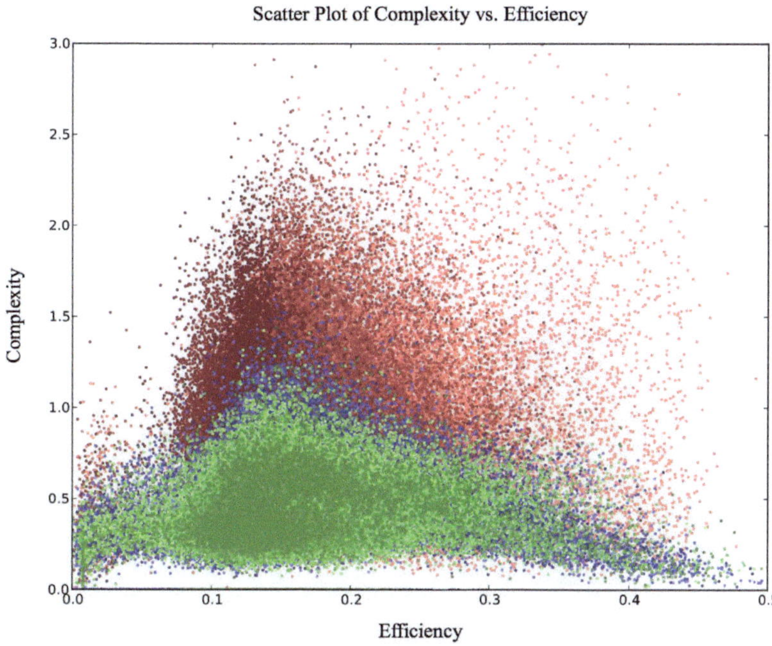

Fig. 15.12 Neural complexity as a function of global efficiency. (For a discussion of color-coding of this data, see Section 15.9 of the text.)

towards high complexity, any agent whose genome produces a network topology with a large value of SWI is unlikely to exhibit sufficient complexity and will thus be at an evolutionary disadvantage. This produces an indirect, but strong evolutionary pressure on SWI towards 1.0 in the driven runs. In contrast, the random walk is under no such pressure and quickly produces high values of SWI (with correspondingly low values of complexity). Thus SWI grows more rapidly, at first, in the passive runs. Over time, an evolutionary pressure towards higher complexity in the driven runs produces values of SWI that are still close to 1.0, while the passive runs merely drift thus producing large numbers of random connections that yield low values of CC and SWI, thus inverting the relationship between driven and passive trends.

As for why complexity is tied so tightly to values of SWI near 1.0, we can only speculate at this stage, but there are reasons to think this is reasonable. For SWI to adopt a value much larger than one, either CC/CC_{rand} must grow substantially larger than one or CPL/CPL_{rand} must fall to values substantially smaller than one, or both. But in the case of extremely high levels of clustering, integration is likely to suffer, as clusters begin to behave almost or entirely independently. At the opposite extreme, shortest possible path lengths will likely imply extreme levels of integration, at the expense of segregation, as networks work essentially in lockstep.

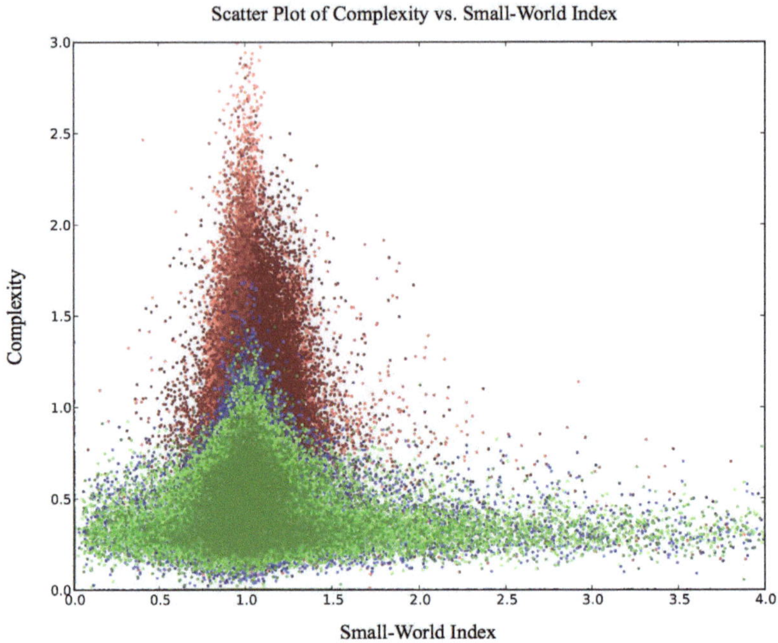

Fig. 15.13 Neural complexity as a function of small-world index. (For a discussion of color-coding of this data, see Section 15.9 of the text.)

So neither extreme exhibits the balance of integration and segregation that is necessary for high complexity. Also, since small-world graphs already exhibit low values of *CPL*, by definition, it is unlikely that CPL/CPL_{rand} *can* grow very much smaller than one. Accordingly, in retrospect, it seems reasonable to expect complexity to peak for small values of *SWI*, though, absent data, we might have expected a value somewhat larger than 1.0.

Given this tight correlation between complexity and *SWI*, one might expect the closest parallel to complexity's period of statistical significance (between driven and passive), which, as we saw in Fig. 15.9, proves to be the case.

Despite the sharpness of this *SWI* peak, here again, possession of an optimal value of the graph metric is no guarantor of high complexity. However, it appears unlikely that a network of high complexity will be found with an *SWI* outside of a modest range around 1.0.

15.10 Broader Applicability

It is reasonable to ask how these results are impacted by the nature of the simulations used to produce them. Are there specific or general constraints on the underlying mechanisms of the simulator that might constrain network structure or function and thus produce the observed effects?

For some graph metrics, the actual numerical values of these optimal ranges may be scaled by the magnitude of the weights in our simulations, since distances between nodes are interpreted as the inverse of the synaptic strength between those nodes. However, it seems unlikely that varying the maximum weight—the only free parameter in this regard—would eliminate this ubiquitous "optimal-range" behavior, since this weight only establishes an upper bound on synaptic strength, and thus a lower bound on edge length.

These numerical values may also depend on simple network scaling parameters, such as the maximum number of neural groups and the maximum number of neurons per group, but, here again, we expect variations in these parameters to, at most, shift the optimal ranges, not eliminate them. We also expect that a more normalized metric such as SWI will be largely insensitive to these kinds of maximum weight and node count parameters, since what is being measured is a deviation from a comparable random graph of equal weight distributions, node counts, and edge counts. In fact, we suspect that normalizing all of the graph metrics by their values in comparable random graphs, a common practice in network science, may pull in the outliers and tighten the distributions of the other metrics.

We have seen a degree of variation in structure/function relationships from run to run for some graph metrics. Of course, varied outcomes from natural selection or a genetic algorithm in a complex, chaotic system are to be expected. But by examining not only multiple runs with different seeds, but multiple *types* of runs—driven, passive, and fitness—in which the evolutionary pressures on structure and function are very different (and basically non-existent for passive runs), we have some confidence that we have captured the majority of the space of possible dynamics, given the neural model and genetic encoding of network structure we utilize.

And that genetic encoding is, deliberately, extremely general. As discussed earlier, specific architectures are not specified or selected for. Rather, the scale of the network and the statistics of connectivity are all that evolution has available to operate upon. Internal neural group counts simply range from zero to some maximum number (five, plus some fixed input and output/behavior neural groups, in these experiments). Neuron counts within each group simply vary between one and 16. The number of synaptic connections between neurons is determined by a genetically specified connection density, and specific connections are determined stochastically, in a more or less ordered sequence, depending on a genetically determined "topological disorder" gene. Only self-terminations are disallowed. So essentially any network topology is possible, within the limits of group and neuron counts.

Given that all graph topologies are possible and we have explored their structure and function in multiple settings under very different selection pressures (including non-existent), there is reason to think these results may be robust.

We cannot rule out the possibility that an alternative neural activation model—continuous-time or spiking, for example—might produce sufficiently different dynamics that the current results do not apply. However, a small degree of experimentation with continuous-time and Izhikevich spiking models (Izhikevich 2003, 2004) in our system has produced mostly comparable evolutionary outcomes in agent behaviors, so there is at least some hope that these results will extend even to different models of neural activation.

15.11 Discussion and Conclusions

Four key components—a computational ecosystem capable of carrying out synchronized natural-selection and null-model simulations, a resulting wide array of evolved neural network topologies and time series, an information-theoretic metric for measuring neural complexity, and a collection of powerful graph theoretical tools—have allowed us to ask questions and observe trends that are difficult to pose and assess by any other means.

By "replaying the tape of life" (Gould 1989) in these novel ways, we have been able to investigate evolutionary trends in complexity, a subject much discussed and debated since Darwin. Our results demonstrate a clear natural selection for complexity, in a driven, biased fashion. But they also show a tendency to weakly stabilize complexity at a "just good enough" level. Though our work does not specifically address it, there is little doubt that evolution will, under the right circumstances, select, in a driven sense, to reduce complexity, such as when dark-dwelling organisms in a cave give up their eyes to avoid wasting metabolic energy. All of which suggests, then, that at the scale of individuals and species evolution always guides trends in complexity.

But the scale of the discussion very much matters. Gould (1996) and Dawkins (1997) have argued strongly for passive and driven evolutionary trends, respectively. However, much of their disagreement seems to stem from an issue of scale. Our work agrees with Dawkins's claim that evolution is always driven—at the level of individuals comprising a species. But integrating over an entire biome's collection of increasing, decreasing, and stabilized trends in complexity is likely to produce a process that at least appears random, at the larger scale. In one of the earliest attempts to model evolution computationally in order to address these kinds of questions, Raup et al. (1973) observed that fully deterministic trends at small scales may very well be at the base of larger scale trends, even if those larger scale trends appear to be passive. Our current simulations reinforce this notion of natural selection driving the evolution of complexity at small scales, but driving it in all directions—up, down, and stable. So larger scale trends are likely to be obfuscated by these opposing trends at the smaller scale, and the often conflicting and inconclusive evidence for trends in the paleobiological record is unavoidable and to be expected.

By looking at a specific adaptation—the evolution towards behaviors resulting in an Ideal Free Distribution—we were able to demonstrate a correlation between behavioral adaptation and our measures of neural complexity and graph topology. But results from the passive and complexity-as-fitness runs make it clear that this correlation only applies when complexity gains are the result of natural selection.

There is an implicit assumption in all this work that neural complexity is a valuable commodity—something worth identifying and fostering. Indirect evidence from biology and simulation exists for this opinion, in which the kinds of small-world networks found in biological brains are produced by evolutionary selection for complex neural dynamics (Sporns et al. 2000). We believe our results indicate that the need to produce viable, adaptable behavioral strategies in the face of complex, changing environments is sufficient to *require* complex neural dynamics and to *produce* them, given sufficient time for evolution to work its course.

In many ways our experiments with the use of an information theoretic measure of neural complexity to guide agent evolution and behavior, using complexity as a fitness function in a standard genetic algorithm (GA) sense, is closest in spirit to some of the staples of the field of Guided Self-Organization, such as homeokinesis (Der et al. 1999, 2001), predictive information (Der et al. 2008), and empowerment (Klyubin et al. 2005). In some unpublished work we have attempted to guide these complexity-as-fitness evolutionary trajectories towards behaviors more traditionally associated with biological evolutionary fitness, such as foraging and reproduction. To this end we have tried applying filters to the neural activation time series used to compute complexity, suppressing their contribution to complexity outside of temporal windows around behaviorally meaningful events, such as the consumption of food and successful (virtual[6]) reproduction. There have been some hints that this approach might help guide the GA in the desired directions, but so far the results are inconclusive. We have considered other methods for combining traditional evolutionary fitness with complexity as fitness, but not yet had the opportunity to explore them. So we have yet to reliably join complexity as a fitness function with evolutionarily useful behaviors.

This disconnect between complexity and behavioral fitness is at odds with the very direct coupling of behavior and complexity found in a task discussed by Sporns and Lungarella (2006), in which a robotic arm attempts to grasp a randomly moving block. We believe the discrepancy is due to significant differences in the range of possible behaviors, environments, and sensory inputs in these two experiments. In the work by Sporns and Lungarella, evolving the robot for maximum complexity produced block-grasping behaviors as efficient as when the robot was evolved to perform the block-grasping task directly. But in their experiment, aside from the block, the environment was empty or consisted of random noise, and the only way to provide non-random sensory input to the controlling network, and thus increase the complexity metric, was to attend to that block. By contrast, in Polyworld the environment contains appearing (growing) and disappearing (eaten) food, other structural artifacts (the barriers), and, most significantly, many other agents behaving according to their own network dynamics, providing a rich and varied source of visual complexity. Each Polyworld agent also expresses control over a suite of seven primitive behaviors, most of which directly affect that agent's experience of the world. All these additional sensory inputs and multiple low-level behaviors produce a complicated, non-unique relationship between the complexity of the agent's

[6] In GA mode reproduction is handled by the algorithm, not agent behaviors, so we devised a method of recognizing events corresponding to the activations of mating behaviors by collocated agents that would normally have produced an offspring, and recording these as "virtual births".

neural dynamics and any resulting higher order behaviors. It is the extreme difference between the agents' sensoria and possible sources of complexity in the two works that we believe accounts for the marked difference in observed correlation between complexity and behavior.

One curious, but sweeping conclusion one might draw from the contrasting relationship between neural complexity and behavior when natural selection is present and when it is not (in the null-model and complexity-as-fitness runs) is that it may only be possible to study biologically meaningful neural complexity in what are typically thought of as "artificial life" models. That is, absent a computational ecosystem in which neurally controlled behaviors dictate survival and reproduction there appears to be no correlation between complexity and evolutionarily useful behaviors or what is generally deemed evolutionary fitness, despite clear evidence of such a correlation when natural selection of neurally controlled behaviors is at work in our model. It remains to be seen whether such a correlation exists in biological organisms and, unfortunately, rigorously establishing such a link is likely to remain difficult for the foreseeable future. Techniques do not currently exist to quantitatively assess neural complexity in biological organisms and, of course, all extant biological organisms have been subject to natural selection. However, structural information about biological neural systems is accruing and increasing in detail rapidly, and it may be possible to relate structural complexity to the richness of behavioral repertoires before too much longer. We may consider it a prediction and partial confirmation of our model if indicators of behavioral complexity prove to be correlated with neural topologies approaching those we have identified here as most correlated with neural complexity and evolutionarily adaptive behavior.

Our model also demonstrates and predicts evolutionary selection for neural complexity during behavioral adaptation of a species to a complex environment. We suggest that it is reasonable to expect results from the model to apply to biological systems, since, even though Polyworld provides only an abstract model of biological evolution, the principle in action is, at its base, simply natural selection—an amplification of advantageous traits. The mechanisms of action and behavior clearly differ between model and biology, and there will be differences in the specific costs and benefits between any biological agents and artificial agents, but such differences exist between any one biological agent or niche and another. There will always be cost/benefit trade-offs in competition and niche exploitation for any evolutionary ecosystem, real or artificial, and it is these trade-offs that evolution explores. As long as increases in complexity improve the ability to survive and reproduce by agents so advantaged, evolution will actively select for those complexity increases. The subsequent plateau in driven complexity seen here, due to the widespread adoption of a "good enough" strategy, that acts as a kind of local optimum, demonstrates how evolution can act to serve as a weak stabilizing force for complexity, in accord with Dennett's (1996) observation, "The cheapest, least intensively designed system will be 'discovered' first by Mother Nature, and myopically selected." Yet when variation produces a sufficiently improved network and corresponding set of behaviors, our system exhibits a punctuated equilibrium, as so often observed in nature.

Of course one expects neural function to be heavily dependent upon neural structure. Indeed, prior work has demonstrated a correlation between neural complexity and network small-world-ness (Sporns et al. 2000). However, our results show that at least when small-world-ness is assessed by SWI, that correlation is decidedly non-monotonic. Indeed, all our data suggest that the complexity of a neural network's function is maximized within a fairly modest range of optimal graph theoretical measures of the network.

It has been suggested that evolution may strongly select for a combination of maximum efficiency and minimum communication cost in biological brains (Bullmore and Sporns 2009), and there is evidence for a genetic component to this kind of cost-efficiency in the organization of human cortical networks (Fornito et al. 2011). In biological systems, evolutionary pressures due to physical constraints, such as wiring length (Mitchison 1991), brain volume (Murre and Engelhardt 1995), and fast response time (Lago-Fernández et al. 2000), naturally lead to reduced communication costs. Combined with a need for efficiency, it is reasonable to speculate that brains have long been under an evolutionary pressure towards small-world graph structures (Sporns et al. 2000; Bullmore and Sporns 2009).

Here we see evidence for an evolutionary pressure towards network efficiency and small-world structures—exhibiting high clustering and short path lengths[7]—based purely on network functionality, absent all physical constraints. This bolsters the argument for evolutionary pressures on efficiency and suggests that low-cost, small-world structures provide a purely functional benefit, independent of physical constraints. Indeed, though physical and functional constraints could, in principle, have produced evolutionary pressures that were independent or even conflicting, our work shows that they are aligned and convergent, working in combination to produce the kinds of network structures that foster complex, adaptive behavior. We suggest that humans and all biological organisms with even modestly complex nervous systems are the fortunate beneficiaries of these convergent and synergistic physical and functional constraints.

We have also identified limited ranges amongst several graph theoretical metrics within which high-complexity networks are to be found. It is not currently possible to predict network function purely from structure—each of these limited ranges of metrics may be seen as necessary but not sufficient to guarantee complexity. However, one wonders if perhaps the addition of more refined graph theoretical metrics, such as (signed) motif distributions, modularity, w-cores (weighted k-cores), excess entropy of graph links, and so on, might enable such a feat. Regardless, knowledge of the parameter regimes in which neural networks are most capable of producing complex dynamics can be used to guide evolutionary search for the most capable networks. Armed with a successful characterization of those network structures most likely to confer dynamical complexity, we may be able to close the loop on

[7] Technically, the evidence is for *moderately* high clustering and *moderately* short path lengths, but the requirement that both integration and segregation be present to achieve high complexity, together with the fundamentally opposing tension between integration and segregation, is exactly mirrored and driven by this trade-off between path length and clustering, meaning neither can be so extreme that the other is degraded overly much.

guided self-organization and apply our learned structural constraints to the more rapid, more far-reaching evolution of artificially intelligent agents.

Acknowledgements. Thanks to Olaf Sporns for early complexity calculations, ongoing moral support, and, together with Rolf Kötter, Mikail Rubinov, and others, the original MATLAB-based Brain Connectivity Toolbox. Thanks to Virgil Griffith for his work on the "passive" null model in Polyworld, the original C++ version of the complexity tools, and many scripts (first shell, then Python) for automating the data analysis. Thanks to Steven Williams for his efforts as technical lead in the development and deployment of bct-cpp, the C++ version of the Brain Connectivity Toolbox, that made the network structure calculations possible. Thanks to Xin Shuai for his prototype network analysis experiments with NetworkX. Thanks to Jaimie Murdock for the insights gleaned from his genetic clustering analysis of Polyworld agents. And ongoing thanks to Sean Dougherty for his continued and extensive enhancement of Polyworld. Work carried out while on faculty of School of Informatics & Computing at Indiana University, Bloomington and on sabbatical in London with Daniel Polani at U. of Hertfordshire. Chapter written (on weekends) while at Google. Some support provided by the National Academies / Keck Futures Initiative grant number NAKFI CS22.

References

Adami, C.: What is complexity? BioEssays 24, 1085–1094 (2002)

Adami, C., Ofria, C., Collier, T.: Evolution of biological complexity. PNAS 97(9), 4463–4468 (2000)

Arthur, B.: On the evolution of complexity. In: Cowan, G.A., et al. (eds.) Complexity: Metaphors, Models, and Reality, pp. 65–81. Addison-Wesley, Reading (1994)

Avery, J.: Information Theory and Evolution. World Scientific Publishing Company (2003)

Ay, N., Olbrich, E., Bertschinger, N., Jost, J.: A unifying framework for complexity measures of finite systems, Working Paper 06-08-028. In: Proceedings of ECCS 2006, Oxford, UK, pp. 6–8 (2006)

Bedau, M., Snyder, E., Brown, C., Packard, N.: A Comparison of Evolutionary Activity in Artificial Evolving Systems and in the Biosphere. In: Husbands, P., Harvey, I. (eds.) Proceedings of the Fourth European Conference on Artificial Life, pp. 125–134. MIT Press, Cambridge (1997)

Bonner, J.: The Evolution of Complexity by Means of Natural Selection. Princeton Univ. Press, Princeton (1988)

Bullmore, E., Sporns, O.: Complex brain networks: graph-theoretical analysis of structural and functional systems. Nature Reviews Neuroscience 10, 186–198 (2009)

Carroll, S.: Chance and necessity: the evolution of morphological complexity and diversity. Nature 409, 1102–1109 (2001)

Chaitin, G.J.: Toward a mathematical definition of "life". In: Levine, R.D., Tribus, M. (eds.) The Maximum Entropy Formalism, pp. 477–498. MIT Press (1979)

Channon, A.: Passing the ALife test: Activity statistics classify evolution in Geb as unbounded. In: Kelemen, J., Sosík, P. (eds.) ECAL 2001. LNCS (LNAI), vol. 2159, pp. 417–426. Springer, Heidelberg (2001)

Cope, E.D.: The method of creation of organic forms. Proc. Am. Phil. Soc. 12, 229–263 (1871)

Crutchfield, J.P., Feldman, D.P.: Regularities unseen, randomness observed: Levels of entropy convergence. Chaos 13(1), 25–54 (2003)

Darwin, C.: On the Origin of Species by Means of Natural Selection, or the Preservation of Favoured Races in the Struggle for Life. John Murray, London (1859)

Darwin, C.: The Descent of Man, and Selection in Relation to Sex. Princeton Univ. Press, Princeton (1871)

Dawkins, R.: Human Chauvinism (a review of Gould's Full House). Evolution 51(3), 1015–1020 (1997)

Der, R., Güttler, F., Ay, N.: Predictive information and emergent cooperativity in a chain of mobile robots. In: Bullock, S., Noble, J., Watson, R., Bedau, M.A. (eds.) Artificial Life XI: Proceedings of the Eleventh International Conference on the Simulation and Synthesis of Living Systems, pp. 166–172. MIT Press, Cambridge (2008)

Der, R., Steinmetz, U., Pasemann, F.: Homeokinesis–a new principle to back up evolution with learning. Computational Intelligence for Modelling, Control, and Automation in Concurrent Systems Engineering Series 55, 43–47 (1999)

Der, R., Steinmetz, U., Pasemann, F.: Self-organized acquisition of situated behavior. Theory in Biosciences 120, 179–187 (2001)

Farmer, J.D., Griffith, V.: On the viability of self-reproducing machines (2007) (unpublished)

Fornito, A., Zalesky, A., Bassett, D.S., Meunier, D., Ellison-Wright, I., Yücel, M., Wood, S.J., Shaw, K., O'Connor, J., Nertney, D., Mowry, B.J., Pantelis, C., Bullmore, E.T.: Genetic Influences on Cost-Efficient Organization of Human Cortical Functional Networks. The Journal of Neuroscience 31(9), 3261–3270 (2011)

Fretwell, S.D.: Populations in a seasonal environment. Princeton Univ. Press, Princeton (1972)

Fretwell, S.D., Lucas, H.L.: On territorial behavior and other factors influencing habitat distribution in birds. Acta Biotheoretica 19, 16–36 (1970)

Gould, S.J.: Wonderful Life: The Burgess Shale and the Nature of History. Norton, New York (1989)

Gould, S.J.: The evolution of life on earth. Scientific American 271(4), 62–69 (1994)

Gould, S.J.: Full House. Harmony Books, New York (1996)

Grassberger, P.: Toward a quantitative theory of self-generated complexity. Int. J. Theor. Phys. 25(9), 907–938 (1986)

Griffith, V., Yaeger, L.S.: Ideal Free Distribution in Agents with Evolved Neural Architectures. In: Rocha, L., Yaeger, L.S., Bedau, M., Floreano, D., Goldstone, R., Vespignani, A. (eds.) Artificial Life X: Proceedings of the Tenth International Conference on the Simulation and Synthesis of Living Systems, pp. 372–378. MIT Press (Bradford Books), Cambridge (2006)

Han, T.S.: Nonnegative entropy measures of multivariate symmetric correlations. Information and Control 36, 133–156 (1978)

Heylighen, F.: Evolutionary Transitions: How do levels of complexity emerge (2000), `http://citeseer.ist.psu.edu/375313.html`

Humphries, M.D., Gurney, K., Prescott, T.J.: The brainstem reticular formation is a small-world, not scale-free, network. Proc. R. Soc. B 273, 503–511 (2006)

Huynen, M.A.: Exploring phenotype space through neutral evolution. Journal of Molecular Evolution 43(3), 165–169 (1996)

Izhikevich, E.M.: Simple Model of Spiking Neurons. IEEE Transactions on Neural Networks 14, 1569–1572 (2003)

Izhikevich, E.M.: Which Model to Use for Cortical Spiking Neurons? IEEE Transactions on Neural Networks 15, 1063–1070 (2004)

Jaynes, E.T.: Information Theory and Statistical Mechanics. Physical Review 106, 620–630 (1957a)
Jaynes, E.T.: Information Theory and Statistical Mechanics, II. Physical Review 108, 171–190 (1957b)
Katz, M.J.: Is evolution random? In: Raff, R.A., Raff, E.C. (eds.) Development as an Evolutionary Process, pp. 285–315. Alan R. Liss, New York (1987)
Kimura, M.: The Neutral Theory of Molecular Evolution. Cambridge University Press, Cambridge (1983)
Klyubin, A., Polani, D., Nehaniv, C.: All else being equal be empowered. In: Capcarrère, M.S., Freitas, A.A., Bentley, P.J., Johnson, C.G., Timmis, J. (eds.) ECAL 2005. LNCS (LNAI), vol. 3630, pp. 744–753. Springer, Heidelberg (2005)
Knoll, A., Bambach, R.K.: Directionality in the History of Life: Diffusion from the Left Wall or Repeated Scaling of the Right? Paleobiology 26(4) (supplement), 1–14 (2000)
Lago-Fernández, L.F., Huerta, R., Corbacho, F., Sigüenza, J.A.: Fast Response and Temporal Coherent Oscillations in Small-World Networks. Phys. Rev. Lett. 84, 2758–2761 (2000)
Latora, V., Marchiori, M.: Efficient Behavior of Small-World Networks. Phys. Rev. Lett. 87(19), 198701 (2001)
Latora, V., Marchiori, M.: Economic Small-World Behavior in Weighted Networks. Europ. Phys. Journ. B 32, 249–263 (2003)
Lizier, J.T., Piraveenan, M., Pradhana, D., Prokopenko, M., Yaeger, L.S.: Functional and Structural Topologies in Evolved Neural Networks. In: Kampis, G., Karsai, I., Szathmáry, E. (eds.) ECAL 2009, Part I. LNCS, vol. 5777, pp. 140–147. Springer, Heidelberg (2011)
Lungarella, M., Pegors, T., Bulwinkle, D., Sporns, O.: Methods for quantifying the information structure of sensory and motor data. Neuroinformatics 3(3), 243–262 (2005)
Maynard Smith, J.: Time in the evolutionary process. Studium Generale 23, 266–272 (1970)
McShea, D.W.: Evolutionary Change in the Morphological Complexity of the Mammalian Vertebral Column. Evolution 47(3), 730–740 (1993)
McShea, D.W.: Mechanisms of large-scale evolutionary trends. Evolution 48, 1747–1763 (1994)
McShea, D.W.: Metazoan complexity and evolution: Is there a trend? Evolution 50, 477–492 (1996)
McShea, D.W.: The minor transitions in hierarchical evolution and the question of a directional bias. J. Evol. Biol. 14, 502–518 (2001)
McShea, D.W.: The evolution of complexity without natural selection, a possible large-scale trend of the fourth kind. Paleobiology 31(2) (supplement), 146–156 (2005)
McShea, D.W., Brandon, R.N.: Biology's First Law: The Tendency for Diversity and Complexity to Increase in Evolutionary Systems. The University of Chicago Press (2010)
Mitchison, G.: Neuronal branching patterns and the economy of cortical wiring. Proceedings of the Royal Society of London. Series B: Biological Sciences 245(1313), 151–158 (1991)
Murdock, J., Yaeger, L.S.: Genetic clustering for the identification of species. In: Krasnogor, N., Lanzi, P.L., Engelbrecht, A., Pelta, D., Gershenson, C., Squillero, G., Freitas, A., Ritchie, M., Preuss, M., Gagne, C., Ong, Y.S., Raidl, G., Gallager, M., Lozano, J., Coello-Coello, C., Silva, D.L., Hansen, N., Meyer-Nieberg, S., Smith, J., Eiben, G., Bernado-Mansilla, E., Browne, W., Spector, L., Yu, T., Clune, J., Hornby, G., Wong, M.-L., Collet, P., Gustafson, S., Watson, J.-P., Sipper, M., Poulding, S., Ochoa, G., Schoenauer, M., Witt, C., Auger, A. (eds.) GECCO 2011: Proceedings of the 13th Annual Conference Companion on Genetic and Evolutionary Computation, pp. 29–30. ACM, Dublin (2011a)

Murdock, J., Yaeger, L.S.: Identifying species by genetic clustering. In: Lenaerts, T., Giacobini, M., Bersini, H., Bourgine, P., Dorigo, M., Doursat, R. (eds.) Advances in Artificial Life: Proceedings of the Eleventh European Conference on Artificial Life (ECAL 2011), pp. 565–572. MIT Press, Paris (2011b)

Murre, J.M.J., Engelhardt, D.P.F.: The connectivity of the brain: multi-level quantitative analysis. Biological Cybernetics 73, 529–545 (1995)

Newman, M.E.J., Engelhardt, R.: Effects of neutral selection on the evolution of molecular species. Proc. R. Soc. London B, 1333–1338 (1998)

Raup, D.M., Gould, S.J.: Stochastic Simulation and Evolution of Morphology–Towards a Nomothetic Paleontology. Systematic Zoology 23(3), 305–322 (1974)

Raup, D.M., Gould, S.J., Schopf, T.J.M., Simberloff, D.S.: Stochastic Models of Phylogeny and the Evolution of Diversity. Journal of Geology 81, 525–542 (1973)

Rechsteiner, A., Bedau, M.A.: A Generic Neutral Model for Quantitative Comparison of Genotypic Evolutionary Activity. In: Floreano, D., Mondada, F. (eds.) ECAL 1999. LNCS, vol. 1674, pp. 109–118. Springer, Heidelberg (1999)

Rensch, B.: Evolution above the species level. Columbia Univ. Press, New York (1960a)

Rensch, B.: The laws of evolution. In: Tax, S. (ed.) The Evolution of Life, pp. 95–116. Univ. of Chicago Press, Chicago (1960b)

Saunders, P.T., Ho, M.W.: On the increase in complexity in evolution. J. Theor. Biol. 63, 375–384 (1976)

Shannon, C.E.: A mathematical theory of communication. Bell System Technical Journal 27, 379–423, 623–656 (1948)

Sporns, O., Lungarella, M.: Evolving coordinated behavior by maximizing information structure. In: Rocha, L. (ed.) Artificial Life X: Proceedings of the Tenth International Conference on the Simulation and Synthesis of Living Systems, pp. 323–329. MIT Press (Bradford Books), Cambridge (2006)

Sporns, O., Tononi, G., Edelman, G.: Theoretical Neuroanatomy: Relating Anatomical and Functional Connectivity in Graphs and Cortical Connection Matrices. Cerebral Cortex 10, 127–141 (2000)

Swenson, R.: Emergent attractors and the law of maximum entropy production: Foundations to a theory of general evolution. Systems Research 6(3), 187–197 (1989)

Thomas, R.D.K., Reif, W.-E.: The Skeleton Space: A Finite Set of Organic Designs. Evolution 47(2), 341–360 (1993)

Tononi, G., Edelman, G., Sporns, O.: Complexity and coherency: integrating information in the brain. Trends in Cognitive Sciences 2(12), 474–484 (1998)

Tononi, G., Sporns, O., Edelman, G.: A measure for brain complexity: Relating functional segregation and integration in the nervous system. Proc. Nat. Acad. Sci. 91, 5033–5037 (1994)

Turney, P.: Increasing Evolvability Considered as a Large-Scale Trend in Evolution. In: Wu, A. (ed.) Proceedings of the Workshop on Evolvability at the 1999 Genetic and Evolutionary Computation Conference (GECCO 1999), pp. 43–46 (1999)

Turney, P.: A simple model of unbounded evolutionary versatility as a largest-scale trend in organismal evolution. Artificial Life 6, 109–128 (2000)

Valentine, J.W., Collins, A.G., Meyer, C.P.: Morphological complexity increase in metazoans. Paleobiology 20(2), 131–142 (1994)

Waddington, C.H.: Paradigm for an evolutionary process. In: Waddington, C.H. (ed.) Towards a Theoretical Biology, Aldine, Chicago, vol. 2, pp. 106–128 (1969)

Wagner, P.J.: Contrasting the Underlying Patterns of Active Trends in Morphologic Evolution Evolution. Evolution 50(3), 990–1007 (1996)

Watts, D.J., Strogatz, S.H.: Collective dynamics of 'small-world' networks. Nature 393(6684), 440–442 (1998)

Yaeger, L.S.: Computational Genetics, Physiology, Metabolism, Neural Systems, Learning, Vision, and Behavior or Polyworld: Life in a New Context. In: Langton, C.G. (ed.) Proceedings of the Artificial Life III Conference, pp. 263–298. Addison-Wesley, Reading (1994)

Yaeger, L.S.: How evolution guides complexity. HFSP 3(5), 328–339 (2009)

Yaeger, L.S.: Identifying neural network topologies that foster dynamical complexity. Advances in Complex Systems 16(02n03), 1350032 (2013)

Yaeger, L.S., Griffith, V., Sporns, O.: Passive and Driven Trends in the Evolution of Complexity. In: Bullock, S., Noble, J., Watson, R., Bedau, M.A. (eds.) Artificial Life XI: Proceedings of the Eleventh International Conference on the Simulation and Synthesis of Living Systems, pp. 725–732. MIT Press, Cambridge (2008)

Yaeger, L.S., Sporns, O.: Evolution of Neural Structure and Complexity in a Computational Ecology. In: Rocha, L., Yaeger, L.S., Bedau, M., Floreano, D., Goldstone, R., Vespignani, A. (eds.) Artificial Life X: Proceedings of the Tenth International Conference on the Simulation and Synthesis of Living Systems, pp. 330–336. MIT Press (Bradford Books), Cambridge (2006)

Yaeger, L.S., Sporns, O., Williams, S., Shuai, X., Dougherty, S.: Evolutionary Selection of Network Structure and Function. In: Fellerman, H., Dörr, M., Hanczyc, M.M., Laursen, L.L., Maurer, S., Merkle, D., Monnard, P.-A., Støy, K., Rasmussen, S. (eds.) Artificial Life XII: Proceedings of the Twelfth International Conference on the Simulation and Synthesis of Living Systems, pp. 313–320. MIT Press, Cambridge (2010)

Chapter 16
Clustering and Modularity in Self-Organized Networks

Somwrita Sarkar and Peter A. Robinson

16.1 Introduction

Many biological, artificial, and social systems are self-organized. Though an overarching, exhaustive definition of self-organization is elusive, there is general agreement on many of the properties that self-organized systems can be characterized by: they are global systems, composed of many, usually identical, micro level components. These components interact locally, while the system shows emergence of global dynamics not directly observable, measurable, quantified, or defined at the local level (Prokopenko 2009).

The dynamics of self-organizing processes can be *guided*, either by design, or as a result of factors in the environment that affect the system. For example, the human brain can be considered as an example of a guided self-organized system, because while the brain grows from infancy to adulthood in a self-organized manner, the interaction of the organism with its external environment guides this self-organization (Sporns 2011). In artificial systems, guided self-organization can occur through design. For example, in artificial life, evolutionary dynamics, and game theoretic experiments, explicitly altering the variables, constraints, or sets of rules for an evolutionary process can result in the emergence of different forms of self-organization (Prokopenko 2009). In the design and engineering domain, the design of complex engineered systems (consider examples such as aircraft, spaceships,

Somwrita Sarkar
Design Lab, Faculty of Architecture, Design, and Planning,
University of Sydney, NSW 2006, Australia

Somwrita Sarkar · Peter A. Robinson
Complex Systems Group, School of Physics, University of Sydney, NSW 2006, Australia

Peter A. Robinson
Brain Dyanmics Center, Sydney Medical School,
University of Sydney, Westmead, NSW 2145, Australia

M. Prokopenko (ed.), *Guided Self-Organization: Inception*,
Emergence, Complexity and Computation 9,
DOI: 10.1007/978-3-642-53734-9_16, © Springer-Verlag Berlin Heidelberg 2014

and aero-engines) is frequently thought of as centrally planned (as opposed to self-organized). However, it is worth noting that the process of designing itself is, on the one hand, driven by the system or project objectives and constraints, and on the other hand the organization's internal and external environments (Sosa et al. 2003, 2004). This "guiding process" results in a guided self-organized system of teams of human designers and engineers working together to achieve the design (Sosa et al. 2003, 2004; Sarkar et al. 2013a). Thus, looked at in one way, every self-organized system would be guided, either partially or fully, by the environment in which the system is situated, either by design or by other contextual factors.

16.2 Modularity of Self-Organized Systems

In studying the evolution of such systems, one fundamental organizational property is *modularity*, *community formation*, or *cluster formation*. Modularity is defined as *the organization of system elements in clusters or tightly interacting modules, with elements of the same cluster densely connected to each other, and comparatively sparsely connected with elements from other clusters* (Fortunato 2010; Newman 2010). Further, it has been shown that many real world systems show *hierarchical modularity*: the presence of nested smaller modules in larger modules, at many levels of organization.

Why modularity or hierarchical modularity is observed in systems is an open research question, and is not particularly well-understood. One possibility is that the clusters are functionally formed: the elements of a system that all perform similar functions, as a group, are densely connected together. From this perspective, modularity is a functional phenomenon, and modules are a measure of the functional or behavioral complexity of the system. For example, the human cortex has visual or auditory or somatosensory areas, and engineered systems have sub-systems and components that are connected based on specific functions such as heat or material transfer.

Another possibility discussed in the literature is that the modularity in the system can be a result of optimizing factors such as path lengths for information communication, total wiring length for volume optimization, or any other such global optimization factor (Sporns 2011; Fortunato 2010).

The question of modular and hierarchical organization is especially important for guided self-organized systems. Some studies have shown that guiding the self-organization process towards optimizing certain definite sets of behaviors results in modular architectures (Kashtan and Alon 2005). This has particular implications for the evolutionary design of engineered systems, as well as understanding under what conditions in a guided self-organization process can modularity emerge, for biological and natural systems. In other words, the presence or absence of modularity can signal the types of guided self-organization or self-organization occurring in a system.

Modularity appears to be a fundamental organizational property of self-organized systems. Therefore, methods for its detection become important: Given a system, and its description in terms of its elements and their inter-relationships, can we detect the naturally occurring (not artificially defined) modules? Answering this detection question can help to answer the deeper questions discussed above: why do self-organized systems organize themselves into modules? If we detect the modules of a certain system, can we then predict the "guiding" process or principle that results in the formation of these modules?

Detecting and characterizing modularity or communities in real world systems has proved to be difficult. Despite being a significant research focus in multiple disciplines, the answer to how to detect and optimally describe the modular structure of a system remains elusive.

Self-organized complex systems can be represented as graphs or networks of interaction between its component elements: nodes represent system components and links represent the interaction between these components (Newman 2010; Fortunato 2010). In various domains, the nodes and links can have various representations. The study of complex systems as networks has attracted widespread attention in physics, computer science, biology and many other disciplines, with networks representing a wide array of social, biological, technological, or economic systems.

16.3 Chapter Summary

In this Chapter, we focus on spectral methods of modularity finding in networks (Fortunato 2010; Newman 2010; Sarkar and Dong 2011b; Sarkar et al. 2013b). Spectral methods involve studying the eigenvalue spectra of classes of matrices related to the networks. In general, they provide a powerful lens to study modularity, in comparison with other methods of modularity finding, for two main reasons. First, many methods focus on the local properties of nodes and their connected neighbors, to produce locally greedy algorithms to detect modules (for a review of methods see (Fortunato 2010)). As opposed to local methods, spectra provide a powerful global method of looking at the connectivity structure, without losing the local information. Second, eigenvalue spectra of connectivity matrices and graphs are also often used to understand the dynamics of a system, specifically with regards to properties such as stability and instability of system dynamics (Rajan and Abbott 2006; Gray et al. 2009; Robinson et al. 2009; Arenas et al. 2006). Relating the dynamical systems view to the network modularity can therefore provide a powerful mechanism to understand how system structure and system dynamics are co-dependent. The relationships between eigenvalues and the system dynamics arises from underlying behavioral models, and are therefore, subject to model assumptions. But they still provide an insight into the structure-behavior connection. However, methods that look for only structural modularity, on the other hand, will fail to make this structure-behavior connection.

In Sec. 16.4, we first review some background on the use of spectral methods for understanding network modularity. In our previous work, we have shown that eigenvalue spectra of graph adjacency matrices can help to fingerprint the modular structure of networks (Sarkar et al. 2009; Sarkar and Dong 2011b,a; Sarkar et al.

2013b,a). In this work, we show how the adjacency matrix provides a powerful mechanism to study hierarchical modularity and modular overlaps. In particular, we review how traditional spectral clustering methods dependent on using the Laplacian matrix have, in general, not captured hierarchical organization and modular overlaps. In Sec. 16.5, we then present a generalized algorithm for modularity detection, developing on the specific methods presented in our recent work (Sarkar and Dong 2011b; Sarkar et al. 2013b). We demonstrate the efficiency of the algorithm on synthetic networks generated using stochastic block model type network models (Sarkar et al. 2013b,a).

16.4 Background

In this section, we present a critical review of the prevalent approaches to spectral modularity detection in graphs, and some of their limitations that we address in our work.

16.4.1 Spectra and Graph Structure

One research direction involving the use of network spectra to make inferences about network structure stems from a long history in graph theory, random matrix theory and quantum mechanics. In graph theory, a general problem is to infer relations between the spectral and structural properties of a graph by studying the eigenvalues and eigenvectors of a matrix associated with the graph (Biggs 1993; Cvetkovic et al. 1995, 1997, 2010). An unweighted network of n nodes is represented by its symmetric square adjacency matrix $\mathbf{A}$, where

$$A_{ij} = \begin{cases} 1 \text{ if an edge exists between nodes } i \text{ and } j, \\ 0 \text{ otherwise.} \end{cases} \tag{16.1}$$

The *spectrum* of the network (Biggs 1993) is the set of eigenvalues of $\mathbf{A}$ together with their multiplicities. If $\mathbf{A}$ has distinct eigenvalues $\lambda_0 > \lambda_1 > \ldots > \lambda_{n-1}$ with respective multiplicities $m(\lambda_0), m(\lambda_1), \ldots, m(\lambda_{n-1})$, the spectrum is

$$S = \begin{bmatrix} \lambda_0 & \lambda_1 & \ldots & \lambda_{n-1} \\ m(\lambda_0) & m(\lambda_1) & \ldots & m(\lambda_{n-1}) \end{bmatrix}. \tag{16.2}$$

Starting with the work of Wigner (Wigner 1955, 1957, 1958), random matrix theory (Mehta 2004) and the spectral density of a network have been used to classify complex network types (Farkas et al. 2001; Goh et al. 2001). The *spectral density* of a network is defined to be the density of the eigenvalues of its adjacency matrix. When the number of nodes n is finite, this is expressed as a sum of Dirac δ functions:

$$\rho(\lambda) = \frac{1}{n} \sum_{j=0}^{n-1} \delta(\lambda - \lambda_j), \tag{16.3}$$

where λ is the eigenvalue at which the spectral density ρ is computed. It has been shown (Farkas et al. 2001) that random, non-modular small world, and scale-free models have distinct spectra. When the spectral density $\rho(\lambda_i)$ is plotted against the eigenvalues $\lambda_j, i = 0 \ldots n-1$, [see Figs. 1-4 in (Farkas et al. 2001)], as n grows large, the spectral density for an ER random graph approaches a semi-circular form, the spectral density of a Watts-Strogatz (WS) small world network shows multiple sharp peaks, and the spectral density of a Barabasi-Albert scale free network shows a triangular form. The spectral density approach was then extended to study modularity, hierarchy, and dynamics (de Aguiar and Bar-Yam 2005) using a scale-free hierarchical modular network as defined by (Ravasz et al. 2002). However, this fingerprinting of network types was not extended to determine the number of modules, the number of hierarchical levels, and the exact modular composition.

16.4.2 Spectral Clustering and Partitioning Approaches

A second principal direction in the use of spectra to infer network structure, and specifically look at graph partitioning and finding network modularity stems from work in computer science and mathematics (Chung 1997; Mieghem 2011; Cvetkovic et al. 1995, 1997; Newman 2010). In general called *spectral clustering* or *graph partitioning*, these approaches use the information contained in the eigenvectors and eigenvalues of a suitable matrix representation of a graph. Most used is the graph Laplacian (Pothen et al. 1990), that we now define. We have defined earlier that **A** represents the adjacency matrix of a graph G with n nodes. Further, let **D** be the degree matrix, where

$$D_{ij} = \begin{cases} d_i \text{ degree of node } i \text{ when } i = j, \\ 0 \text{ when } i \neq j. \end{cases} \tag{16.4}$$

Then the Laplacian matrix is

$$\mathbf{L} = \mathbf{D} - \mathbf{A} \tag{16.5}$$

with

$$L_{ij} = \begin{cases} d_i \text{ when } i = j, \\ -1 \text{ when } i \neq j \text{ and } i \text{ is adjacent to } j, \\ 0 \text{ otherwise.} \end{cases} \tag{16.6}$$

Almost all spectral clustering or partitioning approaches operate on **L** (or its other variants) to partition the graph recursively, each time finding an optimal bisection (Newman 2010; Pothen et al. 1990). The basic approach followed is that first an optimization function is defined with respect to the graph properties, such as minimizing the number of cuts needed to partition the graph, or minimizing the distance needed to travel across the graph. Usually, the graph is initially partitioned into two modules with respect to an optimization function, followed by a recursive reapplication of the bisection step to find more modules.

As we have discussed in (Sarkar and Dong 2011b), and now demonstrate more extensively in this work, this approach fails to capture many of the important characteristics important for modularity detection in real world networks. We specifically show that the strict partitioning assumption, (the assumption that the graph be partitioned into two disjointed parts), is a cornerstone of the methods and rests on the specific use of the second leading eigenvector of the Laplacian. However, this same assumption renders the method incapable of identifying other natural patterns fundamental to modularity detection, such as modular overlaps and hierarchical organization, unless the algorithms are specifically modified. One way of doing this is presented in this chapter.

The original spectral bi-partitioning algorithm (Pothen et al. 1990) defines and minimizes the *cut size* R, defined as the number of edges running between two groups of vertices into which a cut partitions a graph. Following (Newman 2010), an index vector is defined, $\mathbf{s} = \{s_1, s_2, ..., s_n\}$, with each s_i = +1 or -1 depending on which of the two modules vertex i is assigned to, with the normalization condition $\mathbf{s}^T\mathbf{s} = 1$. The cut size R can then be defined in terms of the index vector $\mathbf{s}$ and the Laplacian L follows, and this becomes the function to be minimized:

$$\text{Min } R = \frac{1}{4}\mathbf{s}^T\mathbf{L}\mathbf{s}. \tag{16.7}$$

Following a parallel formulation, Newman's spectral approach (Newman 2010) presents a *modularity function* Q that is maximized to find a partition that optimally divides the network into two modules. A modularity matrix B is defined that measures the difference between the actual number of edges existing between a pair of vertices and an expected number derived from an equivalent random graph with the same number of vertices and the same degree distribution but with no community structure (Newman 2010). Then, the modularity function that is to be maximized is defined in terms of the modularity matrix B, the partition vector $\mathbf{s}$ as in Eq. (16.7), and the number of edges m in the network:

$$\text{Max } Q = \frac{1}{4m}\mathbf{s}^T\mathbf{B}\mathbf{s}. \tag{16.8}$$

We note here the equivalence between the forms of the two optimization functions in (16.7) and (16.8). Usually, solving these optimization problems, (minimizing Eq. (16.7) or maximizing Eq. (16.8)), are NP-hard (Newman 2010). However, it can be shown (Pothen et al. 1990; Newman 2010) that an approximate solution can be found if $\mathbf{s}$ is chosen proportional to the eigenvector corresponding to the second smallest eigenvalue of $\mathbf{L}$ or the leading eigenvector of $\mathbf{B}$. Thus, both these approaches represent $\mathbf{s}$ as a linear combination of the eigenvectors $\mathbf{v}_i$ of the Laplacian matrix $\mathbf{L}$ in Eq. (16.7) or the modularity matrix $\mathbf{B}$ in Eq. (16.8), as

$$\mathbf{s} = \sum_{i=1}^{n} a_i\mathbf{v}_i. \tag{16.9}$$

However, since $\mathbf{s}$ is, by definition, constrained to take on discrete $+1$ or -1 values, an approximate solution is resorted to in both cases, with $s_i = +1$ if the ith element of the corresponding eigenvector is positive and -1 if negative. This corresponds to the strict partition and bisection assumption: vertices can belong to only one of two communities. For more than two communities, the vector **s** is replaced by an $n \times k$ matrix **S** with k communities, with $i = 1$ to n and $j = 1$ to k, such that:

$$S_{ij} = \begin{cases} 1 \text{ when vertex } i \text{ belongs to community } j, \\ 0 \text{ otherwise.} \end{cases} \tag{16.10}$$

The strict partitioning assumption still holds. Thus, *a priori* assumptions built into the definition of a community do not permit identification of overlapping communities, even when the eigenvectors and eigenvalues may implicitly contain this information.

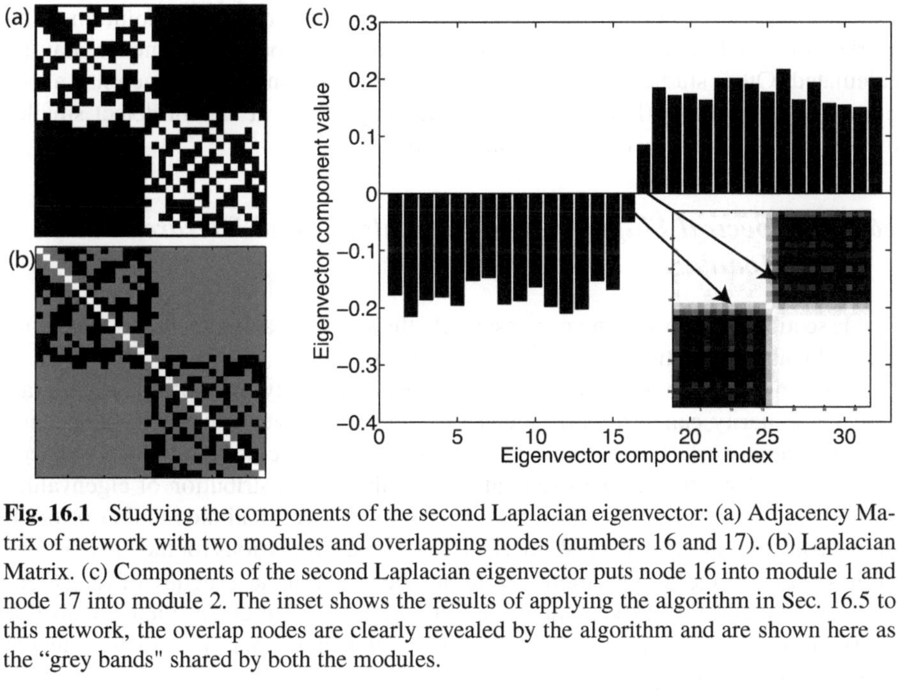

Fig. 16.1 Studying the components of the second Laplacian eigenvector: (a) Adjacency Matrix of network with two modules and overlapping nodes (numbers 16 and 17). (b) Laplacian Matrix. (c) Components of the second Laplacian eigenvector puts node 16 into module 1 and node 17 into module 2. The inset shows the results of applying the algorithm in Sec. 16.5 to this network, the overlap nodes are clearly revealed by the algorithm and are shown here as the "grey bands" shared by both the modules.

We show a detailed example to demonstrate the above claim. Figure 16.1 shows an example 32 node network with two modules with node numbers 16 and 17 deliberately connected with both modules; i.e., they share dense interactions with both modules, and are therefore, overlapping nodes. It would be incorrect to assign them into independent modules, because they connect equally strongly with both modules. With the algorithm described above, however, we see that the relevant eigenvector component for node 16 is negative and node 16 is therefore placed in module 1 by the algorithm. Similarly, the relevant eigenvector component for node 17 is positive, and is therefore placed in module 2 by the algorithm. Clearly, forcing an approximate solution may throw away useful information contained in the eigenvectors and eigenvalues that can potentially be used to shed more light onto the community structure existing in the the graph: the approach we demonstrate later in this work. The inset in Fig. 16.1(c) shows the results of applying the algorithm presented in this paper in Sec. 16.5 to identify the modules. The white portions of the matrix show the two principal modules, and the overlap nodes clearly emerge as gray bands that are "shared" between the two modules.

Further, the spectral bi-partitioning algorithm (Pothen et al. 1990) asks the user to choose the number and relative sizes of communities beforehand, and always provides a solution, whether or not a clear community structure exists in the graph. Newman's modified approach (Newman 2010) addresses most of these limitations by using the modularity matrix instead of the Laplacian. However, his approach maintains the strict partitioning assumption. Thus, overlapping modules are not revealed. The calculation of the modularity metric also requires a solution to be worked out, and nodes to be assigned to communities, before the metric may be calculated. Other studies have also shown that that Newman's approach has a resolution limit problem (Fortunato and Barthelemy 2006): it cannot detect smaller hidden sub structures inside larger communities.

16.4.3 Spectral Fingerprints of Modularity and Hierarchical Modularity: Adjacency Matrix

In this section, we show, using our past work, the second major gap: that of detecting hierarchical organization of modules in networks.

In recent work, we have shown that network modularity, and specifically hierarchical modularity, can be fingerprinted using the largest eigenvalues of the adjacency matrix of a network (Sarkar et al. 2013b). Gaps between clusters of closely spaced large eigenvalues that are well separated from the bulk distribution of eigenvalues around the origin reveal the number of hierarchical levels and the number of modules at each hierarchical level. We derived analytical expressions for the mean values of these largest eigenvalues, thereby relating the hierarchical structure of typical hierarchical stochastic block model type networks and matrices to their eigenvalue distributions. This provided an algorithm-independent manner of characterizing the hierarchical modularity of networks.

Hierarchical modular graphs have "modules nested inside modules". The probability of an edge inside the lowest level (smallest) module is the highest, and progressively decreases as the level of hierarchy increases. We follow the typical stochastic block model form for constructing a hierarchical network (Sarkar et al. 2013b). A hierarchical modular network is constructed by recursively placing random matrix blocks with decreasing levels of connectivity between nodes in hierarchical levels in a block diagonal form. We consider the matrix

$$\mathbf{A}' = \begin{bmatrix} \mathbf{A} & \mathbf{P} \\ \mathbf{P} & \mathbf{A} \end{bmatrix}, \tag{16.11}$$

where **A** is a binary random network of size s and edge probability p, and **P** is a random network of size s and edge probability pq. Here, the parameter q sets the level of decrease in connectivity between the various levels of hierarchy. That is, q is a numeric parameter that is varied to define the connectivity of the first level hierarchy of off-diagonal networks or embedded modules represented by **P**. Thus, the lower the value of q, the stronger the hierarchical modular structure, and higher the value of q (to 1), the weaker the hierarchical modular structure.

We now define the second level of perturbation $\mathbf{A}''$ as

$$\mathbf{A}'' = \begin{bmatrix} \mathbf{A}' & \mathbf{P} \\ \mathbf{P} & \mathbf{A}' \end{bmatrix}, \tag{16.12}$$

where $\mathbf{A}'$ is the matrix defined in Eq. (16.11) and **P** is a random network or matrix of size $2s$ and edge probability pq^2. Note here the second hierarchical level: $\mathbf{A}'$ already has the first level of hierarchy built in as described previously, with the first level off-diagonal blocks having connectivity pq and the diagonal blocks having connectivity p, with $pq < p$. Now, the second level off-diagonal blocks, represented by matrix **P**, have connectivity pq^2 with $pq^2 < pq < p$. In general, the matrix **P** defines each successive level L of perturbations of increasing size $(s, 2s, 4s, \ldots, N/2)$ and decreasing probability of connection $(pq, pq^2, \ldots, pq^{L-1})$, producing an extra level of hierarchical modular structure with each perturbation level. Figure 16.2 [inset] shows an example 1024 node hierarchical network adjacency matrix with 5 hierarchical levels.

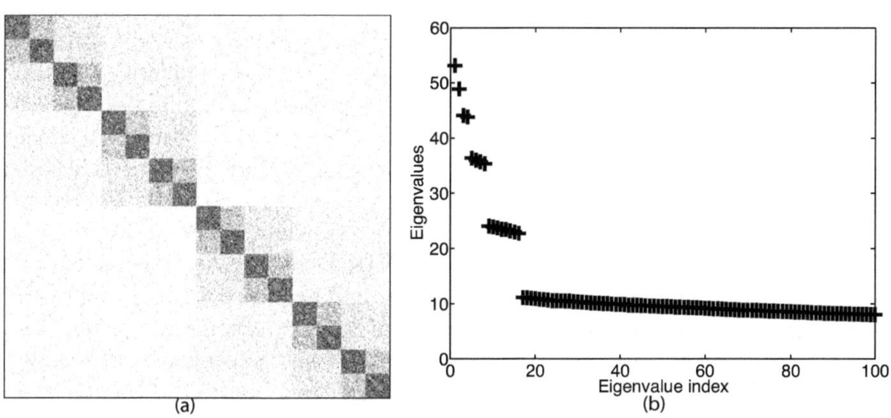

Fig. 16.2 Hierarchical networks and their adjacency matrix spectra clearly bring out the hierarchical organization; (a) the adjacency matrix for a hierarchical network (b) plot shows eigenvalues arranged in decreasing order. Clusters of eigenvalues indicate the hierarchical structure.

We have proved in (Sarkar et al. 2013b), that for L hierarchical levels, the expected values of the largest eigenvalues (those separated from the bulk of the eigenvalues) of a hierarchical network $\mathbf{A}_L$, along with their algebraic multiples, are:

$$Sp(\mathbf{A}_L) = \begin{bmatrix} s\left[p+pq\Sigma_{i=0}^{L-1}(2q)^i\right] & 1 \\ s\left[p+pq\left\{\left[\Sigma_{i=0}^{L-2}(2q)^i\right]-(2q)^{L-1}\right\}\right] & 1 \\ \dots & \dots \\ s\left[p+pq(1-2q)\right] & 2^{L-2} \\ s\left[p-pq\right] & 2^{L-1} \end{bmatrix}. \tag{16.13}$$

Figure 16.2 shows the spectrum correctly predicting the hierarchical structure: the coarsest network module of 1024 nodes, 2 modules of 512 nodes each, 4 modules of 256 nodes each, 8 modules of 128 nodes each, and 16 modules of 64 nodes each. Note the gaps in eigenvalues after the 2nd, 4th, 8th, and 16th eigenvalues.

These findings tell us that considering other eigenvalues and eigenvectors may tell us more about the structure of the graph, and therefore, must be explored further in a systematic way. In the next section, we present a generalized algorithm for detecting the modular organization at multiple hierarchical levels, using the information contained in the eigenvalues and eigenvectors of the adjacency matrix.

16.5 Detecting the Modular Structure

In this section, we present a generalized algorithm based on work presented (Sarkar et al. 2009; Sarkar and Dong 2011b; Sarkar et al. 2013a), and briefly review the main steps of the algorithm. The main difference between the approach we present and the optimization approaches discussed in relation to the Laplacian and modularity matrices in Sec. 16.4.2 is that instead of stating the modularity detection problem as an optimization problem, we state it as a pattern recognition based unsupervised clustering problem. We have already seen that eigenvalues and corresponding eigenvectors beyond the second ones have important information encoded about the structure of the system.

The method contains three aspects. First, the eigenvalue decomposition (EVD) of the adjacency matrix is computed as $\mathbf{A} = \mathbf{V}\mathbf{D}\mathbf{V}^T$. If $\mathbf{A}$ has n nodes, then $\mathbf{V}$ is the $n \times n$ orthonormal matrix of its eigenvectors and $\mathbf{D}$ is an $n \times n$ diagonal matrix of its eigenvalues. The eigenvalues are arranged in a decreasing order, along with the corresponding eigenvectors. The main idea is to now express the connectivity of each node with other nodes as a vector in space, as a linear combination of an eigenvector component times the corresponding eigenvalue as

$$\mathbf{a}_i = \left[v_{i1}\lambda_0, v_{i2}\lambda_1, \dots v_{in}\lambda_{n-1}\right], \tag{16.14}$$

as in (Sarkar and Dong 2011b). Thus, connectivity is now expressed as a function of position in space.

Second, we perform a dimensionality reduction on the original adjacency matrix by preserving the k largest eigenvectors and eigenvalues to produce a reduced approximation of the node vectors as

$$\mathbf{a}_i^{(k)} = [v_{i1}\lambda_0, v_{i2}\lambda_1, \ldots v_{ik}\lambda_k]. \quad (16.15)$$

The heuristic choice on k is guided by the findings in the previous section: we identify the largest gaps in the spectrum between successive eigenvalues, and then choose k (or $k+1$) based on the these largest gaps. If the graph has a hierarchical modular structure, there will be many such gaps. Using different k values (corresponding to each of the large eigengaps) will bring out the modularity structure at different hierarchical levels, since each k value corresponds to the number of modules at a specific hierarchical level in the system. Usually, the more pronounced the modularity structure in the network, the sharper the gaps, and the weaker the modularity structure in the network or higher the noise, the smaller the gaps.

The dimensionality reduction step identifies data redundancy in the adjacency matrix. Each of the n nodes can be thought of as a separate dimension. However, the pattern in the data is the number of modules, which are always going to be lesser than n. Therefore, all the n dimensions are not needed to capture the pattern: the number of needed dimensions corresponds to the number of modules. In a module, two nodes with many common neighbors or the same set of neighbors are likely to fall in the same module, while two nodes that do not share common neighbors are likely to fall in separate modules. If two nodes have exactly the same set of neighbors, then there are dependent rows and columns in the matrix. If they share many common neighbors, then their vertex vector representations will share high dot products. In either case, this redundancy in the matrix means that there is a lower number of dimensions that can be used to represent the modular organization of the system. This redundancy in the graph matrix is used to compute a linear least squares, optimal, lower dimensional approximation of the original matrix by retaining the k largest eigenvectors and eigenvalues. These, when arranged in decreasing order, capture the relative information content that each orthogonal dimension contains about modular organization.

Finally, to find the modules, dot products are computed between all the k reduced vector representations of nodes, resulting in a dot product or cosine matrix. The higher the cosine between two node vectors, the higher the probability that they belong to the same module. The lower the cosine, the higher the probability that they belong to different modules. With the cosine matrix suitably reordered to reveal the highly connected groups of nodes along the block diagonal. We have previously provided an algorithm for reordering (Sarkar and Dong 2011b), we can identify the modular hierarchical-overlapping organization in the network, as well as the overlap nodes. The overlap nodes appear in the cosine matrix as "bands" of high dot products: nodes that fall in an overlap between multiple modules share high dot products with nodes of all those respective modules. In Fig. 16.1(c), the inset shows that the two overlap nodes in the network are clearly revealed as "white-grey" bands (high cosine similarity) values, shared by both the modules (shown in white).

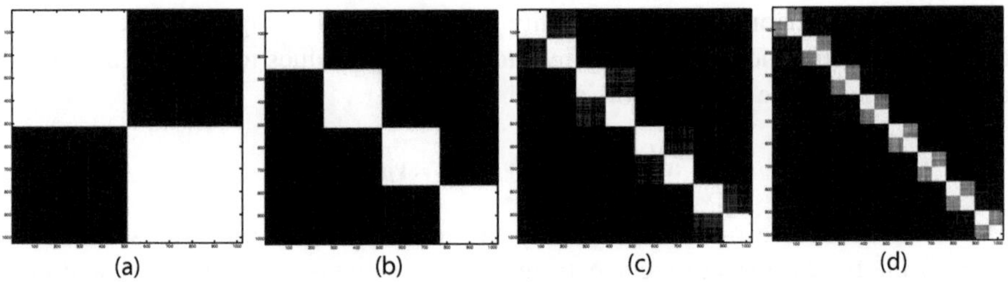

Fig. 16.3 Using the algorithm to bring out hierarchical structure. Dot product matrices computed to reveal the modules for 1024 node hierarchical network of Fig. 16.2 using: (a) $k = 2$, (b) $k = 4$, (c) $k = 8$, and (d) $k = 16$. Modules at separate hierarchical levels are clearly brought out.

Fig.16.3 shows the dot product matrices after performing the modularity identification using $k = 2, 4, 8, 16$, respectively for the 1024 hierarchical network shown in Fig. 16.2. Note that when $k = 2$, the coarsest hierarchical arrangement is visible, when $k = 4, 8$, the second and third hierarchical levels are visible, and when $k = 16$ the finest hierarchical level is visible.

16.6 Discussion and Conclusion

We have presented spectral methods for modularity and hierarchy detection in complex self-organized systems represented as graphs or networks. The area of modularity detection is a very important one, as it has been shown to be deeply connected with self-organizing processes in systems. We reviewed the traditional spectral clustering approaches from the literature, and then showed how incorporating the information in the full spectrum and reformulating the manner of modularity detection can lead to the detection of modular overlaps and hierarchical organization. Our main aim in this chapter was to demonstrate the methods and their comparison with previous methods in the literature. Elsewhere, we have also tested these methods on large classes of synthetic, benchmark, and real world networks, with very accurate modularity detection results (Sarkar and Dong 2011b; Sarkar et al. 2013b,a).

This work is at the preliminary stages, and can be extended in a number of other directions, especially with regard to deepening our understanding of self-organizing processes in systems. We can use the spectra to chart the evolution of self-organizing systems, both for studying emergence of modularity in naturally self-organizing systems, or those that are being guided by design. In evolving systems, we can study the spectra of temporal snapshots of a system. If modularity emerges at any stage, this will be clearly indicated and can be studied using the spectra.

Second, we can use the spectra to understand, explore, and map the relationships between the self-organization processes and the system structure. For example, in designed systems, several versions of self-organizing, optimization or evolutionary rules can be specified, and charting the evolution of the system under the effect of these various sets of generative rules can be performed using the spectra. If modularity or hierarchical organization emerges, the spectra can be used as a classification tool to understand which processes lead to emergence of modularity.

Finally, there are a number of methodological directions. The literature shows various types of matrices used for modularity detection and classification, ad we have explored preliminary relationships between them, as reported in this paper. The work here can be extended to better understand the relationships between the various matrix classes, and to explore what type of information on modularity is captured (or not captured) in specific matrix forms.

Acknowledgements. The Seed Staff Support Grant, Faculty of Architecture, Design, and Planning, University of Sydney, Australian Research Council and Westmead Millenium Institute supported this work.

References

Arenas, A., Díaz-Guilera, A., Pérez-Vicente, C.J.: Synchronization reveals topological scales in complex networks. Phys. Rev. Lett. 96(11), 114102 (2006)

Biggs, N.: Algebraic Graph Theory, 2nd edn. Cambridge University Press (1993)

Chung, F.: Spectral Graph Theory. In: CBMS Regional Conference Series in Mathematics (1997)

Cvetkovic, D., Doob, M., Sachs, H.: Spectra of Graphs: Theory and Applications. Johann Ambrosius Barth Verlag (1995)

Cvetkovic, D., Rowlinson, P., Simic, S.: Eigenspaces of Graphs. Encyclopedia of Mathematics and Its Applications, vol. 66. Cambridge University Press (1997)

Cvetkovic, D., Rowlinson, P., Simic, S.: An introduction to the theory of graph spectra, London Mathematical Society Texts, vol. 75. Cambridge University Press (2010)

de Aguiar, M.A.M., Bar-Yam, Y.: Spectral analysis and the dynamic response of complex networks. Phys. Rev. E 71(1), 016106 (2005)

Farkas, I., Derényi, I., Barabási, A., Vicsek, T.: Spectra of "real-world" graphs: Beyond the semicircle law. Phys. Rev. E 64(2), 026704 (2001)

Fortunato, S.: Community detection in graphs. Phys. Rep. 486, 75–174 (2010)

Fortunato, S., Barthelemy, M.: Resolution limit in community detection. Proc. Natl. Acad. Sci. 104(1), 36–41 (2006)

Goh, K.-I., Kahng, B., Kim, D.: Spectra and eigenvectors of scale-free networks. Phys. Rev. E 64(5), 051903 (2001)

Gray, R.T., Fung, C.K.C., Robinson, P.A.: Stability of random brain networks with excitatory and inhibitory connections. Neurocomp. 72, 1565–1574 (2009)

Kashtan, N., Alon, U.: Spontaneous evolution of modularity and network motifs. Proceedings of the National Academy of Sciences 102(39), 13773–13778 (2005)

Mehta, M.L.: Random Matrices, 3rd edn. Elsevier (2004)

Mieghem, P.V.: Graph Spectra for Complex Networks. Cambridge University Press, Cambridge (2011)

Newman, M.E.J.: Networks: An Introduction. Oxford University Press (2010)

Pothen, A., Simon, H., Liou, K.: Partitioning sparse matrices with eigenvectors of graphs. SIAM J. Matrix Anal. Appl. 11, 430–452 (1990)

Prokopenko, M.: Guided self-organization. HFSP Journal 3(5), 287–289 (2009)
Rajan, K., Abbott, L.F.: Eigenvalue spectra of random matrices for neural networks. Phys. Rev. Lett. 97, 188104 (2006)
Ravasz, E., Somera, A., Mongru, D., Oltvai, Z., Barabasi, A.: Hierarchical organization of modularity in metabolic networks. Science 297, 1551 (2002)
Robinson, P.A., Henderson, J.A., Matar, E., Riley, P., Gray, R.T.: Dynamical reconnection and stability constraints on cortical network architecture. Phys. Rev. Lett. 103, 108104 (2009)
Sarkar, S., Dong, A.: Characterizing modularity, hierarchy, and module interfacing in complex design systems. In: Proceedings of the American Society of Mechanical Engineers (ASME) 2011 International Design Engineering and Technical Conference and Computers and Information in Engineering Conference (DETC/CIE), pp. DETC2011–DETC47992. ASME Press, New York (2011a)
Sarkar, S., Dong, A.: Community detection in graphs using singular value decomposition. Phys. Rev. E 83(4), 046114 (2011b)
Sarkar, S., Dong, A., Gero, J.S.: Design optimization problem reformulation using singular value decomposition. Journal of Mechanical Design 131(8), 081006 (2009)
Sarkar, S., Dong, A., Henderson, J.A., Robinson, P.A.: Spectral characterization of hierarchical modularity in product architectures. Journal of Mechanical Design (2013a) (accepted for publication)
Sarkar, S., Henderson, J.A., Robinson, P.A.: Spectral characterization of hierarchical network modularity and limits of modularity detection. PLoS One 8(1), e54383 (2013b)
Sosa, M.E., Eppinger, S.D., Rowles, C.M.: Identifying modular and integrative systems and their impact on design team interactions. Journal of Mechanical Design 125, 240–252 (2003)
Sosa, M.E., Eppinger, S.D., Rowles, C.M.: The misalignment of product architecture and organizational structure in complex product development. Management Science 50(12), 1674–1689 (2004)
Sporns, O.: Networks of the brain. MIT Press (2011)
Wigner, E.P.: Characteristic vectors of bordered matrices with infinite dimensions. Ann. Math. 62(3), 548–564 (1955)
Wigner, E.P.: Characteristic vectors of bordered matrices with infinite dimensions ii. Ann. Math. 65(2), 203–207 (1957)
Wigner, E.P.: On the distribution of the roots of certain symmetric matrices. Ann. Math. 67(2), 325–327 (1958)

Index

ε-machines 122
"topological disorder" gene 447

action switching 310
action-perception loop 5
active information storage 327, 347, 349, 395
actuator 74, 263, 268, 283, 395, 398
  state 68, 269
adaptation 22, 72, 240, 367, 379, 418, 426
  task-less 403
adaptive behavior 300, 450
adaptivity 317
adjacency matrix 436, 459
algorithmic complexity 419
ANOVA 375
arrow of complexity 417
artificial chemistry 369
artificial evolution 412
Artificial Intelligence 309
artificial life 418, 457
artificial neural network 200, 398, 418, 421
Ashby's law 47, 326
attractor 22, 124, 212, 306, 326
  basin 306
autaptic connection 202
autistic vehicle 206
autocatalytic adaptive response 346
autonomous behavior 377
autonomous robot 194
autopoiesis 20
autotroph 26
Avida 426
axiomatic 8

backprojection 205
backpropagation 205, 322
Bayesian network 264
behavioral complexity 458
behavioral trajectory 312
best matching unit 334
bifurcation 197, 307, 322, 329
biodiversity 426
Blahut-Arimoto algorithm 77
boids 369
Bonferroni's test 375
bootstrapping 204, 215
brain-body-environment 297, 330
Braitenberg vehicle 194
branching clade lineage 424

categorical perception 304
causal Bayesian network 77
causal effect 76, 264
causal information flow 277
causal network 264
causality 264
cellular automata 33, 116, 381, 396
  blinkers 117
  collisions 117
  density classification 145
  Game of Life 125
  gliders 131, 396
  particles 117
  Sierpinski Gasket 125
channel capacity 5, 68
chaos control 231

chaotic phase 25, 63
chaotic regime 330
chaotic system 447
characteristic path length 436, 437
chemosensor 308
chemotaxis 303, 307
circular pattern 208
classification 334, 469
closeness centrality 85
cluster formation 458
clustering algorithm 432
clustering coefficient 436
clustering method 460
coherence 149, 345, 381, 395, 434
collective behavior 346, 352
collective communications 345
collective memory 345
collective motion 197
collision-based computing 116
communication 408, 409, 439, 458
  channel 24
  referential 304
community formation 458
community structure 462
complexity 12, 19, 53, 116, 212, 224, 318, 336
complexity growth 438
computation
  distributed 115
  universal 115
condensed matter physics 367
connectivity 421, 459
conservation laws 56
continuous-time recurrent neural network 304, 329
control parameter 53
control rule 412
control theory 54
controller 204, 225, 268
  quasi-linear, 224
coordinated behavior 9, 264, 300, 301, 391, 403
coordination pattern 299
correlational importance 168
cortex 458
cortical network 451
coupled oscillators 356
critical regime 25, 125, 228
critical state 55
cross-entropy 324
cross-motor teaching 248

decentralized control 197
description complexity 419
deterministic systems 142
differential equations 53, 206
diffusive control 54
dimensionality reduction 467
directed acyclic graph 265
directed graph 435
directed information 277
distal learning 231
distributed cognition 297
distributed computation 12, 345, 347, 396
dual total correlation 420
dynamic behavior 380
dynamic coordination 355
dynamic stability 322
dynamical complexity 418
dynamical hypothesis 300
dynamical interaction 408
dynamical system 22, 56, 71, 193, 224, 322
dynamical systems 4, 298, 459

echo state network 322
ecological interaction 379
ecological web 419
ecospace 422
ecosystem 33
  computational 418
edge of chaos 228
effective measure complexity 119, 286, 338, 420
eigenvalue decomposition 466
eigenvalue spectra 459
embedding vector 119
embodied agent 391
embodied cognition 69, 297
embodied intelligence 263
embodied robotics 70
embodiment 193, 243, 287, 412
emergence 20, 367, 395, 417, 468
emergent behavior 227
emerging dynamics 199
empowerment 5, 67, 194, 264, 395, 396, 449
  closed-loop 81

context free 80
continuous 92
multi-agent 91
open-loop 81
state-dependent 80
enactive cognition 297
energy cost 422
energy functional 57
entropy 5, 24, 60, 75, 117, 160, 202, 276, 325, 393, 417, 419
entropy production 363
entropy rate 118
equilibrium 198, 213, 302
ethology 394
Evolino approach 322
evolution 12, 33, 54, 72, 369, 417
driven trend 424
passive trend 424
evolutionary activity 424
evolutionary algorithm 322, 398
evolutionary convergence 379
evolutionary design 369
evolutionary dynamics 457
evolutionary innovation 380
evolutionary learning 231
evolutionary method 391
evolutionary model 426
evolutionary optimisation 264
evolutionary pressure 422
evolutionary theory 417
excess entropy 4, 119, 263, 395, 420
exponential distribution 325
exterioception 213
extinction rate 425

fault tolerance 439
feed-forward neural network 200
Finsler metric 85
firing rate 60
Fisher information 46, 326
Fisher memory curve 326
Fisher memory matrix 326
fitness function 369, 391, 395, 422
fitness landscape 422
fixed point 196, 226, 306, 329
foraging behaviour 394
free energy 194, 203
free energy principle 4
functional complexity 419
functional segregation 394

Gödel's Theorem 123
generalization capability 327
generating functional 54
genetic algorithm 305, 373, 422, 429, 449
tournament selection 430
genetic complexity 422
genetic consistency 426
genetic drift 424
genetic encoding 421
genetic regulatory networks 25
genome 421
genotype 397
genotypic information 383
global efficiency 436, 438
global integration 394
global pattern 412
goal-directed behavior 307
goal-oriented behavior 229
gradient descent 226, 322
gradient method 200
Granger causality 279
graph
optimal bisection 461
partitioning 461
graph theory 4, 421, 460
Grassberger's forecast complexity 338
group behaviour 391

halting problem 123
haploid agent 421
HEXAPOD 195, 205, 207, 215, 216, 218, 219
hidden layer 322
hierarchical modular graph 464
homeokinesis 71, 194, 203, 223, 449
homeostasis 20, 57, 71, 224
HUMANOID 194, 195, 197, 198, 205, 207, 212, 215, 218, 219
hyper-interactive evolutionary computation 372

ideal free distribution 435
imitation learning 235
Infomax principle 5, 200, 264, 325
information 4, 19, 69, 116, 120, 276, 324, 346, 395, 420, 467
coherence 117
dynamics 116

modification 116
redundant 160
synergistic 159
transfer 116
unique 162
information bottleneck 69
information cascades 12 346, 353
information dynamics 326, 345
information flow 5, 68, 76, 263, 324, 394
information gain 330
information geometry 160
information loss 330
information processing 345
capacity 328
information storage 326
information structure 395
information theory 4, 20, 68, 116, 159, 263, 324, 347, 392, 417
information transfer 326, 346–348, 350, 395
information-driven evolution 396
informational redundancy 69
Infotaxis 396
input entropy 122
input-driven network 322
integration 419, 437
interactive evolutionary computation 372
narrowly defined 372
simulated breeding 372
intervention 76, 270
intrinsic excitability 325
intrinsic motivation 70
intrinsic plasticity 202, 325
learning 331
invariance 218
inverse minimum path length 438
irreducibility 20
irreversible process 24

Jacobian matrix 201, 226
Java Information Dynamics Toolkit 131
Jung's Monte Carlo Integration 96

Kolmogorov complexity 338
Kullback-Leibler divergence 60, 325, 383

Lagrange function 56
Laplacian matrix 460
leaky integrator 58, 398
learning 4, 70, 194, 223, 263, 303, 322, 391, 396
anti-Hebbian 201
Hebbian 201, 421
learning rate 421
learning rule 201
unsupervised 201
Legrende polynomials 329
limit cycle 207 306
limnology 20
linear separation 327
local efficiency 439
locomotion behavior 252
Long Short Term Memory network 322
loop motifs 128
Lyapunov exponent 330

Mackey-Glass time series 335
Maximum Entropy Production Principle 6, 71
mean square error 324
memory capacity 326
metalearning 54
MIMO channel 100
Minimum Description Length 338
minimum path length 438
model acquisition 108
modular architecture 458
modular composition 461
modular overlap 460
modular robot 115, 395
modular structure
hierarchical 465
modularity 451, 458
hierarchical 458
modularity detection 460
modularity function 462
modularity matrix 462
morphogenesis 377
morphological computation 12, 193, 287
morphology 418
motif distribution 451
motion pattern 199
motor 102, 195, 200, 225, 249, 303, 399
acceleration 103
activation 404
continuous activity 246
control 92
coordinated motor behavior 300

force 198
in-phase activity 249
pattern 249
servo 197, 227
space 232
state 391
synchronized 251
teaching signal 233, 235
movement pattern 224
multi-information 279, 419
mutual information 5, 76, 118, 160, 202, 264, 325, 393, 419

natural selection 422
negentropy 363
network dynamics 10
network modularity 459
network topology 418
neural activation
continuous-time 447
spiking 447
neural activation model 447
neural complexity 394, 418
neural controller 391, 421
neural dynamics 419
neural gain 54
neural group 421
neural network 418
neural structure 115
neural topology 427
NeuralGas network 333
neuromodulator 54
neuron 159, 196, 256, 304, 321, 325, 398, 399
excitatory 421
firing rate 421
firing-rate 202
inhibitory 421
neuronal circuit 322
neuroprosthetics 255
neuroscience 394
neutral mutation 422

objective function 58, 194, 224, 263, 347
online learning 244, 329
open loop control 269
open-ended evolution 387, 422
optimization 461
order parameter 367
ordered phase 25
origin of life 424
orthonormal matrix 466
oscillation 213, 409
oscillatory phase 198
oscillatory behaviour 392
oscillatory movement 409

paleobiology 417
paleontology 424
Partial Information Decomposition 160
partial information diagram 160
particle swarm 369
path length 458
pattern formation 395
pattern recognition 466
pattern spin-off effect 218
perception-action loop 4
perpetual motion machine 196
perturbation 25, 145, 196, 213, 215, 224, 239, 306, 330, 346, 354, 372, 465
environmental 22, 385
exogenous 381
post-synaptic potential 201
phase space 12, 53, 60, 306
phase transition 46, 326, 347
phenotypic trait 159
pheromone 394
plasticity 5
polyhomeostasis 57
Polyworld 418
population statistics 427
post-synaptic potential 201
predictive information 4, 71, 119, 194, 200, 263, 286, 396, 449
principal component analysis 315
principle of common cause 278
problem-specific error function 233
punctuated equilibrium 450

quantum mechanics 460
quasi-periodic structure 211

random Boolean networks 20
random graph 440
random matrix theory 460
random variation
anagenetic 424
cladogenetic 424
random walk 424

reaction-diffusion 329
reactive behavior 225
reactive controller 270
recurrent neural network 202, 287, 322, 413
Recurrent Self-Organizing Map 333
redundancy 263, 280, 467
reinforcement learning 231, 244
relational categorization 304
relevant information 74, 264
reservoir computing 12, 322
reward measure 391
reward signal 257
reward-based guidance 258
robot 75, 193, 195, 204, 206, 223, 225, 227, 229, 287, 292, 369, 391, 392, 397, 402, 410, 449
  AIBO 75
  armband 251
  chain-like 227
  control 54
  hexapod 227
  humanoid 197
  quadruped 227
  snake-like 227
  wheeled 227
robotics 193, 223, 394
robustness 244, 311, 425

saddle manifold 306
scale-free network 461
segregation 419, 436
selective attention 304
self-amplification 198, 228
self-destruction 196
self-localization 212
self-model matrix 205
self-motivated learning 231
self-organized chaos 202
Self-Organizing Map 333
self-organizing recurrent network 331
Self-Organizing Reservoir 333
self-regulation 5, 22
self-repair 377
self-reproducing machine 425
sensitivity 225
sensor 54, 74, 225, 268, 311, 395, 449
  ambient light 402
  angle 198
  combinations 206
  distribution 288
  dynamics 201, 234
  information 290, 434
  infrared 403
  motion 243
  noise 228
  proprioceptive 196, 213
  proximity 409
  space 245, 254
  state 68, 244, 269, 395
  teaching signal 235
  visual 239
sensor process 284
sensor-to-motor coupling 216
sensorimotor coordination 224, 241, 395
sensorimotor interaction 314
sensorimotor loop 4, 196, 224, 263, 396
sensors morphology 396
sensory-motor feedback 301
separable information 145
short path length 436
Singular Value Decomposition 100
situated cognition 297
slow point 329
small-world index 436, 440
small-world network 436, 461
social environment 408
social learning 347
soliton dynamics 116
somatosensory area 458
spatial aggregation 345
spatial formation 410
spatio-temporal pattern 3, 369
spatio-temporal representation 360
speciation 425
specific information 330
spectral bi-partitioning algorithm 462
spectral clustering 461
spectral density 460
spectral method 459
spectrum 460
spike-time dependent plasticity 331
spontaneous evolution 377
stable regime 330
statistical complexity 122
statistical learning theory 324
statistical mechanics 4, 367
stochastic block model 460

stochastic dependence 273
structural information 273
structure-behavior connection 459
structure/function relationship 447
Student's T-test 430
subcritical regime 252
super-criticality 199
supercritical regime 252
swarm 10, 391
swarm behaviour 391
swarm chemistry 369
swarm dynamics 345
swarm robotics 412
symmetry breaking 12, 194, 226
synaptic connection 398, 447
synaptic efficacy 421
synaptic normalization rule 331
synaptic plasticity 54
synaptic strength 421, 435
synaptic value 200
synchronization 115, 403
synchronized movement 392
synchrony 199, 413
synergistic mutual information 170
synergy 159, 195, 211, 256
system dynamics
  instability 459
  stability 459
system-environment loop 4

task oriented behavior 287
task-independent objective 391
teaching signal 233
theory of computation 4
thermodynamics 4, 24, 417
three-zone swarming model 348
Tierra 381
time-loop error 194, 203, 224
topological distortion 421
topology
  binary 421
Trafalgar effect 346
transfer entropy 5, 141, 263, 277, 326, 347, 350, 395
  thermodynamic interpretation 139, 277, 363
transient 326
Tsallis entropy 46
TSE Complexity 418
Tukey's test 375
TWOWHEELED 205–207, 210, 212, 216, 218, 219

ultrastability 22
union-information 171
unsupervised clustering 466
unsupervised training 334
utility function 68

virtual ecosystem 387

weighted k-cores 451
world model 276

YARS simulator 288

Zero Force Evolutionary Law 424

MIX
Papier aus verantwortungsvollen Quellen
Paper from responsible sources
FSC® C105338

If you have any concerns about our products,
you can contact us on
ProductSafety@springernature.com

In case Publisher is established outside the EU,
the EU authorized representative is:
Springer Nature Customer Service Center GmbH
Europaplatz 3, 69115 Heidelberg, Germany

Printed by Libri Plureos GmbH
in Hamburg, Germany